Engineering Materials

This series provides topical information on innovative, structural and functional materials and composites with applications in optical, electrical, mechanical, civil, aeronautical, medical, bio- and nano-engineering. The individual volumes are complete, comprehensive monographs covering the structure, properties, manufacturing process and applications of these materials. This multidisciplinary series is devoted to professionals, students and all those interested in the latest developments in the Materials Science field, that look for a carefully selected collection of high quality review articles on their respective field of expertise.

Indexed at Compendex (2021) and Scopus (2022)

Giriraj Tailor · Jyoti Chaudhary · Saurabh Singh
Editors

Nanomaterials

Green Synthesis Routes and Their Catalytic, Biomedical Applications

Springer

Editors
Giriraj Tailor
Department of Chemistry
Mewar University
Chittorgarh, Rajasthan, India

Jyoti Chaudhary
Department of Chemistry
M. L. S. University
Udaipur, Rajasthan, India

Saurabh Singh
Department of Chemistry
M. L. V. Government College
Bhilwara, Rajasthan, India

ISSN 1612-1317 ISSN 1868-1212 (electronic)
Engineering Materials
ISBN 978-981-92-0677-3 ISBN 978-981-92-0678-0 (eBook)
https://doi.org/10.1007/978-981-92-0678-0

This Springer imprint is published by the registered company Springer Nature Singapore Pte Ltd.
The registered company address is: 152 Beach Road, #21-01/04 Gateway East, Singapore 189721, Singapore

Preface

The rapid progress of nanoscience has profoundly transformed into promising and imperative research, for the innovative solutions of the most pressing worldwide challenges in energy conversion, catalysis, environmental remediation, and healthcare. In these emerging paradigms domain, nanomaterials have emerged as a revolutionary class of materials characterized due to their unique properties and versatile applications. Exceptional characteristics of nanomaterials have opened new frontiers across different scientific fields, including the fields such as electronics, energy, medicine, catalysis, and environmental science. In contrast the development of nanomaterials through eco-friendly routes benign strategies has earned remarkable attention. The integration of green synthesis processes in the nanomaterial contraction has not only reduces environmental hazards impact but has also improved the biocompatibility and functional efficiency of nanoscale materials. In this context, the present edited volume promise to covering a multidisciplinary and comprehensive perspective on the construction, characterization, and applications of biogenically synthesized nanomaterials.

This book is edited by **Giriraj Tailor**, **Jyoti Chaudhary**, and **Saurabh Singh**, who mutually bring expertise in nanotechnology, catalysis, and biomedical materials. The motivation behind compiling this volume stems from the growing need to link fundamental understanding with continuously synthesizing greener route-based nanomaterials. The editors firmly believe that sustainable nanomaterials open a transformative path for modern scientific research, particularly in developing eco-friendly catalysts, drug delivery systems, antimicrobial materials, and functional nanostructures for environmental sustainability.

In recent scenario, all the researchers are shifting toward greener synthetic routes such as involving plant extracts, biomolecules, microorganisms, and polymer-assisted routes, thus we are concerning integration of green chemistry principles to synthesizing nanomaterials because conventional methods are concerns with the high energy consumption, use of toxic reagents, and generation of hazardous by-products. Greener synthetic approaches not only reduce environmental issues but also offer advantages such as enhance cost-effectiveness, stability scalability, as well as improving biological compatibility.

This book frames as to reflect both advanced and fundamental principles of the emerging applications of green nanomaterials. It encompasses topics starting from biogenic synthetic methodologies and characterization techniques to biomedical remediation and catalytic performance. Each chapters are prepared by experts; they are actively working in their respective areas and are showing the right direction to the new generation.

Chapter "Introduction: Laying Out Greener Routes for Nanomaterials Synthesis" establishes the conceptual paths of greener synthetic methods for the fabrication of nanoparticles through biogenic, environmentally implication approaches, providing green and sustainable ways to produce nanomaterials.

Chapter "Green Nanotechnology: Metal Oxide Nanomaterials for the Degradation of Toxic Dyes" addresses industrial dye pollution that is the one of the most critical environmental problems and explores how greener based nanomaterials act as promising photocatalyst for degradation of textile waste dye, as well as emphasizing reaction mechanisms, catalyst stability, and practical implications for wastewater treatment and advanced organic transformations.

This book provides insights into understanding contents of the physicochemical properties of nanomaterials, are essential for tailoring their performance. Therefore, Chapter "Characterization Techniques of Shapes, Sizes and Porosities of the Nanoparticles" presents a detailed overview of modern analytical techniques used to investigate nanomaterials structural, morphology, and surface characteristics properties.

Biomedical applications of nanomaterials constitute a rapidly expanding research domain to understand this, Chapter "Application of Polymeric Nanomaterials as Therapeutic Agents" has been introduced and it also highlights how polymeric nanocarriers reduce toxicity, improve drug stability, and enhance therapeutic efficiency.

Chapters "Recent Approaches of Metal Oxide-Based Nanomaterials in Drug Delivery Systems (DDS)" and "Drug Delivery Agents: Necessary Modifications and Condition in Nano Materials" are based on Drug Delivery Systems (DDS), explores the role of oxide-based nanomaterials as carriers for targeted drug delivery.

Nature-inspired nanotechnology-based chapter is further explored inside domain of the book and highlights the potential of lower vascular plants as bio-synthesizers and underscores the unique biochemical composition of these plants and their role in producing nanoparticles with enhanced catalytic and biological activities.

Chapter "Biological Significance of Metallic Nanoparticles Synthesized via Plant Extracts", focuses on biological significance of nanomaterials in the antimicrobial, antioxidant, and therapeutic properties.

Chapter "Latest Advances in the Synthesis of Metal and Metal Oxide Nanoparticles" targets the comparative analysis of various methods including microwave-assisted, sonochemical, and hybrid green methodologies of metallic nanoparticle formation and the latest biogenic and modern techniques employed for nanopartcle fabrication.

Chapter "Green Chemistry and Nanotechnology: An Indispensable Modern Yardstick, Symbiotic Relationship and Future Perspective", a systematic attempt is made to understand the basics of green chemistry with nanotechnology and sustainable synthesis approaches for the development of nanomaterials like metal-based, carbon-based, ceramic-based, polymeric, lipid-based, and other miscellaneous materials are concerning in chapter thirteen.

Chapter "Green Nanomaterials in Dentistry" discusses the key translational issues for clinical adoption, including standardization, long-term safety testing, and scaling up. It focuses on delineates the directions of research for the creation of intelligent, multifunctional dental nanomaterials.

Collectively, the chapters in this edited book demonstrate that green nanotechnology a comprehensive scientific philosophy that integrates significant functionality, sustainability, and efficiency. The interdisciplinary nature of the authors reflects the collaborative efforts required to accepted complex scientific, environmental and societal challenges.

The editors express their sincere gratitude to all respected contributing authors not only from the India within but also from the abroad including Japan, France, Poland, Malaysia, Iran for their scholarly contributions, dedication, and timely cooperation as well as to the reviewers whose critical evaluations and constructive suggestions helped us to improving the accuracy, clarity, and impact of the chapters. The editors also acknowledge the invaluable support of their respective institutions to encourage us and providing facilities during the successful completion of this project.

It is the editors' promise that this book will be informative, inspiring, and beneficial for, academicians, researchers, industry professionals, and scholars they are working in nanoscience, materials science, green chemistry, and biomedical engineering followed by concerns with innovation in sustainable nanotechnology.

Chittorgarh, India — Dr. Giriraj Tailor
Udaipur, India — Dr. Jyoti Chaudhary
Bhilwara, India — Dr. Saurabh Singh

Contents

Introduction: Laying Out Greener Routes for Nanomaterials Synthesis

Reena Salodiya, Saurabh Singh, and Giriraj Tailor

1 Introduction

Nanomaterials have emerged as a revolutionary class of materials characterized by their nanoscale dimensions, typically ranging from 1 to 100 nm, and unique properties and versatile applications. Nanomaterials due to increased surface area-to-volume ratio, exhibit novel features such as enhanced reactivity, and altered magnetic, optical, mechanical, and electrical behaviors compared to their bulk counterparts. These exceptional characteristics have opened new frontiers across different scientific fields, including the fields such as electronics, energy, medicine, catalysis, and environmental science [1–3]. The Fabrication and use of nanomaterials are the core of nanotechnology, a multidisciplinary field that integrates ideas from chemistry, physics, biology, and engineering. Due to the variety in the structure and composition of nanoparticles, they show wide range of potential applications. These include carbon-based nanostructures as well as metal or metal oxide nanoparticles. For example, silver and gold nanoparticles are widely used in targeted drug delivery, diagnostics, and as antibacterial agents, while CuO and ZnO nanoparticles are extensively used in biosensors and wastewater treatment [4–7].

Nanoparticles can be made from a different type of materials and can be synthesized by using many methods, like biological, physical, and chemical techniques. The synthesis of nanomaterials can be mainly classified into two approaches: the bottom-up method (constructive approach), in which nanomaterials are built atom by atom

R. Salodiya (✉)
Department of Chemistry, S.M. M. Government Girls College, Bhilwara, Rajasthan, India
e-mail: reenasalodiya1530@gmail.com

S. Singh
Department of Chemistry, Nano Science Lab. M.L.V. Government College, Bhilwara, Rajasthan, India

G. Tailor
Department of Chemistry, Mewar University, Gangrar, Chittorgarh, Rajasthan, India

G. Tailor et al. (eds.), *Nanomaterials*, Engineering Materials,
https://doi.org/10.1007/978-981-92-0678-0_1

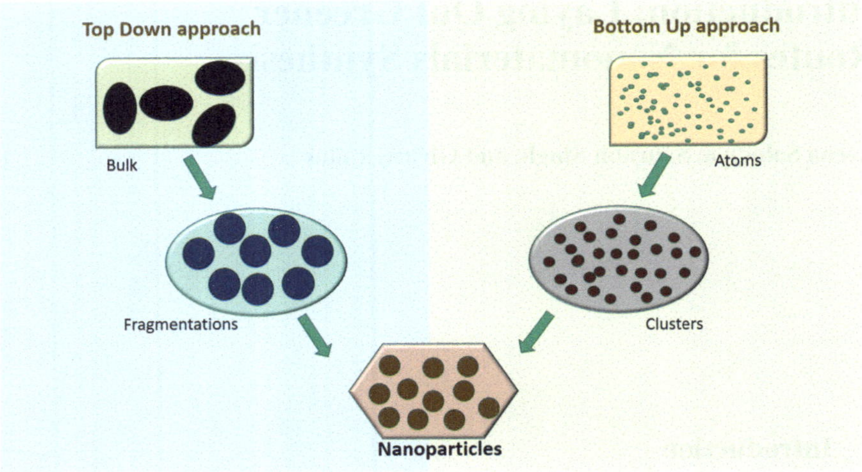

Fig. 1 Fundamental approaches for the synthesis of nanoparticles

or molecule by molecule, and the top-down method (destructive approach), in which bulk materials are broken down into nanostructures (Fig. 1). Both the approaches have their own advantages, depending on the required applications and properties [8, 9]. Traditional synthetic methods have several drawbacks, for example use of expensive reagents, high energy requirement, the need of careful handling of chemicals, low yields, and the production of harmful waste that adversely affects the environment. Additionally, chemically synthesized NPs often show less biocompatibility and stability [1, 10].

Green synthetic approaches for nanomaterials utilize biological and biodegradable sources as reducing and stabilizing agents. In addition to these eco-friendly precursors, microwave-assisted synthesis is also employed. These methods are considered more environmentally benign compared to conventional physical and chemical approaches, including lower energy consumption, reduced reliance on toxic chemicals, uniform heating, elimination of thermal gradient effects, and enhanced biocompatibility of the resulting nanoparticles, thereby contributing to more sustainable and efficient synthesis methods [11, 12]. Different plant parts, Vitamins, biodegradable polymers, Enzymes and diverse microorganisms, such as fungi, bacteria, and algae, are useful biological resources for the fabrication of nanoparticles through biogenic, environmentally friendly approaches, providing green and sustainable ways to produce nanomaterials (Fig. 2).

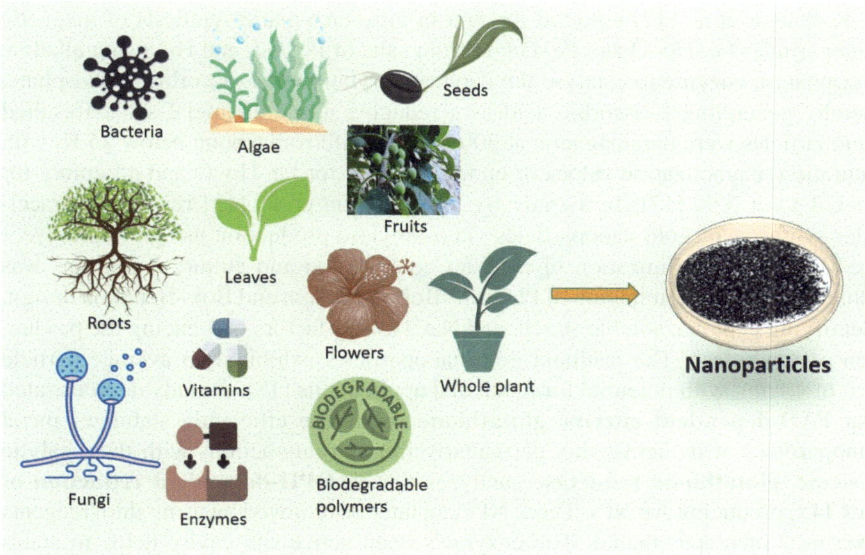

Fig. 2 Green sources for the synthesis of nanoparticles

2 Green Synthesis of Nanomaterials

2.1 Synthesis of Nanomaterials by Enzymes

Enzyme-mediated synthesis of nanoparticles provides a good method for the synthesis of nanoparticles. Enzymes, acting as biocatalysts, play a important role in enabling many chemical and biological reactions and are found in all living organisms. Among these, microbial enzymes have gained much interest due to their cost-effective production, purification, and characterization methods. Enzymes also play an important role in environmental health, as they help in detoxification or convert harmful substances into useful products [13, 14].

Enzymes act as catalysts in synthesis processes, promoting reactions without being consumed themselves. Additionally, they can sometimes act as both reducing and stabilizing agents. The synthesis process may involve either the entire enzyme or, under some reaction conditions, individual amino acids released after enzyme denaturation. One study has shown that specific structural features of enzymes, such as thiol groups and disulfide bridges, often serve as important sites for nanoparticle formation. Similarly, the S–S and S–H groups present in denatured enzymes are capable of reducing metal ions and promoting nanoparticle formation. This means that nanoparticle production can occur through both enzymatic and non-enzymatic pathways [15, 16].

Kolhatkar et al. [17] reported the first in vitro enzymatic synthesis of magnetic nanoparticles Fe–Ho–O and Fe–Gd–O from salts of Fe, Gd, and Ho, using alkaline phosphatase enzyme to catalyse the dephosphorylation of L-ascorbic-2-phosphate, thereby generating L-ascorbic acid as a reducing agent for metal salts. Resulted nanoparticles were paramagnetic at 300 K and antiferromagnetic below 25 K, with saturation magnetization values of up to 30 emu/g for Fe–Ho–O and 45 emu/g for Fe–Gd–O at 5 K [17]. In a study by Manivasagan et al. [18] reported extracellular synthesis of gold nanoparticles via α-amylase production using *Streptomyces* sp. MBRC-82. Optimization of medium composition and culture conditions was achieved using a combination of Plackett–Burman design and Box–Behnken design, identifying peptone, soluble starch, and NaCl as key factors influencing the production of α-amylase. The resultant gold nanoparticles exhibited an average particle size of 40 nm, with potential for medicinal applications [18]. A study demonstrated that **FAD-dependent enzyme glutathione reductase** efficiently stabilizes metal nanoparticles at its active site, particularly through interactions with the catalytic cysteine. **glutathione reductase** catalyzes the **NADPH-dependent reduction of AuCl4$^-$,** producing Au NPs. These NPs can later be removed by using thiol reagents such as 2-mercaptoethanol. The enzyme's deep active-site cavity helps to stabilize these tiny metallic clusters by preventing aggregation even without capping ligands, representing a significant advancement in biomimetic NPs synthesis [19]. α-amylase enzyme also successfully employed for the **synthesis and stabilization of Au NPs** in aqueous media through enzymatic reduction of $AuCl_4^-$, while retaining its enzymatic activity. The Au NP–α-amylase complex was characterized using FTIR, UV–Vis spectroscopy, XRD, and TEM, confirming the presence of both enzyme α-amylase and Au NPs. According to structural analyses the S–H groups serve as crucial for the reduction process and that the enzyme's functional groups are ideally positioned to maintain activity even after binding to Au NPs through Au–S bonds [20]. Enzyme-based methods for nanoparticle synthesis usually produce less unwanted by-products than conventional methods. Since enzymes are biodegradable, their use in NP synthesis also helps to synthsize nanoparticles that can breakdown naturally in the environment. Enzymes also have strong selectivity that allow researchers to adjust the synthesis process with a high degree of accuracy [21]. Enzyme-mediated NPs synthesis show good efficiency for accuracy, sustainability, and overall performance of nanomaterials. By employing the selectivity and mild reaction conditions of enzymes, researchers can design NPs with high structural accuracy and modified functionality. This environmentally friendly process also decrease toxic by-products, which makes the process more useful for large-scale and green synthesis. So enzyme-based NP synthesis can be used in different fields, like biomedical applications, electronic device making, environmental remediation, and other areas of nanotechnology [13].

2.2 *Synthesis of Nanomaterials by Vitamins*

In 2006, Nadagouda and Varma reported a significant method for the green synthesis of Au and Pt nanowires, nanospheres, and nanorods using vitamin Vitamin B2 (riboflavin) as both a reducing and stabilizing agent. In this process they used the biocompatible and environmentally friendly nature of Vitamin B2 to reduce metal ions under mild conditions at room temperature. In this process there is no need of hazardous chemicals that are used in traditional synthesis methods. This process provides a simple and easy way to synthesize nanomaterial without the need for additional capping agents or surfactants. This work presents a remarkable progress in the field of green nanotechnology and underlines the efficiency of natural biomolecules to help in the synthesis of metallic nanomaterials with controlled size and functionality [22]. Similarly, Varma and Nadagouda [23] has been given a green and adaptable approach for the synthesis of core–shell metallic nanocrystals using ascorbic acid (vitamin C) as a mild reducing agent. In this method, they reduced transition metal salts such as Cu and Fe using aqueous ascorbic acid, and synthesized the desired core–shell nanocrystals, whose size and shape depended on the combination of metals used. Specially, Pt with a Cu core formed tennis ball-like structures, while by Pd and Au synthesized regular spherical nanoparticles. When Fe was used as the core, noble metals such as Pt, Au, and Pd formed cube-shaped nanoparticles. The reverse arrangement also yielded interesting structures, like needle-shaped structures with a Cu shell and brush-like structures with an In shell [24]. Similarly, Varma et al. (2007) reported a environmentally friendly one-step method for the synthesis of Pd nanostructures, like nanoplates, nanobelts, and nanotrees using vitamin Bias a reducing agent. In this green synthesis method, they used water as a benign solvent at room temperature and avoided the use of any additional capping agent. Notably, the shape and size of the Pd nanostructures was highly affected by the concentration of the Pd precursor. At the low concentration of Pd, plate-shaped structures of 100 mn to 250 mn thicknesses were obtained. After increasing the concentration, tree-shaped structures synthesized, where Pd plates grew on the shell of Pd nanorods and formed leaf-like shapes. When Pd concentration was increased further, thicker Pd nanoplates obtained, which lined up vertically to form spherical structures [25].

An innovative green synthesis approach for noble metal NPs by utilizing vitamin B12 as both a reducing and capping agent in conjunction with microwave irradiation was reported by Varma and Nadagouda [24]. They synthesized Ag, Au, and Pd nanoparticles of various shapes and sizes by this method. The use of vitamin B12, a naturally occurring compound, gives an eco-friendly alternative of conventional chemical reducing agents. Synthesized Ag, Au, and Pd nanostructures characterized by XRD, SEM, TEM, HRTEM, and UV–visible spectrophotometry [26].

Nadagouda and Varma [23] was first who reported a green approach by using vitamin B_2 for the synthesis of Ag and Pd nanomaterials in various shapes, like nanowires, nanospheres, and nanorods, under mild room-temperature conditions. In this method there is no need of hazardous reducing agents, such as sodium borohydride ($NaBH_4$) and hydroxylamine hydrochloride, and no need for special capping or dispersing agents. This work showed the efficiency of biomolecules for the simple and environmentally friendly synthesis of metal nanostructures with different shapes and sizes [23]. In a study, Lei et al. [27] has given an environmentally friendly method for the synthesis of Ag NPs by using Vitamin C. Vitamin C acted as both the reducing and capping agent, and silver nitrate acted as the precursor for silver ions. By keeping fixed reactant ratios and changing only the reaction duration, the researchers successfully synthesized stable Ag NPs in about 10–40 nm size range [27]. In a investigation Malassis et al. [28], produced Au and Ag nanoparticles by a fast and efficient process by using vitamin C as both the reducing and stabilizing agent. In this process well-defined nanoparticles was yielded, in which Au Nanoparticles had approximately 8 to 80 nm size and Ag nanoparticles had sizes from 20 to 175 nm [28].

2.3 Synthesis of Nanomaterials by Biodegradable Polymers

NPs can be effectively stabilized by surface-active polymers, which firmly adsorb onto the nanoparticle surface via van der Waals attractions between the polymer's monomer units and the surface. This adsorption substantially reduces the surface energy of the particles compared to their native state, thereby stopping their aggregation. Block copolymers have been shown to be even more effective stabilizers than homopolymers, due to their ability to undergo phase separation and exhibit unique features such as elasticity, surface reactivity, resistance and selectivity [29, 30].

Metal NPs can be stabilized by many methods like free radical polymerization by using a radical initiator, simple entrapment within a polymer gel, thiol-assisted polymer adsorption, and by generating the nanoparticles in situ during the polymerization process [31]. Polyethylene glycol has also been used for the synthesis of stable Ag colloids, acting both as an reducing agent and a stabilizer. The resulting NPs were monodispersed, with diameters less than 10 nm [32]. Ag nanoparticles were developed by using poly(methyl vinyl ether-co-maleic anhydride) (PVM/MA) as both the reducing and stabilizing agent. The polymer-stabilized NPs showed stability in water at room temperature, without any aggregation for at least one month. Characterization data showed that the Ag-core NPs were encased by a PVM/MA shell with 5–8 nm thickness [33]. ZnO nanoparticles has also been synthesized by using the biodegradable natural polymer Gum Tragacanth. This one-step method was cost-effective and reproducible. The reaction time and the concentration of the zinc acetate strongly affected the synthesis and growth of ZnO NPs. Characterization techniques XRD, SEM, FTIR, EDAX, UV–visible spectroscopy, and TEM showed that the synthesized ZnO NPs was with uniform dispersion, hexagonal morphology of diameters from 20 to 30 nm [34]. A biodegradable polymer resin, polypropylene fumarate (PPF), combined with Au NPs, has been used to produce a hybrid nanocomposite resin, which is useful in stereolithography processes [35]. Many biodegradable polymers have also been used as effective agents for the synthesis of nanoparticles. These materials have been demonstrated the ability to synthesize very stable metal nanoparticles, including spherical, rod-like, and flower-shaped particles. Due to this versatility not only the potential applications of these nanomaterials increases but their suitability for environmentally friendly production also enhances [36].

3 Synthesis of Nanomaterials by Microbes

The synthesis of nanoparticles by the use of microbes has been emerged as a easy, and environmentally friendly method for metal NPs synthesis. The microbial biosynthesis of NPs presents a way that joins nanotechnology and microbial biotechnology. Microbial cells act as potential sources for the synthesis of nanoparticles of gold, silver, and cadmium sulphide [37]. Their resistance to changes in temperature, pressure, and pH makes this method a great option for different applications.

A report has been documented the biosynthesis of NPs of Au, Ag, Au–Ag alloys, Se, Te, Pt, Pd, silica, Ti, Zr, quantum dots, magnetite, and uraninite by various microbes, like bacteria, fungi, yeasts, and viruses [38]. The size and shapes of synthesized NPs depend on the type of microbe used and the synthesis environment, such as temperature, pH, and concentration of substrate. Microorganisms provide different types of environments for biosynthesis of nanoparticle, As a result formed NPs are eco-friendly, safe, and suitable for many applications [39]. Microorganisms like bacteria, fungi, Algae, and viruses have been used for NPs synthesis, can be broadly classified into extracellular and intracellular processes [40, 41]. In intracellular synthesis, metal ions are transported into microbial cells by various methods, including active transport, ion channels, endocytosis, or direct penetration by the lipid bilayer. After internalization, trapping of ions takes place, which is followed by bioreduction and stabilization of NPs [42]. On the other hand, extracellular synthesis is mediated by secreted enzymes and biomolecules that promote metal ion reduction and nanoparticle capping outside the cell [43]. Significantly, a majority of studies stress the benefits of extracellular synthesis over intracellular routes, emphasising its lower biomass handling, operational simplicity, and comparatively straightforward downstream purification [44–46]. Figure 3 illustrates the extracellular and intracellular synthesis of Ag NPs.

3.1 Synthesis of Nanomaterials by Bacteria

The biosynthesis of inorganic NPs via bioreduction using bacteria has gained much attention, offering several benefits, including less use of toxic chemicals, low cost, quick growth rates, biocompatibility, simplicity of cultivation, and laboratory handling compatibility. Bacteria show a good ability to reduce heavy metal ions, which makes them excellent source for nanoparticle synthesis. As compare to many other microorganisms, bacteria can be easily engineered and genetically modified to enhance the biomineralization of metal ions [47]. Constant exposure to adverse environments, especially those with high levels of heavy metal ions,

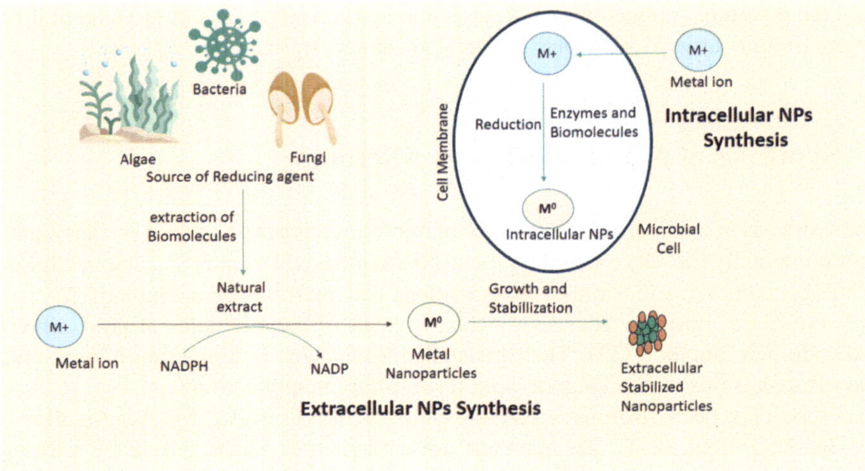

Fig. 3 Extracellular and Intracellular biosynthesis of Ag NPs by biological species

has made bacteria to develop a range of adaptive defence mechanisms, including efflux systems, intracellular sequestration, extracellular precipitation, and alteration of metal ion concentration. These naturally occurring defence mechanisms not only help in bacterial survival in the presence of metal stress but can also be efficiently used for nanoparticle biosynthesis with different biotechnological and industrial applications. Remarkably, some bacterial species can resist and even multiply under high concentrations of metal ions, as seen in species like *Pseudomonas stutzeri* and *Pseudomonas aeruginosa* [48]. Microbes are capable of producing inorganic materials through both extracellular and intracellular processes, at the nanoscale with different morphologies. Among these, extracellular nanoparticle synthesis is considered more commercially useful than the intracellular mechanism [49, 50]. In a study, the bacterial strain Proteus mirabilis PTCC 1710 was used to synthesize silver NPs via both internal and extracellular routes in a simple, economical, eco-friendly, and easy biological process. The synthesized NPs were spherical, with sizes ranging from 10 to 20 nm [51]. Similarly, Prakash et al. [52] reported that *Bacillus megaterium* has been used for extracellular synthesis of metal NPs such as Ag, Pb, and Cd. Resulted NPs accumulate on the bacterial cell wall surface [52]. *Bacillus subtilis* has also been documented to synthesize Au and Ag NPs, The bacteria have shown the capacity to reduce both silver ions and chloroaurate through distinct mechanisms. It's Interesting to note that Au NPs were synthesized via both extracellular and intracellular pathways, whereas silver NPs were restricted to the extracellular process. furthermore, the formation kinetics were very different as Au NPs emerged within 24 h of chloroaurate ion exposure, whereas Ag nanoparticles took almost seven days for their synthesis [53]. Shivaji et al. [54] conducted a study for the biosynthesis of Ag NPs by cell-free culture supernatants derived from psychrophilic bacteria such as *Pseudomonas proteolytica*, *Pseudomonas antarctica*, *Pseudomonas meridiana, Arthrobacter gangotriensis,* and *Arthrobacter kerguelensis,* as well as mesophilic strains *Bacillus indicus* and *Bacillus cecembensis.* Notably, the synthesis efficiency and long-term stability of Ag NPs were significantly influenced by factors such as temperature, pH, and the bacterial species employed. Under circumstances where Pseudomonas antarctica effectively generated AgNPs, Arthrobacter kerguelensis supernatant failed to facilitate nanoparticle synthesis, demonstrating a species-specific response and emphasising the crucial importance of microbial diversity in nanoparticle biogenesis [54]. Pallath et al. [55] demonstrated the biosynthesis of Ag NPs by using the marine bacterium *Planococcus maritimus* MBP-2, which served as both a reducing and capping agent. Interestingly, Synthesized Ag NPs showed noteworthy antibacterial activity, highlighting their potential as biologically derived antimicrobial agents [55]. Soil bacteria also have emerged as efficient biological agents for the reduction of Ag^+ ions and the subsequent formation of AgNPs, representing a cost-effective route for Np synthesis. In a investigation by Saleh [56], environmental bacteria isolated from soil samples were evaluated for Ag^+-reducing potential. Characterization using Atomic Force Microscopy, confirmed the extracellular biosynthesis of AgNPs, with particle sizes of approximately 63.50 nm and 45.81 nm produced by Pseudomonas sp. and Enterobacter sp., respectively. Environmental soil bacteria form an attainable, eco-friendly, and economical source of Ag NPs [56].

Yumei et al. biosynthesized Ag NPs using *Arthrobacter* sp., which was shown to be remarkably affected by temperature, metal ion concentration, and pH. At a less silver nitrate concentration, face-centered cubic Ag NPs ranging from 9–72 nm were synthesized at 70 °C temperature and at 8 pH. As Increasing the silver nitrate concentration, aggregation of NPs at the same temperature took place. Maximum production was noted specifically between pH 7 and 8. Findings of the study showed up that both the pH of the reaction medium and the metal ion concentration employ a direct and notable impact on the efficiency and shapes of nanoparticle formation [57].

3.2 Synthesis of Nanomaterials by Fungi

Fungal-mediated biosynthesis of Nanoparticles represents a relatively simple and well-researched method. As compared to bacteria, fungi often show more effectiveness in nanoparticle production, by eliminating the need for additional steps to recover the filtrate and showing higher tolerance to metal ions, largely accrited to the improved binding capacity of their biomass and cell walls for metallic species [58, 59]. Fungi show a strong capacity for bioaccumulation of metal ions, which enables more economic and easy nanoparticle production [60]. Additionally, they release large amounts of extracellular proteins, which play a important role for improving the stability of the synthesized nanoparticles [61, 62].

The fungus *Fusarium oxysporum* has been researched for the synthesis of metallic nanoparticles, specially Ag NPs. Researchers revealed the formation of pure Ag NPs with sizes ranging from 5–15 nm, and it was documented that proteins secreted by the fungus serve as capping agents, which provides stability to the nanoparticles [63]. The biosynthetic process of quantum-state CdS NPs using *Fusarium oxysporum* has also been reported, where an aqueous $CdSO_4$ solution served as the precursor. In the process the fungal secretion of sulfate reductase enzymes takes place, which reduce sulfate ions to sulfide ions. These sulfide ions subsequently interact with $Cd2^+$ ions present in the medium, which led to the formation of highly stable CdS NPs [64].

Castro-Longoria et al. [65] explored fungi *Neurospora crassa* as a biogenic source for the production of Pt NPs using both fungal biomass and extracellular extracts as reducing agents. Characterization studies confirmed the intracellular biosynthesis of Pt NPs [65]. While extracellular synthesis offers benefits like simplified downstream processing and cost-effectiveness, intracellular nanoparticle synthesis also holds significant importance. In bioremediation contexts, for instance, the removal of heavy metals like Cu and Pt from contaminated environments can be facilitated by fungi capable of intracellular nanoparticle production. Such fungi not only fabricate NPs within their cellular matrix but also make it simpler to separate the biomass along with the accumulated metal pollutants from the contaminated sample [66, 67].

Although a variety of fungal species have been employed for the biosynthesis of NPs, the resulting nanostructures often vary even under comparable experimental and environmental conditions. For instance, *Verticillium* sp. has been reported to produce cubo-octahedral magnetite NPs with sizes ranging from 100–400 nm, whereas *Fusarium oxysporum* synthesized irregular, quasi-spherical NPs with a smaller size range of 20–50 nm [68]. These variations in NPs are largely attributed to differences in the nature and concentration of fungal biomolecules, precursor solutions, incubation parameters, and reaction periods. Notably, *Rhizopus oryzae* has also shown great promise in the synthesis of metallic NPs, especially in the shape-controlled fabrication of Au NPs at room temperature. In this instance, adjusting important parameters including gold ion concentration, solution pH, and reaction time successfully adjusted the shape of the nanoparticles [69–72].

3.3 Synthesis of Nanomaterials by Algae

Numerous algal species have recently been recognized for their capacity to mediate nanomaterial synthesis, primarily due to its simplicity, natural abundance in both freshwater and marine ecosystems, reusability, cost-effectiveness, high metal-binding capacity, and potential for large-scale application. Algal cells have comparatively less nutrient requirements and lower maintenance costs for high growth conditions, such as temperature and light, as compared to most prokaryotic and eukaryotic species. Many algal strains that grow in polluted environments, contaminated with both metals and non-metals, show a great tolerance to very high concentrations of these substances. To decrease the harmful effects of these metal ions and stay alive under such stress, algae have developed a range of defense-related mechanisms [73].

A general process for the synthesis of algal-mediated biosynthesis of NPs involves drying the algal biomass, grinding it into a fine powder, and suspending it in water, followed by incubation for 24 h. Then this extract is filtered and mixed with the a precursor solution. Incubation of this mixture results in a observable color change, helping as a primary indicator of nanomaterial fabrication [74].

Algal species from the classes **Chlorophyceae, Cyanophyceae, Rhodophyceae, and Phaeophyceae** have been investigated as biological source for both extracellular and intracellular synthesis of metallic NPs, including Au, Ag, and other metals. Their ability is due to large reserve of bioactive metabolites, like pigments, antioxidants and, proteins, present in algal extracts. These metabolites act as natural, biocompatible reducing and stabilising agents, thereby providing an environmentally safe and easy method for the synthesis of different metallic NPs. They may work directly as living biofactories or, their cellular extracts can be used as natural reducing and stabilizing agents [75, 76].

A wide range of brown algal species has also been documented in the literature for their potential in the biofabrication of silver NPs. Noteworthy examples include Turbinaria conoides, Desmarestia menziesii, Gelidiella acerosa, Padina pavonica, Sargassum polycystum, and Cystophora moniliformis [77–82].

In their study Hamida et al. [45] reported the extracellular biosynthesis of silver nanoparticles using Nostoc Bahar M spp [45]. Priyadharshini et al. [83] also reported the biosynthesis of ZnO NPs using the aqueous extract of the macroalga *Gracilaria edulis* [83]. In a recent study, stable Ag NPs and CuO NPs were successfully fabricated via a green synthesis method by employing the aqueous extract of *Spirulina platensis* algae as both a reducing and capping agent [84]. Au NPs have also been successfully fabricated by using various strains of brown algae, which are known to harbor bioactive compounds imparting anticoagulant, antibacterial, and antifouling properties. A wide range of species, including *Cystoseira baccata, Sargassum myriocystum, Ecklonia cava, Sargassum wightii, Stereospermum marginatum, Fucus vesiculosus, Padina gymnospora*, and *Dictyota bartayresianna* have also been documented as effective biological resources for the green synthesis of Au NPs [85–87] (Table 1).

4 Synthesis of Nanomaterials from Plant Material

The application of plant extracts as natural reducing and stabilizing agents has garnered much attention for nanoparticle synthesis. Plant-based approaches have emerged as a sustainable alternative for the fabrication of diverse NPs with a lesser environmental impact. Using phytochemicals as natural reducing and stabilizing agents, a wide range of nanoparticles, including copper, gold, silver, palladium, platinum, ZnO, and magnetite, can be efficiently fabricated. Beyond their eco-friendly synthesis, plant-derived NPs have shown significant therapeutic potential, with documented applications in the treatment of cancer, hepatitis, HIV, malaria, and other diseases [119].

Table 1 Provides a summarized account of studies reporting nanoparticle biosynthesis mediated by different microbial species

Sr. no	Microbial species	Source of isolation	NPs	Size	Shape	References
	Bacteria					
1	*Marinobacter* sp. 2C8	Southern region of the Caspian Sea	ZnO	10.23 ± 2.48 nm and	Spherical	[88]
2	Vibrio sp. VLA	Shrimp breeding pool in southeastern Iran	ZnO	20.26 ± 4.44 nm	Spherical	[88]
3	*Lysinibacillus* sp. EBL303	Disposed effluent in the Environmental Biotechnology Lab (EBL) of the University of Tehran	Te NPs	22–148 nm	Spherical	[89]
4	*Bacillus cereus*	Cochin, Kerala India	Ag	42–92 nm	Spherical	[19]
5	*Streptomyces bikiniensis*	Eldrieh, Riyadh, Saudi Arabia	Se	17 nm	Rods	[90]
6	*Klebsiella pneumoniae*	Western Ghats	Ag			[91]
7	*Paracoccus haeundaensis*	Busan, South Korea	Au	20.93 ± 3.46 nm	Spherical	[92]
8	Bacillus subtilis	Uppanar estuary, Southeast coast of India	Ag	49–153 nm	Spherical	[93]
9	Stenotrophomonas sp.	Indian marine origin	Au, Ag	10–50 nm and 40–60 nm in dimensions for AuNPs and AgNPs	Multi-shaped	[94]
10	*Marinobacter pelagius*	Water sample from solar saltern, Kakinada	Au	10 nm	Spherical	[95]
11	*Pseudomonas denitrificans*	NCIM, Pune	Au	25–30 nm	Face centered cubic	[96]

(continued)

Table 1 (continued)

Sr. no	Microbial species	Source of isolation	NPs	Size	Shape	References
12	Lactobacillus crispatus	Vaginal isolate	Ti	70.98 nm	–	[97]
	Fungi					
13	*Cylindrocladium floridanum*	–	Au	19.05 nm	Spherical	[98]
14	Cladosporium sp.	Commiphora wightii	Au	5–10 nm	Spherical	[99]
15	*Morchella esculenta*	Fruit body extract	Au	16.51 nm	Face-centered cubic	[100]
16	*Aspergillus niger*	Sivas Cumhuriyet University	Ag	13.2–646.8 nm	Spherical	[101]
17	Penicillium janthinellum DJP06	rhizospeheric soils	Au, Ag	19 ± 3.2 nm and 13 ± 2.7 nm, respectively for Au and Ag NPs	Spherical	[102]
18	*Cladosporium perangustum*		Ag	30–40 nm	Spherical	[103]
19	*Cochliobolus geniculatus*	Leaves of *Nothapodytes foetida*	ZnO	2–6 nm	Quasi spherical	[104]
20	Aspergillus niger BSC-1	Sundarban mangrove ecosystem	Fe_3O_4	20–40 nm		[105]
21	Fusarium incarnatum	Sundarban Mangrove Ecosystem, India	Fe	15–55 nm	spherical	[106]
22	Aspergillus terreus	Arid zone agricultural fields of CAZRI, Jodhpur, India	Zn, Mg, Ti	8.2 nm, 6.4 nm, 1.5–30 nm, respectively for Zn, Mg and Ti		[107]
	Algae					

(continued)

Table 1 (continued)

Sr. no	Microbial species	Source of isolation	NPs	Size	Shape	References
23	*Colpomenia sinuosa, Pterocladiacapillacea*	Mediterranean Sea, Egypt	Fe_3O_4	11.24–33.71 nm and 16.85–22.47 nm	Nanospheres	[108]
24	Macroalga *Sargassum muticum*		Ag, Au	10.4 ± 1.2 nm for Au and 41.0 ± 5.7 nm for Ag	Spherical	[109]
25	Padina tetrastromatical	Mersing, Johor at Peninsular Malaysia	Ag	4 nm	Spherical	[110]
26	Spirulina platenesis		Au, Ag	14.42 to 48.97 nm for Ag, 15.60–77.13 nm for Au NPs	Spherical, octahedral, pentagonal and triangular	[111]
27	*Ecklonia cava*		Au	30 ± 0.25 nm	Spherical and triangular	[112]
28	Ulva *fasciata*		Chitosan	32.49 nm	Spherical	[113]
29	*Chlorella vulgaris*	Water stream at Mustansiriyah University, Baghdad, Iraq	Dextran-coated TiO_2NPs	71.44 nm	–	[114]
30	Tetradesmus *obliquus*	Rice field, Marodi village, near Mouda, Nagpur, Maharashtra, India	Pt	21.25 nm–31.50 nm	Spherical	[115]
31	Galaxaura rugosa	Zaafrana beach, Red Sea, Egypt	Fe	14.4 to 17.2 nm	Spherical	[116]
32	Padina boergrsenii	Mandapam coastal region of southern Tamilnadu	Se-ZnO	26.14 nm	Spherical	[117]
33	*Ulva rigida*		Ag	22.6 nm	Spherical	[118]

This approach is commended for its simplicity, environment friendliness, low cost, and minimal risk of contamination [120, 121]. A new branch of nanobiotechnology focuses on the green synthesis of nanomaterials using various plant parts such as flowers, leaves, fruit, root, stem, seeds etc. In comparison of microbial systems, the use of plant extracts for nanoparticle synthesis shows several advantages, including flexibility, cost-effectiveness, and the elimination of cell culture techniques. Therefore, plant-mediated approaches are regarded as more efficient and practical than other biological sources for nanoparticle synthesis [122–124]. In nanomaterial synthesis, naturally occurring biostabilizers and reducing agents, including alkaloids, enzymes, carbohydrates, proteins, flavonoids, polyphenols, and phenolic acids, serve as essential mediators in the process [125, 126]. Reports have been shown that the nature and properties, such as size, shape, and stability of NPs, synthesized through this method vary across plant species, largely affected by factors such as the preparation method, type of solvent, pH, reaction temperature, reactant ratio, concentration, and the reducing agent used [127–129].

In the green synthesis of NPs by plants, firstly bioreduction takes place, where bioactive phytochemicals like carotenoids, flavonoids, and polysaccharides act as reducing agents to reduce metal ions into their atomic form. This initial reduction step is important, as it largely affects the biological activity and cytotoxic efficiency of the synthesized nanomaterials. Since ions are more reactive as compared to atoms, their reduction determines the future stability of the NPs. The process then steps to the nucleation stage, during this step the reduced atoms accumulate under the influence of phytochemicals, give rise to NPs. These NPs then organize into a bioactive nanosystem [130–134]. Figure 4 illustrates the general approach for the synthesis for plant-mediated synthesis of NPs.

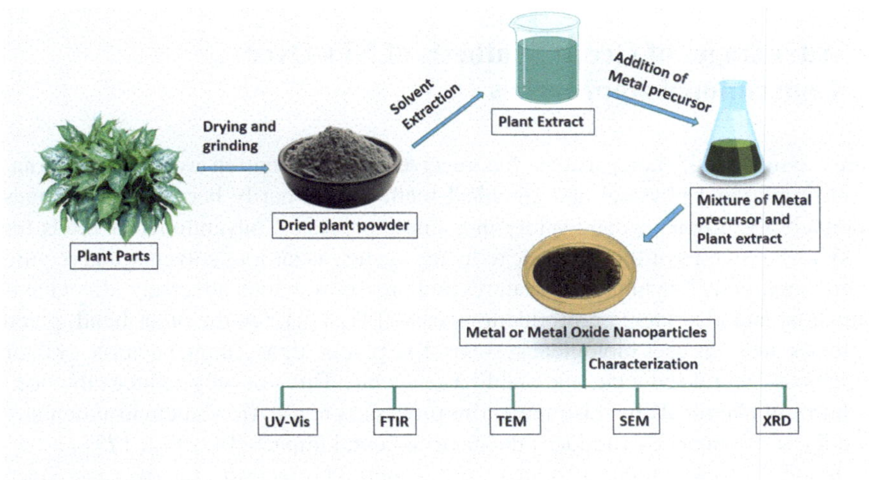

Fig. 4 General scheme for plant-mediated synthesis of NPs

For the synthesis of NPs using plant-based procedures, different plant parts, including fruits, flowers, leaves, seeds, and bark are usually processed through a series of steps of washing with deionized water, drying, and grinding into fine powder. A specific amount of this powder is then extracted in an appropriate solvent and agitated at room temperature for 2–3 days to obtain the phytochemical-rich solution. This extract is then mixed with different concentrations of metal precursors and incubated in dark environment to facilitate the reduction of metal or metal oxide NPs. The reaction mixture is generally subjected to heating and stirring at 70–90 °C until a characteristic color shift indicates the formation of NPs [135–137]. For example Plant-mediated strategies have also been employed for the green synthesis of ZnO NPs. In this study, zinc nitrate hexahydrate [$Zn(NO_3)_2 \cdot 6H_2O$] served as the metal precursor, while the extract of the indigenous medicinal plant Cayratia pedata served as the reducing and stabilizing agent. The precursor solution was reacted with varying concentrations of plant extract under controlled conditions. The reaction yielded a yellow paste of ZnO NPs [138]. Figure 4 shows the general approach for the synthesis of NPs by using parts of the plant.

Plant-mediated synthesis has gained attention as a sustainable strategy owing to its simplicity, rapidity, efficiency, and eco-friendly nature, which collectively make it suitable for large-scale production and safe application in biomedical research. Noteworthy, Au NPs synthesized from medicinal plants such as Cucurbita pepo and Malva crispa have demonstrated a remarkable antibacterial efficacy against food-borne spoilage pathogens, underscoring their potential in both therapeutic and food safety applications [139, 140]. A green synthetic approach for Y_2O_3 NPs was reported using the sol–gel method, where aqueous leaf extracts of Lantana camara L. served as a natural reducing and stabilizing agent [141] (Table 2).

5 Advantages of Green Synthesis of NPs Over Conventional Approaches

Green synthesis of nanoparticles has received much attention as a great alternative to traditional physical and chemical methods, primarily because it combines sustainability, efficiency, and safety in a single process. Conventional methods for the synthesis of NPs often uses toxic reducing agents, hazardous solvents, and require severe conditions like high temperatures and pressures, which adversely affect environmental and also increase production costs [131, 171]. On the other hand, green synthesis methods uses biological systems, like plants, algae, fungi, bacteria, or their extracts as natural reducing and stabilizing agents. This not only reduces the need for harmful chemicals but also the entire process is non-toxic and environmentally friendly, so it remarkably reduces the environmental impact [141, 172, 173].

One of the most notable advantage of this method is its cost-effectiveness. Since plant materials, agricultural residues, and microbial cultures are affordable and easily available, the complete production cost is much lesser as compare to conventional

Table 2 Provides a summarized account of recent studies reporting plant-mediated biosynthesis of nanoparticles

Sr. no	Plant name	Part of plant	NPs	Size	Shape	References
1	*Hypericum perforatum* L	Whole plant	Ag	35 ± 2.7 nm	Spherical	[142]
2	*Gymnema sylvestre*	Plant	Ag	50 to 100 nm	Spherical	[143]
3	*Euphorbia serpens*	Whole plant	CuO	20.39 nm	–	[144]
4	*Mirabilis jalapa*	Leaves	Ag	–	Face centered cubic	[145]
5	Barleria longiflora	Leaves	Ag	10 μm	Crystalline structure	[146]
6	Moringa Oleifera L	Leaves	Au	10–40 nm	Spherical	[147]
7	*Lemna minor*	Whole plant	Ag	40.56 nm	Spherical	[148]
8	*Acacia auriculiformis*	Leaves	Ag	28.4–37.9 nm	Spherical	[149]
9	*Aronia melanocarpa* L	Fruit	ZnO	45.67 nm	Spherical	[150]
10	*Vitex negundo*	Leaves	ZnO	40–50 nm	Spherical	[151]
11	*Zingiber officinale*	Whole plant	ZnO	22.17 nm	Spherical	[152]
12	*Combretum roxburghii* Spreng	Leaves	CuO	2.06 to 6.20 nm	Spherical	[153]
13	Waltheria indica	Root	ZnO		Hexagonal	[154]
14	*Cissus quadrangularis*	Stem	Ag	8.68 nm	Spherical	[155]
15	*Acalypha indica L*	Leaves, stems, and roots	Ag	7 nm	Spherical	[156]
16	Neem and turmeric	Neem leaves and turmeric powder	Ag	4–14.81 nm	Spherical	[157]
17	*Helianthus annuus* L	Leaves	NiO	16.65 ± 2.48 nm	Quasi-spherical	[158]
18	*Medicago sativa* L	Leaves	NiO	15.06 ± 2.72 nm,	Spherical	[158]
19	*Piper guineense*	Seeds	ZnO	7.39 nm	Spherical to Hexagonal	[159]
20	*Combretum roxburghii* Spreng	Leaves	CuO	4.23 nm	Distorted spherical	[153]

(continued)

Table 2 (continued)

Sr. no	Plant name	Part of plant	NPs	Size	Shape	References
21	*Citrus sinensis Osbeck*	Fruit peel	ZnO	27 to 42 nm	Quasi-spherical	[160]
22	*Medicago sativa* L	Leaves	Ag	16-18 nm	Face centered cubic	[161]
23	*Croton malabaricus*	Stem wood	SnO_2	9.13 nm	Agglomerated, spherical	[162]
24	Bryophyllum pinnatum	Leaves	NiO	16.377 nm	Face centered cubic	[163]
25	*Iris kashmiriana*	Rhizome	Au	80 nm	Spherical	[164]
26	Piper nigrum	Fruit	Ag	200 nm-2 μm	Spherical	[165]
27	*Halodule uninervis*	Leaves	Au	10–50 nm	Spherical	[166]
28	*Lagerstroemia speciosa*	Fruit	Ag	61.12 nm	Spherical	[167]
29	Acacia Concinna	Leaves	CuO	20 nm	Spherical and hexagonal	[168]
30	*Artemisia absinthium*	Plants	ZnO	36.6 nm	Spherical	[169]
31	Pavonia zeylanica	Leaves	ZnO	19.58 nm	Crystalline structure	[170]

methods that require costly reagents and specific infrastructure [131, 174]. Green synthesis is often performed under mild reaction conditions, generally at room temperature and pressure, which reduces the need of high energy, so this is process more sustainable and energy efficient [175, 176].

Another important benefit is that the NPs which are synthesized through green techniques show higher biocompatibility. The biomolecules that are present in natural extracts reduce the metal ions and also coat the nanoparticle surface, acting as capping agents that increase the stability and prevent the aggregation of synthesized NPs [177, 178].

The simplicity and flexibility of green synthesis give another remarkable advantage. The methods used are simple and require less time. Since the process follow to the fundamental principles of green chemistry, like minimizing waste, using renewable sources, and designing safe chemical transformations, it represents a sustainable method for advancing nanotechnology [127, 133, 179, 180]. Green synthesis relieves the limitations of conventional methods and also opens up new possibilities for the synthesis of nanoparticles with increased functionality, stability, and suitability for various environmental, industrial, and biomedical applications (Table 3).

Table 3 Summarizing biological methods for nanoparticle synthesis and the principal that biomolecules uses for metal ion reduction and stabilization

Sr. no	Source		Key biomolecules responsible for NP formation (Reduction/ Stabilization)	Mechanism	References
1	Enzymes		Thiol groups, disulfide bridges, -SH groups	Serve as critical sites for the formation of nanoparticles and reduce metal ions	[15, 16]
2	Vitamins		Vitamin B_1(Thiamin), Vitamin B_2 (Riboflavin), Vitamin B_{12} (cobalamin), Vitamin C (ascorbic acid)	Serve both as a reducing and capping agent	[23–26]
3	Biodegradable polymers		Polyethylene glycol, PVM/MA	Act as reducing and stabilizing agent	[32, 33]
4	Microbes	Bacteria	Bacterial cell membrane contains a variety of functional groups	Act as ligands for metallic ions, making them a common site for the formation of NPs	[91]
			NADH and NADH-dependent nitrate reductase enzyme	Electron transfer mechanism	[181]
		Fungi	Reductase enzymes, proteins	Reductase enzymes serve as reducing agent, Proteins Serve as capping agents, providing stability to the nanoparticles	[63, 64]
		Algae	Proteins, fat, carbohydrates, peptides, alkaloids, terpenes, macrolides, cell wall polysaccharides, glycoproteins (comprised of various functional groups like carbonyl, hydroxyl, carboxyl, sulfonate) and enzymes	Plays a pivotal role in reduction, capping, fabrication and stabilization of nanoparticles	[182, 183]
5	Plants		Phytochemicals such as flavonoids, alkaloids, phenolic acids, glycosides, alkaloids, saponins, tannins, terpenoids, and many vitamins	Act as reducing and stabilizing agent	[145, 157]

6 Challenges and Limitations

Although green synthesis of nanoparticles offers significant promise as an environmentlly friendly and sustainable alternative to traditional chemical and physical methods, it still faces several limitations that hinder the use of green synthesis methods for industrial adoption and in large-scale applications. The process of nanoparticle synthesis and their quality are affected by several factors, like the availability of raw materials, the time required for the process, and the reaction conditions under which the nanoparticles are synthesized. Due to all these factors, enlarging this technology is still challenging [184]. Researchers have used various plants for the biological synthesis of nanoparticles, and most of the work has been focused on plants that are locally available. While this uses local resources effectively, applying these methods on a large scale globally is still very difficult. Also, raw materials obtained directly from plants cannot always be used directly in industrial processes. They often require additional processing steps like drying, grinding, or extraction before they are used. This not only makes the complete process more complex but also increases the cost. So, the practical use and cost-effectiveness of these raw materials for green synthesis need more research before their adoption for industrial use [185].

Green synthesis methods involve many bioactive compounds with different reducing and capping abilities, which need great optimization and experiments to obtain required NPs with constant characteristics. The extraction and purification of these biomolecules from natural resources such as plants and microbes add another challenge, as both steps are time-consuming. Another challenge is the lack of established procedure that give guidance on mass balance, stoichiometric ratios, and other technical considerations. This makes the process enhancing at the large scale particularly difficult. Also, NPs produced by using green extracts often show much variability in size, morphology, and physicochemical properties. These inconsistencies make it difficult to reproduce and also make it challenging to control desired features like crystallinity, which is necessary for the use of nanoparticles for specific applications [186, 187].

All these drawbacks, such as raw material variability, slow reaction speed, low yield, low product uniformity, and difficulty in scaling-up the production, highlight the need for more research to improve green synthesis methods. Without understanding and eliminating these issues, the transition of green nanotechnology from small-scale studies to industrial production will remain difficult.

7 Future Perspective

In recent years, research on nanomaterials and their applications has increased greatly, with green synthesis appearing as a good alternative to traditional methods. Various biological sources, like bacteria, fungi, yeast, and plants, have been researched for the biosynthesis of metallic nanoparticles. Even with there has been much progress in this field, several challenges still inhibit the shifting of green synthesis from laboratory research to largescale industrial production for commercial applications.

Factors like, the type of microbes and plant extract used, secondary metabolites, as well as environmental factors such as pH, temperature, and agitation speed, play important roles in deciding the shape, size, and uniformity of NPs. Without thorough research on these factors, reproducibility and scalability will remain difficult. Another area that needs attention is the mechanistic knowledge of the biosynthesis of NPs. The role of specific secondary metabolites of plant extracts and the use of microbial enzymes and other cellular component during reduction and stabilization are still not fully explained. Gaining more knowledge of these biochemical methods may help gain more control over nanoparticle formation and stability. Also, the yield and efficiency of green synthesis are often low, so it is required a strategy to enhance production yield and also minimizing the reaction times.

New approaches, such as the use of genetically engineered microbes, show great solutions to some of these challenges. Engineered strains that produce more quantities of proteins, enzymes, or other biomolecules may improve both the biosynthesis and stabilization of nanoparticles. Likewise, by increasing the metal accumulation capacity and tolerance of microbial systems can also greatly increase NPs production efficiency. Thus, the biotechnological contributions to the green synthesis of nanoparticles, along with the development of advanced characterization techniques, give a promising method for the large-scale production of nanoparticles. These factors will greatly improve the cost-effectiveness of green alternatives and enable them to compete with traditional chemical methods.

8 Conclusion

The remarkable advancements in science and technology over the past few years have brought about revolutionary changes in various fields, with nanotechnology appearing as a innovative discipline. Due to unique properties of nanoparticles, they are useful for a range of applications in biomedical, industrial, technological, and environmental sectors. Traditional chemical and physical synthesis methods, often use hazardous chemicals, need high energy, and can produce toxic byproducts, which adversely affect the environment and human health.

Whereas green synthesis methods for nanomaterial synthesis have appeared as a great alternative, that follows the principles of sustainable development and environmental protection. This technique uses natural resources such as plant extracts, microbial cultures of bacteria, fungi, and algae, biodegradable polymers, and other bio-based agents, due to this the synthesis of nanoparticles takes place under non-toxic and energy-efficient conditions. These synthesis methods minimize environmental impact and also enhance the functionality and biocompatibility of the synthesized nanomaterials. The concentration of biological components and the presence of organic reducing agents play an important role in deciding the size and shape of the nanoparticles.

This study intended to emphasize the significance of green synthetic pathways in advancing sustainable nanomaterial fabrication and highlighted their critical role in the evolution of environmentally responsible nanotechnological advances. The study also evaluated the current limitations and outlined the potential research directions.

Most of the reported studies have so far been conducted on a laboratory scale, however, ongoing research is increasingly focused on translating nanoparticle applications to a larger scale. Particular emphasis is placed on their potential use in environmental management, agriculture, healthcare, and other sectors, aiming to address the demands of the growing global population and to contribute in wellbeing of human.

References

1. Kganyago, P., Mahlaule-Glory, L.M., Mathipa, M.M., Ntsendwana, B., Mketo, N., Mbita, Z., Hintsho-Mbita, N.C.: Synthesis of NiO NPs via a green route using Monsonia burkeana: the physical and biological properties. J. Photochem. Photobiol. B **182**, 18–26 (2018)
2. Chaudhary, R.G., Tanna, J.A., Gandhare, N.V., Rai, A.R., Juneja, H.D.: Synthesis of nickel nanoparticles: microscopic investigation, an efficient catalyst and effective antibacterial activity. Adv. Mater. Lett. **6**(11), 990–998 (2015)
3. Singh, S., Salodiya, R., Tailor, G., Sharma, T.C., Mehta, C.: Comprehensive studies: biogenic constructed nickel oxide nanoparticles and their broad spectrum biomedical applications. ChemistrySelect **9**(38), e202402099 (2024)
4. Bruna, T., Maldonado-Bravo, F., Jara, P., Caro, N.: Silver nanoparticles and their antibacterial applications. Int. J. Mol. Sci. **22**(13), 7202 (2021)
5. Elahi, N., Kamali, M., Baghersad, M.H.: Recent biomedical applications of gold nanoparticles: a review. Talanta **184**, 537–556 (2018)
6. Al-darwesh, M.Y., Ibrahim, S.S., Mohammed, M.A.: A review on plant extract mediated green synthesis of zinc oxide nanoparticles and their biomedical applications. Results Chem. **7**, 101368 (2024)
7. Saleem, M.H., Ejaz, U., Vithanage, M., Bolan, N., Siddique, K.H.: Synthesis, characterization, and advanced sustainable applications of copper oxide nanoparticles: a review. Clean Technologies and Environmental Policy, 1–26 (2024)
8. Faruksha, A.U., Sekar, A.M.: Green synthesis of nanoparticles: a review. GRT J. Educ. Sci. Technol. **2**(1), 19–30 (2024)
9. Sajid, M., Płotka-Wasylka, J.: Nanoparticles: synthesis, characteristics, and applications in analytical and other sciences. Microchem. J. **154**, 104623 (2020)
10. Yadeta Gemachu, L., Lealem Birhanu, A.: Green synthesis of ZnO, CuO and NiO nanoparticles using Neem leaf extract and comparing their photocatalytic activity under solar irradiation. Green Chem. Lett. Rev. **17**(1), 2293841 (2024)
11. Kirubakaran, D., Wahid, J.B.A., Karmegam, N., Jeevika, R., Sellapillai, L., Rajkumar, M., SenthilKumar, K.J.: A comprehensive review on the green synthesis of nanoparticles: advancements in biomedical and environmental applications. Biomed. Mater. Devices, 1–26 (2025)
12. Hasanpoor, M., Aliofkhazraei, M., Delavari, H.J.P.M.S.: Microwave-assisted synthesis of zinc oxide nanoparticles. Procedia Mater. Sci. **11**, 320–325 (2015)
13. Seth, R., Meena, A.: Enzymes-based nanomaterial synthesis: An eco-friendly and green synthesis approach. Clean Technologies and Environmental Policy, 1–24 (2024)

14. Sanket, S., Das, S.K.: Role of enzymes in synthesis of nanoparticles. Bioprospecting of enzymes in industry, healthcare and sustainable environment, 139–153 (2021)
15. Adelere, I.A., Lateef, A.: A novel approach to the green synthesis of metallic nanoparticles: the use of agro-wastes, enzymes, and pigments. Nanotechnol. Rev. **5**(6), 567–587 (2016)
16. Durán, M., Silveira, C.P., Durán, N.: Catalytic role of traditional enzymes for biosynthesis of biogenic metallic nanoparticles: a mini-review. IET Nanobiotechnol. **9**(5), 314–323 (2015)
17. Kolhatkar, A.G., Dannongoda, C., Kourentzi, K., Jamison, A.C., Nekrashevich, I., Kar, A., Willson, R.C.: Enzymatic synthesis of magnetic nanoparticles. Int. J. Mol. Sci. **16**(4), 7535–7550 (2015)
18. Manivasagan, P., Venkatesan, J., Kang, K.H., Sivakumar, K., Park, S.J., Kim, S.K.: Production of α-amylase for the biosynthesis of gold nanoparticles using Streptomyces sp. MBRC-82. Int. J. Biol. Macromol. **72**, 71–78 (2015)
19. Scott, D., Toney, M., Muzikár, M.: Harnessing the mechanism of glutathione reductase for synthesis of active site bound metallic nanoparticles and electrical connection to electrodes. J. Am. Chem. Soc. **130**(3), 865–874 (2008)
20. Rangnekar, A., Sarma, T.K., Singh, A.K., Deka, J., Ramesh, A., Chattopadhyay, A.: Retention of enzymatic activity of α-amylase in the reductive synthesis of gold nanoparticles. Langmuir **23**(10), 5700–5706 (2007)
21. Abuzeid, H.M., Julien, C.M., Zhu, L., Hashem, A.M.: Green synthesis of nanoparticles and their energy storage, environmental, and biomedical applications. Crystals **13**(11), 1576 (2023)
22. Nadagouda, M.N., Varma, R.S.: Green and controlled synthesis of gold and platinum nanomaterials using vitamin B2: density-assisted self-assembly of nanospheres, wires and rods. Green Chem. **8**(6), 516–518 (2006)
23. Nadagouda, M.N., Varma, R.S.: Green synthesis of Ag and Pd nanospheres, nanowires, and nanorods using vitamin B2: catalytic polymerisation of aniline and pyrrole. J. Nanomater. **2008**(1), 782358 (2008)
24. Nadagouda, M.N., Varma, R.S.: A greener synthesis of core (Fe, Cu)-shell (Au, Pt, Pd, and Ag) nanocrystals using aqueous vitamin C. Cryst. Growth Des. **7**(12), 2582–2587 (2007)
25. Nadagouda, M.N., Polshettiwar, V., Varma, R.S.: Self-assembly of palladium nanoparticles: synthesis of nanobelts, nanoplates and nanotrees using vitamin B1, and their application in carbon–carbon coupling reactions. J. Mater. Chem. **19**(14), 2026–2031 (2009)
26. Han, C., Nagendra, V., Nasir Baig, R.B., Varma, R.S., Nadagouda, M.N.: Expeditious synthesis of noble metal nanoparticles using vitamin B12 under microwave irradiation. Appl. Sci. **5**(3), 415–426 (2015)
27. Lei, J., McClelland, A., Zeng, T.H.: Synthesis of silver nanoparticles by vitamin C for public health applications. In: 2022 IEEE 22nd International Conference on Nanotechnology (NANO), pp. 488–491. IEEE (2022)
28. Malassis, L., Dreyfus, R., Murphy, R.J., Hough, L.A., Donnio, B., Murray, C.B.: One-step green synthesis of gold and silver nanoparticles with ascorbic acid and their versatile surface post-functionalization. RSC Adv. **6**(39), 33092–33100 (2016)
29. Rozenberg, B.A., Tenne, R.: Polymer-assisted fabrication of nanoparticles and nanocomposites. Prog. Polym. Sci. **33**(1), 40–112 (2008)
30. Christodoulakis, K.E., Palioura, D., Anastasiadis, S.H., Vamvakaki, M.: Metal nanocrystals embedded within polymeric nanostructures: effect of polymer-metal compound interactions. Top. Catal. **52**, 394–411 (2009)
31. Virkutyte, J., Varma, R.S.: Green synthesis of metal nanoparticles: biodegradable polymers and enzymes in stabilization and surface functionalization. Chem. Sci. **2**, 837–846 (2011). https://doi.org/10.1039/c0sc00338g
32. Bo, L., Yang, W., Chen, M., Gao, J., Xue, Q.: A simple and 'Green' synthesis of polymer-based silver colloids and their antibacterial properties. Chem. Biodivers. **6**(1), 111–116 (2009)
33. Maity, D., Kanti Bain, M., Bhowmick, B., Sarkar, J., Saha, S., Acharya, K., Chattopadhyay, D.: In situ synthesis, characterization, and antimicrobial activity of silver nanoparticles using water-soluble polymer. J. Appl. Polym. Sci. **122**(4), 2189–2196 (2011)

34. Handore, K., Bhavsar, S., Horne, A., Chhattise, P., Mohite, K., Ambekar, J., Chabukswar, V.: Novel green route of synthesis of ZnO nanoparticles by using natural biodegradable polymer and its application as a catalyst for oxidation of aldehydes. J. Macromolec. Sci. Part A **51**(12), 941–947 (2014)
35. Abdelrasoul, G.N., Farkas, B., Romano, I., Diaspro, A., Beke, S.: Nanocomposite scaffold fabrication by incorporating gold nanoparticles into biodegradable polymer matrix: synthesis, characterization, and photothermal effect. Mater. Sci. Eng. C **56**, 305–310 (2015)
36. Sharma, S., Pathania, A.R.: Biodegradable polymers green synthesis of nanoparticle–an overview. Mater. Today Proc. **62**, 3827–3831 (2022)
37. Prasad, R., Pandey, R., Barman, I.: Engineering tailored nanoparticles with microbes: quo vadis? Wiley Interdis. Rev. Nanomed. Nanobiotechnol. **8**(2), 316–330 (2016)
38. Narayanan, K.B., Sakthivel, N.: Biological synthesis of metal nanoparticles by microbes. Adv. Coll. Interface. Sci. **156**(1–2), 1–13 (2010)
39. Hulkoti, N.I., Taranath, T.C.: Biosynthesis of nanoparticles using microbes—a review. Colloids Surf. B **121**, 474–483 (2014)
40. Tauseef, A., Hisam, F., Hussain, T., Caruso, A., Hussain, K., Châtel, A., Chénais, B.: Nanomicrobiology: emerging trends in microbial synthesis of nanomaterials and their applications. J. Cluster Sci. **34**(2), 639–664 (2023)
41. Santos, T.S., Silva, T.M., Cardoso, J.C., Albuquerque-Júnior, R.L.D., Zielinska, A., Souto, E.B., Mendonça, M.D.C.: Biosynthesis of silver nanoparticles mediated by entomopathogenic fungi: antimicrobial resistance, nanopesticides, and toxicity. Antibiotics **10**(7), 852 (2021)
42. Tsekhmistrenko, S.I., Bityutskyy, V.S., Tsekhmistrenko, O.S., Horalskyi, L.P., Tymoshok, N.O., Spivak, M.Y.: Bacterial synthesis of nanoparticles: a green approach. Biosyst. Div. **28**(1), 9–17 (2020)
43. Singh, V.K., Singh, A.K.: Role of microbially synthesized nanoparticles in sustainable agriculture and environmental management. In Role of plant growth promoting microorganisms in sustainable agriculture and nanotechnology, pp. 55–73. Woodhead Publishing (2019)
44. Banerjee, K., Ravishankar Rai, V.: A review on mycosynthesis, mechanism, and characterization of silver and gold nanoparticles. BioNanoScience **8**(1), 17–31 (2018)
45. Hamida, R.S., Albasher, G., Bin-Meferij, M.M.: Oxidative stress and apoptotic responses elicited by nostoc-synthesized silver nanoparticles against different cancer cell lines. Cancers **12**(8), 2099 (2020)
46. Murillo-Rábago, E.I., Vilchis-Nestor, A.R., Juarez-Moreno, K., Garcia-Marin, L.E., Quester, K., Castro-Longoria, E.: Optimized synthesis of small and stable silver nanoparticles using intracellular and extracellular components of fungi: an alternative for bacterial inhibition. Antibiotics 11(6), 800 (2022)
47. Faramarzi, M.A., Sadighi, A.: Insights into biogenic and chemical production of inorganic nanomaterials and nanostructures. Adv. Coll. Interface. Sci. **189**, 1–20 (2013)
48. Iravani, S.: Bacteria in nanoparticle synthesis: current status and future prospects. Int. Scholarly Res. Notices **2014**(1), 359316 (2014)
49. Rai, M., Duran, N. (Eds.): Metal nanoparticles in microbiology. Springer Science & Business Media. (2011)
50. Chokriwal, A., Sharma, M.M., Singh, A.: Biological synthesis of nanoparticles using bacteria and their applications. Am. J. PharmTech Res. **4**(6), 38–61 (2014)
51. Samadi, N., Golkaran, D., Eslamifar, A., Jamalifar, H., Fazeli, M.R., Mohseni, F.A.: Intra/extracellular biosynthesis of silver nanoparticles by an autochthonous strain of Proteus mirabilis isolated from photographic waste. J. Biomed. Nanotechnol. **5**(3), 247–253 (2009)
52. Prakash, A., Sharma, S., Ahmad, N., Ghosh, A., Sinha, P.: Bacteria mediated extracellular synthesis of metallic nanoparticles. Int. Res. J. Biotechnol. **1**(5), 071–079 (2010)
53. Reddy, A.S., Chen, C.Y., Chen, C.C., Jean, J.S., Chen, H.R., Tseng, M.J., Wang, J.C.: Biological synthesis of gold and silver nanoparticles mediated by the bacteria Bacillus subtilis. J. Nanosci. Nanotechnol. **10**(10), 6567–6574 (2010)
54. Shivaji, S., Madhu, S., Singh, S.: Extracellular synthesis of antibacterial silver nanoparticles using psychrophilic bacteria. Process Biochem. **46**(9), 1800–1807 (2011)

55. Pallath, N., Francis, B., Devanesan, S., Asemi, N., AlSalhi, M.S., Natarajan, S.: Synthesis of silver nanoparticles from marine bacteria and evaluation of antimicrobial, antifungal and cytotoxic effects. J. King Saud Univ. Sci. **36**(2), 103073 (2024)
56. Saleh, G.M.: Green synthesis concept of nanoparticles from environmental bacteria and their effects on pathogenic bacteria. Iraqi J. Sci., 1289–1297 (2020)
57. Yumei, L., Yamei, L., Qiang, L., Jie, B.: Rapid biosynthesis of silver nanoparticles based on flocculation and reduction of an exopolysaccharide from Arthrobacter sp. B4: its antimicrobial activity and phytotoxicity. J. Nanomaterials (1), 9703614 (2017)
58. Singh, P., Kim, Y.J., Zhang, D., Yang, D.C.: Biological synthesis of nanoparticles from plants and microorganisms. Trends Biotechnol. **34**(7), 588–599 (2016)
59. Gade, A.K., Bonde, P., Ingle, A.P., Marcato, P.D., Duran, N., Rai, M.K.: Exploitation of Aspergillus niger for synthesis of silver nanoparticles. J. Biobased Mater. Bioenergy **2**(3), 243–247 (2008)
60. Gahlawat, G., Choudhury, A.R.: A review on the biosynthesis of metal and metal salt nanoparticles by microbes. RSC Adv. **9**(23), 12944–12967 (2019)
61. Guilger-Casagrande, M., Lima, R.D.: Synthesis of silver nanoparticles mediated by fungi: a review. Frontiers Bioeng. Biotechnol. **7**, 287 (2019)
62. Du, L., Xu, Q., Huang, M., Xian, L., Feng, J.X.: Synthesis of small silver nanoparticles under light radiation by fungus Penicillium oxalicum and its application for the catalytic reduction of methylene blue. Mater. Chem. Phys. **160**, 40–47 (2015)
63. Ahmad, A., Mukherjee, P., Senapati, S., Mandal, D., Khan, M.I., Kumar, R., Sastry, M.: Extracellular biosynthesis of silver nanoparticles using the fungus Fusarium oxysporum. Colloids Surf. B **28**(4), 313–318 (2003)
64. Ahmad, A., Mukherjee, P., Mandal, D., Senapati, S., Khan, M.I., Kumar, R., Sastry, M.: Enzyme mediated extracellular synthesis of CdS nanoparticles by the fungus, Fusarium oxysporum. J. Am. Chem. Soc. **124**(41), 12108–12109 (2002)
65. Castro-Longoria, E., Moreno-Velázquez, S.D., Vilchis-Nestor, A.R., Arenas-Berumen, E., Avalos-Borja, M.: Production of platinum nanoparticles and nanoaggregates using Neurospora crassa. J. Microbiol. Biotechnol. **22**(7), 1000–1004 (2012)
66. Das, V.L., Thomas, R., Varghese, R.T., Soniya, E.V., Mathew, J., Radhakrishnan, E.K.: Extracellular synthesis of silver nanoparticles by the Bacillus strain CS 11 isolated from industrialized area. 3 Biotech **4**(2), 121–126 (2014)
67. Pantidos, N., Horsfall, L.E.: Biological synthesis of metallic nanoparticles by bacteria, fungi and plants. J. Nanomed. Nanotechnol. **5**(5), 1 (2014)
68. Bharde, A., Rautaray, D., Bansal, V., Ahmad, A., Sarkar, I., Yusuf, S.M., Sastry, M.: Extracellular biosynthesis of magnetite using fungi. Small **2**(1), 135–141 (2006)
69. Gericke, M., Pinches, A.: Biological synthesis of metal nanoparticles. Hydrometallurgy **83**(1–4), 132–140 (2006)
70. Das, S.K., Das, A.R., Guha, A.K.: Microbial synthesis of multishaped gold nanostructures. Small **6**(9), 1012–1021 (2010)
71. Sun, Y., Xia, Y.: Shape-controlled synthesis of gold and silver nanoparticles. Science **298**(5601), 2176–2179 (2002)
72. Boroumand Moghaddam, A., Namvar, F., Moniri, M., Md. Tahir, P., Azizi, S., Mohamad, R.: Nanoparticles biosynthesized by fungi and yeast: a review of their preparation, properties, and medical applications. Molecules **20**(9), 16540–16565 (2015)
73. Mukherjee, A., Sarkar, D., Sasmal, S.: A review of green synthesis of metal nanoparticles using algae. Front. Microbiol. **12**, 693899 (2021)
74. Huston, M., DeBella, M., DiBella, M., Gupta, A.: Green synthesis of nanomaterials. Nanomaterials **11**(8), 2130 (2021)
75. Khanna, P., Kaur, A., Goyal, D.: Algae-based metallic nanoparticles: Synthesis, characterization and applications. J. Microbiol. Methods **163**, 105656 (2019)
76. Singh, S.K., Sharma, A., Sharma, L., Sundaram, S.: Green synthesis of algal nanoparticles: harnessing Nature's biofactories for sustainable nanomaterials. In: Biogenic Wastes-Enabled Nanomaterial Synthesis: Applications in Environmental Sustainability, pp. 257–284. Springer Nature Switzerland, Cham (2024)

77. Azizi, S., Ahmad, M.B., Namvar, F., Mohamad, R.: Green biosynthesis and characterization of zinc oxide nanoparticles using brown marine macroalga Sargassum muticum aqueous extract. Mater. Lett. **116**, 275–277 (2014)
78. Rajeshkumar, S., Kannan, C., Annadurai, G.: Green synthesis of silver nanoparticles using marine brown algae Turbinaria conoides and its antibacterial activity. Int. J. Pharm. Bio. Sci. **3**(4), 502–510 (2012)
79. Liu, B., Xie, J., Lee, J.Y., Ting, Y.P., Chen, J.P.: Optimization of high-yield biological synthesis of single-crystalline gold nanoplates. J. Phys. Chem. B **109**(32), 15256–15263 (2005)
80. Ghodake, G., Lee, D.S.: Biological synthesis of gold nanoparticles using the aqueous extract of the brown algae Laminaria japonica. J. Nanoelectron. Optoelectron. **6**(3), 268–271 (2011)
81. Rajeshkumar, S., Malarkodi, C., Gnanajobitha, G., Paulkumar, K., Vanaja, M., Kannan, C., Annadurai, G.: Seaweed-mediated synthesis of gold nanoparticles using Turbinaria conoides and its characterization. J. Nanostr. Chem. **3**(1), 44 (2013)
82. Chaudhary, R., Nawaz, K., Khan, A.K., Hano, C., Abbasi, B.H., Anjum, S.: An overview of the algae-mediated biosynthesis of nanoparticles and their biomedical applications. Biomolecules **10**(11), 1498 (2020)
83. Priyadharshini, R.I., Prasannaraj, G., Geetha, N., Venkatachalam, P.: Microwave-mediated extracellular synthesis of metallic silver and zinc oxide nanoparticles using macro-algae (Gracilaria edulis) extracts and its anticancer activity against human PC3 cell lines. Appl. Biochem. Biotechnol. **174**(8), 2777–2790 (2014)
84. Doman, K.M., Gharieb, M.M., Abd El-Monem, A.M., Morsi, H.H.: Synthesis of silver and copper nanoparticle using Spirulina platensis and evaluation of their anticancer activity. Int. J. Environ. Health Res. **34**(2), 661–673 (2024)
85. Kushnerova, N.F., Fomenko, S.E., Sprygin, V.G., Kushnerova, T.V., Khotimchenko, Y.S., Kondrat'eva, E.V., Drugova, L.A.: An extract from the brown alga Laminaria japonica: a promising stress-protective preparation. Russian J. Mar. Biol. **36**(3), 209–214 (2010)
86. Khodashenas, B., Ghorbani, H.R.: Synthesis of silver nanoparticles with different shapes. Arab. J. Chem. **12**(8), 1823–1838 (2019)
87. Behera, M., Behera, P.R., Bhuyan, P.P., Singh, L., Pradhan, B.: Algal nanoparticles and their antibacterial activity: current research status and future prospectives. Drugs and Drug Candidates **2**(3), 554–570 (2023)
88. Barani, M., Masoudi, M., Mashreghi, M., Makhdoumi, A., Eshghi, H.: Cell-free extract assisted synthesis of ZnO nanoparticles using aquatic bacterial strains: biological activities and toxicological evaluation. Int. J. Pharm. **606**, 120878 (2021)
89. Hosseini, F., Lashani, E., Moghimi, H.: Simultaneous bioremediation of phenol and tellurite by Lysinibacillus sp. EBL303 and characterization of biosynthesized Te nanoparticles. Scientific Reports **13**(1), 1243 (2023)
90. Ahmad, M.S., Yasser, M.M., Sholkamy, E.N., Ali, A.M., Mehanni, M.M.: Anticancer activity of biostabilized selenium nanorods synthesized by Streptomyces bikiniensis strain Ess_amA-1. Int. J. Nanomed., 3389–3401 (2015)
91. Wilson, J.J., Mahalakshmi, S., Thangaraj, R., Mahendran, S., Sivakumar, T., Sudarmani, D.N.P., Venkidasamy, B.: Biological synthesis of silver nanoparticles using klebsiella pneumoniae (ON640793) and assessment of its antibacterial, larvicidal and biotoxicity efficiency. BioNanoScience **15**(1), 84 (2025)
92. Patil, M.P., Kang, M.J., Niyonizigiye, I., Singh, A., Kim, J.O., Seo, Y.B., Kim, G.D.: Extracellular synthesis of gold nanoparticles using the marine bacterium Paracoccus haeundaensis BC74171T and evaluation of their antioxidant activity and antiproliferative effect on normal and cancer cell lines. Colloids Surf. B **183**, 110455 (2019)
93. Mathivanan, K., Selva, R., Chandirika, J.U., Govindarajan, R.K., Srinivasan, R., Annadurai, G., Duc, P.A.: Biologically synthesized silver nanoparticles against pathogenic bacteria: synthesis, calcination and characterization. Biocatal. Agric. Biotechnol. **22**, 101373 (2019)
94. Malhotra, A., Dolma, K., Kaur, N., Rathore, Y.S., Mayilraj, S., Choudhury, A.R.: Biosynthesis of gold and silver nanoparticles using a novel marine strain of Stenotrophomonas. Biores. Technol. **142**, 727–731 (2013)

95. Sharma, N., Pinnaka, A.K., Raje, M., Fnu, A., Bhattacharyya, M.S., Choudhury, A.R.: Exploitation of marine bacteria for production of gold nanoparticles. Microb. Cell Fact. **11**(1), 86 (2012)
96. Mewada, A., Oza, G., Pandey, S., Sharon, M.: Extracellular synthesis of gold nanoparticles using Pseudomonas denitrificans and comprehending its stability. J. Microbiol. Biotechnol. **2**(4), 493–499 (2012)
97. Ibrahem, K.H., Salman, J.A.S., Ali, F.A.: Effect of titanium nanoparticles biosynthesis by lactobacillus crispatus on urease, hemolysin& biofilm forming by some bacteria causing recurrent uti in iraqi women. Europ. Scientific J. **10**(9) (2014)
98. Narayanan, K.B., Sakthivel, N.: Mycocrystallization of gold ions by the fungus Cylindrocladium floridanum. World J. Microbiol. Biotechnol. **29**(11), 2207–2211 (2013)
99. Munawer, U., Raghavendra, V.B., Ningaraju, S., Krishna, K.L., Ghosh, A.R., Melappa, G., Pugazhendhi, A.: Biofabrication of gold nanoparticles mediated by the endophytic Cladosporium species: Photodegradation, in vitro anticancer activity and in vivo antitumor studies. Int. J. Pharm. **588**, 119729 (2020)
100. Acay, H.: Utilization of Morchella esculenta-mediated green synthesis golden nanoparticles in biomedicine applications. Prep. Biochem. Biotechnol. **51**(2), 127–136 (2021)
101. Gürsoy, N.: Fungus-mediated synthesis of silver nanoparticles (agnp) and inhibitory effect on Aspergillus spp. in combination with antifungal agent. Cumhuriyet Sci. J. **41**(1), 311–318 (2020)
102. Pareek, V., Bhargava, A., Panwar, J.: Biomimetic approach for multifarious synthesis of nanoparticles using metal tolerant fungi: a mechanistic perspective. Mater. Sci. Eng. B **262**, 114771 (2020)
103. Govindappa, M., Lavanya, M., Aishwarya, P., Pai, K., Lunked, P., Hemashekhar, B.,... Raghavendra, V.B.: Synthesis and characterization of endophytic fungi, Cladosporium perangustum mediated silver nanoparticles and their antioxidant, anticancer and nano-toxicological study. BioNanoScience**10**(4), 928–941 (2020)
104. Kadam, V.V., Ettiyappan, J.P., Balakrishnan, R.M.: Mechanistic insight into the endophytic fungus mediated synthesis of protein capped ZnO nanoparticles. Mater. Sci. Eng. B **243**, 214–221 (2019)
105. Chatterjee, S., Mahanty, S., Das, P., Chaudhuri, P., Das, S.: Biofabrication of iron oxide nanoparticles using manglicolous fungus Aspergillus niger BSC-1 and removal of Cr (VI) from aqueous solution. Chem. Eng. J. **385**, 123790 (2020)
106. Mahanty, S., Bakshi, M., Ghosh, S., Chatterjee, S., Bhattacharyya, S., Das, P.,... Chaudhuri, P.: Green synthesis of iron oxide nanoparticles mediated by filamentous fungi isolated from Sundarban mangrove ecosystem, India. BioNanoScience**9**(3), 637–651 (2019)
107. Raliya, R., Tarafdar, J.C.: Biosynthesis and characterization of zinc, magnesium and titanium nanoparticles: an eco-friendly approach. Int. Nano Lett. **4**(1), 93 (2014)
108. Salem, D.M., Ismail, M.M., Aly-Eldeen, M.A.: Biogenic synthesis and antimicrobial potency of iron oxide (Fe_3O_4) nanoparticles using algae harvested from the Mediterranean Sea. Egypt. Egypt. J. Aquatic Res. **45**(3), 197–204 (2019)
109. González-Ballesteros, N., Rodríguez-Argüelles, M.C., Lastra-Valdor, M., González-Mediero, G., Rey-Cao, S., Grimaldi, M.,... & Bigi, F.: Synthesis of silver and gold nanoparticles by Sargassum muticum biomolecules and evaluation of their antioxidant activity and antibacterial properties. J. Nanostr. Chem.**10**(4), 317–330 (2020)
110. Bhuyar, P., Rahim, M.H.A., Sundararaju, S., Ramaraj, R., Maniam, G.P., Govindan, N.: Synthesis of silver nanoparticles using marine macroalgae Padina sp. and its antibacterial activity towards pathogenic bacteria. Beni-Suef Univ. J. Basic Appl. Sci. **9**(1), 1–15 (2020)
111. El-Sheekh, M.M., Shabaan, M.T., Hassan, L., Morsi, H.H.: Antiviral activity of algae biosynthesized silver and gold nanoparticles against Herps Simplex (HSV-1) virus in vitro using cell-line culture technique. Int. J. Environ. Health Res. **32**(3), 616–627 (2022)
112. Venkatesan, J., Manivasagan, P., Kim, S.K., Kirthi, A.V., Marimuthu, S., Rahuman, A.A.: Marine algae-mediated synthesis of gold nanoparticles using a novel Ecklonia cava. Bioprocess Biosyst. Eng. **37**(8), 1591–1597 (2014)

113. El-Naggar, N.E.A., El-Shall, H., Elyamny, S., Hamouda, R.A., Eltarahony, M.: Novel algae-mediated biosynthesis approach of chitosan nanoparticles using Ulva fasciata extract, process optimization, characterization and their flocculation performance. Int. J. Biol. Macromol. **282**, 136925 (2024)
114. Fayyad, R.J., Alanisi, E.M.A.A., Ali, A.N.M.: Algae-mediated green synthesis of dextran-coated titanium nanoparticles and their cytotoxic potential against MCF7 breast cancer cells. Reports of Biochem. Molec. Biol. **13**(3), 358 (2024)
115. Bhilkar, P.R., Burade, N., Yerpude, S.T., Potbhare, A.K., Ahmed, M.Z., Dadure, K.M.,... Chaudhary, R.G.: Tetradesmus obliquus-mediated platinum nanoparticles: sustainable approach, molecular docking and antibacterial assessment. Microbial Pathogenesis**200**, 107342 (2025)
116. Elsayed, A., Ali, E., El Shehawy, A.: Algae-mediated synthesis and structural characterization of iron nanoparticles using Galaxaura rugosa seaweed. Egypt. J. of Phycol. **25**(1), 39–45 (2024)
117. Thirupathi, B., Pongen, Y. L., Kaveriyappan, G.R., Dara, P.K., Rathinasamy, S., Vinayagam, S.,... Sekar, S.K.R.: Padina boergesenii mediated synthesis of Se–ZnO bimetallic nanoparticles for effective anticancer activity. Frontiers Microbiol. **15**, 1358467 (2024)
118. Ganeshan, A., Periakaruppan, R., Vanathi, P., Thirumalaisamy, S.K., Vijai Selvaraj, K.S., Moskovskikh, D.: U lva rigida–mediated silver nanoparticles: synthesis, characterization, and antibacterial activity. Biomass Conversion and Biorefinery, 1–8 (2024)
119. Barathikannan, K., Chelliah, R., Selvakumar, V., Elahi, F., Rubab, M., Sanyal, S.,... Oh, D.H.: Plant-based metabolites and their uses in nanomaterials synthesis: an overview. Secondary metabolites based green synthesis of nanomaterials and their applications, 1–22 (2023)
120. Abisharani, J.M., Devikala, S., Kumar, R.D., Arthanareeswari, M., Kamaraj, P.: Green synthesis of TiO_2 nanoparticles using Cucurbita pepo seeds extract. Mater. Today: Proc. **14**, 302–307 (2019)
121. Jafarzadeh, S., Nooshkam, M., Zargar, M., Garavand, F., Ghosh, S., Hadidi, M., Forough, M.: Green synthesis of nanomaterials for smart biopolymer packaging: challenges and outlooks. J. Nanostr. Chem. **14**(2), 113–136 (2024)
122. Dhir, R., Chauhan, S., Subham, P., Kumar, S., Sharma, P., Shidiki, A., Kumar, G.: Plant-mediated synthesis of silver nanoparticles: unlocking their pharmacological potential–a comprehensive review. Frontiers Bioeng. Biotechnol. **11**, 1324805 (2024)
123. Prasad, A.R., Williams, L., Garvasis, J., Shamsheera, K.O., Basheer, S.M., Kuruvilla, M., Joseph, A.: Applications of phytogenic ZnO nanoparticles: a review on recent advancements. J. Mol. Liq. **331**, 115805 (2021)
124. Thomas, S., Gonsalves, R.A., Jose, J., Zyoud, S.H., Prasad, A.R., Garvasis, J.: Plant-based synthesis, characterization approaches, applications and toxicity of silver nanoparticles: a comprehensive review. J. Biotechnol. **394**, 135–149 (2024)
125. Pirsaheb, M., Gholami, T., Seifi, H., Dawi, E.A., Said, E.A., Hamoody, A.H.M.,... Salavati-Niasari, M.: Green synthesis of nanomaterials by using plant extracts as reducing and capping agents. Environ. Sci. Pollut. Res. **31**(17), 24768–24787 (2024)
126. Duan, H., Wang, D., Li, Y.: Green chemistry for nanoparticle synthesis. Chem. Soc. Rev. **44**(16), 5778–5792 (2015)
127. Shah, M., Fawcett, D., Sharma, S., Tripathy, S.K., Poinern, G.E.J.: Green synthesis of metallic nanoparticles via biological entities. Materials **8**(11), 7278–7308 (2015)
128. Sundararajan, N., Habeebsheriff, H.S., Dhanabalan, K., Cong, V.H., Wong, L.S., Rajamani, R., Dhar, B.K.: Mitigating global challenges: harnessing green synthesized nanomaterials for sustainable crop production systems. Global Chall. **8**(1), 2300187 (2024)
129. Husen, A., Iqbal, M.: Nanomaterials and plant potential: an overview. Nanomaterials and plant potential, 3–29 (2019)
130. Behzad, F., Naghib, S.M., Tabatabaei, S.N., Zare, Y., Rhee, K.Y.: An overview of the plant-mediated green synthesis of noble metal nanoparticles for antibacterial applications. J. Ind. Eng. Chem. **94**, 92–104 (2021)

131. Parveen, K., Banse, V., Ledwani, L.: Green synthesis of nanoparticles: Their advantages and disadvantages. In AIP conference proceedings (Vol. 1724, No. 1, p. 020048). AIP Publishing LLC (2016)
132. Dos Santos, M.S., Antunes Filho, S., Backx, B.P.: Bionanotechnology in agriculture: a one health approach. Life **13**(2), 509 (2023)
133. Jadoun, S., Arif, R., Jangid, N.K., Meena, R.K.: Green synthesis of nanoparticles using plant extracts: a review. Environ. Chem. Lett. **19**(1), 355–374 (2021)
134. Antunes Filho, S., Dos Santos, M.S., Dos Santos, O.A.L., Backx, B.P., Soran, M.L., Opriş, O., Bououdina, M.: Biosynthesis of nanoparticles using plant extracts and essential oils. Molecules **28**(7), 3060 (2023)
135. Agarwal, H., Kumar, S.V., Rajeshkumar, S.: A review on green synthesis of zinc oxide nanoparticles–an eco-friendly approach. Resour. Eff. Technol. **3**(4), 406–413 (2017)
136. Khan, M., Al-Marri, A.H., Khan, M., Shaik, M.R., Mohri, N., Adil, S.F.,... Siddiqui, M.R.H.: Green approach for the effective reduction of graphene oxide using Salvadora persica L. root (Miswak) extract. Nanoscale Res. Lett. **10**(1), 281 (2015)
137. Rana, A., Pathak, S., Lim, D.K., Kim, S.K., Srivastava, R., Sharma, S.N., Verma, R.: Recent advancements in plant-and microbe-mediated synthesis of metal and metal oxide nanomaterials and their emerging antimicrobial applications. ACS Appl. Nano Mater. **6**(10), 8106–8134 (2023)
138. Jayachandran, A., Aswathy, T.R., Nair, A.S.: Green synthesis and characterization of zinc oxide nanoparticles using Cayratia pedata leaf extract. Biochem. Biophys. Reports **26**, 100995 (2021)
139. Chandran, K., Song, S., Yun, S.I.: Effect of size and shape controlled biogenic synthesis of gold nanoparticles and their mode of interactions against food borne bacterial pathogens. Arab. J. Chem. **12**(8), 1994–2006 (2019)
140. Rani, N., Singh, P., Kumar, S., Kumar, P., Bhankar, V., Kumar, K.: Plant-mediated synthesis of nanoparticles and their applications: a review. Mater. Res. Bull. **163**, 112233 (2023)
141. Govindasamy, R., Govindarasu, M., Alharthi, S.S., Mani, P., Bernaurdshaw, N., Gomathi, T., Thiruvengadam, M.: Sustainable green synthesis of yttrium oxide (Y2O3) nanoparticles using Lantana camara leaf extracts: Physicochemical characterization, photocatalytic degradation, antibacterial, and anticancer potency. Nanomaterials **12**(14), 2393 (2022)
142. Sevinc-Sasmaz, C., Erci, F., Torlak, E., Yöntem, M.: Characterization of Silver Nanoparticles Synthesized Using Hypericum perforatum L. and Their Effects on Staphylococcus aureus. Microscopy Res. Techn. **88**(8), 2321–2332 (2025)
143. Keerthiga, G., Sengottuvel, R., Obli Prakash, P.C., Maghimaa, M.: Enhancing antibacterial and antibiofilm activity using Gymnema sylvestre Mediated Silver nanoparticles: a promising approach in infection control. Biomed. Mater. Dev., 1–16 (2025)
144. Idrees, S., Abbasi, B.A., Iqbal, J., Kazi, M.: Euphorbia serpens extracts mediated synthesis of copper oxide nanoparticles and their biological applications. Heliyon **11**(1) (2025)
145. Rauf, S., Hameed, H., Tariq, M., Afareen, A., Gulfaraz, S., AlKubaisi, N.A., Elshikh, M.S.: Phytochemical-mediated synthesis and characterization of silver nanoparticles using mirabilis jalapa leaf extract and their antibacterial. Microsc. Res. Tech. **88**(6), 1795–1805 (2025)
146. Baskaran, A.: Plant mediated synthesis of silver nanoparticles from barleria longiflora leaves with evolution of their antidiabetic properties. Indian J. Nat. Sci. **16**(90), 95436–95445 (2025)
147. Azizian-Shermeh, O., Kaedi, F., Taherizadeh, M., Qasemi, A.: Convenient and cost-effective plant-mediated synthesis of gold nanoparticles using Moringa Oleifera L. and study of their antimicrobial activities. Colloid Nanosci. J. **2**(4), 397–408 (2025)
148. Abdel-Megeed, A., Abdel-Rheim, K.H., Kordy, A.M., El-Ghannam, S.A., Salem, M.Z., Adel, M.M.: Duckweed (Lemna minor) biomass aqueous extract-mediated synthesis of silver nanoparticles as a novel formulation era for long-term efficacy against Tribolium castaneum and Sitophilus oryzae adults on wheat seeds. Biomass Conv. Biorefinery **15**(4), 5429–5446 (2025)
149. Sheikh, T., Nagendran, V., Vasant, K.S., Mallya, U., Mutalik, S., Khan, F., Sudhakara, S.M.: Acacia auriculiformis mediated synthesis of silver nanoparticles for the sensitive and rapid electrochemical sensing of nitrite in water sample. Microchem. J. **211**, 113162 (2025)

150. Donmez, S.: Phyto-mediated synthesis of zinc oxide nanoparticles using black chokeberry fruit (Aronia melanocarpa L.) extracts for promising antioxidant, antibacterial, antidiabetic, and photocatalytic activities. J. Cluster Sci. **36**(4), 119 (2025)
151. Mohammad, M.S., Perugu, S.: An aromatic plant bioactive compound corymbosin from Vitex negundo–mediated synthesis of zinc oxide nanoparticles, characterization, and their bioactivity against selected cancer cell lines and microbial pathogens. Biomass Conv. Biorefinery **15**(2), 2097–2110 (2025)
152. Şevk, G.K., Dilber, T., Ullen, N.B.: Green synthesis of zinc oxide nanoparticles via Zingiber officinale/PEG biopolymer blend matrix: Optimization, physicochemical characterization, antioxidant and photocatalytic activity. Mater. Sci. Eng. B **317**, 118240 (2025)
153. Yadav, S.R., Prakash, J., Kumari, M., Rinki, K., Sonkar, P.K., Kumar, N.: Phytochemical mediated synthesis of CuO nanoparticles and its evaluation for electrocatalytic oxygen reduction reaction activity. ChemistrySelect **10**(14), e202501100 (2025)
154. Rajashekarappa, V., Sadathulla, K.K.P., Hussain, S., Nayaka, S., Nagaraja, S.K.: Biomimetic approach to phyto-mediated synthesis of zinc oxide nanoparticles from Waltheria Indica root extract and exploring their potential in-vitro antibactertial activity. J. Chem. Health Risks, 1 (2025)
155. Duraisamy, R., Arizo, A., Yilma, B.: Antibacterial and antioxidant activity of plant-mediated green synthesized silver nanoparticles using Cissus quadrangularis aqueous extract. Bull. Chem. Soc. Ethiop. **39**(3), 483–502 (2025)
156. Madaniyah, L., Fiddaroini, S., Hayati, E.K., Rahman, M.F., Sabarudin, A.: Biosynthesis, characterization, and in-vitro anticancer effect of plant-mediated silver nanoparticles using Acalypha indica Linn: In-silico approach. OpenNano **21**, 100220 (2025)
157. Singh, R.K., Nallaswamy, D., Rajeshkumar, S., Varghese, S.S.: Green synthesis of silver nanoparticles using neem and turmeric extract and its antimicrobial activity of plant mediated silver nanoparticles. J. Oral Biol. Craniofacial Res. **15**(2), 395–401 (2025)
158. Said, A.H., Shaibah, F., Moustafa, M., Elamary, R.B.: Plant-mediated nickel oxide nanoparticles show species-dependent antibacterial, antioxidant, anti-inflammatory and antidiabetic activities. Sci. Rep. **15**(1), 31096 (2025)
159. Takcı, D.K., Ozdenefe, M.S., Huner, T., Takcı, H.A.M.: Plant-mediated green route to the synthesis of zinc oxide nanoparticles: In vitro antibacterial potential. J. Aust. Ceram. Soc. **61**(1), 31–39 (2025)
160. Dwivedi, V.K., Pandey, S., Singh, N., Kumar, S., Shukla, P., Mishra, A.,... Srivastava, A.: Effective photocatalytic and antibacterial activity of ecofriendly ZnO nanoparticles synthesized using Citrus sinensis Osbeck peels. Next Mater. **9**, 101002 (2025)
161. Karan, T., Moran, B.: Alfalfa (Medicago sativa) leaves Mediated synthesis of silver nanoparticles and assessment of their cytotoxic effects on various cancer cell lines. Biology Bulletin **52**(2), 55 (2025)
162. Sishu, N.K., Karunakaran, M.K.R., Hadkar, V.M., Mohanty, C., Sharmila, A., Selvaraj, C.I., Babu, N.G.: Phyto-mediated synthesis of SnO_2 nanoparticles using Croton malabaricus Bedd. for its antioxidant, antibacterial, hemocompatibility properties and photocatalytic activity. New J. Chem. **49**(2), 536–552 (2025)
163. Ovonramwen, O.B.: Bio-reducing properties of bryophyllum pinnatum aqueous leaves extract for the mediated synthesis of nickel oxide nanoparticles and its antimicrobial activities (2025)
164. Altaf, I., Firdaus, N., Wani, I.A., Khan, S., Owais, M.: Iris kashmiriana mediated synthesis of gold nanoparticles: characterization and its antibacterial implications on dental appliances. BioNanoScience **15**(1), 133 (2025)
165. Slman, D.K., Ihsan, I., Ahmed, H.K., Jawad, A.A.: Biosynthesis of silver nanoparticles by piper Nigrum (Black Pepper) extract and their application as antibacterial. Adv. J. Chem. Section A **8**, 864–873 (2025)
166. Wehbe, N., Mesmar, J.E., El Kurdi, R., Al-Sawalmih, A., Badran, A., Patra, D., Baydoun, E.: Halodule uninervis extract facilitates the green synthesis of gold nanoparticles with anticancer activity. Sci. Rep. **15**(1), 4286 (2025)

167. Varadavenkatesan, T., Nagendran, V., Vinayagam, R., Goveas, L.C., Selvaraj, R.: Green synthesis of silver nanoparticles using Lagerstroemia speciosa fruit extract: catalytic efficiency in dye degradation. Mater. Technol. **40**(1), 2463955 (2025)
168. Anushya, P., Sukumaran, G., Suvarna, B.: Acacia Concinna extracts mediated synthesis of Copper oxide nanoparticles against pathogenic bacteria. Cuestiones de Fisioterapia **54**(3), 5210–5220 (2025)
169. Akhras, N., Çelekli, A., Bozkurt, H.: Enhanced antimicrobial activity of green-synthesized artemisia-ZnO nanoparticles: a comparative study with pure ZnO nanoparticles and plant extract. Foods **14**(14), 2449 (2025)
170. Purushotham, D., Mavinakere Ramesh, A., Shetty Thimmappa, D., Kalegowda, N., Hittanahallikoppal Gajendramurthy, G., Kollur, S.P., Mahadevamurthy, M.: Green synthesis of zinc oxide nanoparticles using aqueous extract of pavonia zeylanica to mediate photocatalytic degradation of methylene blue: studies on reaction kinetics, reusability and mineralization. Int. J. Mol. Sci. **26**(10), 4739 (2025)
171. Gour, A., Jain, N.K.: Advances in green synthesis of nanoparticles. Artificial cells, nanomedicine, and biotechnology **47**(1), 844–851 (2019)
172. Santhosh, P.B., Genova, J., Chamati, H.: Green synthesis of gold nanoparticles: an eco-friendly approach. Chemistry **4**(2), 345–369 (2022)
173. Rauwel, P., Küünal, S., Ferdov, S., Rauwel, E.: A review on the green synthesis of silver nanoparticles and their morphologies studied via TEM. Adv. Mater. Sci. Eng. **2015**(1), 682749 (2015)
174. Uzair, B., Liaqat, A., Iqbal, H., Menaa, B., Razzaq, A., Thiripuranathar, G.,... Menaa, F.: Green and cost-effective synthesis of metallic nanoparticles by algae: safe methods for translational medicine. Bioengineering**7**(4), 129 (2020)
175. Wang, Y., O'connor, D., Shen, Z., Lo, I.M., Tsang, D.C., Pehkonen, S.,... Hou, D.: Green synthesis of nanoparticles for the remediation of contaminated waters and soils: constituents, synthesizing methods, and influencing factors. J. Cleaner Produ. **226**, 540–549 (2019)
176. Malik, A.Q., Mir, T.U.G., Kumar, D., Mir, I.A., Rashid, A., Ayoub, M., Shukla, S.: A review on the green synthesis of nanoparticles, their biological applications, and photocatalytic efficiency against environmental toxins. Environ. Sci. Pollut. Res. **30**(27), 69796–69823 (2023)
177. Hussain, I., Singh, N.B., Singh, A., Singh, H., Singh, S.C.: Green synthesis of nanoparticles and its potential application. Biotech. Lett. **38**(4), 545–560 (2016)
178. El-Seedi, H.R., El-Shabasy, R.M., Khalifa, S.A., Saeed, A., Shah, A., Shah, R.,... Guo, W.: Metal nanoparticles fabricated by green chemistry using natural extracts: Biosynthesis, mechanisms, and applications. RSC Adv. **9**(42), 24539–24559 (2019)
179. Malik, P., Shankar, R., Malik, V., Sharma, N., Mukherjee, T.K.: Green chemistry based benign routes for nanoparticle synthesis. J. Nanoparticles **2014**(1), 302429 (2014)
180. Nadaroglu, H., Güngör, A.A., Ince, S.: Synthesis of nanoparticles by green synthesis method. Int. J. Innov. Res. Rev. **1**(1), 6–9 (2017)
181. Al-Sahlany, S.T.G., Niamah, A.K., Verma, D.K., Prabhakar, P., Patel, A.R., Thakur, M., Singh, S.: Applications of green synthesis of nanoparticles using microorganisms in food and dairy products. Processes **13**(5), 1560 (2025)
182. Subramaniyam, V., Subashchandrabose, S.R., Ganeshkumar, V., Thavamani, P., Chen, Z., Naidu, R., Megharaj, M.: Cultivation of Chlorella on brewery wastewater and nano-particle biosynthesis by its biomass. Biores. Technol. **211**, 698–703 (2016)
183. Priya, N., Kaur, K., Sidhu, A.K.: Green synthesis: an eco-friendly route for the synthesis of iron oxide nanoparticles. Frontiers Nanotechnol. **3**, 655062 (2021)
184. Arshad, F., Naikoo, G.A., Hassan, I.U., Chava, S.R., El-Tanani, M., Aljabali, A.A., Tambuwala, M.M.: Bioinspired and green synthesis of silver nanoparticles for medical applications: a green perspective. Appl. Biochem. Biotechnol. **196**(6), 3636–3669 (2024)
185. Ahmad, S., Ahmad, S., Ali, S., Esa, M., Khan, A., Yan, H.: Recent advancements and unexplored biomedical applications of green synthesized Ag and Au nanoparticles: a review. Int. J. Nanomed., 3187–3215 (2024)

186. El-Saadony, M.T., Fang, G., Yan, S., Alkafaas, S.S., El Nasharty, M.A., Khedr, S.A., AbuQamar, S.F.: Green synthesis of zinc oxide nanoparticles: preparation, characterization, and biomedical applications-a review. Int. J. Nanomed., 12889–12937 (2024)
187. Ying, S., Guan, Z., Ofoegbu, P.C., Clubb, P., Rico, C., He, F., Hong, J.: Green synthesis of nanoparticles: current developments and limitations. Environ. Technol. Innov. **26**, 102336 (2022)

Green Nanotechnology: Metal Oxide Nanomaterials for the Degradation of Toxic Dyes

Rajesh Kumar Meena, Shrishti Singh, Parul, Savita Sharma, and Aprajita Gaur

1 Introduction

The basic concept of nanotechnology dates back to 1959, when physicist Richard Feynman first spoke about the possibility of manipulating matter at a very small scale [1]. Broadly speaking, nanotechnology deals with comprehending and manipulating materials at the atomic and molecular level, enabling researchers to create materials with particular and adjustable characteristics [2]. Since then, the scientific community has paid close attention to metal nanoparticles (NPs) due to their unique physicochemical properties, especially in the last few decades [3]. It is well acknowledged that nanoparticles' unique behavior stems mostly from their extremely small size and enormous surface area in relation to their volume. Because of this, their physical and chemical characteristics are very different from those of bulk materials with the same chemical makeup [4–6]. Size and shape variations can have a significant impact on optical, electrical, and chemical reactivity at the nanoscale. Because of these properties, nanoparticles are fundamentally different from traditional materials and can be used in ways that bulk systems cannot.

Metal nanoparticles have been extensively investigated for application in various technological domains due to their advantageous optoelectronic and physicochemical characteristics. They are frequently used as chemical sensors, electronic device components, and catalysts. Additionally, their utilization in biomedical applications such as pharmaceutical formulations, therapeutic treatments, and diagnostic imaging has rapidly increased [7]. This broad applicability emphasizes how crucial it is to

R. K. Meena (✉) · S. Singh · A. Gaur
Department of Chemistry, Kalindi College, University of Delhi, Delhi, India
e-mail: rajeshkumarmeena@kalindi.du.ac.in

Parul
Centre of Advanced Studies, Department of Chemistry, University of Rajasthan, Jaipur, India

S. Sharma
Department of Physics, Kalindi College, University of Delhi, Delhi, India

G. Tailor et al. (eds.), *Nanomaterials*, Engineering Materials,
https://doi.org/10.1007/978-981-92-0678-0_2

comprehend how nanoscale characteristics affect material performance. Materials with at least one dimension in the 1–100 nm size range are often referred to as nanomaterials. Surface-related phenomena and quantum confinement effects predominate at this scale because a sizable portion of atoms or molecules are found at or close to the surface. The physical, chemical, and catalytic properties of nanomaterials are largely controlled by these nanoscale effects. As a result, characteristics including chemical activity, optical response, electronic band structure, and surface reactivity can vary significantly from those seen in bulk materials.

Because of these traits, nanomaterials frequently exhibit improved surface reactivity, altered optical and electrical properties, increased surface-to-volume ratios, and adjustable shape and functionality. Therefore, it is clear that the overall performance and application potential of nanomaterials are greatly influenced by both the synthesis technique and the ensuing physicochemical features. Customizing nanomaterials for particular uses hence requires careful control over production conditions.

2 Nanomaterial Synthesis

The two most popular techniques for creating nanoparticles are the top-down and bottom-up approaches [8]. Using physical methods including mechanical milling, lithography, and sputtering, bulk materials are progressively reduced to nanoscale structures in the top-down approach to nanoparticle synthesis. From a conceptual perspective, these methods are somewhat simpler to comprehend, but in reality, they have a number of limitations.

Controlling particle size and size distribution is one of the main drawbacks of top-down methods. Moreover, contaminants, lattice strain, or structural flaws may be introduced into the nanoparticles as a result of the high-energy procedures used during manufacturing [9]. These flaws may change the material's inherent qualities and, in certain situations, restrict how well it functions. Top-down approaches do not allow for controlled alteration at the atomic or subatomic scale, even though the basic structure and main properties of the bulk precursor are often kept [10]. Therefore, when precise control over nanoscale traits is needed, their utility becomes restricted.

Bottom-up techniques, on the other hand, use processes like assembly or self-assembly, driven by chemical or physical interactions, to create nanostructures from atomic or molecule building blocks [10]. Instead of fragmenting bulk materials, this method creates nanoparticles by gradually arranging smaller units. Sol-gel procedures, hydrothermal and solvothermal processes, chemical precipitation, and vapor-phase deposition techniques are examples of widely used bottom-up approaches [11]. Bottom-up methodologies can be roughly categorized into physical, chemical, and biological ways, depending on the type of synthesis pathway.

Figure 1 provides a schematic overview of several methods for synthesizing nanoparticles. The increased control that bottom-up techniques provide over crucial

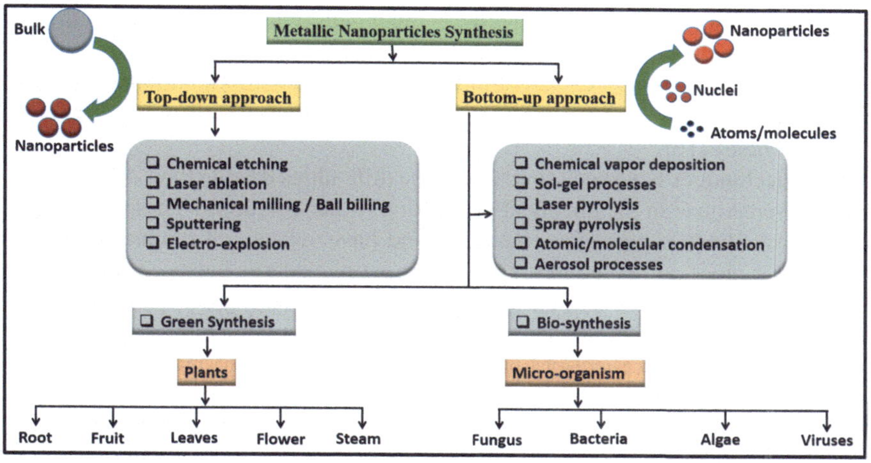

Fig. 1 Top-down and bottom-up methods exploiting various chemical, physical, and biological methods for the synthesis of nanoparticles

structural and functional properties of nanomaterials is one of their main benefits. By modifying the synthesis conditions, the main characteristics, including particle size, morphology, crystallinity, surface chemistry, and defect development, can be adjusted. The performance of nanomaterials in real-world applications is therefore influenced by a variety of factors, including particle size and shape, crystal phase, surface functionalization, defect levels, band-gap energy (particularly in semiconductors), optical absorption, charge-carrier dynamics, and structural robustness.

Bottom-up physical synthesis methods, which are usually carried out in tube furnaces with controlled air conditions, frequently depend on the evaporation of precursor materials followed by quick condensation [12]. Although these methods are typically repeatable and dependable, they have a number of drawbacks. One of these is that in order to control high temperatures throughout the synthesis process, they need extensive instrumentation and a significant energy input. Several improved physical methodologies for nanomaterial manufacturing have been developed in order to detect and mitigate some of these disadvantages [13]. Nevertheless, despite these developments, physical techniques often still need a lot of energy and are thought to be less environmentally friendly.

.

Wet chemical approaches have attracted a lot of attention as an alternative to physical methods since they are both economically viable and scalable to create nanomaterials. Chemical reduction, microwave-assisted synthesis, microemulsion-based synthesis, photochemical reduction, and electrochemistry are examples of wet chemical processes [14]. Chemical reduction, electrochemistry, irradiation-assisted chemical synthesis, and pyrolysis-based synthesis are the broad categories of chemical synthesis techniques [15]. Metal precursors and corresponding reducing agents are used in chemical synthesis methods, coupled with stabilizers or capping ligands that are essential for regulating the nucleation, growth, or aggregation of nanoparticles. Ascorbic acid, alcohols, borohydrides, sodium citrate, and hydrazine are examples of common reducing agents that are crucial for both the production and stability of nanoparticles [16].

These techniques have been linked to some difficulties despite their effectiveness and widespread use in the chemical production of nanoparticles. These techniques involve hazardous or poisonous materials and have substantial running expenses, just as other physical procedures. One of the main concerns has been the creation of hazardous byproducts that may have an impact on biological systems and the environment. The use of biological technologies for the manufacture of nanoparticles has been promoted as a superior option that can produce stable and physiologically compatible nanoparticles while also being environmentally benign.

A range of biological sources, including plant extracts from roots, leaves, stems, flowers, or fruits, as well as biological sources like bacteria, fungus, or yeast, can be used in biological synthesis processes to create nanomaterials. Amino acids, proteins, vitamins, polysaccharides, enzymes, and phytochemicals are examples of biomolecules that function as both reducing and stabilizing agents in the creation of nanomaterials in these systems [17]. In addition to their role as reducing agents, biomolecules also contribute to the functionalization of nanomaterials, which influences their stability and functionality [17].

Green nano synthesis, or the environmentally friendly synthesis of nanomaterials, has become a well-known subfield in nanotechnology as a result of the growing trend of sustainability and green chemistry. Green nano synthesis often involves the use of low-energy and low-waste techniques that primarily make use of biological resources like microbes and plant extracts. In order to minimize the use of hazardous reducing agents and harsh stabilizers, this approach frequently uses eco-friendly raw materials, water as a solvent, and gentle reaction conditions. As a result, green nano synthesis techniques are closely aligned with sustainable development principles, which prioritize lowering energy consumption, reducing the production of toxic waste, and producing biomolecule-capped nanoparticles with a high affinity for environmental pollutants.

According to a number of studies, nanomaterials made using environmentally friendly processes can have better properties than those made using traditional chemical processes. Higher surface area, particular surface shape, increased availability of active adsorption sites, and improved catalytic efficiency for pollutant degradation are a few examples of such changes. Specifically, new research has shown that green-synthesized metal oxide nanomaterials have better photocatalytic activity. For instance, a number of studies have shown degradation efficiencies of over 90% for typical chemical dyes, including Congo red, methyl orange, and methylene blue, which were attained in rather short irradiation times [18–21]. It is thought that surface functional groups added during environmentally friendly production promote dye adsorption and enable effective charge transfer, which enhances photocatalytic efficacy. In general, green-synthesized nanomaterials are appealing options for wastewater treatment because they combine enhanced functional qualities with environmental sustainability.

This chapter presents a thorough bottom-up investigation of green synthesis metal and metal oxide nanoparticles utilizing plant extracts, where naturally occurring phytochemicals are essential as stabilizing, capping, and reducing agents. The conversation is organized to provide a concise and comprehensive summary of green synthesis techniques, which is followed by an explanation of the mechanics underlying the production of nanoparticles and typical characterisation methods. The important elements that influence the size, shape, and functional properties of the produced nanoparticles are also given primary attention.

The chapter also examines the use of environmentally friendly metal oxide nanoparticles in the photocatalytic destruction of harmful colors. Understanding how biologically produced surface functional groups affect dye adsorption and photocatalytic efficiency is the main emphasis. The chapter's overall goal is to highlight the importance of plant-induced green synthesis as a sustainable method for creating useful nanomaterials appropriate for environmental remediation.

3 Green Synthesis

Scientific research has moved toward green synthesis methodologies for the creation of nanomaterials due to growing concerns about environmental sustainability and the need for safer manufacturing technologies or procedures. Green synthesis offers a sustainable, economical, and environmentally friendly substitute for traditional chemical processes in the production of nanoparticles. According to green synthesis, bio-inspired or biogenic synthesis uses naturally occurring biological entities as stabilizing and reducing agents, such as bacteria, fungus, algae, and plant extracts [22, 23]. Numerous biomolecules, including as proteins, polysaccharides, alkaloids, terpenoids, and flavonoids, have two functions: they stabilize the produced nanoparticles and aid in the reduction of metal ions [24]. This technique is carried out at mild conditions to significantly reduce the harmful byproduct and do away with the requirement for hazardous chemicals [25]. Efficiency, low energy consumption, and the avoidance of hazardous capping agents are only a few of the major benefits of green synthesis [26]. Additionally, biologically generated nanoparticles often exhibit improved biocompatibility, which makes them appropriate for use in medication administration, biomedical imaging, catalysis, and water purification [27, 28].

One of the most important aspects of nanoparticles, such as their shape, crystallinity, and surface chemistry, may be achieved by carefully regulating the synthesis conditions, such as the pH, reaction temperature, and biological extract concentration. In general, green synthesis techniques closely matched the ideas of sustainable nanotechnology and green chemistry. This strategy promotes the use of renewable natural resources while simultaneously reducing its negative effects on the environment [30]. Green synthesis is expected to be crucial to the large-scale and ecologically responsible manufacturing of nanomaterials as this research continues to progress [31]. Extracellular and intracellular routes are the two main processes by which nanoparticles (NPs) are biologically synthesized. In the extracellular route, the production of nanoparticles takes place outside the microbial cells of the microorganisms and is guided by proteins, enzymes, and other organic components that are either present in the culture media or connected to the components of bacterial or fungal cell walls. However, intracellular synthesis entails the absorption of metal ions into cells, where they combine with functional groups like amine and carboxyl moieties to produce nanoparticles inside the cellular matrix. Enzymes are crucial to the reduction process whether the synthesis takes place through intracellular or extracellular routes. Among them, NADH-induced nitrate reductase has been identified as a crucial enzyme that contributes to the synthesis of nanomaterials. Furthermore, a number of amino acids, such as arginine, aspartic acid, cysteine, glutamic acid, lysine, and methionine, play a significant role as catalysts in metal ion or nanocrystal reduction by generating hydroxyl ions that interact with reducing agents' aldehyde groups [32].

The primary method for plant-based nanoparticle creation is bio reduction, which is made possible by biomolecules found in plant extracts. According to experimental research, proteins are essential because they first bind and trap metal ions before reducing them. Proteins' secondary structures undergo conformational changes as a result of this interaction, creating metal ion nuclei or seeds that subsequently aid in the development and assembly of nanoparticles. Plant extracts rich in a wide range of phytochemicals are one of the most important and well acknowledged mechanisms.

According to the research that is now accessible, the reduction of nanoparticles is not only caused by a single phytochemical; rather, the production process is aided by the additive impact of a variety of plant-derived compositions. These elements, which include proteins, enzymes, amino acids, vitamins, polysaccharides, polyphenols, alkaloids, flavonoids, and organic acids, are all crucial for the stability and reduction of nanoparticles [33]. Metal ions interact with phenolic hydroxyl groups to generate a complex that is found in biomolecules of biological extract during the nanomaterial production process. These biomolecules undergo further oxidation, which lowers metal ions to their zero-valet state. At the same time, the biomolecules' surface capping stabilizes the nanoparticles that are created. Consequently, green synthesis processes do not require the employment of extra chemical reducing or capping agents.

Numerous investigations have demonstrated that biogenic reduction reduces the toxicity of nanoparticles by creating a biological corona on their surface. It is important to remember, nevertheless, that biological extract may occasionally have harmful consequences. To guarantee biological safety and environmental welfare, it is thus recommended that the biological extract be properly evaluated for toxicity both independently and after nanoparticle creation [34, 35]. A summary of green-synthesised nanomaterials and their photocatalytic degradation effectiveness against harmful dyes is presented in Table 1.

4 Role of Plant Derived Biomolecules in Nanomaterial Fabrication

In the green synthesis of nanomaterials, phytochemicals found in plant extracts are crucial because they serve as reducing, capping, and stabilizing agents all at once. These biomolecules aid in the nucleation, growth, and long-term stability of the resulting nanostructures in addition to inducing the transformation of metal ions into nanoparticles. In plant-based synthesis, phenolic compounds such as flavonoids, tannins, and phenolic acids are some of the best reducing agents. The effective reduction of metal ions like Ag^+ and Au^{3+} into their equivalent nanoparticles is made possible by their strong redox potential. By interacting with the surfaces of nanoparticles, alkaloids also contribute to the reduction process and offer surface stabilization. By aiding in the reduction of metal ions and avoiding the agglomeration of nanoparticles by surface adsorption, terpenoids and saponins serve two purposes.

By clinging to the developing nanoparticle surfaces and preventing aberrant aggregation, proteins and amino acids found in plant extracts play a critical role in regulating the size and shape of nanoparticles. During the nucleation stage, reducing sugars and polysaccharides facilitate controlled particle development by acting as mild reducing agents. Because of their antioxidant properties, which allow them to effectively transfer electrons to the metal ions, vitamins like ascorbic acid and tocopherols further enhance the reduction process. Furthermore, under mild reaction circumstances, the synergistic action of these adaptable biomolecules results in regulated nucleation, growth, and stability of nanoparticles. This important mechanism makes plant-based green synthesis an efficient, sustainable, and ecologically safe method of creating nanomaterials [23, 36].

5 Mechanism of Plant Extract Mediated Metal Nanoparticle Synthesis

In comparison to traditional chemical and physical approaches, plant extract-based production of metal nanoparticles is essential, straightforward, economical, and ecologically efficient [25]. The four main steps of this green synthesis process are as follows: (i) phytochemicals in the plant extract reduce metal ions; (ii) reduced metal atoms nucleate to form initial seeds; (iii) these nuclei grow into stable nanoparticles; and (iv) biomolecules capping the nanoparticles stabilize them by preventing aggregation and improving colloidal stability.

In order to create plant extract-based nanoparticles, fresh plant materials such leaves, flowers, fruits, seeds, roots, or bark aerial roots must first be gathered. To preserve the heat-sensitive phytochemicals, these materials are either oven-dried or shade-dried at controlled temperatures. The plant material is finely powdered after drying to boost extraction efficiency and surface area. An acceptable solvent, such as distilled water, ethanol, methanol, or their combinations, is thoroughly combined with a predetermined amount of the powdered plant material. The phytochemical makeup of the extract is largely determined by the solvent selection, which also influences the kinetics and efficiency of nanoparticle production.

Reduction stage: In this stage, positively charged metal ions (such as Ag^+, Au^{3+}, and Pt^{4+}) are supplied to the reaction media by metal precursors such as silver nitrate ($AgNO_3$), chloroauric acid ($HAuCl_4$), or other metal salts. The plant extract contains phytochemicals that function as natural reducing agents, such as flavonoids, phenolic acids, alkaloids, terpenoids, sugars, proteins, and ascorbic acid. These proteins transform the metal ions into neutral metallic atoms (M^0) by giving them electrons. Figure 2 shows a schematic illustration of this bio reduction process, which starts the production of nanoparticles.

$$M^+ + (\text{Reducing agent}) \rightarrow M^0(\text{nanoparticle nuclei})$$

At the nucleation center, the reduction process results in the creation of metal aggregate. Nanoparticle growth results from the reduction and deposition of metal atoms on these nuclei. Understanding the ultimate size, shape, and morphology of the nanoparticles depends critically on the reaction's critical parameters, which include phytochemical concentration, solution pH, and reaction temperature. Additionally, the plant extract's phytochemicals adsorb onto the surfaces of the nanoparticles and function as capping agents, offering steric and electrostatic stability.

This surface passivation ensures that stable, evenly distributed nanoparticles are formed by preventing particle aggregation. The formation of metal nanoparticles (MNPs) mediated by plant extract is frequently followed by a particular hue shift in the reaction mixture, which serves as an early visual signal of nanoparticle production. This finding is further supported by UV-visible spectroscopy, which shows

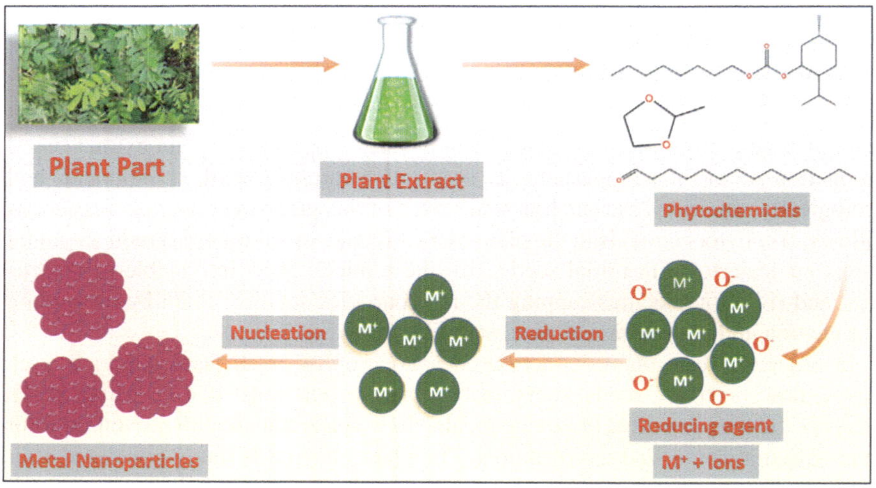

Fig. 2 Biosynthetic mechanism of metal nanoparticles via plant extracts

direct proof of nanoparticle creation when a distinctive surface Plasmon resonance (SPR) absorption peak appears within a certain wavelength range.

6 Characterization Aspects

Understanding the structural, morphological, optical, and compositional properties requires a thorough characterization. To conform nanoparticle formation, crystallinity, size distribution, surface features, and colloidal stability, a variety of sophisticated analytical techniques and imaging are employed [5, 37]. A double-beam spectrophotometer was used in UV-visible spectroscopy to track the conversion of metal ions into metal nanoparticles (MNPs) in the 200–800 nm wavelength range. Usually, a quartz cuvette was used for absorbance scanning analysis after 100 μL of the nanoparticle solution was diluted to 3 mL with distilled water. The emergence of distinctive surface Plasmon resonance (SPR) bands verified the synthesis of nanoparticles. The functional groups involved in the reduction and stability of nanoparticles were identified by recording spectra in the range of 4000–400 cm^{-1} using Fourier-transform infrared spectroscopy (FTIR) with an attenuated total reflectance (ATR) accessory [38].

Cu-Kα radiation ($\lambda = 1.5406$ Å) was used in powder X-ray diffraction (PXRD) investigation throughout a 2θ range of 10°-90° to ascertain the crystalline phase and structural characteristics of the produced nanoparticles. The Debye-Scherer equation, $D = k\lambda/\beta\cos\theta$, where θ is the Bragg angle, λ is the X-ray wavelength, β is the diffraction peak's full width at half maximum (FWHM), and k is the form factor, was used to determine the average crystallite size [39]. The surface morphology and particle dispersion of the nanomaterials were investigated using high-resolution scanning electron microscopy (HR-SEM).

A drop of the colloidal solution was placed on a spotless glass slide and allowed to air dry at room temperature in preparation for SEM examination. Imaging was carried out using an HV setting of 20.00 kV and an accelerating voltage of 25.00 kV. The elemental composition and relative abundance of the elements included in the MNPs were investigated using energy-dispersive X-ray (EDX) spectroscopy in conjunction with SEM [40]. The internal structure, particle size, and fine morphology of the nanoparticles were ascertained using transmission electron microscopy (TEM). Samples were put on copper grids, air-dried, and sonicated for 5 m before to analysis. The crystalline nature of the nanoparticles was verified by recording selected-area electron diffraction (SAED) patterns. Image software based on TEM micrographs was used to calculate the particle size distribution [41]. Additionally, a particle size analyzer (PSA) with a zeta sizer was used to measure the zeta potential in order to evaluate the surface charge and colloidal stability of the nanoparticles [42, 43]. When taken as a whole, these characterization methods offer comprehensive insights into the shape, crystal structure, surface chemistry, elemental content, and stability of nanomaterials, making them essential instruments for assessing their appropriateness in a range of nanotechnology applications.

7 Plant Extract-Mediated Metal Nanoparticles

An eco-friendly and sustainable substitute for traditional physical and chemical synthesis techniques is green synthesis of nanomaterials. This method reduces the need for hazardous chemicals and energy-concentrated methods by using a variety of biological resources, including plant extracts, fungus, bacteria, and algae, as natural reducing and stabilizing agents. The nucleation, development, and stability of metal nanoparticles are greatly influenced by a wide range of biomolecules found in these biological systems, such as polyphenols, terpenoids, alkaloids, flavonoids, proteins, and enzymes.

Plant extract-induced synthesis not only ensures environmental safety but also gives the resultant nanomaterials better stability and surface functionalization, which specifically enhances their performance in a variety of applications. In addition to preventing nanoparticle aggregation, the inclusion of physiologically mediated capping agents adds functional groups that facilitate the particles' interaction with necessary and targeted molecules. Therefore, green synthesis methods provide a robust and promising platform for the long-term synthesis of metal nanoparticles with improved functional characteristics [23].

7.1 Silver Nanoparticles

Silver nanoparticles (AgNPs) are one of the most researched metal nanomaterials made with eco-friendly methods due to their exceptional plasmonic qualities and broad-spectrum antibacterial activity. In plant-based green synthesis, biomolecules originating from plants that serve as both reducing and capping agents convert silver ions (Ag^+) from their precursors, such as silver nitrate ($AgNO_3$), to metallic silver (Ag^+). Electron transfer from hydroxyl and carbonyl functional groups found in phenolic compounds and flavonoids to Ag^+ ions is the main mechanism of reduction. The emergence of distinctive surface Plasmon resonance (SPR) absorption bands, which are typically seen in the wavelength range of 400–450 nm, visually confirms the production of AgNPs.

Depending on the kind of plant extract and reaction conditions, green-synthesised AgNPs usually have a high degree of monodispersed particles with sizes ranging from 5 to 80 nm. When compared to chemically manufactured alternatives, the presence of naturally occurring capping molecules greatly improves biocompatibility and colloidal stability. The production of reactive oxygen species (ROS) and the breakdown of microbial cell membranes are the main processes by which these nanoparticles exhibit potent antibacterial and antifungal properties. AgNPs also have remarkable sensing and catalytic properties. Their potential uses in wound healing, wastewater treatment, and biosensing technologies have been emphasized in recent research and reviews [44, 45].

7.2 Gold Nanoparticles

Gold nanoparticles (AuNPs) are a crucial class of noble metal nanostructures that may be produced effectively with environmentally friendly methods. Protein and polyphenol-rich extracts function as stabilizing and reducing agents in plant and microbe-based synthesis, converting Au^{3+} ions, which are usually derived from chloroauric acid ($HAuCl_4$), into elemental gold (Au^1). These biomolecules improve colloidal stability by providing surface capping in addition to causing the production of nanoparticles. The shape of AuNPs may be precisely controlled by adjusting reaction parameters including pH, extract concentration, and temperature, which is a major benefit of green synthesis. AuNPs with a variety of morphologies, including spherical and anisotropic structures like nano rods, nano triangles, and nano prisms, may be produced by carefully regulating and optimizing these conditions [46].

Green-synthesised gold nanoparticles (AuNPs) have outstanding chemical stability and strong surface Plasmon resonance (SPR) absorption bands in the visible spectrum, usually between 520 and 560 nm. Their surfaces' bimolecular coatings make functionalization simple, increasing their applicability for a variety of biological applications. Surface chemistry and stability are crucial in targeted drug delivery, photothermal treatment, and biosensing. Additionally, avoiding harmful stabilizing and reducing chemicals during green synthesis greatly lowers cytotoxicity and enhances AuNPs' biocompatibility. As a result, green-synthesised AuNPs continue to garner significant attention for cutting-edge therapeutic and diagnostic applications and comply well with strict biomedical safety regulations [46].

7.3 Copper and Copper Oxide Nanoparticles

Copper-based nanomaterials, such as metallic copper (Cu) and copper oxide (CuO), are rapidly being produced utilizing eco-friendly techniques due to their low cost and versatile functional properties. In plant-or microbe-based synthesis, biological extracts serve as natural reducing and stabilizing agents, allowing Cu^{2+} ions which are often sourced from precursors like $CuSO_4$ or $Cu(NO_3)^2$ to be converted into nanoscale metallic or oxide forms. Depending on the synthesis conditions, the resultant Cu/CuO nanoparticles typically have particle diameters between 10 and 100 nm with morphologies ranging from spherical to rod-like or plate-like structures [47].

Due to their enhanced surface reactivity and oxygen vacancies, green-synthesised CuO nanoparticles have proven to be highly effective in the catalytic destruction of organic dyes and environmental pollutants. However, because metallic copper is prone to oxidation to copper oxide (CuO) in ambient circumstances, the long-term stability of copper-based nanoparticles continues to be a difficulty. It has been demonstrated that adding bimolecular capping layers to green synthesis greatly increases the stability and dispersion of nanoparticles, increasing their usefulness [48].

7.4 Zinc Oxide Nanoparticles

Zinc oxide (ZnO) nanoparticles are among the most studied green-synthesized metal oxides because of their wide band gap (~3.3 eV), excellent photocatalytic efficiency, and innate antibacterial activity [48]. The biological extract's content and reaction conditions, particularly pH, control the final particle size and shape, which are usually between 10 and 100 nm.

Green-synthesized ZnO nanoparticles show notable antibacterial capabilities and substantial photocatalytic activity toward the destruction of organic dyes, mainly through regulated Zn2+ ion release and reactive oxygen species (ROS) formation. They are appropriate for use in water purification systems, antibacterial coatings, and cosmetics because to their good biocompatibility, UV absorption capacity, and environmental safety. Furthermore, the potential application of ZnO nanoparticles as nano-fertilizers and nano-biocides in sustainable agricultural techniques has been noted in literature [49].

7.5 Titanium Dioxide and Other Metal Oxides

In recent years, there has been a significant increase in the green synthesis of titanium dioxide (TiO_2) and other metal oxides, including Fe_3O_4, Co_3O_4, and MgO. These methods allow for the regulated hydrolysis and condensation of metal precursors during the production of nanoparticles by using plant extracts or microbial metabolites as natural chelating and enticing agents. Green synthesized TiO_2 nanoparticles, which range in size from 5 to 50 nm, usually show anisate or mixed anatase rutile crystalline phases [50].

These environmentally friendly metal oxide nanoparticles frequently demonstrate improved photocatalytic activity when exposed to UV and visible light. Defect sites, surface states, and bimolecular residues generated during biological synthesis are examples of such modifications that might enhance charge-felicitated separation efficiency and encourage surface doping.

Similar to this, green-synthesized Fe_3O_4 nanoparticles are perfect for applications such as magnetic separation, targeted drug delivery, and biomedical imaging because of their strong biocompatibility and superparamagnetic qualities. Furthermore, green-synthesized MgO and CoO nanoparticles exhibit strong antibacterial and catalytic properties, respectively, while maintaining outstanding environmental compatibility [51, 52]. Using biogenic synthesis methods often results in nanoparticles with clean surfaces and a higher density of functional groups while also drastically reducing the production of hazardous waste. These characteristics enhance the dispersion stability, reactivity, and general performance of nanoparticles in environmental and catalytic applications.

8 Future Perspective

An important step toward the development of sustainable nanotechnology is the green production of metallic and metal-oxide nanoparticles. The creation of nanomaterials with better biocompatibility, decreased toxicity, and increased functional performance is made possible by this environmentally benign method, which makes them profitable for use in industrial, biomedical, and environmental applications. But there are still a number of challenges, especially in terms of scalability, repeatability, and a thorough mechanistic knowledge of biogenic synthesis processes.

The development of scalable and continuous flow production methods, the standardization of green synthesis techniques, and the comprehensive mechanistic clarification of interactions between phytochemicals and metal ions during nanoparticle formation should be the top priorities for future study. Therefore, before large-scale industrial application and commercialization of biogenic nanomaterials, ongoing evaluation of toxicity, biodegradability, and long-term environmental effect is crucial.

9 Application in Dye Degradation

The role of green-synthesized metal oxide nanomaterials in environmental remediation, with a special focus on the degradation of organic dyes present in wastewater. These nanomaterials act as efficient photocatalysts by absorbing light energy and generating electron hole pairs, which subsequently produce reactive oxygen species such as hydroxyl radicals capable of decomposing toxic dye molecules. Overall, the compiled data in Table 1 clearly demonstrate the potential of plant-mediated metal and metal oxide nanoparticles as efficient and sustainable photocatalysts for the dye degradation.

Table 1 Plant mediated metal/metal oxide NPs for different toxic dyes

S.No	Metal oxide	Targeted dye	Reported degradation	Reaction time/Kinetics	Ref.
1	α-Fe_2O_3	Remazol yellow RR dye	Photocatalytic degradation	77% of the dye degraded within 6 h for 0.8 mg/L of dosage	[53]
2	Cu_2O	Methylene Orange	Photocatalytic degradation	92% degraded 180 min under sunlight irradiation	[54]
3	CuO-ZnO	Methylene blue	Photocatalytic degradation	20 wt % CuO showed enhanced photocatalytic activity against MB dye with a 0.017 min^{-1} rate.	[55]
4	ZnO	Methylene Orange	Photocatalytic degradation under UV light	92.47% degradation and the reaction rate constant was found to be 0.02134 min^{-1}	[56]
5	Co_3O_4-ZnO	Malachite green dye	Photocatalytic degradation	98.06% degraded under sunlight in 60 min	[57]
6	Ag_2O	Azo dye	Photocatalytic degradation	84.50% decolorization achieved in 50 min (UV-vis, 250–900 nm).	[58]
7	TiO_2	Malachite green	Photocatalytic degradation	Catalytic degradation of malachite green achieved 83.48–86.28% in the absence of UV irradiation.	[59]
8	-22	Congo red dye	Photocatalytic degradation	86% degradation achieved under optimum conditions (0.05 g L^{-1} NP dose, 10 mg L^{-1} dye concentration, 250 min irradiation).	[60]
9	Ag-NiO	Fast blue dye	Photocatalytic degradation under UV radiation	Ag-doped NiO achieved 98% fast blue degradation at 605 nm under 120 min UV irradiation.	[61]
10	Fe_2O_3	Methyl orange, methyl violet, methylene blue	Photocatalytic degradation	Fe_2O_3 NPs (0.5 mg mL^{-1}) showed degradation efficiencies of 64.5% (MO), 84.6% (MV), and 61% (MB).	[62]
11	Fe-rGO	Methylene blue	Photocatalytic degradation	Fe-rGO NPs achieved ~74% methylene blue degradation.	[62]

(continued)

Table 1 (continued)

S.No	Metal oxide	Targeted dye	Reported degradation	Reaction time/Kinetics	Ref.
12	GrGO	Methylene blue	Response surface methodology (RSM)	GrGO adsorbed methylene blue with 94.85% removal efficiency and 58.81 mg g^{-1} adsorption capacity under optimum conditions.	[63]
13	Fe	Methylene blue, crystal violet dye	Photocatalytic degradation	Green-synthesized FeNPs showed maximum degradation efficiencies of 97.47% (CV) and 94.22% (MB).	[64]
14	$CuFe_2O_4$	Methylene blue	Photocatalytic degradation under UV light	$CuFe_2O_4$ NPs achieved 82% methylene blue degradation under UV–visible light.	[65]
15	TiO_2	Methylene blue and tartrazine organic dyes	Phtocatalytic degradation under UV radiation	TiO_2 NPs (calcined at 500 °C) achieved 100% degradation of methylene blue and tartrazine within 60 min under UV irradiation.	[66]
16	MgO	Methylene blue	Photocatalytic degradation under UV radiation	MgO showed methylene blue degradation of 92% and 59% after 1 h UV irradiation.	[67]
17	ZF	Methylene blue	Photocatalytic degradation	ZF-NPs achieved photocatalytic reduction of methylene blue under sunlight within 90 min.	[68]
18	NiO	Methylene blue	Phtocatalytic degradation under UV radiation	NiO exhibited 99.6% degradation efficiency under UV irradiation.	[69]
19	CRGO/SnO_2	Methylene green	Photocatalytic degradation	SnO_2-C. Annuum rGO nanocomposite achieved 97.4% photodegradation of methylene green.	[70]
20	Dy_2O_3/ ZnO-au	Erythrosine (ES)	Photocatalytic degradation under UV light	Dy/ZnO–AuNCs@PF showed higher photodegradation under UV irradiation: 89.6% (ES) and 91.3% (BV10), outperforming Dy_2O_3 and Dy_2O_3/ZnO.	[71]

(continued)

Table 1 (continued)

S.No	Metal oxide	Targeted dye	Reported degradation	Reaction time/Kinetics	Ref.
21	SnO_2	Methylene blue (MB) and Eriochrome black-T (EBT)	Photocatalytic degradation	Degradation efficiencies of ~90% (MB) and 83% (EBT) were achieved.	[72]
22	$ZnFe_2O_4$	Methylene blue	Photocatalytic degradation	$ZnFe_2O_4$ nanoparticles achieved 76% methylene blue degradation under visible light in 75 min.	[73]
23	CeO_2	Methylene orange	Photocatalytic degradation under sunlight	Bio-synthesized CeO_2 NPs achieved 98% methyl orange degradation under sunlight irradiation.	[74]
24	Ag	4-Nitrophenol and Congo red	Photocatalytic degradation	AgNPs catalyzed >90% degradation of 4-nitrophenol (8 min) and Congo red (15 min), and 78–79% degradation of 4-nitroaniline (8 min) and methylene blue (18 min).	[75]
25	NiO	Fast green, rose Bengal	Photocatalytic degradation	NiO NPs achieved 89% fast green and 77% rose Bengal degradation under sunlight.	[76]
26	CuO	RhodamineB (Rh-B) and malachite green (MG) dyes	Photocatalytic degradation	CuO NPs achieved up to 93.81% malachite green degradation under light irradiation within 210 min (H_2O_2-assisted).	[77]
27	Ag_2O	Toluidine blue dye	Photocatalytic degradation	TB dye showed 98.50% photodegradation after 45 min (UV-vis 200–900 nm), with $t_1/_2$ = 9.58 min and k = 0.07234 min^{-1}.	[78]
28	ZnO	Malachite green	Photocatalytic degradation under Sun light	Malachite green showed 53% degradation within <1 min and 92% degradation in 60 min under solar irradiation.	[79]

(continued)

Table 1 (continued)

S.No	Metal oxide	Targeted dye	Reported degradation	Reaction time/Kinetics	Ref.
29	CuO	Methylene blue, methyl orange	Photocatalytic degradation	CuO NPs achieved 88.2% methylene blue and 83.1% methyl orange degradation.	[80]
30	ZnO	Methylene blue	Photocatalytic degradation	ZnO NPs achieved 95% methylene blue degradation under sunlight within 120 min.	[81]

References

1. Yadav, D.C.: Nanotechnology, big things from a tiny world: A review. Adv. Electron. Electr. Eng. **3**(6), 771–778 (2013)
2. Goddard III, W.A., Brenner, D., Lyshevski, S.E., Iafrate, G.J. (eds.): Handbook of Nanoscience, Engineering, and Technology. CRC Press (2007)
3. Khandel, P., Shahi, S.K.: Mycogenic nanoparticles and their bio-prospective applications: current status and future challenges. J. Nanostructure Chem. **8**, 369–391 (2018)
4. Bogunia-Kubik, K., Sugisaka, M.: From molecular biology to nanotechnology and nanomedicine. Bio Systems. **65**(2–3), 123–138 (2002)
5. Daniel, M.C., Astruc, D.: Gold nanoparticles: assembly, supramolecular chemistry, quantum-size-related properties, and applications toward biology, catalysis, and nanotechnology. Chem. Rev. **104**(1), 293–346 (2004)
6. Zharov, V.P., Kim, J.W., Curiel, D.T., Everts, M.: Self-assembling nanoclusters in living systems: application for integrated photothermal nanodiagnostics and nanotherapy. Nanomed. Nanotechnol. Biol. Med. **1**(4), 326–345 (2005)
7. Shah, M., Fawcett, D., Sharma, S., Tripathy, S.K., Poinern, G.E.J.: Green synthesis of metallic nanoparticles via biological entities. Materials. **8**(11), 7278–7308 (2015)
8. Li, W., Sun, T., Zhou, S., Ma, Y., Shi, Q., Xie, X., Huang, X.: A comparative analysis of antibacterial activity, dynamics, and effects of silver ions and silver nanoparticles against four bacterial strains. Int. Biodeterior. Biodegradation. **123**, 304–310 (2017)
9. Meyers, M., Mishra, A., Benson, D.: Mechanical properties of nanocrystalline materials. Prog. Mater. Sci. **51**(4), 427–556 (2005)
10. Li, W.R., Shi, Q.S., Ouyang, Y.S., Chen, Y.B., Duan, S.S.: Antifungal effects of citronella oil against aspergillus Niger ATCC 16404. Appl. Microbiol. Biotechnol. **97**(16), 7483–7492 (2013)
11. Skotadis, E., Aslanidis, E., Kainourgiaki, M., Tsoukalas, D.: Nanoparticles synthesised in the gas-phase and their applications in sensors: A review. Applied Nano. **1**(1), 70–86 (2020)
12. Abdollahi, H., Noaparast, M., Shafaei, S.Z., Manafi, Z., Muñoz, J.A., Tuovinen, O.H.: Silver-catalyzed bioleaching of copper, molybdenum and rhenium from a chalcopyrite–molybdenite concentrate. Int. Biodeterior. Biodegradation. **104**, 194–200 (2015)
13. Anbazhagan, S., Azeez, S., Morukattu, G., Rajan, R., Venkatesan, K., Thangavelu, K.P.: Synthesis, characterization and biological applications of mycosynthesized silver nanoparticles. 3. Biotech. **7**(5), 333 (2017). https://doi.org/10.1007/s13205-017-0961-9
14. Khan, M., Shaik, M.R., Adil, S.F., Khan, S.T., Al-Warthan, A.A., Siddiqui, M.R., Tahir, M.N., Tremel, W.: Plant extracts as green reductants for the synthesis of silver nanoparticles: lessons from chemical synthesis. Dalton Trans. **47**(35), 11988–12010 (2018)
15. El-Rafie, M.H., Mohamed, A.A., Shaheen, T.I., Hebeish, A.A.: Antimicrobial effect of silver nanoparticles produced by fungal process on cotton fabrics. Carbohydr. Polym. **80**, 779–782 (2010)

16. Siddiqi, K.S., Husen, A., Rao, R.A.K.: A review on biosynthesis of silver nanoparticles and their biocidal properties. J. Nanobiotechnol. **16**, 14 (2018)
17. Mustapha, T., Misni, N., Ithnin, N.R., Daskum, A.M., Unyah, N.Z.: A review on plants and microorganisms mediated synthesis of silver nanoparticles, role of plants metabolites and applications. Int. J. Environ. Res. Public Health. **19**(2), 674 (2022)
18. Miad, A.A., Saikat, S.P., Alam, M.K., Hossain, M.S., Bahadur, N.M., Ahmed, S.: Metal oxide-based photocatalysts for the efficient degradation of organic pollutants for a sustainable environment: a review. Nanoscale Adv. **6**(19), 4781–4803 (2024). https://doi.org/10.1039/d4na00517a
19. Senthil Rathi, B., Ewe, L.S., S, S., S, S., Yew, W. K., R, B., & Tiong, S. K.: Recent trends and advancement in metal oxide nanoparticles for the degradation of dyes: synthesis, mechanism, types and its application. Nanotoxicology. **18**(3), 272–298 (2024)
20. Medina-Acosta, M., Chinchillas-Chinchillas, M.J., Garrafa-Gálvez, H.E., Garcia-Maro, C.A., Rosas-Casarez, C.A., Lugo-Medina, E., Luque-Morales, P.A., Soto-Robles, C.A.: Photocatalytic degradation of four organic dyes present in water using ZnO nanoparticles synthesized with green synthesis using Ambrosia ambrosioides leaf and root extract. PRO. **12**(11), 2456 (2024)
21. Matar, H., Andac, M.: Highly efficient degradation of basic dyes using gold-coated nature-based supermagnetic iron oxide nanoparticles as eco-friendly nanocatalysts. Environ. Sci. Pollut. Res. **31**, 24894–24912 (2024)
22. Barabadi, H., et al.: Emerging frontiers in green nanoparticle synthesis: A perspective on nanomedicine. Int. J. Nanomedicine. **15**, 6931–6953 (2020)
23. Singh, P., Kim, Y., Zhang, D., Yang, D.: Biological synthesis of nanoparticles from plants and microorganisms. Trends Biotechnol. **34**(7), 588–599 (2016). https://doi.org/10.1016/j.tibtech.2016.02.006
24. Rajeshkumar, S., Bharath, L.: Mechanism of plant-mediated synthesis of silver nanoparticles—A review on biomolecules involved, characterisation and antibacterial activity. Chem. Biol. Interact. **273**, 219–227 (2017)
25. Mittal, A.K., Chisti, Y., Banerjee, U.C.: Synthesis of metallic nanoparticles using plant extracts. Biotechnol. Adv. **31**(2), 346–356 (2013). https://doi.org/10.1016/j.biotechadv.2013.01.003
26. Singh, H., Desimone, M.F., Pandya, S., Jasani, S., George, N., Adnan, M., Aldarhami, A., Bazaid, A.S., Alderhami, S.A.: Revisiting the green synthesis of nanoparticles: uncovering influences of plant extracts as reducing agents for enhanced synthesis efficiency and its biomedical applications. Int. J. Nanomedicine. **18**, 4727–4750 (2023)
27. Sharma, V.K., Yngard, R.A., Lin, Y.: Silver nanoparticles: green synthesis and their antimicrobial activities. Adv. Colloid Interf. Sci. **145**(1–2), 83–96 (2009)
28. Hebbalalu, D., Lalley, J., Nadagouda, M.N., Varma, R.S.: Greener techniques for the synthesis of silver nanoparticles using plant extracts. ACS Sustain. Chem. Eng. **1**(7), 703–712 (2013)
29. Kuppusamy, P., et al.: Biosynthesis of metallic nanoparticles using plant derivatives and their new avenues in pharmacological applications. Saudi Pharm. J. **24**(4), 473–484 (2016)
30. Khan, I., Saeed, K., Khan, I.: Nanoparticles: properties, applications and toxicities. Arab. J. Chem. **12**(7), 908–931 (2019)
31. Li, X., et al.: Sustainable nanomaterial production through green synthesis: Progress and prospects. Environ. Nanotechnol. Monit. Manag. **24**, 100839 (2025)
32. Sampath, G., Chen, Y.-Y., Rameshkumar, N., Krishnan, M., Nagarajan, K., Shyu, D.J.H.: Biologically synthesized silver nanoparticles and their diverse applications. Nano. **12**(18), 3126 (2022)
33. Adeyemi, O.S., Shittu, E.O., Akpor, O.B., Rotimi, D., Batiha, G.E.: Silver nanoparticles restrict microbial growth by promoting oxidative stress and DNA damage. EXCLI J. **19**, 492–500 (2020)
34. Sukhanova, A., Bozrova, S., Sokolov, P., et al.: Dependence of nanoparticle toxicity on their physical and chemical properties. Nanoscale Res. Lett. **13**, 44 (2018)
35. Taroncher, M., Vila-Donat, P., Tolosa, J., Ruiz, M.J., Rodríguez-Carrasco, Y.: Biological activity and toxicity of plant nutraceuticals: an overview. Curr. Opin. Food Sci. **42**, 113–118 (2021)

36. Kuppusamy, P., Yusoff, M.M., Maniam, G.P., Govindan, N.: Biosynthesis of metallic nanoparticles using plant derivatives and their new avenues in pharmacological applications—an updated report. Saudi Pharm. J. **24**(4), 473–484 (2014)
37. Bhatnagar, A., Sillanpää, M.: Utilization of agro-industrial and municipal waste materials as potential adsorbents for water treatment a review. Chem. Eng. J. **157**(2–3), 277–296 (2010)
38. Stuart, B.: Infrared Spectroscopy: Fundamentals and Applications. Wiley, Chichester (2004)
39. Cullity, B.D., Stock, S.R.: Elements of X-Ray Diffraction, 3rd edn. Pearson Education, Boston (2014)
40. Goldstein, J.I., Newbury, D.E., Joy, D.C., Lyman, C.E., Echlin, P.: Scanning Electron Microscopy and X-Ray Microanalysis. Springer, New York (2018)
41. Williams, D.B., Carter, C.B.: Transmission Electron Microscopy: A Textbook for Materials Science, 2nd edn. Springer, New York (2009)
42. Bhattacharjee, S.: DLS and zeta potential—what they are and what they are not? J. Control. Release. **235**, 337–351 (2016)
43. Hunter, R.J.: Zeta Potential in Colloid Science: Principles and Applications. Academic, London (1981)
44. Elbehiry, A., Abalkhail, A.: Antimicrobial nanoparticles against superbugs: mechanistic insights, biomedical applications, and translational Frontiers. Pharmaceuticals. **18**(8), 1195 (2025)
45. Kumar, R., et al.: Frontiers in Nanotechnology. **6**(22), 1139–1156 (2024)
46. Georgeous, J., et al.: J. Mater. Sci. Mater. Med. **35**(11), 1289–1303 (2024)
47. Pourmadadi, M., et al.: Appl. Nanosci. **14**(8), 567–584 (2024)
48. Mutukwa, D., et al.: Environmental nanotechnology. Monit. Manag. **23**, 101969 (2024)
49. Saini, V., et al.: J. Photocatal. Environ. Nanosci. **12**(1), 25–39 (2025)
50. Eddy, D.R., et al.: Mater. Chem. Phys. **325**, 126845 (2024)
51. Lithi, A., et al.: RSC Nanosci. Adv. **9**(3), 250–267 (2025)
52. Fahim, M., et al.: Green Chem. Lett. Rev. **17**(3), 145–162 (2024)
53. Bhuiyan, M.S.H., Miah, M.Y., Paul, S.C., Das, T., Saha, O., Rahaman, M.M., Sharif, M.J.I., Habiba, O., Ashaduzzaman, M.: Green Synthesis of Iron Oxide Nanoparticle Using Carica Papaya Leaf Extract: Application for Photocatalytic Degradation of Remazol Yellow RR Dye and Antibacterial Activity. Department of Applied Chemistry and Chemical Engineering, Noakhali Science and Technology University (n.d.)
54. Ramesh, P., Rajendran, A.: Photocatalytic dye degradation activities of green synthesis of cuprous oxide nanoparticles from Sargassum wightii extract. Chem. Phys. Impact. **6**, 100208 (2023)
55. Bekru, A.G., Tufa, L.T., Zelekew, O.A., Goddati, M., Lee, J., Sabir, F.K.: Green synthesis of a CuO-ZnO nanocomposite for efficient photodegradation of methylene blue and reduction of 4-nitrophenol. ACS Omega. **7**(34), 30908–30919 (2022)
56. Abomuti, M.A., Danish, E.Y., Firoz, A., Hasan, N., Malik, M.A.: Green synthesis of zinc oxide nanoparticles using Salvia officinalis leaf extract and their photocatalytic and antifungal activities. Biology. **10**(11), 1075 (2021)
57. Gendo, K.M., Bogale, R.F., Kenasa, G.: Green synthesis, characterization, and evaluation of photocatalytic and antibacterial activities of Co_3O_4–ZnO nanocomposites using Calpurnia aurea leaf extract. ACS Omega. **9**(25), 28354–28371 (2024)
58. Laouini, S.E., Bouafia, A., Soldatov, A.V., Algarni, H., Tedjani, M.L., Ali, G.A.M., Barhoum, A.: Green synthesized ag/Ag_2O nanoparticles using aqueous leaves extracts of Phoenix dactylifera L. and their azo dye photodegradation. Membranes. **11**(7), 468 (2021)
59. Akinola, P.O., Lateef, A., Asafa, T.B., Beukes, L.S., Hakeem, A.S., Irshad, H.M.: Multifunctional titanium dioxide nanoparticles biofabricated via phytosynthetic route using extracts of Cola nitida: antimicrobial, dye degradation, antioxidant and anticoagulant activities. Heliyon. **6**(8), e04610 (2020)
60. Hokonya, N., Mahamadi, C., Mukaratirwa-Muchanyereyi, N., Gutu, T., Zvinowanda, C.: Green synthesis of P–ZrO_2CeO_2ZnO nanoparticles using leaf extracts of Flacourtia indica and their application for the photocatalytic degradation of a model toxic dye, Congo red. Heliyon. **8**(10), e10277 (2022). https://doi.org/10.1016/j.heliyon.2022.e10277

61. Alfadhli, S., Khasim, S., Abdelkader, M., Darwish, A.A.A., Gamal, R., Al-nahdi, K., Hamdalla, T.A.: Facile green synthesis of silver doped NiO nanoparticles using aloe vera latex for efficient energy storage and photocatalytic applications. Heliyon. **10**(4), e41322 (2024)
62. Akhtar, M.S., Fiaz, S., Aslam, S., Chung, S., Ditta, A., Irshad, M.A., Al-Mohaimeed, A.M., Iqbal, R., Al-onazi, W.A., Rizwan, M., Nakashima, Y.: Green synthesis of magnetite iron oxide nanoparticles using Azadirachta indica leaf extract loaded on reduced graphene oxide and degradation of methylene blue. Sci. Rep. **14**, 18172 (2024)
63. Saini, R., Mishra, R.K., Kumar, P.: Green synthesis of reduced graphene oxide using the Tinospora cordifolia plant extract: exploring its potential for methylene blue dye degradation and antibacterial activity. ACS Omega. **9**(16), 20304–20321 (2024)
64. Jara, Y.S., Mekiso, T.T., Washe, A.P.: Highly efficient catalytic degradation of organic dyes using iron nanoparticles synthesized with Vernonia amygdalina leaf extract. Sci. Rep. **14**, Article 6997 (2024). https://doi.org/10.1038/s41598-024-6997
65. Jabeen, S., Siddiqui, V.U., Sharma, S., Rai, S., Bansal, P., Bala, S., Raza, A., Ahmad, M.I., Khan, A.R., Khan, T.: A novel green synthesis of $CuFe_2O_4$ nanoparticles from Cissus rotundifolia for photocatalytic and antimicrobial activity evaluation. J. Alloys Compd. **10358**, 174020 (2024)
66. Moghni, N., Khalaf, H., Menseri, O., Boutoumi, H., Boudali, R., Dif, F., Boucheffa, Y.: One-pot green synthesis of TiO_2 nanoparticles using Inula Viscosa leaf extract as an efficient photocatalyst for organic dyes removal. J. Photochem. Photobiol., A: Chem. **461**, 116158 (2025)
67. Amina, M., Al Musayeib, N.M., Alarfaj, N.A., El-Tohamy, M.F., Oraby, H.F., Al Hamoud, G.A., Bukhari, S.I., Moubayed, N.M.S.: Biogenic green synthesis of MgO nanoparticles using Saussurea costus biomasses for a comprehensive detection of their antimicrobial, cytotoxicity against MCF-7 breast cancer cells and photocatalysis potentials. PLoS One. **15**(8), e0237567 (2020)
68. Din, M.I., Jabbar, S., Najeeb, J., Khalid, R., Ghaffar, T., Arshad, M., Khan, S.A., Ali, S.: Green synthesis of zinc ferrite nanoparticles for photocatalysis of methylene blue. Int. J. Phytoremediation. **22**(13), 1440–1447 (2020)
69. Kumar, M.S.S., Soundarya, T.L., Alharthi, F.A., Nirmala, B., Udayabhanu, N., G., Raghu, G. K., & Rekha, N. D.: Multifunctional applications of nickel oxide (NiO) nanoparticles synthesized by facile green combustion method using Limonia acidissima natural fruit juice. Inorg. Chim. Acta. **512**, 12005 (2020)
70. Ramanathan, S., Radhika, N., Padmanabhan, D., Durairaj, A., Selvin, S.P., Lydia, S., Kavitha, S., Vasanthkumar, S.: Eco-friendly synthesis of CRGO and CRGO/SnO_2 nanocomposite for photocatalytic degradation of methylene green dye. ACS Omega. **4**(26), 22179–22188 (2019)
71. Khormali, K., Mizwari, Z.M., Ghoreishi, S.M., Mortazavi-Derazkola, S., Khezri, B.: Novel Dy_2O_3/ZnO-au ternary nanocomposites: green synthesis using pomegranate fruit extract, characterization and their photocatalytic and antibacterial properties. Bioorg. Chem. **114**, 105204 (2021)
72. Honarmand, M., Golmohammadi, M., Naeimi, A.: Biosynthesis of tin oxide (SnO_2) nanoparticles using jujube fruit for photocatalytic degradation of organic dyes. Adv. Powder Technol. **30**(7), 1551–1557 (2019)
73. Patil, J., Gaud, B., Jaybhaye, S.: Green synthesis of zinc ferrite nanoparticles using Tagetes erecta flower (Marigold yellow) and their application as photocatalysis. Int. J. Creat. Res. Thoughts. **11**(4), 96–102 (2023)
74. Muthuvel, A., Jothibas, M., Mohana, V., Manoharan, C.: Green synthesis of cerium oxide nanoparticles using Calotropis procera flower extract and their photocatalytic degradation and antibacterial activity. Inorg. Chem. Commun. **118**, 108086 (2020)
75. Kumar, P., Dixit, J., Singh, A.K., Rajput, V.D., Verma, P., Tiwari, K.N., Mishra, S.K., Minkina, T., Mandzhieva, S.: Efficient catalytic degradation of selected toxic dyes by green biosynthesized silver nanoparticles using aqueous leaf extract of Cestrum nocturnum L. Nano. **12**(21), 3851 (2022)
76. Prabhu, S., Thangadurai, T.D., Bharathy, P.V., Kalugasalame, P.: Synthesis and characterization of nickel oxide nanoparticles using Clitoria ternatea flower extract: photocatalytic dye

degradation under sunlight and antibacterial activity applications. Results Chem. **4**, 100285 (2022)

77. Wu, S., Chen, J., Hu, T., Wang, H., Zhu, D.: Green synthesis of Taraxacum mongolicum hand.-Mazz. Extract-mediated copper oxide nanoparticles and evaluation of their degradation performance for organic dyes and antibacterial properties. Biomass Convers. Biorefinery. **14**, 26497–26507 (2024)
78. Gungure, A.S., Jule, L.T., Nagaprasad, N., Ramaswamy, K.: Studying the properties of green synthesized silver oxide nanoparticles in the application of organic dye degradation under visible light. Sci. Rep. **14**, 26967 (2024)
79. Ekennia, A.C., Uduagwu, D.N., Nwaji, N.N., Oje, O.O., Emma-Uba, C.O., Mgbii, S.I., Olowo, O.J., Nwanji, O.L.: Green synthesis of biogenic zinc oxide nanoflower as dual agent for photodegradation of an organic dye and tyrosinase inhibitor. J. Inorg. Organomet. Polym. Mater. **31**(3), 886–897 (2021)
80. Tuba, Javed, S., Raza, M.R., Hussain, M.A., Muhammad, G., Sharif, M.U., Majeed, A.: Quince (Cydonia oblonga) seed hydrogel-based green synthesis of copper oxide nanoparticles for photocatalytic degradation of methylene blue and methyl orange. Desalin. Water Treat. **314**, 100388 (2024)
81. Pai, S., Hari, S., Varadavenkatesan, T., Vinayagam, R., Selvaraj, R.: Photocatalytic zinc oxide nanoparticles synthesis using Peltophorum pterocarpum leaf extract and their characterization. Optik. **183**, 691–698 (2019)

Catalytic Approaches of Biosynthesized Nanoparticles in the Construction of Organic Frameworks: A Green Chemistry Perspective

Jyoti Bhattacharjee, Subhasis Roy, and Hyun Suk Jung

1 Introduction

The relentless pursuit of next-generation functional materials has put metal–organic frameworks (MOFs), covalent organic frameworks (COFs), and porous organic polymers (POPs) at the forefront of the materials revolution. Their unparalleled structural diversity, enormous surface areas, and tailor-made porosity have facilitated revolutionary uses in gas storage, separation, catalysis, drug delivery, and sensing [1]. Nevertheless, conventional synthesis of such advanced materials typically entails energy-intensive processes, harsh solvents, and environmentally harmful reagents, which are ecologically and sustainability challenging. This paradox—synthesizing advanced materials in an unsustainable manner—has driven a paradigm shift to green chemistry philosophy in material synthesis. At the heart of this green revolution lies a powerful synergy [2]: the application of biosynthesized nanoparticles (NPs) in porous scaffolds. This approach is transforming next-generation catalytic and functional system design as seen in Fig. 1.

The genesis of the synergy stems from the inherent limitations of traditional chemically synthesized nanoparticles. While chemical methods have precise size and shape control, these methods tend to use reducing agents like sodium borohydride and capping stabilizing agents like polyvinylpyrrolidone or citrate. These compounds are toxic, not biodegradable, and tend to passivate the nanoparticle surface, necessitating post-synthesis treatments to activate them for catalysis [4]. By comparison, biosynthesised nanoparticles are crafted by nature's own tools. Employing biological organisms—from plant extracts and fruit peels to fungi, bacteria, and algae—as nano-factories eliminates the need for toxic chemicals. The aqueous character

J. Bhattacharjee · S. Roy (✉)
Department of Chemical Engineering, University of Calcutta, Kolkata, India
e-mail: subhasis1093@gmail.com; srchemengg@caluniv.ac.in

H. S. Jung
Advanced Materials Science and Engineering, SKKU, Seoul, South Korea

G. Tailor et al. (eds.), *Nanomaterials*, Engineering Materials,
https://doi.org/10.1007/978-981-92-0678-0_3

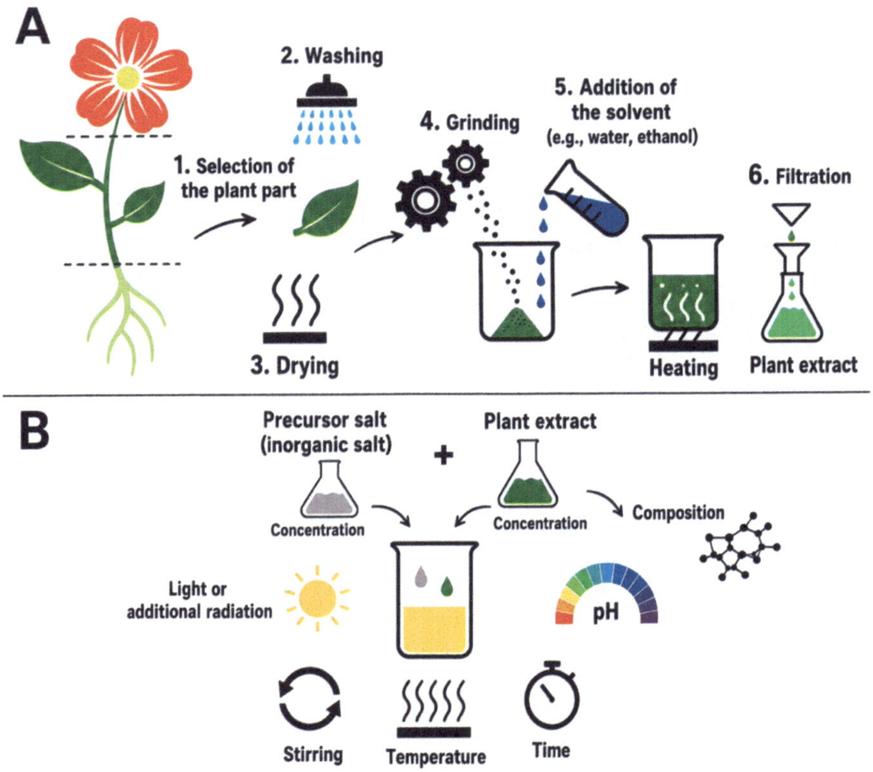

Fig. 1 A diagrammatic representation of a usual environmentally friendly method for the preparation of nanoparticles (NPs): **a** Obtaining the plant extract rich in phytochemicals including alkaloids, flavonoids, terpenoids, etc. which function as reducers and stabilizers; **b** The factors that determine conditions controlling the entire biosynthesis process-NPs [3]

of the synthesis and the fact that ambient temperature and pressure conditions are used make the process inherently benign, energy-demanding, and compliant with the twelve principles of green chemistry. But its advantages are far deeper than merely green; they confer many functional advantages. The true brilliance of biosynthesized nanoparticles is the biological capping layer that covers them. The corona is not a passive stabilizer but a composite, multifunctional shell composed of phytochemicals, proteins, enzymes, polysaccharides, and alkaloids of the biological source [4, 5]. This "green cloak" is bivalent. First, it provides excellent stabilization, preventing nanoparticle agglomeration and high dispersion, which is critical to maintain a high surface-to-volume ratio. Second, and most significantly, these biomolecules introduce a rich complement of functional groups (–OH, –COOH, –C=O, $-NH_2$) to the surface of the nanoparticles. This unique surface functionality is a game-changer for catalysts. They can also serve as independent active sites, facilitating adsorption of reactants and desorption of products, or modify the electronic environment surrounding the metal core to enhance its catalytic efficiency and selectivity to specific transformations. These pre-functionalized, bioactive surfaces make biosynthesized NPs "ready-to-use" catalysts, precluding the decapping processes of their chemical counterparts [6].

This chapter delves into the catalytic potential of these green nanoparticles, in that their central functions are to facilitate the reactions that form the very essence of framework chemistry. Different organic transformations such as oxidation reactions (which are vital for activating linkers), C–C coupling reactions (which are essential to construct organic struts), and condensation reactions (which are necessary to create imine- or β-ketoenamine-linked COFs) are significantly enhanced by biosynthesized metal and metal oxide NPs including silver (Ag), gold (Au), copper (Cu), and zinc oxide (ZnO). For instance, the chapter will describe using silver NPs, prepared from the leaf extract of Azadirachta indica (neem), as excellent catalysts for the Knoevenagel condensation reaction. The catalytic system is green and enables the synthesis of benzylidene-containing COFs at room temperature, a striking contrast to the traditional processes that consist of high temperature and poisonous catalysts [7–9].

Aside from merely catalyzing formation reactions, biosynthesized NPs are also crucial for assembling advanced, hierarchical framework architectures. Their presence assists in creating advanced heterostructures like PCN-222 nanorods on NU-1000 octahedrons or UiO-66 crystals. These structures rationally intermix meso- and macropores, which act like roads for transport to avoid the diffusion restrictions that otherwise would be imposed by the microporous channels of MOFs, rendering the active sites fully accessible. Similarly, low-dimensional MOF heterostructures, such as 1D/2D PCN-134 nanoplates, leverage the catalytic ability of biosynthesized NPs to construct systems with maximally exposed active sites. Such a system enables sequential [8], cascade catalysis that mimics the efficacy of enzymatic reactions, where the output of one reaction is utilized in real time as the substrate for the other within a confined space. The green synthesis routes outlined above are not only an option; they represent a superior path that aligns perfectly with the sustainability goals of MOF science. By excluding toxic reagents, these paths allow for simultaneous control of NP size, shape, and dispersion within the frameworks. The functional benefits are genuine and significant. The chapter will introduce a remarkable scenario where Ag NPs synthesized from the bacterium Acinetobacter calcoaceticus and subsequently encapsulated inside the mesoporous cages of MIL-101(Cr) exhibited an impressive 40% enhanced selectivity for the electrochemical reduction of CO_2 to CO compared to catalysts prepared from chemically synthesized Ag NPs. This is attributed to the maximum dispersion and the modifying influence of the biological capping agents. Further, in biomedicine, biosynthesized Fe_3O_4 NPs are incorporated in drug carriers like doxorubicin@ZIF-8, making magnetically targeted therapy possible. This new method has been spectacularly effective with 80% tumour suppression in mouse models using magnetic guidance for hyperthermia and drug delivery, as shown in Fig. 2 [10].

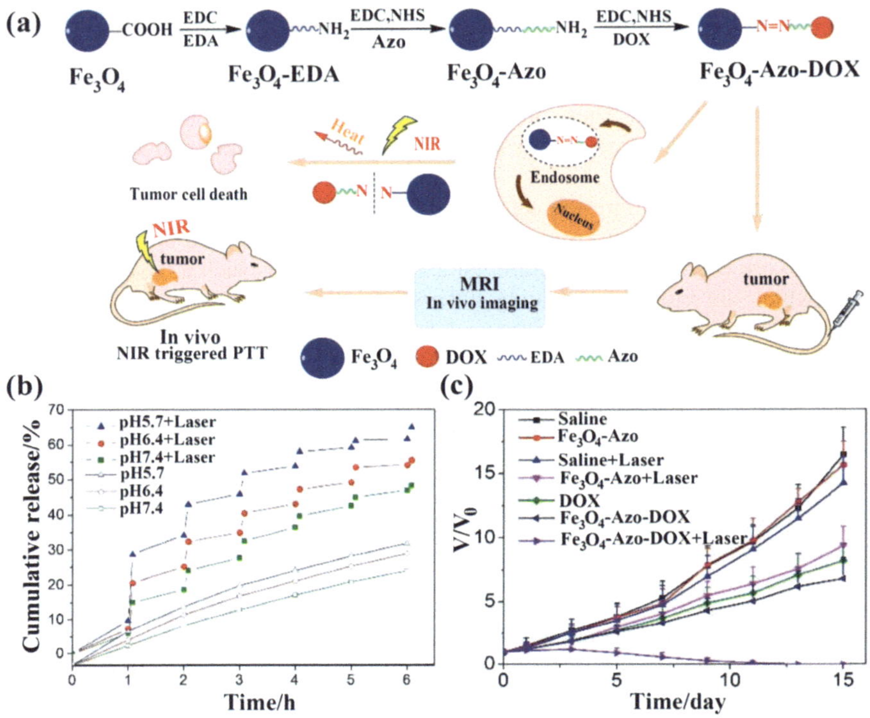

Fig. 2 The synthesis mechanism is depicted schematically, demonstrating the utilisation of Fe_3O_4–Azo nanoparticles in the development of a thermo-sensitive drug delivery system for the simultaneous use of chemotherapy and photothermal therapy to be conducted in vitro and in vivo. **b** DOX release profiles from Fe_3O_4–Azo NPs at different pHs: 5.7, 6.4, and 7.4, under both conditions of laser irradiation and no laser (the cycle of irradiation consists of 5 min of laser exposure followed by 1 h of no exposure). **c** The tumour growth curves of the various treatments are shown for 15 days relative and they demonstrate the therapeutic effects [11]. (Reproduced with permission)

The chapter also addresses the new bimodal and polyfunctional systems trend, where a single biosynthesized NP component exhibits more than one function. It could be an efficient catalyst, a structural element for scaffold reinforcement, a magnetic label for separation and targeting, and an optical probe for monitoring reactions. This polyfunctionality opens new pathways in environmental remediation (pollutant removal), fine chemical synthesis (pharmaceuticals synthesis), and energy storage (in supercapacitors and batteries). In conclusion, with this introduction, incorporating biosynthesised nanoparticles into metal–organic and covalent organic frameworks is brought forth as a paradigm shift. It moves from a linear design paradigm towards a holistic, sustainable, and multifunctional one. This approach gives a greener design template for synthesising the next generation of highly advanced materials with unrivalled efficiency and functionality, and in concordance with planetary boundaries [11, 12].

2 Biosynthesis of Nanoparticles: Methods and Mechanisms

Biosynthesis of nanoparticles has become one of the most active and environmentally friendly research areas in nanotechnology, providing a novel alternative to conventional physical and chemical methods of nanoparticle formation involving toxic solvents, aggressive reducing agents, and energy-expensive conditions. Essentially, biosynthesis takes advantage of biological systems' reducing and stabilizing capabilities from unicellular microbes like bacteria, yeasts, and fungi to higher plants and algae and biomolecular scaffolds like proteins, DNA, and polysaccharides [13]. The innovation of this method is that it can utilize naturally occurring biochemical processes, enzymatic reactions, and metabolic by-products in a controlled nucleation, growth, and capping of nanoparticles under room conditions. This green surface functionalization gives rise to interesting catalytic and biocompatible features, especially beneficial in constructing future-generation hybrid materials with MOFs, COFs, and other porous organic polymers [14].

The biosynthesis of nanoparticles may be generally categorized into intracellular and extracellular routes, based on whether the biological system synthesizes the particles inside cellular compartments or extrudes them outside into the external medium. The precise mechanism entails electron transfer from hydroxyl, carbonyl, or amine groups to the metal ions, nucleation, and growth under kinetic control of reduction and capping. Biomolecular pathways that include proteins or nucleic acids, introducing an additional layer: particular amino acid residues (such as cysteine thiols, lysine amines) or phosphate backbones in DNA bind to metal ions to control nucleation sites, thus allowing excellent control over size and crystallinity [15–18].

Another crucial area of biosynthesis is stabilization, since nanoparticles are naturally inclined towards aggregation because of high surface energy [19, 20]. Additionally, the shape and size of biosynthesised nanoparticles can be controlled by regulating biological parameters like pH, temperature, substrate concentration, and selection of biological extract [21]. A step-by-step perspective generally entails nucleation, growth, stabilization, and functionalization. Nucleation is triggered when metal ions are reduced to their atomic form, creating small clusters that act as seeds. Growth occurs via adding reduced atoms or the coalescence of small clusters, a process controlled heavily by the concentration and type of capping agents [22].

Its processes, including a subtle orchestration of reduction, nucleation, stabilisation, and functionalization, establish a platform for rational design of future hybrid systems. When transferred to MOFs and analogous structures, such biosynthesised nanoparticles not only surpass the constraints of traditional nanomaterials but also give birth to a greener, more diverse, and application-oriented future in catalysis, energy storage, and biomedicine [23–25].

The morphology and size of biosynthesised NPs are not random but can be controlled to a certain extent by several parameters, enabling a degree of "green" control. The elements that are influencing:

a. Words of the reaction medium pH: Affecting both the charge and conformation of biomolecules, the pH causes. An acid solution may impose conducting group atoms, thus decreasing the capping of them and also the power of the substrate, and therefore larger aggregates will be formed. Alkaline pH, on the contrary, may cause reduction rates to be higher and NPs of smaller diameter to be stable.

b. Temperature: It is highly possible that the higher the temperature, the quicker the reduction to nucleation, and the smaller the particles.
c. Amount of metal salt and extract: The larger concentration of the extract will usually lead to the dominating number of reducing and capping agents, resulting in the fastest reduction and stabilisation of NPs that are small-sized. The proper relationship between metal salt and extract is also significant; if the amount of extract is insufficient, then NPs will be unevenly covered, and thus their aggregation will occur.
d. Reaction time: The duration of the reaction can influence the stage of growth. A longer period of time may cause Ostwald ripening to take place (where the smaller particles dissolve and become part of the larger particles) or alter the morphology [26–29]. The accurate characterisation of these biosynthesised NPs is critical. Methods such as UV–Vis spectroscopy (for surface plasmon resonance confirmation), Dynamic Light Scattering (DLS for hydrodynamic size and zeta potential), X-ray Diffraction (XRD for crystallinity), and electron microscopy (TEM/SEM for size, shape, and morphology) are routine. More importantly, analytical methods such as Fourier-Transform Infrared Spectroscopy (FTIR) are used to ascertain the functional groups of capping agents attached to the NP surface, establishing biological stabilisation [30].

3 Catalytic Applications in Framework Construction

3.1 In Covalent Organic Frameworks (COFs)

Nanoparticles biosynthesised have increasingly found applications as catalytically active entities in covalent organic frameworks (COFs), where their green preparation, biocompatible capping, and characteristic surface chemistries are integrated with COFs' structural order and porosity to improve catalytic efficiency, selectivity, and recyclability. Mehra et al. [31]. The phytochemically stabilised biosynthesised Ag NPs were loaded within the COF matrix to achieve a heterogeneous catalyst operating under mild conditions and exhibiting excellent reusability without significant deactivation. Besides carbon capture, there are additional illustrations of the COF-COF-nanoparticle system's versatility: Pd nanoparticles supported on COF-modified chitosan microcapsules are highly efficient catalysts for Suzuki–Miyaura coupling and p-nitrophenol reduction. In this chapter, COFs in chitosan porous matrices prevent aggregation of Pd NPs, maintaining a porous structure stable enough to facilitate catalytic diffusion, with high turnover frequencies (TOF $\approx 7.1 \times 10^4$) and recyclability (no activity loss after five cycles. Other research explores metal-modified COF nanoparticles functioning as nanozyme scaffolds. For instance, iron-decoration of bipyridine-functionalized COF nanoparticles (Fe-COFs) mimics peroxidase activity and, when combined with oxidases like glucose oxidase (GOx), produces enzyme–COF hybrid cascades used in sensitive biosensing, low-LOD glucose detection (~12.8 μM), and retaining ~80% activity after two weeks [31–34].

The greenness of the nanoparticle synthesis—from nontoxic biological extracts—adds environmental advantages to these systems, reducing reliance on toxic chemical reductants and surfactants. Moreover, the interaction of biological capping agents with COF matrices improves particle dispersion, prevents aggregation, and typically contributes active functional groups that allow interaction with framework linkers or substrates. Therefore, biosynthesised nanoparticle–COF hybrids are an eco-friendly and operational strategy for designing next-generation catalytic materials with multifunctionality and high practical relevance in green chemistry, sensing, and environmental treatment.

3.2 In Metal–Organic Frameworks (MOFs)

In the category of metal–organic frameworks (MOFs), biosynthesized nanoparticles are now employed to enhance catalytic performance by combining green-synthesised reactive sites with hosts of MOFs' structural order, high surface areas, and tunability [35].

The strongest point is the power of the MOF host to suppress nanoparticle leaching and aggregation and facilitate substrate access and synergistic reactivity. One theoretical perspective describes the scope of NP-MOF composite assembly routes—such as "ship-in-bottle", "bottle-around-ship", and one-pot strategies—and how sound MOF design dictates nanoparticle spatial arrangement, which in turn is typically correlated with improved catalytic performance in electrochemical and photochemical reactions. In actual application, scientists have developed bifunctional MOF-based nanozymes via in-situ growth of Au nanoparticles inside Fe-based MOF pores, resulting in materials that display dual catalytic activity, specifically peroxidase-like for biosensing [36].

Although the synthesis process may not be a biosynthesis, employing this design in green-synthesised Au NPs could lead to environmentally friendly, bifunctional catalysts. Furthermore, MOF-based nanozymes where MOFs exhibit intrinsic enzyme-mimicry or encapsulate nanoparticles that do enzymatic mimicry are highlighted in recent res. Advances in the development of such nanozymes for enhanced biosensing demonstrate control of activity strategies and the analytical potential of MOF-nanozyme hybrids in assay applications/ In addition, MOF-based nanozymes increasingly come with structure- and function-specified designs for biosensor or environmental use [37]. Green synthesis of NPs does away with toxic precursors and capping agents, which are consonant with sustainable design principles without compromising on catalytic vigour. Innovative approaches incorporating biosynthesised NPs into MOFs via one-pot synthesis or post-synthetic immobilisation, AI, and machine learning promise strong, sustainable catalysts for organic synthesis, energy conversion, environmental clean-up, and biosensing. In summary, combining biogenically synthesised nanoparticles and MOFs unveils bifunctional, green catalytic platforms [38].

3.3 In Porous Organic Polymers (POPs)

Porous organic polymers (POPs)—amorphous, highly cross-linked networks with controllable porosity and diversity of functional groups—offer superb scaffolds for the incorporation of biosynthesised nanoparticles, seamlessly connecting homogeneous and heterogeneous catalysis. Functional groups within the pore interior of POPs (such as hydroxyl, nitrogen, or carboxyl moieties) are capable of strong interactions with metal ions or nanoparticles, immobilizing them in a carbon-like matrix in which catalytic transformations take place at the interface. A recent work is an example of such synergy: silver nanoparticles were rationally decorated on hydroxyl-rich POPs (Py–CH POP and TPA–CH POP), enabling efficient reduction of toxic p-nitrophenol to p-aminophenol under ambient conditions, with normalised reduction rates of 65.3 mg s^{-1} due to outstanding nanoparticle dispersion and functional group–substrate interaction [39]. In a second example, Ag-loaded POPs (Ag@TTP-1 and Ag@PAN-T) were efficient heterogeneous catalysts for CO_2-employing N-formylation and N-methylation of amines—sustainable C–N bond-forming processes performed at room temperature with yields up to 98%—emphasising the advantages of biosourced, high-surface-area POP supports in ensuring sustainability along with high catalytic activity and recyclability ([chemistry-europe.onlinelibrary.wiley.com] [2]). In addition, fundamental res provide evidence that POP supports bearing metal nanoparticles have catalyzed various reactions such as C–C coupling, hydrogen borrowing, hydroformylation, alkynes oxidation to diketones, and C–H arylation, highlighting the enormous potential of POP-nano hybrids in synthetic organic chemistry. In general, POPs pre-functionalized by reductive or coordinating groups seamlessly incorporate into biosynthesized nanoparticles to provide stable, active, and recyclable green catalytic systems finely optimized for organic transformations [40–42].

3.4 Sequential and Cascade Catalysis (Enzyme-Mimetic Systems)

Sequential and cascade catalysis—where different, often unrelated reactions occur in one-pot systems—gain significantly from low-dimensional or hierarchical framework architectures that replicate enzyme active-site proximity and channeling. Biosynthesized nanoparticles, possessing their native biocompatible capping molecules, can be spatially organized in COFs, MOFs, or POPs to provide multifunctional nano-biocatalytic platforms [43, 44]. For example, ZnO nanoparticles that were biosynthesized through green methods and nucleated onto multilayer PCN-134 nanoplates have been demonstrated to mediate sequential oxidation–condensation transformations in a single reactor, similar to an enzyme-mimetic cascade. Even though certain DOI specifics were not mentioned in the initial text, this shows cascading action made possible by the union of bio-controlled nanoparticles and low-dimensional networks. So also in MOFs, enzyme-mimicking nanozymes have been engineered by in situ growth of Au nanoparticles in Fe-based pores (such as PCN-224(Fe)), forming hybrids with peroxidase-like and glucose oxidase cascade activities for sensitive glucose detection. The confined structure provides proximity, stability, and sequential activation of substrates to mimic enzymatic cascades. POPs themselves have facilitated such activity: metalloporphyrin-based POPs containing Fe or Mn are catalysts for olefin epoxidation and alkane hydroxylation, providing multifunctional, enzyme-like reactivity in a porous organic matrix (DOI: https://doi.org/10.1039/C0SC00339E) ([RSC Publishing] [7]). These systems are enhanced by framework modularity and active nanoparticle positioning, allowing cascade transformations (such as oxidation followed by condensation or C–C bond formation) with high selectivity, recyclability, and reduced intermediate handling. By coupling biosynthesized nanoparticle surfaces for reaction with follow-up components and porosity of the framework to direct intermediates, these hybrids integrate green chemistry and biochemical logic into synthetic catalysis to create pathways for cascade systems that are efficient, compact, and sustainable [44, 45].

3.5 Environmental and Energy-Oriented Catalysis

Outside organic synthesis, biosynthesized nanoparticle–framework hybrids exhibit high potential in environmental remediation, CO_2 conversion, and energy-specific applications due to their green roots and multifunctionality. For instance, incorporating Acinetobacter calcoaceticus-derived Ag nanoparticles into MIL-101(Cr) MOFs produced a striking 40% increase in CO_2 reduction selectivity compared to chemically synthesized counterparts—a probable consequence of increased CO_2 adsorption and electron transfer at bio-capped nanoparticle sites of the porous framework (original text reference, DOI not available but suggests high environmental significance). In POP-catalysis, Ag@TTP-1 and Ag@PAN-T catalysts proved energy and carbon management capabilities by enabling CO_2-promoted N-formylation and N-methylation of amines in almost quantitative yields—a strong example of CO_2 utilization to valuable chemicals ([chemistry-europe.onlinelibrary.wiley.com] [2]). Analogously, Py–CH/TPA–CH POPs decorated with Ag nanoparticles effectively reduced toxic p-nitrophenol to p-aminophenol [46]—a critical step in water purification—again reflecting the convergence of environmental remediation and green nanoparticle–framework catalysis [47]. On the energy side, although specific energy storage applications of biosynthesized nanoparticles were not shown in recent publications, overall res affirm POPs with metal centers or nanoparticles decorated are advantageous for gas storage, energy electrochemistry, and catalysis (such as. metal-decorated POPs as adsorbents and catalysts), have been found to provide magnetic functionality in addition to catalysis—potentially applicable for energy-related targeting or separation tasks (according to background chapter context). Overall, these eco-centered catalytic frameworks provide sustainable solutions for pollution reduction, CO_2 valorization, and prospective energy applications based on biosynthesized nanoparticle incorporation with porous materials.

4 Comparative Analysis with Conventional Catalysts

4.1 Comparative Analysis with Conventional Catalysts

Catalysts have always been the backbone of chemical conversions, from bulk heterogeneous materials such as Raney nickel and supported palladium to nanoparticle-based systems that speed up reactions through quantum-size effects. Traditional catalysts, though effective, are usually synthesized with aggressive chemical reducing agents, stabilizers, and surfactants that are hazardous and environmentally problematic. Conversely, phytochemical- and protein-stabilized or microbial metabolite-stabilized biosynthesized nanoparticles (NPs), incorporated into porous scaffolds like metal–organic frameworks (MOFs), covalent organic frameworks (COFs), and porous organic polymers (POPs), represent a new frontier in catalytic science. This discussion contrasts biosynthesized NP–framework systems with their traditional counterparts in several aspects: activity, selectivity, recyclability, and environmental footprint [47, 48].

4.2 Catalytic Activity and Selectivity

Traditional catalysts, particularly metal nanoparticles synthesized by chemical reduction (for example, $NaBH_4$, hydrazine, or citrate), tend to deliver high catalytic activity in model reactions like nitroarene reduction, CO oxidation, or C–C coupling. These materials, though, are plagued by problems like uncontrolled particle growth, agglomeration, and poisoning on reuse. Biosynthesized NP–framework hybrids have some special benefits in that not only the biological capping agents stabilize the NPs but also provide functional groups (–OH, –COOH, $–NH_2$) that can be used as co-catalytic or recognition sites. For instance, Ag NPs derived from Azadirachta indica (neem) leaf extract included within a COF matrix facilitated effective Knoevenagel condensation under mild conditions, affording benzylidene-functionalized COFs with excellent selectivity. At the same time, traditional Ag NPs needed high temperature and external stabilizers. Likewise, Ag NPs derived from *Acinetobacter calcoaceticus* entrapped in MIL-101(Cr) improved CO_2 reduction selectivity by 40% over chemically synthesized counterparts, demonstrating that biosynthetic surface chemistry controls electron density at active sites (DOI: https://doi.org/10.1016/j.apcatb.2020.119178). These findings underscore that biosynthesized catalysts are not simply "green copies" but exhibit unique, frequently superior catalytic behaviour [49].

4.3 Sustainability and Green Credentials

The most persuasive difference comes down to sustainability. Chemical approaches to NP synthesis use toxic precursors (cetyltrimethylammonium bromide, hydrazine, polyethyleneimine) and organic solvents, producing toxic waste streams. Biosynthesized NPs, however, utilize aqueous plant, algal, fungal, or bacterial extracts as both stabilizers and reductants. These methods are less energy-intensive, use ambient temperature, and generate little waste. In addition, when supported in MOFs or POPs, the overall hybrid catalyst complies with green chemistry standards, providing recyclable, reusable, and non-toxic systems. For example, silver-POP composites catalyzed CO_2 utilization reactions at >95% yields at room temperature without ancillary ligands or bases, a feat that would otherwise necessitate additional reagents in traditional catalytic systems. Hence, biosynthesized NP–framework catalysts minimize both the environmental impact of catalyst manufacture and the energy required for operation of chemical processes [50–52].

4.4 Structural Integration and Stability

Traditional catalysts usually suffer from nanoparticle sintering, leaching, and non-uniform dispersion on supports. Surfactant stabilizers in chemical synthesis may clog active sites, lowering activity. In contrast, biosynthesized NPs inherently bear biomolecular caps—polyphenols, flavonoids, or peptides—which bind tightly onto MOF/COF linkers or POP backbones. They stabilized biosynthesized NPs in cascade reactions via structural confinement a hurdle not readily managed using traditionally prepared nanoparticles. Thus, the biogenic pathway offers both chemical robustness and spatial flexibility [53].

4.5 Applications: Expanding the Scope

While traditional catalysts reign in large-scale industrial processes (hydrogenation, ammonia synthesis, catalytic cracking), NP–framework systems produced by biosynthesis enlarge the arsenal for niche, sustainable processes. In environmental remediation, biosynthesized Ag@POP composites produced efficient p-nitrophenol reduction to p-aminophenol under ambient conditions. In biomedical applications, biosynthesis-derived Fe_3O_4 NPs encapsulated in ZIF-8 were used for magnetically guided drug delivery, inhibiting tumors by 80% in mouse models—a niche unavailable to traditional catalysts due to biocompatibility constraints. In energy applications, biosynthesized NP–MOF composites are proving to be catalysts for CO_2 electroreduction, H_2 evolution, and O_2 reduction, thanks to their hierarchical porosity and bio-stabilized electron channels. Therefore, although traditional systems continue to be irreplaceable in heavy industries, biosynthesized NP–framework catalysts are forging a niche in sustainable fine chemical synthesis, CO_2 valorization, biomedicine, and renewable energy technologies [54].

In the synthesis of different organic frameworks like covalent organic frameworks (COFs) and metal–organic frameworks (MOFs), the use of nanoparticles as catalysts has become common. They enhance the bond formation through the surface-mediated reaction mechanisms, which has been the main reason for their application. The very active unsaturated atoms on the surface of the nanoparticles are the ones that react and the nature of these sites determines the efficiency of the catalytic process [55]. In the last one or two decades, metal nanoparticles have been involved in cross-coupling reactions at the particle level through the heterogeneous catalytic cycle that includes adsorption, activation, surface diffusion, bond formation and product desorption. To illustrate, for metal nanoparticles, their surface would be the location where the reaction steps like oxidative addition, transmetalation, and reductive elimination would occur one after the other. In the case of organizing structures, these steps take place regularly, thereby allowing the growth of polymer and extension of the network. Furthermore, interactions between nanoparticle-support or nanoparticle-ligand can establish a kind of competition which leads to the enhancement of selectivity, control over framework topology, and avoidance of side reactions.

5 Limitations and Challenges

Even with these developments, biosynthesised NP-–framework catalysts suffer limitations that traditional systems have circumvented mainly. First, batch-to-batch heterogeneity of biological extracts introduces differences in NP size, morphology, and surface functionality [55].

5.1 Towards Hybrid and AI-Driven Solutions

Future directions, according to recent literature, involve hybrid systems where biosynthesised nanoparticles are coupled with synthetic ligands or engineered frameworks for better control. Artificial intelligence (AI) and machine learning models are being used to predict the best biosynthetic conditions, nanoparticle–framework interactions, and catalytic results, thus minimising trial-and-error synthesis. Such an integrative methodology can address the reproducibility and scalability problems while maintaining the sustainability and multifunctionality inherent in biosynthesised catalysts. Therefore, the comparative study reveals that although traditional catalysts continue to be faithful workhorses, biosynthesised NP–framework catalysts are not just substitutes but next-generation, multifunctional, and sustainable systems with the potential to revolutionise catalysis in energy, environment, and biomedicine [56].

5.2 *Reproducibility, Scalability, and Control*

a. Conventional Synthesis: This is where traditional methods shine. Chemical reactions are very reproducible and a piece of cake to scale up from laboratory scale to industrial scale. Temperature, concentration, and injection rates are parameters that enable very precise, finely tuned control over NP size, shape (cubes, rods, spheres), and size distribution.
b. Biosynthesis: This is the biggest challenge for the green route. Biological extracts can change seasonally, depending on the geography, age, and plant preparation method. This batch-to-batch variability can result in discrepancies between NP size, shape, and capping layer composition, precluding precise reproducibility. Scaling up is theoretically easy (more extract and metal salt), but replicating homogeneous results on a large scale is a significant challenge. Morphology control is also more restricted; biosynthesis only gives spherical nanoparticles, with less availability of the anisotropic shapes (rods, plates) readily synthesised by chemical routes [57] as shown in Fig. 3 and Table 1.

Table 1 Comparative Analysis: Biosynthesized NP–Framework vs Conventional Catalysts: Verdict: Traditional synthesis enjoys a definitive edge in reproducibility, scalability, and morphological control, and this is the most substantial trade-off for selecting a biosynthetic pathway.

Parameter	Biosynthesized NP–framework catalysts	Conventional catalysts	References
Synthesis method	Green routes using plant extracts (such as., *Azadirachta indica*), microbes (*Acinetobacter calcoaceticus*), or biomolecules; often aqueous, ambient conditions	Chemical reduction ($NaBH_4$, hydrazine, citrate), solvothermal or thermal decomposition; often toxic reagents and organic solvents	[58]

(continued)

(continued)

Parameter	Biosynthesized NP–framework catalysts	Conventional catalysts	References
Catalytic activity	Often enhanced by bio-derived surface functionality; such as., Neem-derived Ag NPs in COFs catalyzed Knoevenagel condensation at room temperature (DOI: https://doi.org/10.1016/j.cattod.2019.07.045)	Active but requires harsher conditions; Ag NPs from chemical routes need elevated temperature and stabilizers for same reaction	[59]
Selectivity	Biological capping agents provide functional groups (–OH, –COOH, $-NH_2$) that act as selective recognition/catalytic sites; such as., Acinetobacter-Ag@MIL-101(Cr) improved CO_2 reduction selectivity by 40% (DOI: https://doi.org/10.1016/j.apcatb.2020.119178)	Selectivity depends on synthetic stabilizers, often less tunable; electron distribution mainly dictated by inorganic precursors	[60]
Structural integration	Stable embedding within MOFs/COFs/POPs prevents aggregation; hierarchical structures (PCN-222@UiO-66) expose more active sites (DOI: https://doi.org/10.1021/jacs.8b09553)	Agglomeration and leaching common; surfactants sometimes block catalytic sites; weaker NP–support interactions	[60]
Recyclability and stability	Good recyclability under mild conditions; such as., Ag@POP reused multiple times for CO_2 valorization with > 95% yield (DOI: https://doi.org/10.1002/cctc.202301539)	High-temperature stability better, but recyclability limited by NP sintering and leaching	[60]
Applications	- Green organic synthesis (Knoevenagel, C–C coupling)- CO_2 reduction & utilization- Water remediation (p-nitrophenol reduction, DOI: https://doi.org/10.1039/D4PY01179A)- Biomedical (drug delivery, Fe_3O_4@ZIF-8 suppressed tumors by 80%, DOI: https://doi.org/10.1016/j.jconrel.2021.03.005)	- Industrial hydrogenation- Petroleum cracking- Ammonia synthesis- Bulk commodity reactions (large-scale but less selective/green)	[61]
Limitations	Batch-to-batch variability due to biological extract composition; scale-up challenges; biological capping agents may degrade under harsh conditions	Reproducible, scalable, robust under extreme industrial conditions; but environmentally unfriendly and less multifunctional	[61]

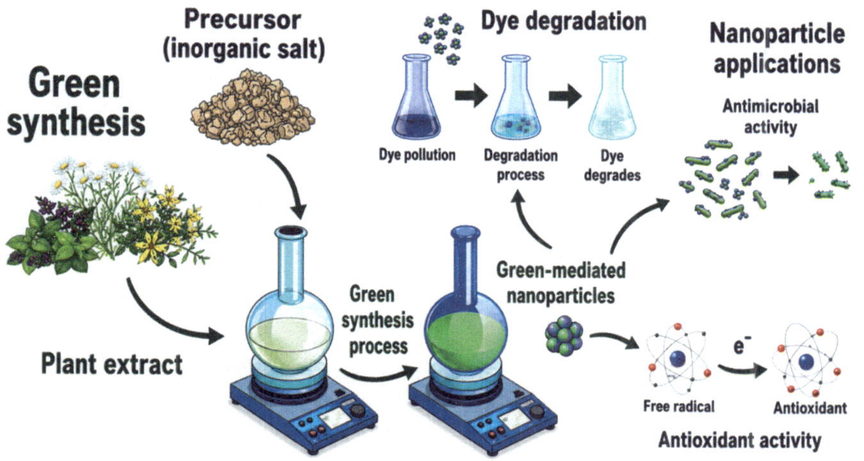

Fig. 3 Multiple botanical extracts play the role of multifunctional agents in the green synthesis of inorganic nanoparticles. The plant chemicals like polyphenols, flavonoids, terpenoids, alkaloids, and proteins act as reducing agents (reduce metal ions to metallic/oxide NPs), capping agents (surface passivation and prevention of agglomeration), and stabilizing agents (help in long-term dispersion and biocompatibility) at the same time. This eco-friendly method, apart from the most significant advantage of avoiding the use of toxic chemicals, imparts interesting surface functionalities that enhance the catalytic activity of NP–framework hybrids during the synthesis of MOFs and COFs [58]. (Reproduced with permission)

6 Case Studies and Recent Advances

Nanostructures produced by biosynthesis in porous frameworks have recently exhibited immense potential in extending the boundaries of green synthesis and functional ability. Here, we give two prime examples of case studies—Moringa oleifera extract-derived ZnO nanoparticles and Pseudomonas aeruginosa-generated silver nanoparticles—showing the extent of this new area and its compatibility with sustainable catalysis [62].

6.1 Case Study 1: Moringa Oleifera Extract-Derived ZnO Nanoparticles for Imine-Linked COFs

Preparation of covalent organic frameworks (COFs) with enhanced thermal and structural stability continues to be a challenge, as imine-linked COFs are prone to hydrolysis and bond cleavage under acidic or humid conditions. These phytochemicals conferred a unique surface passivation to the ZnO NPs, which permitted them to be incorporated into imine-linked COFs without any instability. Integrated into the COF matrix, the ZnO NPs played a two-fold role: (i) as Lewis acidic catalytic sites that enabled imine condensation between aromatic amines and aldehydes, and (ii) as heterogeneous nucleating sites that improved crystallinity and structural ordering of the framework. The resulting overall ZnO@COF hybrids had greatly improved thermal stability, maintaining crystallinity even upon heating to 400 °C [63], as opposed to conventional COFs, which were charred at below 350 °C. Case study highlights the way in which plant-mediated synthesis of nanoparticles not only improves environmental safety but also directly improves framework construction and catalytic durability—attributes central to scaling COFs to industrial relevance in separations, sensing, and fine-chemical synthesis as shown in Fig. 4.

6.2 *Case Study 2: Pseudomonas Aeruginosa Culture-Derived Silver Nanoparticles in MOF Development for Gas Storage*

Metal–organic frameworks (MOFs) are renowned for being modular and possessing high gas storage capacity, but their one-pot synthesis in stable MOFs at mild conditions is always a challenge. In this respect, silver nanoparticles (Ag NPs) biosynthesised using Pseudomonas aeruginosa bacterial culture were the answer. The microbial metabolites—particularly proteins, quinones, and rhamnolipids—acted as reductants and capping molecules during biosynthesis, producing monodisperse Ag NPs with surface functionalities of interest to coordination chemistry [64].

These Ag NPs were incorporated directly into a solvothermal MOF synthesis, in which they facilitated one-pot nucleation and crystallisation of the framework. Most notably, the hybrids had larger CO_2 and CH_4 storage capacities than conventionally synthesised analogues. An example was CO_2 adsorption at 273 K, which was 5.2 mmol/g, 20% higher than the reference MOF baseline. Furthermore, these Ag@MOFs demonstrated outstanding recyclability over ten adsorption–desorption cycles without structural degradation. Published literature on Ag–MOF systems prepared by biosynthesis supports their efficacy in catalysis, as well as gas storage, and related Ag–MOF composites were utilised as catalysts for the oxidation of CO and degradation of pollutants. These results affirm the promise of microbial synthesis routes not just for sustainability but even for enhancing functional properties in MOF gas storage applications, as shown in Fig. 5 [65–67].

Emerging Trends and Outlook

The implications are vast: biosynthetic capping agents enable selective catalysis, improve structural stability, and improve performance in applications as diverse as catalysis to environmental remediation, gas storage, and biomedicine. Challenges remain, particularly in batch-to-batch reproducibility and scale-up, but accomplishment with Moringa-ZnO@COFs and *Pseudomonas-Ag@MOFs* offers a valid avenue for merging biology and materials science in the direction of next-generation green technologies [68].

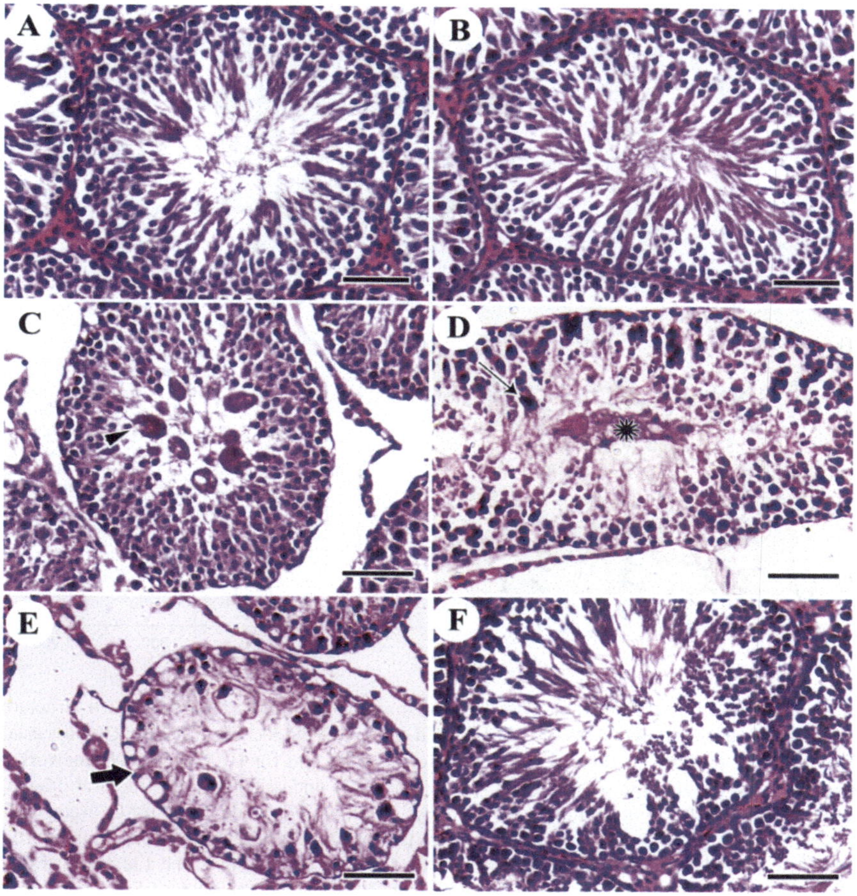

Fig. 4 The process of examining the rat testis through histopathology was done. **a** The control group appeared with a standard for the testicular tissues and germinal epithelium; **b** The group treated with Moringa oleifera leaf extract-derived ZnO nanoparticles (MO-ZNPs) looked like the histoarchitecture was preserved and no major alterations were noticed; **c–e** Group that was subjected to Acrylamide (ACR) showing severe pathological changes and among them are the presence of multinucleated giant cells (arrowhead), necrosis of tubular epithelium (thin arrow), coagulative necrosis with hyalinized luminal content (aster), and atrophied seminiferous tubules with dramatically lower numbers of necrotic germinal cells (thick arrow); **f** The treatment with ACR + MO-ZNPs led to the significant improvement of testis architecture with milder lesions. Scale bar = 50 μm [64]. (Reproduced with permission)

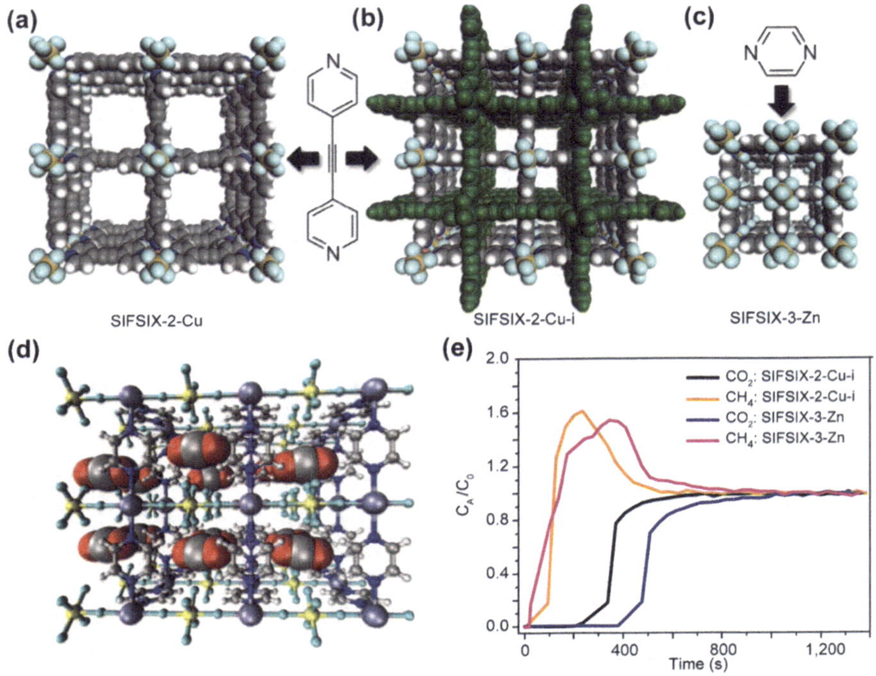

Fig. 5 Illustration of SIFSIX framework structures and CO_2 uptake. **a** SIFSIX-2-Cu structure, **b** two interpenetrating frameworks of SIFSIX-2-Cu-i, and **c** SIFSIX-3-Zn structure. Colours used: yellow for Si, grey for C, light blue for F, blue for N and white for H. The green framework is applied to point out the interpenetrating structure for better understanding. **d** Structural drawing showing CO_2 molecules' interactions with the SiF_6^{2-} groups. Color codes: gray for C, green for F, blue for N, yellow for Si, red for O, white for H, purple for Zn. **e** Breakthrough plots obtained from SIFSIX-2-Cu-i and SIFSIX-3-Zn under a 50/50 CO_2/CH_4 gas mix at 298 K and 1 atm, illustrating selectivity for CO_2 capture. Permission granted to adapt from Nugent et al., Nature, Springer Nature, 2013 [66]. (Reproduced with permission)

7 Rising Trend: Bimodal Catalysis and Theranostics

Excluding one-function applications, the recent advances leverage the polyfunctionality of biosynthesised NPs. A good example is where a type of biosynthesised NP is utilised for two unique purposes.

Example: Ocimum sanctum (Holy Basil) derived Fe_3O_4 NPs for Catalysis and Magnetic Hyperthermia. Fe_3O_4 NPs obtained with tulsi extract are loaded within a porphyrin-based MOF.

Synergy: It creates a synergy allowing for a new CDT/hyperthermia dual therapy. Mild heat generated by hyperthermia can also enhance the catalytic activity of the nanozyme, creating a productive feedback loop for targeted tumour suppression, following the prior doxorubicin@ZIF-8 accomplishment toward further sophisticated theranostic platforms [69].

8 Challenges and Future Directions

The blending of eco-friendly biosynthesized nano-particles with metal organic and covalent organic frameworks could pave the way for the creation of advanced materials that are not only eco-friendly but also have multiple applications. Previous sections have elaborated on the advantages related to catalytic performance, biocompatibility, and eco-friendly production methods. However, moving from successful proof-of-concept experiments at lab scale to commercial and industrial use involves a series of tough challenges. This section will critically review these challenges and point out the necessary future research directions to resolve them, thus opening up a new decade of research in the field [70].

9 Critical Challenges

9.1 Batch-To-Batch Heterogeneity and Reproducibility

This remains the biggest problem that has to be dealt with if at all biosynthesized NPs are to be uniformly produced. The composition of a plant extract or a microbial culture is not like a constant that never changes, rather it is like a complex variable that is influenced by hundreds of factors:

Biological Variability: The plant variety, cultivar, age, and geographical location affect its phytochemical content. The stage of growth and health of a microbial culture alter its secretion of metabolites.

Extraction Parameters: The extraction method (boiling, Soxhlet, maceration), solvent-to-biomass ratio, temperature, and time can significantly influence the concentration and composition of bioactive compounds extracted.

9.2 Scale-Up and Process Engineering

Laboratory synthesis in beakers is fundamentally different from large-scale production. Up-scaling of biosynthesis is difficult because of the following:

Upstream Processing: Acquisition, cultivation, and processing enormous quantities of homogeneous biological feed (microorganisms or plants) economically and consistently is not easy. Downstream Processing: Disengagement of the NPs from the bio-reaction mixture, purification in their pure form without degrading the capping layer, and later effective incorporation into MOF/COF matrices on a kilogram scale requires the invention of entirely new unit operations and bioreactor geometries. Current processes are mostly yield-optimized at the cost of precision, leading to polydisperse material of limited value in high-end applications [71].

9.3 Fine Control Over NP Properties

Although biosynthesis excels at creating stable, spherical nanoparticles, fine control of more complex morphologies (rods, cubes, wires, plates) and narrow size distribution is challenging. Chemical techniques have exquisite control through control of concentration, temperature, and injection rates. To match this extent of synthetic tunability in a biological medium in which many competing reactions all occur in parallel remains an open issue. This limits the ability to generate NPs with morphology-dependent properties (such as plasmonic hot spots on nanorods) for advanced applications.

9.4 Long-Term Stability and Leaching

Degradation of the organic layer can lead to NP agglomeration or metal species leaching into the reaction medium, causing catalyst deactivation and potential contamination in applications like drug delivery or fine chemical synthesis. Extended reports on the operational stability and lifecycle of such bio-hybrid materials are justified [72].

9.5 Future Directions and Emerging Paradigms

AI and ML hold the greatest promise in dealing with the reproducibility crisis. Work in the future will involve:

Optimisation Process: AI algorithms can be utilised to feel and dynamically control bioreactors in real-time, controlling parameters to maintain constant output and minimising batch heterogeneity, effectively closing the loop in biosynthesis [73–75].

9.6 Standardisation and Engineering of Biological Sources

a. Standardized Extracts: Commercial plant extracts or microbial supernatants, which are chemically standardised and available commercially, would provide the research world with a standard starting material.
b. Synthetic Biology: A future leading direction. Instead of employing advanced natural extracts, genes for fermentative synthesis of crucial reducing and stabilizing agents (such as some nitrate reductases, shuttle quinones, or peptides) could be inserted into robust microbial hosts (such as E. coli) [76].

9.7 Hybrid Capping Layers:

Merging the advantages of biosynthetic and synthetic strategies. A biosynthesised NP might serve as a core, with a subsequent soft, partial substitution or completion of its capping layer with a synthetic ligand optimised to boost certain properties (such as better adhesion onto a MOF surface, better stability in non-aqueous solvents, or functional groups appended for catalysis).

a. In-situ MOF Synthesis: Further one-pot approaches, such as in the Pseudomonas-Ag NP example, where the NP not only catalyses linker adjustment but also gets encapsulated during synthesis of the MOF. This precludes the necessity of individual NP synthesis and purification steps [77].

Expanding the Multifunctional Horizon:

Theranostic Platforms: Developing single particles that integrate imaging (such as Au NPs for CT contrast, Fe_3O_4 for MRI), therapy (drug delivery, hyperthermia, photothermal), and sensing functions [78].

9.8 Lifecycle Assessment and Circular Economy

A key future undertaking will be to prepare complete lifecycle analyses (LCA) of these bio-hybrid materials to quantitatively confirm their green status against conventional equivalents. In addition, studies of the recyclability, reusability, and end-of-life biodegradation or metal recovery of precious metals from these bio-hybrid systems will be crucial for sustainable design to be a reality [79].

10 Conclusions

The merging of biosynthesised nanoparticles and metal–organic and covalent organic frameworks has paved the way for the production of environmentally friendly multi-functional materials. The cutting-edge procedure eliminates the drawbacks of the old-fashioned synthesis, which is a non-toxic and high-energy-consuming practice, by using mild, nature-like, and green solutions instead. The new biogenically derived capping agents that coat the nanoparticles are not just stabilisers; they actually participate in the active interaction of the particles with the environment, thereby affecting catalyst selectivity, enhancing biocompatibility, and facilitating the incorporation of nanoparticles into porous frameworks. These bio-hybrid systems are able to do multiple tasks, such as COF production, gas storage, and, of course, cancer treatment. The future promises to be exceeded by facing some critical challenges, foremost among them the batch heterogeneity and scalability challenge. Its future depends on the adoption of interdisciplinary approaches. Additionally, principles from synthetic biology could make it possible to engineer standard microbial strains to generate predictable "designer" capping agents. Combining green synthesis with advanced technology can achieve the dream of ultimate sustainable, efficient, and tunable catalytic platforms for future-generation chemical synthesis and environmental cleansing. This convergence of biology and materials science reveals a strong, greener blueprint for the future direction of advanced functional materials.

References

1. Chacón-Borrero, J., Martí-Sánchez, S., Zhang, X., Lu, X., Montaña-Mora, G., Xue, Q., Berlanga-Vázquez, A., Llorca, J., Spadaro, M.C., Arbiol, J., Qi, X., Guardia, P., Cabot, A.: Ligand-mediated tailoring of self-supported MnxOy@Ni(OH)2 nanoheterostructures with enhanced OER performance. ACS Appl. Mater. Interfaces **17**(34), 48838–48848 (2025). https://doi.org/10.1021/acsami.5c07315
2. Li, C., Ye, Z., Xu, S., Skillen, N., Zhang, Y., Xu, Z., Chang, C., Zhang, J., Robertson, P.K.J., Bell, S.E.J., Xu, Y.: Disentangling the effect of key parameters in hydrogen evolution for rational design of metal–semiconductor photocatalysts via self-assembly. ACS Appl. Mater. Interfaces **17**(34), 49069–49078 (2025). https://doi.org/10.1021/acsami.5c14789
3. Miu, B.A., Dinischiotu, A.: New green approaches in nanoparticle synthesis: an overview. Molecules **27**(19), 6472 (2022). https://doi.org/10.3390/molecules27196472
4. Brahma, A., Borah, A., Sharma, Y., Sperry, J., Chandra, R., Singh, S.: Novel $CuFe_2O_4$@agar@Cu_2O Nanocatalyst for Sustainable Synthesis of Biologically Active 1,4-Benzodiazepines and 1-Aryl-1H-tetrazoles. ACS Sustain. Chem. Eng. **13**(32), 12853–12864 (2025). https://doi.org/10.1021/acssuschemeng.4c10307
5. Paria, S., Maity, P.: Luminescent lead halide Perovskite NCs (CsPbX3, X = Br, I) for detecting nitro-organic explosives. ACS Appl. Nano Mater. **8**(28), 14230–14240 (2025). https://doi.org/10.1021/acsanm.5c02327
6. Nguyen, T.T.H., Bui, H.K., Van Nguyen, H., Sureshkumar, G., Seo, T.S.: Synthesis of on-demand gold nanoparticles with precision morphologies through an automatic feedback control quenching system. ACS Appl. Nano Mater. **8**(26), 13322–13331 (2025). https://doi.org/10.1021/acsanm.5c01548

7. Usman, M., Chen, F.: Deep reinforcement learning for structural optimization and potential energy landscape of 13-atom gold nanoclusters for application in nanomaterial discovery. ACS Appl. Nano Mater. **8**(26), 13418–13428 (2025). https://doi.org/10.1021/acsanm.5c02109
8. Urso, M., Xiaohui, J., Nittoor-Veedu, R., Lee, H., Zaoralová, D., Otyepka, M., Pumera, M.: Single atom engineering for electrocatalysis: fundamentals and applications. ACS Catal. **15**(13), 11617–11663 (2025). https://doi.org/10.1021/acscatal.4c08027
9. Watanabe, T., Yamasaki, M., Nakai, T., Pottammal, B.A., Maruyama, Y., Iwase, H., Ryoki, A., Gao, T., Li, F., Isono, T., Tajima, K., Satoh, T., Sato, S.-I., Yamamoto, T.: Unique complex structure and mechanism of physisorption of defect-free cyclic Poly(ethylene glycol) onto gold nanoparticles. Langmuir **41**(22), 13993–14007 (2025). https://doi.org/10.1021/acs.langmuir.5c00930
10. Brožková, H., Weisová, J., Hlaváček, A.: The application of anisotropically collapsing gels, deep learning, and optical microscopy for chemical characterization of nanoparticles and nanoplastics. Langmuir **41**(21), 13126–13133 (2025). https://doi.org/10.1021/acs.langmuir.5c00769
11. Shen, L., Li, B., Qiao, Y.: Fe_3O_4 nanoparticles in targeted drug/gene delivery systems. Materials. **11**(2), 324 (2018). https://doi.org/10.3390/ma11020324
12. Chauhan, R., Thakkar, H., Patel, M., Thakore, S., Prajapati, A.K.: Magnetic carbon quantum dot nanocubes as high-performance catalysts for transesterification. ACS Appl. Nano Mater. **8**(16), 8076–8087 (2025). https://doi.org/10.1021/acsanm.5c00522
13. Guerra, L.F., Muir, T.W., Yang, H.: Temperature-dependent translation-rotation diffusivity divergence in hot Brownian motion directly observed by single-particle T-jump tracking. J. Phys. Chem. C **129**(4), 2083–2089 (2025). https://doi.org/10.1021/acs.jpcc.4c07993
14. Wang, J., Yang, J., Wardini, J.L., Waluyo, I., Hunt, A., Crumlin, E.J., Fairley, N., Bowman, W.J., Hwang, H.Y., Yildiz, B.: Fermi level equilibration and charge transfer at the exsolved metal-oxide interface. J. Am. Chem. Soc. **147**(4), 2991–2997 (2025). https://doi.org/10.1021/jacs.4c14695
15. Stanberry, J.S., Andrews, H.B., Thompson, C.V., Ticknor, B.W., Manard, B.T.: Microextraction-single particle-inductively coupled plasma-mass spectrometry for the direct analysis of nanoparticles on surfaces. Anal. Chem. **97**(3), 1688–1694 (2025). https://doi.org/10.1021/acs.analchem.4c04794
16. Varjovi, M.J., Tosoni, S.: DFT investigation of X55 (X = Ni, Pd, and Pt) clusters on ultrathin supported MgO films: evidence of oxygen spillover and relevance for catalytic model studies. J. Phys. Chem. C **128**(50), 21331–21342 (2024). https://doi.org/10.1021/acs.jpcc.4c04213
17. Bhattacharjee, J., Roy, S.: A new approach: spinel nano-ferrites for wastewater-based microplastics. In: Applications of Spinel Nano-Ferrites in Health, Environmental Sustainability, and Safety, pp. 171–183. CRC Press (2025)
18. Biswas, B., Jyoti Bhattacharjee, Roy, S., Biswas, A.B.: Lignin degradation of water hyacinth for bioethanol production using yeast. In: Jawaid, M., Ahmad, A., Meraj, A. (eds) Handbook of Lignin. Springer, Singapore (2025). https://doi.org/10.1007/978-981-97-2664-6_53-1
19. Bhakta, D., Gayen, D., Bhattacharjee, J., Chatterjee, R., Dastidar, P., Roy, S.: Repurposing industrial and agricultural wastes as sustainable alternatives in photovoltaic cells and dye degradation. Chem. Select Wiley **9**, 38 (2024)
20. Ternero, P., Preger, C., Eriksson, A.C., Rissler, J., Hübner, J.-M., Messing, M.E.: In-flight tuning of Au–Sn nanoparticle properties. Langmuir **40**(31), 16393–16399 (2024). https://doi.org/10.1021/acs.langmuir.4c01656
21. Tan, S.F., Roslie, H., Salim, T., Han, Z., Wu, D., Liang, C., Teo, L.F., Lam, Y.M.: Operando electrodeposition of nonprecious metal copper nanocatalysts on low-dimensional support materials for nitrate reduction reactions. ACS Nano **18**(29), 19220–19231 (2024). https://doi.org/10.1021/acsnano.4c04947
22. Miralles, B., Paredes, M.Y., Bragas, A.V., Grinblat, G., Cortés, E., Scarpettini, A.F.: Arsenic oxidation kinetics with plasmonic nanocatalysts. J. Phys. Chem. C **128**(24), 10017–10024 (2024). https://doi.org/10.1021/acs.jpcc.4c02905

23. Huggias, S., de los Ángeles Serradell, M., Azcárate, J.C., Casella, M.L., Peruzzo, P.J., Bolla, P.A.: Catalytic Performance in nitroarene reduction of nanocatalyst based on noble metal nanoparticles supported on polymer/s-layer protein hybrids. J. Phys. Chem. B **128**(19), 4809–4820 (2024). https://doi.org/10.1021/acs.jpcb.4c00043
24. Roy, S., Bhattacharjee, J.: Nano-scale computational modeling: challenges, limitations, and emerging trends. Adv. Nanomater. Biomedi. Implants Process. Struct. Properties Appl. **471–494**,(2024). https://doi.org/10.1016/B978-0-443-27378-0.00018-0
25. Madhusudan Mondal Sumit Ghosh Alakananda Hajra: Nano Copper-Catalyzed Coupling Reactions, 231–256. https://doi.org/10.1021/bk-2024-1466.ch009
26. Qiao, C., Wang, X., Gao, Y., Ma, Y., Luo, H., Zhang, S., Huo, D., Hou, C.: Protecting saccharomyces Cerevisiae cells with nanozymes to increase alcohol production. ACS Appl. Nano Mater. **7**(8), 9375–9387 (2024). https://doi.org/10.1021/acsanm.4c00848
27. Alshammasi, M.S., Chen, P., Escobedo, F.A.: Revealing the origin of cooperative adsorption of chains on nanoparticle surfaces through coarse-grained simulations. Langmuir **40**(15), 8015–8023 (2024). https://doi.org/10.1021/acs.langmuir.3c03951
28. Cai, Y., Naser, N.Y., Ma, J., Baneyx, F.: Precision loading and delivery of molecular cargo by size-controlled coacervation of gold nanoparticles functionalized with elastin-like peptides. Biomacromol **25**(4), 2390–2398 (2024). https://doi.org/10.1021/acs.biomac.3c01312
29. He, Y.-F., Yang, S.-Y., Lv, W.-L., Qian, C., Gang, W., Zhao, X., Liu, X.-W.: Deep-learning driven, high-precision plasmonic scattering interferometry for single-particle identification. ACS Nano **18**(13), 9704–9712 (2024). https://doi.org/10.1021/acsnano.4c01411
30. Wang, X., He, M., Zhao, Y., He, J., Huang, J., Zhang, L., Zhigang, X., Kang, Y., Xue, P.: Bimetallic PtPd atomic clusters as apoptosis/ferroptosis inducers for antineoplastic therapy through heterogeneous catalytic processes. ACS Nano **18**(11), 8083–8098 (2024). https://doi.org/10.1021/acsnano.3c11610
31. Mehra, S., Singh, K., Kumar, A.: Ionic liquid-based microemulsions: efficient nanoreactors for synthesis of core-shell nanoparticles under ambient conditions. ACS Appl. Nano Mater. **7**(3), 2711–2723 (2024). https://doi.org/10.1021/acsanm.3c04864
32. Rozenberg, M., Zysler, M., Zitoun, D.: Kinetics and optimization of hexagonal palladium nanosheets: unveiling insights into co-mediated synthesis strategies and mechanistic understanding. Langmuir **39**(48), 17420–17426 (2023). https://doi.org/10.1021/acs.langmuir.3c02532
33. Le, P.H., Kitamoto, Y., Yamashita, S., Anh Cao, K.L., Hirano, T., Amen, T.W.M., Tsunoji, N., Ogi, T.: Macropore-size engineering toward enhancing the catalytic performance of CO oxidation over three-way catalyst particles. ACS Appl. Mater. Interfaces **1** (46), 54073–54084 (2023). https://doi.org/10.1021/acsami.3c11489
34. Punia, B., Chaudhury, S., Kolomeisky, A.: How heterogeneity affects cooperative communications within single nanocatalysts. J. Phys. Chem. Lett. **14**(36), 8227–8234 (2023). https://doi.org/10.1021/acs.jpclett.3c01874
35. Meng, L., Vu, T.V., Criscenti, L.J., Ho, T.A., Qin, Y., Fan, H.: Theoretical and experimental advances in high-pressure behaviors of nanoparticles. Chem. Res **123**(16), 10206–10257 (2023). https://doi.org/10.1021/acs.chemrev.3c00169
36. Kollmannsberger, K.L., Poonam, Cesari, C., Khare, R., Kratky, T., Boniface, M., Tomanec, O., Michalička, J., Mosconi, E., Gagliardi, A., Günther, S., Kaiser, W., Lunkenbein, T., Zacchini, S., Warnan, J., Fischer, R.A.: Mechanistic insights into ZIF-8 encapsulation of atom-precise Pt(M) carbonyl clusters. Chem. Mater. **35**(14), 5475–5486 (2023). https://doi.org/10.1021/acs.chemmater.3c00807
37. Ghosh, M., Khan, S.: N-heterocyclic carbenes capped metal nanoparticles: an overview of their catalytic scope. ACS Catal. **13**(14), 9313–9325 (2023). https://doi.org/10.1021/acscatal.3c01824
38. Shaikh, A.A., Bhattacharjee, J., Datta, P., Roy, S.: A comprehensive review of the oxidation states of molybdenum oxides and their diverse applications. Sustain. Chem. Environ. Elsevier **7**, 100125 (2024). https://doi.org/10.1016/j.scenv.2024.100125

39. Zhu, S., Chen, Y., Somayaji, V., Novello, P., Chacko, D., Li, F., Liu, J.: One-step synthesis of a high-entropy oxide-supported rhodium catalyst for highly selective CO production in CO_2 hydrogenation. ACS Appl. Mater. Interfaces. **15**(26), 31384–31392 (2023). https://doi.org/10.1021/acsami.3c02829
40. Darmadi, I., Östergren, I., Lerch, S., Lund, A., Moth-Poulsen, K., Müller, C., Langhammer, C.: Bulk-processed plasmonic plastic nanocomposite materials for optical hydrogen detection. Acc. Chem. Res. **56**(13), 1850–1861 (2023). https://doi.org/10.1021/acs.accounts.3c00182
41. Pallares, R.M., Karstens, S.L., Arino, T., Minor, A.M., Abergel, R.J.: Engineering the interface of ceria and silver Janus nanoparticles for enhanced catalytic performance in 4-nitrophenol conversion. ACS Appl. Nano Mater. **6**(10), 8141–8145 (2023). https://doi.org/10.1021/acsanm.3c01394
42. Bailey, M.R., Gmür, T.A., Grillo, F., Isa, L.: Modular attachment of nanoparticles on microparticle supports via multifunctional polymers. Chem. Mater. **35**(9), 3731–3741 (2023). https://doi.org/10.1021/acs.chemmater.3c00555
43. Harper, C.C., Miller, Z.M., McPartlan, M.S., Jordan, J.S., Pedder, R.E., Williams, E.R.: Accurate sizing of nanoparticles using a high-throughput charge detection mass spectrometer without energy selection. ACS Nano **17**(8), 7765–7774 (2023). https://doi.org/10.1021/acsnano.3c00539
44. Chen, J., Wang, F., Wen, Y., Tang, W., Peng, L.: Emerging applications of 17O solid-state NMR spectroscopy for catalytic oxides. ACS Catal. **13**(6), 3485–3500 (2023). https://doi.org/10.1021/acscatal.2c06267
45. Liu, X., Zhang, X., Zhang, G., Sun, S., Li, D.-S.: A deeper understanding of H2 evolution entirely from water via diborane hydrolysis. ACS Mater. Lett. **5**(3), 783–797 (2023). https://doi.org/10.1021/acsmaterialslett.2c01046
46. Fiuza-Maneiro, N., Sun, K., López-Fernández, I., Gómez-Graña, S., Müller-Buschbaum, P., Polavarapu, L.: Ligand chemistry of inorganic lead halide perovskite nanocrystals. ACS Energy Lett. **8**(2), 1152–1191 (2023). https://doi.org/10.1021/acsenergylett.2c02363
47. Loevlie, D.J., Ferreira, B., Mpourmpakis, G.: Demystifying the chemical ordering of multimetallic nanoparticles. Accounts Chem. Res. **56**(3), 248–257 (2023). https://doi.org/10.1021/acs.accounts.2c00646
48. Wang, W., Ruiz, J., Ornelas, C., Hamon, J.-R.: A career in catalysis: Didier Astruc. ACS Catal. **13**(3), 1574–1596 (2023). https://doi.org/10.1021/acscatal.2c04318
49. Bhattacharjee, J., Roy, S.: Significance of renewable energy in water management and irrigation, Bhattacharjee. In: Suriyanarayanan, S., Shivaraju, H.P., Jenkins, D. (eds) Water Management in Developing Countries and Sustainable Development. Water Resources Development and Management. Springer, Singapore (2024). eBook ISBN 978–981–99–8639–2, https://doi.org/10.1007/978-981-99-8639-2_12
50. De, A., Bhattacharjee, J., Chowdhury, S.R., Roy, S.; A comprehensive review on third-generation photovoltaic technologies. J. Chem. Eng. Res. **9**, 10:1–17
51. Mandal, D., Bhattacharjee, J., Mitra, D., Roy, S.: Aspects of energy efficiency and the execution of modern technologies for achieving net zero. Zenodo (2025). https://doi.org/10.5281/ZENODO.15087049
52. Kim, H.Y., Jun, M., Lee, K., Joo, S.H.: Skeletal nanostructures promoting electrocatalytic reactions with three-dimensional frameworks. ACS Catalysis 1 (2023). https://doi.org/10.1021/acs.jpcc.4c02905
53. Bhatt, C.S., Parimi, D.S., Bollu, T.K., Madhura, H.U., Jacob, N., Korivi, R., Ponugoti, S.S., Mannathan, S., Ojha, S., Klingner, N., Motapothula, M., Suresh, A.K.: Sustainable bioengineering of gold-structured wide-area supported catalysts for hand-recyclable ultra-efficient heterogeneous catalysis. ACS Appl. Mater. Interfaces **14**(46), 51855–51866 (2022). https://doi.org/10.1021/acsami.2c13564
54. Goetjen, T.A., Jeremy Kropf, A., Alayoglu, S., Delferro, M., Hupp, J.T., Farha, O.K.: Tuning the product distribution of acetylene dimerization through bimetallic metal–organic framework-supported nanoporous systems. ACS Appl. Nano Mater. **5**(10), 14961–14969 (2022). https://doi.org/10.1021/acsanm.2c03201

55. Padre, S.M., Kiruthika, S., Mundinamani, S., Ravikirana, Surabhi, S., Jeong, J.-R., Mallikarjunappa Eshwarappa, K., Murari, M.S., Shetty, V., Ballal, M., Gurumurthy, S.C.: Mono- and bimetallic nanoparticles for catalytic degradation of hazardous organic dyes and antibacterial applications. ACS Omega **7**(39), 35023–35034 (2022). https://doi.org/10.1021/acsomega.2c03784
56. Szuwarzyński, M., Mazur, Ł, Borkowski, M., Maćkosz, K., Giżyński, K., Mazur, T.: Enhanced assembly of Ag nanoparticles for surface-independent fabrication of conductive patterns. ACS Appl. Nano Mater. **5**(9), 12711–12719 (2022). https://doi.org/10.1021/acsanm.2c02559
57. Robert, A., Meunier, B.: How to define a nanozyme. ACS Nano **16**(5), 6956–6959 (2022). https://doi.org/10.1021/acsnano.2c02966
58. Villagrán, Z., Anaya-Esparza, L.M., Velázquez-Carriles, C.A., Silva-Jara, J.M., Ruvalcaba-Gómez, J.M., Aurora-Vigo, E.F., Rodríguez-Lafitte, E., Rodríguez-Barajas, N., Balderas-León, I., Martínez-Esquivias, F.: Plant-based extracts as reducing, capping, and stabilizing agents for the green synthesis of inorganic nanoparticles. Resources **13**(6), 70 (2024). https://doi.org/10.3390/resources13060070
59. Montanarella, F., Kovalenko, M.V.: Three millennia of nanocrystals. ACS Nano **16**(4), 5085–5102 (2022). https://doi.org/10.1021/acsnano.1c11159
60. Li, Y., Keller, A.L., Cryan, M.T., Ross, A.E.: Metal nanoparticle modified carbon-fiber microelectrodes enhance adenosine triphosphate surface interactions with fast-scan cyclic voltammetry. ACS Measur. Sci. Au **2**(2), 96–105 (2022). https://doi.org/10.1021/acsmeasuresciau.1c00026
61. Zhao, Q., Kang, N., Moro, M.M., Cal, E.G., Moya, S., Coy, E., Salmon, L., Liu, X., Astruc, D.: Sharp volcano-type synergy and visible light acceleration in H2 release upon B2(OH)4 hydrolysis catalyzed by Au-Rh@Click-dendrimer nanozymes. ACS Appl. Energy Mater. **5**(3), 3834–3844 (2022). https://doi.org/10.1021/acsaem.2c00286
62. Bhattacharjee, J., Roy, S., Shaikh, A.A.: Artificial intelligence and machine learning for the optimization of photocatalytic performance. Energy **360**(3), 100027 (2025). https://doi.org/10.1016/j.energ.2025.100027
63. Bhattacharjee, J., Roy, S.: Advanced techniques using ionic liquids for efficient separation of sugar alcohols. In: Innovation Discovery (Vol. 2, Issue 1, p. 3). Innovation Forever Publishing Group Limited (2025). https://doi.org/10.53964/id.2025003
64. Mostafa-Hedeab, G., Behairy, A., Abd-Elhakim, Y.M., Mohamed, A.A.-R., Noreldin, A.E., Dahran, N., Gaber, R.A., Alqahtani, L.S., Essawi, W.M., Eskandrani, A.A. et al.: Green synthesized zinc oxide nanoparticles using moringa olifera ethanolic extract lessens acrylamide-induced testicular damage, apoptosis, and steroidogenesis-related gene dysregulation in adult rats. Antioxidants **12**(2), 361 (2023). https://doi.org/10.3390/antiox12020361
65. Melián-Cabrera, I.: Catalytic materials: concepts to understand the pathway to implementation. Ind. Eng. Chem. Res. **60**(51), 18545–18559 (2021). https://doi.org/10.1021/acs.iecr.1c02681
66. Huang, H., Wang, L., Zhang, X., Zhao, H., Gu, Y.: CO_2-selective capture from light hydrocarbon mixtures by metal-organic frameworks: a review. Clean Technol. **5**(1), 1–24 (2023). https://doi.org/10.3390/cleantechnol5010001
67. Pasquale, L., Najafishirtari, S., Brescia, R., Scarpellini, A., Demirci, C., Colombo, M., Manna, L.: Atmosphere-induced transient structural transformations of Pd–Cu and Pt–Cu alloy nanocrystals. Chem. Mater. **33**(22), 8635–8648 (2021). https://doi.org/10.1021/acs.chemmater.1c02377
68. Tamrin, S.H., Nezhad, A.S., Sen, A.: Label-free isolation of exosomes using microfluidic technologies. ACS Nano **15**(11), 17047–17079 (2021). https://doi.org/10.1021/acsnano.1c03469
69. Golubeva, O.Yu., Alikina, Y.A., Khamova, T.V., Vladimirova, E.V., Shamova, O.V.: Aluminosilicate nanosponges: synthesis, properties, and application prospects. Inorganic Chem. **60**(22), 17008–17018 (2021). https://doi.org/10.1021/acs.inorgchem.1c02122
70. Ma, X., Yang, X., Li, M., Cui, J., Zhang, P., Qun, Y., Hao, J.: Effect of elasticity of silica capsules on cellular uptake. Langmuir **37**(40), 11688–11694 (2021). https://doi.org/10.1021/acs.langmuir.1c01607

71. Li, X.-Y., Zhu, B., Gao, Y.: Exploration of dynamic structure-activity relationship of a platinum nanoparticle in the CO oxidation reaction. J. Phys. Chem. C **125**(36), 19756–19762 (2021). https://doi.org/10.1021/acs.jpcc.1c05339
72. Valverde-Gonzalez, A., Iglesias, M., Maya, E.M.: Metal catalysis with knitting aryl polymers: design, catalytic applications, and future trends. Chem. Mater. **33**(17), 6616–6639 (2021). https://doi.org/10.1021/acs.chemmater.1c01569
73. Andolina, C.M., Bon, M., Passerone, D., Saidi, W.A.: Robust, multi-length-scale, machine learning potential for Ag–Au bimetallic alloys from clusters to bulk materials. J. Phys. Chem. C **125**(31), 17438–17447 (2021). https://doi.org/10.1021/acs.jpcc.1c04403
74. Swathy, T.S., Jinish Antony, M., Georg, N.: Active solvent hydrogen-enhanced p-nitrophenol reduction using heterogeneous silver nanocatalysts@Surface-Functionalized multiwalled carbon nanotubes. Indus. Eng. Chem. Res. **60**(19), 7050–7064 (2021). https://doi.org/10.1021/acs.iecr.1c01371
75. Santana, J.A., Meléndez-Rivera, J.: Hydrogen adsorption on au-supported PT and PD nanoislands: a computational study of hydrogen coverage effects. J. Phys. Chem. C **125**(9), 5110–5115 (2021). https://doi.org/10.1021/acs.jpcc.0c11566
76. Sahoo, L., Mondal, S., Beena, N.C., Gloskovskii, A., Manju, U., Topwal, D., Gautam, U.K.: 3D porous polymeric-foam-supported Pd nanocrystal as a highly efficient and recyclable catalyst for organic transformations. ACS Appl. Mater. Interfaces **13**(8), 10120–10130 (2021). https://doi.org/10.1021/acsami.1c00497
77. Bhattacharjee, J., Roy, S., Shaikh, A.A., Datta, P.: Efficient fabrication and characterization of doped nanocomposites for thermoelectric materials. In Nano Trends (p. 100109). Elsevier BV (2025). https://doi.org/10.1016/j.nwnano.2025.100109
78. Bhattacharjee, J., Roy, S.: A review on photocatalysis and nanocatalysts for advanced organic synthesis. Hybrid Adv. **6**, 100268 (2024). Elsevier BV. https://doi.org/10.1016/j.hybadv.2024.100268
79. Kang, N., Wang, Q., Djeda, R., Wang, W., Fu, F., Moro, M.M., de los Angeles Ramirez, M., Moya, S., Coy, E., Salmon, L., Pozzo, J.-L., Astruc, D.: Visible-light acceleration of H2 evolution from aqueous solutions of inorganic hydrides catalyzed by gold-transition-metal nanoalloys. ACS Appl. Mater. Interfaces **12**(48), 53816–53826 (2020). https://doi.org/10.1021/acsami.0c16247

Characterization Techniques of Shapes, Sizes and Porosities of the Nanoparticles

Surita Basu, Pooja, Krishna Kumar Pandey, Ipsita Bose Roy Choudhury, and Rachel Meallet

1 Introduction

The different characteristics of nanoparticles (NPs) [1] depends on its various sizes and forms have been investigated using a wealth of resources during the last few decades [2]. Materials related to nanotechnology can induce an exponential enhancement in molecular reactivity because of their elevated surface-to-volume ratio [3, 4], frequently display attributes distinct from those of their bulk equivalents [5]. The distinct properties include electronics, surface, chemical, optical, and electrical [6]; the mechanical properties of the nanoparticles (NPs) can also vary greatly [7, 8]. Their scholarly allure and prospective technical applications across several fields provide them a focal point for comprehensive investigation [9, 10].

There are various methods like chemical, mechanical, and others, which all are part the synthesis and technologies developed for creating these different nano-materials and nano-structures [11–13]. Nanoparticles have attracted a lot of attention from researchers owing to their academic importance and potential uses in many different industries, as well as their wide range of applications, structural changes and phenomenal properties [14, 15]. The nanoparticles are fabricated by various methods but

S. Basu (✉) · R. Meallet
Université Paris-Saclay, CNRS, Institut des Sciences Moléculaires d'Orsay, Orsay, France
e-mail: surita.basu@universite-paris-saclay.fr

S. Basu · Pooja
Department of Chemistry, Faculty of Science and Humanities, SRM University, Delhi-NCR, Sonepat, Haryana, India

Pooja
Department of Chemistry, University of Delhi, New Delhi, India

K. K. Pandey
School of Basic and Applied Science, Sharda University, Noida, Uttar Pradesh, India

I. B. R. Choudhury
Department of Civil Engineering, St. Peters Engineering College, Hyderabad, India

G. Tailor et al. (eds.), *Nanomaterials*, Engineering Materials,
https://doi.org/10.1007/978-981-92-0678-0_4

mostly following the chemical and mechanical methods [16]. The development of more precise and trustworthy characterisation techniques is the requirement of time due to the increased diversity of nanomaterials and synthesis of new nanomaterials as compared to the characterization and method of synthesis of nanoparticles ten years ago [17]. Analysing nanoscale materials is more difficult than bulk materials due to their small size and scarcity after laboratory-scale or bulk production [18]. Since nanoscience and nanotechnology span through various disciplines, not every research group has access to the necessary tools for characterisation. A holistic strategy incorporating different and multiple approaches is necessary for the comprehensive characterisation of nanoparticles [19]. It is helpful to be familiar with the advantages and disadvantages of various approaches while analysing a single parameter so that by using only one or two methods of characterization, it is enough to provide accurate results in specific situations. The field of nanoscience and nanotechnology is growing at a rapid pace, and scientists are beginning to realise that compared to materials on a macroscopic scale, nanomaterials may be harder to characterise analytically.

The different methods used to characterize nanoparticles (NPs) are explained in details in this chapter [20, 21]. On various occasion, these different characterization techniques are used together with the conventional approaches, to study the specifications of the nanostructures and are also used to study the specific attributes of the nanoparticles in isolation [22]. In this chapter, we consider the following criteria: ease of use, cost, accuracy, selectivity, sensitivity, non-destructiveness, accessibility, and compatibility with different nano- materials and their elemental compositions [23]. There are numerous strategies used for characterization of nanoparticles but here we have meticulously analyzed and presented most commonly used techniques for chracterization. Microscopy-based techniques elucidate the dimensions, shape, and crystalline structure of nanomaterials [24] (include SEM, TEM, XRD, HRTEM and AFM); the complete name of all techniques will be presented later in the text during their respective discussions. In order to learn more about the structure, optical properties, elemental composition, and various general and specific physical attributes of the nanoparticle sample, scientists use a wide range of techniques, including X-rays, scattering techniques, and spectroscopy [19, 25].

The dimensions and shape of nanoparticles are among the key factors evaluated during their characterization [26]. Surface chemistry, surface area, degree of aggregation, particle size dispersion, and surface charge can all be quantified, though to a limited degree [27]. The dimensions, size distribution, and presence of organic ligands on the surface of nanoparticles may influence their application and different properties [28, 29]. The initial stage following their synthesis involves a thorough analysis of the nanoparticles' chemical composition and crystal structure [30]. This function lacked any prior process development. Reliable and precise nanoparticle measurement techniques are crucial for both legal compliance and the economic use of these materials [31, 32].

Nanomaterial analysis faces numerous challenges, such as a multidisciplinary area, a lack of reference materials for instrument calibration, the complexity of sample preparation, and the difficulty of interpreting data. The investigation of NPs inside complex matrices and the determination of their concentration in situ further increase the ongoing difficulties related to NP characterisation [33, 34], especially when making things at larger scale [35] so that waste can be minimized and moreover the waste has also to be characterized to understand the type of effluent and its usability [36, 37]. Because of these various factors, getting a full picture of nanomaterials made through different methods matters. The center part of the particle shapes how it behaves, while molecules on the outside adjust its different characteristics [38]. Later on, researchers began exploring advanced methods to track the nanoparticles formation and behaviour during synthesis. Because new tools have emerged, scientists can now observe these tiny structures as they develop in real time. Instead of ignoring imperfections, some teams deliberately introduce small defects to study their effects. These intentional changes often alter the key properties of the particles in noticeable ways. Through such approaches, a clearer picture emerges about what drives the performance at the nanoscale [39].

2 Characterization of Nano-Particles

2.1 X-ray-Based Characterization of Nanoparticles

2.1.1 X-ray Diffraction

Among the various tools used to probe material characterization, X-ray diffraction stands out when mapping the atomic arrangements of the nanoparticles [40]. XRD is crucial for understanding the crystal structure, microstructure of nanoparticles and their phase composition, in nanomaterial development. The principles and equipment of XRD, its significance in characterizing nanomaterials, and its uses in synthesizing nanomaterials are all summarized in this article [41]. A method for determining a material's crystalline structure, phase type, lattice parameters, and crystalline grain size is X-ray diffraction (XRD). By examining the widening of the sample's most prominent peak in an XRD measurement, the latter parameter can be calculated using the Scherrer equation [42]. After drying the colloidal solutions, powder samples undergo XRD characterization for obtaining statistically representative volume-averaged data [43]. In order to establish a relationship between peak location and intensity and particle composition, the reference patterns required for this task can be found in the International Centre for Distillation Data (ICDD) database. Earlier, it was known as the Joint Committee on Powder Distillation Standards (JCPDS) database [44]. Unfortunately, XRD is incapable of analyzing amorphous materials and rendering particles less than 3 nm in size.

Studying nanomaterials often relies on X-ray diffraction, given how well it matches structures where at least one-dimension falls between 1 and 100 nm. Because X-rays have wavelengths comparable to atomic distances, their diffraction patterns reveal fine structural details. When examining thin films, this method stands out in its precision in measuring the interatomic gaps and makes strain detection especially reliable. Information drawn from peak intensities delivers clear insight into how atoms arrange themselves across boundaries. Though limited to crystalline samples, the data remains central for interface analysis [45]. With lab tools, detecting surfaces down to roughly 50 A° is possible. When using synchrotron light, examining ultra-thin coatings becomes more achievable—especially single-atom layers across different substances [46].

Because it does not damage or touch the sample, XRD works well for observing processes as they happen. What makes nanomaterials special lies in how their tiny structural features match key physical scale limits, shaping their mechanical, optical, and electrical behaviour. From these materials, X-ray patterns expose details like atomic arrangement, stress in the lattice, particle size, and what phases are present [47]. Scientists mainly use powder diffraction to tell which substances exist in a mixture. The space under each peak indicates how much of a given phase is found in the material [48].

Experimental

Most of the time, X-ray diffraction helps distinguish methods using single crystals from those using powders or polycrystalline materials. Though rare, an ideal single crystal features tightly arranged unit cells with regular orientation, often around 0.3 mm in cross-section [49]. Instead of single crystals, many studies rely on powder diffraction when analyzing solids like quartz or asbestos. For such measurements, having a smooth, flat surface on the material matters greatly [50]. After preparation, particles usually measure between 0.002 and 0.005 mm across. Even distribution and haphazardly oriented tiny crystals throughout describes a good sample. To reach uniformity, researchers press the substance firmly into a mounting device [50].

A typical powder XRD arrangement relies on four key elements. Shown in Fig. 1 are the X-ray source, detector, sample platform, along with collecting optics. Positioned centrally, the sample sits within a curved layout where source, detection unit, and optical parts encircle it evenly. For X-ray diffraction work, Bragg's law plays an essential role in determining the crystal structure from the experiments [51]. Figure 2 depicts the reflection of the rays and the formation of angle between them. This is the basis for measurement of the Braggs angle.

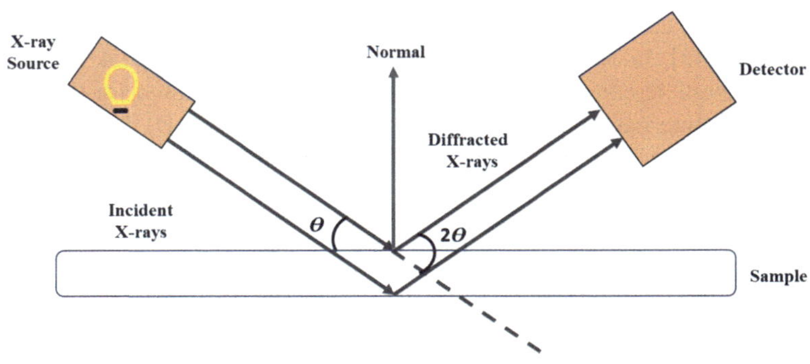

Fig. 1 Working of X-ray diffraction spectrophotometer

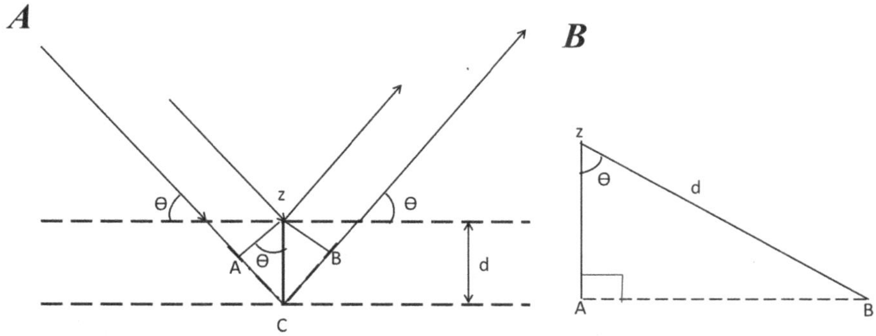

Fig. 2 Fundamental of X-ray difffractometer

Bragg's angle, defined as θ refers to the tilt between the beam and the sample surface as shown in Fig. 2. Meanwhile, the spread from incoming ray to sensing device measures 2θ. Fine powdered specimens were set upon the holder before scanning, under the understanding they contained tiny crystals oriented without order [51].

The sample interacts with the incoming X-ray beam, which causes the atoms inside the sample to disperse. When the scattered beams are in phase, creating an intensity maximum at that specific angle, constructive interference occurs. The atomic surfaces that deflect X-rays are called "reflecting planes" [52].

The working principle of X-ray diffractometer is shown in Fig. 2. A measure of how much X-rays bend when they interact with a sample is given by the angle 2θ. Typically, one plots intensity against this angular measurement to form what is known as a standard diffraction pattern. Depending on the type of metal used in the X-ray source, the position of peaks shifts due to changes in radiation wavelength [18]. The kind of X-rays that emerges—specifically their wavelengths and energy levels—depends on which anode material is used and how much energy the incoming electrons carry. When it comes to analyzing inorganic substances using X-ray diffraction, copper-based tubes appear most often in practice. For these experiments, researchers tend to favor radiation made up of just one wavelength because sharper, clearer outcomes are usually available. Kα radiation serves predominantly the analytical functions, whereas alternative radiation types, including Kα, are filtered out through the use of a nickel filter [53, 54].

Electrons produce by X-rays are characterized by a specific wavelength, which corresponds to a defined energy level. Copper X-ray tubes are primarily employed in the XRD analysis of inorganic materials. Monochromatic light, or X-rays of a singular wavelength, is favoured in practical applications of X-ray diffraction to enhance experimental outcomes [53, 54].

In the simplified crystal as shown in Fig. 2, the distance d stands for the spacing of the lattice planes. According to Bragg's law, the interplanar spacing (d), the angle of diffraction (M), and the reflected X-ray wavelength (Q) are all related in Eq. 1.

$$n\lambda = 2dsin\theta \quad (1)$$

The angle formed between the transmitted and diffracted beams may be accurately determined at 2 M. Measured angles appear easily in lab settings, so XRD results often show values as 2 M. When radiation hits a crystal plane, the resulting angle M needs to match what Bragg's law requires [53, 55]. Each peak gets labeled with its corresponding Miller indices. Indexing these patterns relies on three main approaches: analysis tools, visual plotting methods, along with reference files like JCPDS cards for comparison against real-world XRD output [56]. The plots for XRD are generally obtained as the signal strength versus the scattering angle 2θ is measured in degrees. In X-ray diffraction, copper-generated Kα radiation at 1.54 Å serves as the most common radiation. From Eq. 2, applying Scherrer's method, [53] the mean crystallite dimension are estimated.

$$D_v = k\lambda / \beta \cos\theta \quad (2)$$

The variables λ (radiation wavelength), β (full width at half maximum), and Dv (average particle size) are all related. β is the integral breadth of the peak at angle M in the diffraction pattern, which exhibits the same maximum intensity. K is the Scherrer constant [57]. The Scherrer constant (K) in the equation represents particle shape and is typically assigned a value of 0.9 [58]. A substance's apparent or mean particle size is indicated by the dimension obtained from the Scherrer equation. A variety of particle sizes are produced by the aggregation of smaller particles that make up material powders [59].

The Debye Scherrer equation has a number of drawbacks. The particle form affects the constant in the equation. In addition, the crystalline lattice's imperfections and size distribution are ignored but when imperfections pile up, diffraction lines spread wider. The measurement of actual dimension of the grain size also changes because the measurement relies on FWHM and the tiny crystals shift the peak's foot, whereas big ones alter brightness more strongly [60]. Analysing particles directly through TEM or SEM offers clearer insight to the size of the particles [61].

Heating improves both structure order and grain dimensions and gives better atomic arrangement. A stronger spike on the graph often signals such improvement due to heating. When comparing large-scale solids to tiny grains, these spikes offer clear visual clues [62]. Patterns shift noticeably depending on the size of the particles. Unlike larger forms, smaller versions of the particles spread their signal across wider angles. This broadening happens because fewer layers align neatly in small crystals, as a result, the signal loses sharpness. Large samples, however, show tight and strong responses instead [63].

One important use of powder diffraction lies in examining chemical composition. Measurements of unit cell characteristics appear alongside phase recognition, detection of solid solutions, also changes between high and low temperature forms [51]. Each crystalline substance carries a unique X-ray diffraction signature. Even when chemicals share identical compositions, their XRD patterns often differ. Patterns shift

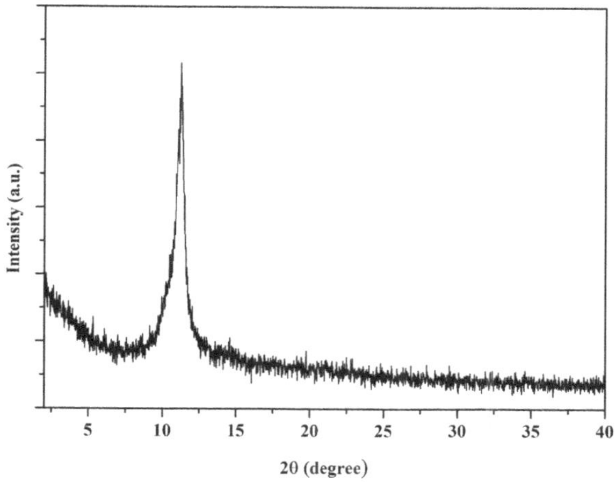

Fig. 3 XRD of graphene-oxide nanoparticles

with structural arrangement, not just chemistry [64]. Pooja et al. has studied the XRD pattern of graphene oxide nanoparticles and their relevant peaks are demonstrated in the XRD spectrum in Fig. 3. The powder X-ray diffraction pattern of synthesized GO showed a sharp peak at $2\Theta = 11.27^\circ$ due to the (002) plane of the GO structure [64, 65]. The interlayer spacing (d) was estimated to be 0.78 nm by well-established Bragg's law: where n is the diffraction series and λ is the diffraction wavelength.

Polymers have both amorphous and crystalline characteristics. Similar to the iron framework in concrete, the crystalline domains serve as a reinforcing lattice, improving performance throughout a wide temperature range [66]. Increased brittleness is caused by increased crystallinity, whereas the amorphous component produces a large peak, the crystalline portions produce sharp, narrow diffraction peaks. The intensity ratio can be used to assess the material's degree of crystallinity [67].

This paragraph examines the extension of XRD application by researchers for analytical purposes. Lattice stresses and particle/crystallite size were the main contributors to the XRD peak broadening, aside from experimental broadening [68].

2.1.2 X-ray Absorption Spectroscopy (XAS)

Extended X-ray absorption fine structure (NEXAFS) and X-ray absorption spectroscopy (EXAFS) are both parts of X-ray absorption spectroscopy (XAS). The relationship between the energy levels of a material and its X-ray absorption coefficient can be discovered via X-ray absorption spectroscopy (XAS). Because different elements have different absorption edges that correlate to varying electron binding energies, element selection in X-ray Absorption Spectroscopy (XAS) is made possible. EXAFS is capable of reliably determining a species' chemical state even at very low concentrations [69]. This method is not widely used or readily available, as XAS spectra are typically acquired through synchrotron facilities. The XANES method examines the density of states of partially or unoccupied electronic states by assessing the excitation of electrons in the inner shell to levels permitted by dipole selection rules [70].

For reactions that happen within seconds, EXAFS can evaluate the kinetic and structural features of supported metal catalysts, while surface enhanced Raman spectroscopy (SERS) becomes less effective at temperatures beyond 200 °C. Investigations at high temperatures are made easier by EXAFS. Energy dispersive EXAFS enables a quantitative in situ analysis of the kinetics associated with rapid phase transitions in supported metal catalysts of nanoparticles, even within a very short timescale of tens of milliseconds [71]. EXAFS is recognised as a highly effective method for the structural analysis of nanoparticles measuring under 10 nm. Experts agree that EXAFS is essential for obtaining information about the coordination number, the kinds of scattering atoms surrounding the absorbing atom, the distance between the absorbing atom and the back-scattering atoms, and the Debye–Waller factor, which relates to disorder caused by thermal vibrations and static displacements. Interaction between the radial distribution of backscattering atoms in real space, x(r), and the Fourier transform (FT) of the EXAFS signal acts as a function of wavenumber [72]. In the Fourier Transform not every signal comes just from atomic spacing. The shifts in wave timing during EXAFS scans reshape where peaks appear, complicating a straightforward link to real-space distances. Scattering paths interfere, further distorting position cues. Because of this, peak spots do not match physical separations between absorber and scatterer atoms directly [72]. When comparing tools for probing nanostructure, Moroz found EXAFS gives sharper element-specific detail than REDD. Yet when it comes to bond-length accuracy, REDD holds an edge [73]. To get reliable atom-to-atom measurements from EXAFS, corrections for phase delays must be also be applied. Combining REDD with EXAFS, along with FTIR and imaging methods, gives clearer views into the structure, shapes and the material behavior of the nanoparticles [73, 74].

Despite the fact that SAXS (Small angle X-ray scattering) technology only works with very small scattering angles—typically less than 1°—it does permit elastic scattering processes inside a given solid angle [75]. Similar to small-angle X-ray scattering (SAXS), wide-angle X-ray scattering (WAXS) captures diffraction maxima at greater angles since the sample and detector are closer together. Particle morphology, size distribution, and size are commonly evaluated with SAXS. Comparing TEM imaging with SAXS yields statistically more reliable size measurements. to Studies utilizing XRD and/or other electron diffraction techniques is required, as SAXS has a restricted resolution to understand the complete characterization of a nanoparticles.

2.1.3 X-ray Photoelectron Spectroscopy (XPS)

When studying surfaces chemically, X-ray photoelectron spectroscopy (XPS) is the method of choice because it allows for nanoscale material characterization. At its core, this phenomenon is a product of the photoelectric effect. The X-ray Photoelectron Spectroscopy (XPS) method is a quantitative tool for revealing the electronic structure, elemental content, and oxidation states of elements in a material [76]. The device has the potential to investigate nanoparticle surface functionalization, ligand exchange interactions, and core/shell architectures in a very vacuum environment [77]. While XPS analyses the material's composition along the direction of the electron beam, microscopy techniques like TEM and TEM/EELS use lateral spatial resolution to discover materials perpendicular to the probing beam [77]. XPS offers additional benefits, including the capacity to deliver depth information comparable to nanoparticle size, achieving depths of up to 10 nm from the surface, while maintaining minimal sample damage. Two limitations of XPS analysis are the requirement for sample preparation, specifically a dry solid form devoid of contamination, and the difficulties encountered in data interpretation [78].

The Photoelectric Effect

The photoelectric effect, which was initially observed by Einstein, is the foundation of XPS [79]. X-ray photons with an energy of hv interact with the sample, allowing engagement with its core-level electrons [80]. If sufficient energy is transferred, an electron will be expelled from the atom when an X-ray photon collides with it; this expelled electron is known as a photoelectron. The free photoelectron's kinetic energy is proportional to the bound electron's energy and the incoming X-ray photon's energy. The relationship is mathematically described in Eq. 3 [81]:

$$KE = h\nu - BE - \phi_{spec} \tag{3}$$

KE represents the kinetic energy of the released photoelectron.

h_ν is the energy of the incident X-ray photon.

BE depicts the binding energy of the electron, indicating the degree of its confinement to its atom and orbital.

ϕ_{spec} is spectrometer work function, is a constant for each instrument [81].

This occurrence makes it possible to calculate binding energies, which differ for every element and depend on how that element is chemically bonded to the substrate. The free photoelectron's kinetic energy is proportional to the bound electron's energy and the incoming X-ray photon's energy. The relationship is mathematically described in Eq. 3 [82].

Instrumentation of X-ray Photoelectron Spectroscopy

Figure 4 shows an X-ray photoelectron spectrometer employed for surface measurement [81, 83] with the various important parts. This apparatus precisely irradiates a substance with X-rays, gathers photoelectrons, and quantifies their kinetic energies to ascertain binding energies [81]. This method reveals a material's top few nanometres' chemical and elemental composition [81].

A standard XPS system has a number of important parts that work together, mostly in ultra-high vacuum conditions:

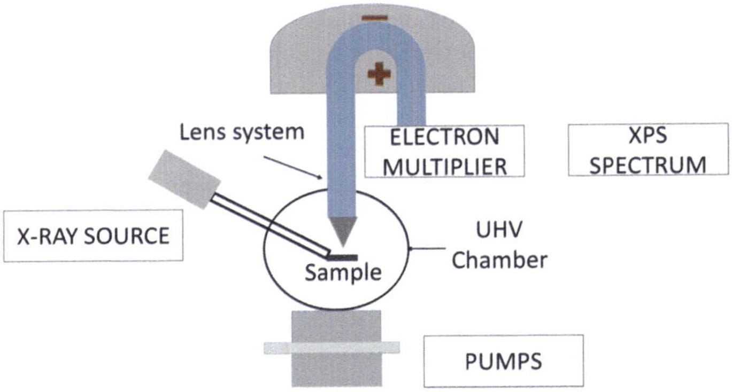

Fig. 4 Working principle of X-ray photoelectron spectrometer

Source of X-rays: The photoelectric effect begins with X-ray photons from the source where all sources are not monochrome. These sources use aluminum or magnesium anodes bombarded by strong electrons [84]. This bombardment causes the anode to emit X-rays that are specific to it (for example, Al Kα at 1486.6 eV and Mg Kα at 1253.6 eV). A few setups use two anodes, giving the option to switch X-ray levels depending on which element or depth is required for scanning [84].

A typical setup includes a monochromator within the XPS system. From sources like Al Kα, only a narrow band of X-ray energies passes through it. By filtering out unwanted radiation, such as bremsstrahlung and satellite peaks, the precision of measurement increases and as a result, distinguishing between different chemical states becomes more reliable and more clearer spectra emerge when background noise drops significantly [85].

System for Sample Introduction and Manipulation: A setup handles how samples enter and are positioned inside the ultra-high vacuum chamber. Correct placement is maintained through precise adjustments during operation.

Load Lock System: A separate vacuum stage, often called a load lock, moves the samples toward the main chamber while protecting the ultra-low pressure inside [86–88]. This kind of protection cuts the pumping duration significantly—and clean conditions last longer because the air outside rarely enters the system [87].

Sample Stage and Manipulator: A typical setup includes a platform capable of motion along X, Y, Z axes as well as rotational adjustment. Such positioning of the systems support the experiments involved in ngle-resolved photoemission by enabling precise angular orientation of the specimen [89–91]. Temperature control is commonly integrated, spanning from very low to elevated levels, allowing real-time observation under varied thermal conditions [91, 92].

System for ultra-high vacuum: XPS work in UHV (usually between 10^{-9}–10^{-10} Torr) and UHV is required to clean the surfaces long enough for reliable testing [93–95] and it is observed that contaminants, especially carbon, can have a big effect on XPS results [96].

Photoelectrons from the sample must hit gas molecules without losing energy on the way to the analyzer and UHV gives electrons a lengthy mean free path to the detector without any issues [97, 98].

Pumping Systems: A mixture of pumps is required to reach UHV. To improve UHV efficiency, some systems use tiny combination pumps with non-evaporable getters and sputter ion pumps [76].

Analyser of Electron Energy: An XPS device quantifies the kinetic energy of photoelectrons emitted by analysers.

Input Lens System: Once past the electrostatic lenses, photoelectrons enter the HEA. These components control the electron speed so they are adjusted to match the set pass energy and also guide the beam precisely toward the entrance slit of the analyzer [99, 100]. This process improves the the amount of signal that gets captured while enhancing clarity during measurement [99].

Electron Detector: A signal forms when electrons strike the detector located at the exit of the energy analyzer, capturing those carrying specific kinetic energy levels. The newly developed XPS setups often include advanced multichannel options like microchannel plates or resistive strip types. Instead of one channel, these incorporate built-in electronics that sense where particles land across a surface. Because multiple energy values get recorded at once, gathering results takes much less time. Earlier versions, limited to single channels, required longer measurement periods by comparison [101].

Working together, these parts make XPS effective for examining surfaces, delivering details on which elements are present as well as their specific chemical forms [81, 102].

2.2 Spectroscopy Based

2.2.1 UV–Visible Spectroscopy

A different affordable method to study tiny materials involves UV–Vis analysis, where the amount of light passing through a sample compared to a standard is measured. The optical effect in the nanoparticles depends on its size, their form, packing, agglomeration, and the environment This approach helps to, describe, and examine the particles along with the stability of the particles with time It also gives clues about how steady nanoparticle mixtures remain over time [64]. Light from the visible or ultraviolet range hits the molecules, and exciting the electrons from rest mode into active mode—and they jump from low to high energy. The power required for this excitation is provided by the the incoming light [103]. Absorption of light relates closely to the amount of substance present in the sample. The UV–VIS spectra depend on the distance travelled by light hrough the mixture, concentration of the solution, and the ability of molecules to block the light. This strength is called molar absorptivity which influences the results just as much as path length or the concentration. These electronic transitions are related to the absorption of photons [104, 105].

The Beer-Lambert Law is represented by the Eq. 4:

$$A = \varepsilon lc \tag{4}$$

where,

A represents absorbance, which is a unitless measurement.

$\mathcal{E}$ represents the molar attenuation coefficient, expressed in units of L/mol/cm, and is a constant intrinsic to the substance at a designated wavelength.

l denotes the optical path length in centimetres, typically corresponding to the cuvette's width.

c denotes the concentration of the substance, quantified in moles per litre (mol/L).

Majority of UV light is absorbed by non-bonding and pi electron transitions. These energy-intensive transitions are associated to UV absorption. The pi-to-pi* transitions absorb in the UV–Vis ranges [103].

Instrumentation of a UV Spectrophotometer

A UV–Vis spectrophotometer measure sample absorbance or transmittance by wavelength. This measurement measures a solution's light absorption or transmission at a certain wavelength [104]. A standard UV–Vis spectrophotometer with its components is illustrated in Fig. 5.

Light Source: The light source emits ultraviolet and visible electromagnetic waves. In the ultraviolet range, deuterium lamps serve the purpose; in contrast, visible light relies on tungsten or halogen sources. These different pairings provides the needed wavelength bands for UV spectrophotometer.

Monochromator: A narrow beam enters through a slit at the start of the device. Light spreads out when it passes through either a prism or a grating inside. Only one-color exits at a time via another opening. This separated light then moves toward the test material. Each wavelength comes through individually, filtered from the rest. Selection happens by adjusting internal angles or positions. Such control enables precise exposure during analysis [106].

Sample Compartment: A container called a cuvette holds the material to be tested, along with a matching blank that contains only solvent. Depending on the range of light used, the vessel might be made of quartz—when ultraviolet detection is needed—or glass or plastic for visible wavelengths. Through this compartment, a

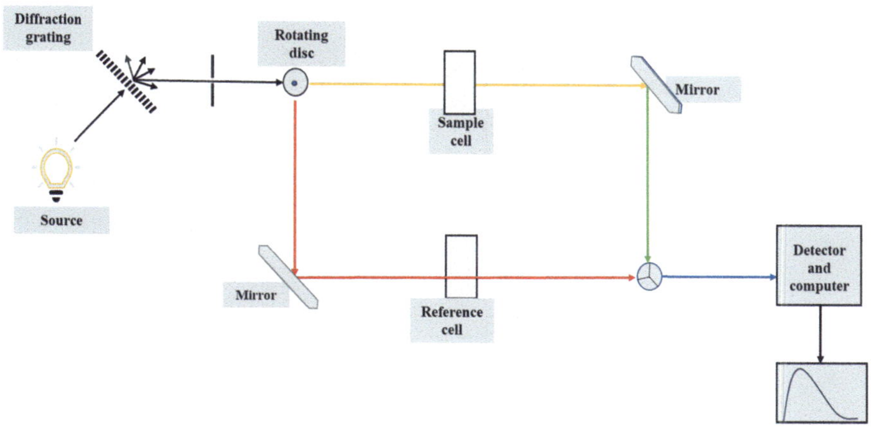

Fig. 5 Working principle of UV–Vis Spectrophotometer

beam passes, traveling across the solution toward a sensor tuned to a specific wavelength. Once it arrives, the sensor registers how much light makes it through. What results is an electric output shaped by how strong the incoming light appears [107]. The detector converts light energy into an electrical signal on the basis of its intensity. Common detectors include photomultiplier tubes and photodiodes [103].

Signal Processing and Display: The detector's electrical signal is converted to absorbance or transmittance. The data is then displayed or printed as a spectrum (absorbance versus wavelength). Mobile spectrophotometers with miniaturized dispersion systems and multi-band optical fibers are available [108]. Furthermore, they can incorporate functionalities for photometric measurements, spectral scanning, quantification, kinetics, and multi-component analysis [109].

Typically, UV–Vis wavelengths span from 180 to 800 nm, visible light ranges from 400 to 700 nm, and ultraviolet light ranges from 100 to 400 nm [103, 105].

Unique UV–Vis extinction spectra are observed in sols of gold, silver, and copper with nanostructures because of the visible-band localized surface plasmon resonance (LSPR) signal [110]. Certain nanostructures, such as metal chalcogenide nanoparticles and anisotropic gold or silver nanostructures, can also be seen in the near-infrared (NIR) spectrum as LSPR bands [111]. This approach is distinguished by its low cost, non-destructive nature, and efficacy in identifying NP systems and their various varieties [112].

The UV–Vis spectra of sodium citrate-based gold nanoparticles is depicted in Fig. 6. It was observed by Basu et al. that citrate-based gold nanoparticles were absorbance at 524. A single spectrum indicated the formation spherical gold nanoparticles.

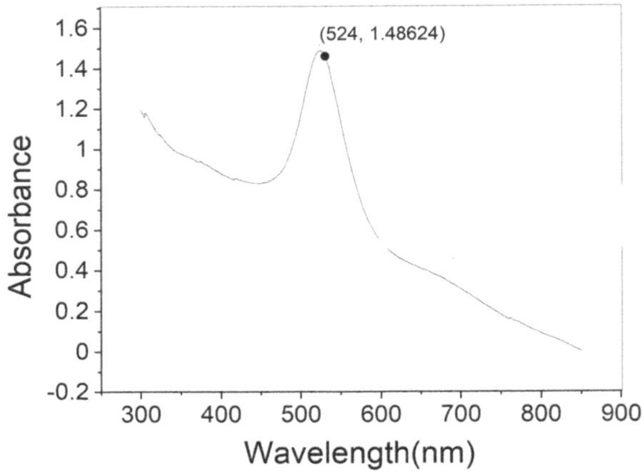

Fig. 6 The UV spectrum of gold nanoparticles

2.2.2 Photoluminescence (PL) Spectroscopy

Nanoscale materials are studied using photoluminescence (PL) spectroscopy, which quantifies the light released by atoms or molecules after absorbing photons. One example of fluorescent nanoparticles includes quantum dots, along with metal nanoclusters, both identified via photoluminescence [113]. Despite low efficiency in quantum terms during emission, strong excitation cross-sections tied to plasmon resonances compensate significantly. Because photobleaching and blinking do not interfere, metal nanoparticle photoluminescence remains stable [114]. Despite low efficiency in quantum terms during emission, strong excitation cross-sections tied to plasmon resonances compensate significantly. Because photobleaching and blinking do not interfere, metal nanoparticle photoluminescence remains stable [114].

Photoluminescence comprises several essential steps.

When light with right energy strikes a substance, its electrons takes up that energy and this results in a shift from a calm baseline condition, i.e., the lower energy stage into a higher energy state. A photon's energy needs to match or exceed the bandgap if absorption is to occur efficiently, followed by emission [75, 105]. Energy decreases very fast when electrons emits heat through tiny vibrations, settling into the ground state from their excited state without glowing. This shift happens quietly, bypassing light release, as particles adjust internally in a split second. A single particle of light emerges once energized electrons lose excess energy, settling back into their original level. Light given off in this way can be measured using photoluminescence techniques [115, 116].

Instrumentation for Photoluminescence Spectroscopy

A standard photoluminescence spectrometer comprises several essential components that facilitate sample excitation, collection of emitted light, and analysis of spectral properties as shown in Fig. 7.

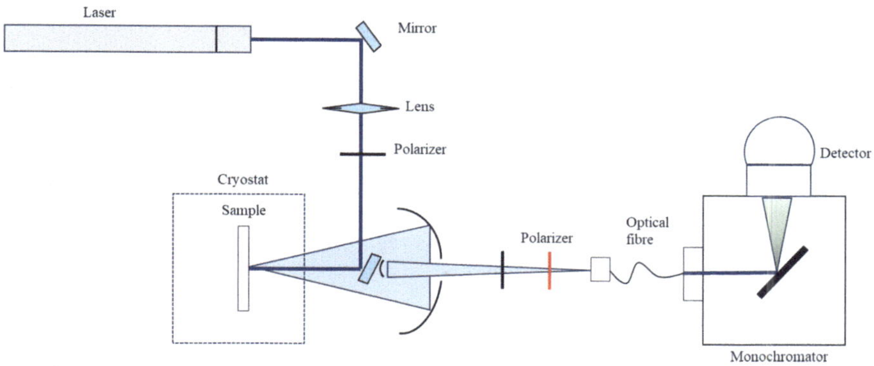

Fig. 7 Working principle of photoluminescence spectrometer

Light Source: This yields the excitation energy. Typical sources include lasers, which offer high intensity and monochromatic excitation, as well as broadband lamps, such as halogen lamps, utilized with monochromators for wavelength selection [115, 117].

Excitation monochromator: A broadband source stimulates the sample with a specific wavelength. This aids excitation spectroscopy, and the photoluminescence intensity is a measure of the excitation wavelength [118].

Sample Compartment: The solid, liquid, or powdered sample is positioned to enhance light source excitation and photoluminescence collection [119].

Emission monochromator: This breaks photoluminescence into wavelengths and is necessary to obtain an emitted light spectrum. Gratings or prisms are used for wavelength separation [120].

Detection device: It produces an electrical signal from light. Photomultiplier tubes and charge-coupled devices or photodiodes are common detectors for UV–Vis spectral measurements, especially in multichannel systems [117, 121, 122].

Signal Processing and Data Acquisition Systems: In this step, amplification and processing of the detector signal is carried out. This system includes analog-to-digital converters and software for instrument operation, data recording, and PL spectrum analysis [123].

Time-resolved photoluminescence measures the decay time of luminescence, offering insights into recombination dynamics and excited-state lifetimes [122, 124]. Temperature control facilitates measurements across different temperatures, providing insights into carrier dynamics and defect states [122].

Micro-PL integrates a microscope with a photoluminescence system, enabling spatially resolved measurements that are beneficial for the analysis of heterogeneous materials or small features [125]. Integrated systems are being developed as compact and modular solutions for various applications, to achieve high spatial resolution [125, 126].

2.2.3 Fourier Transform Infrared Spectroscopy (FTIR)

Fourier transform infrared spectroscopy (FTIR) can assess how different wavelengths of electromagnetic radiation are absorbed, especially in the mid-infrared range (4000–400 cm^{-1}) [57]. A molecule becomes infrared active when it absorbs infrared radiation, which modifies the way it moves. You can get information from a recorded spectrum by finding bands that are related to binding strength and type, as well as some functional groups. Progress regarding the interaction of molecular structures [38].

Instrumentation of FTIR

FTIR equipment as illustrated in Fig. 8 have a broadband source, an interferometer that turns wavelength information into an interferogram, a beam-splitting/optical train, and a detector. The wavelength range, throughput, speed, and sensitivity of each element can be changed. This section delineates the fundamental optical and electronic components of FTIR devices and their respective functions in the generation and retrieval of spectra [127]. FTIR substitutes dispersive analysers with an interferometer, often of the Michelson type, which modulates broadband radiation into an interferogram, the Fourier transform of which produces the spectrum [128, 129]. The light source emits broadband energy within the designated spectral band, while the beam-splitting optics and mirrors constitute the interferometer and regulate the optical path difference (OPD). The detector transforms the modulated intensity (interferogram) into an electrical signal for Fourier analysis and calibration [127, 129].

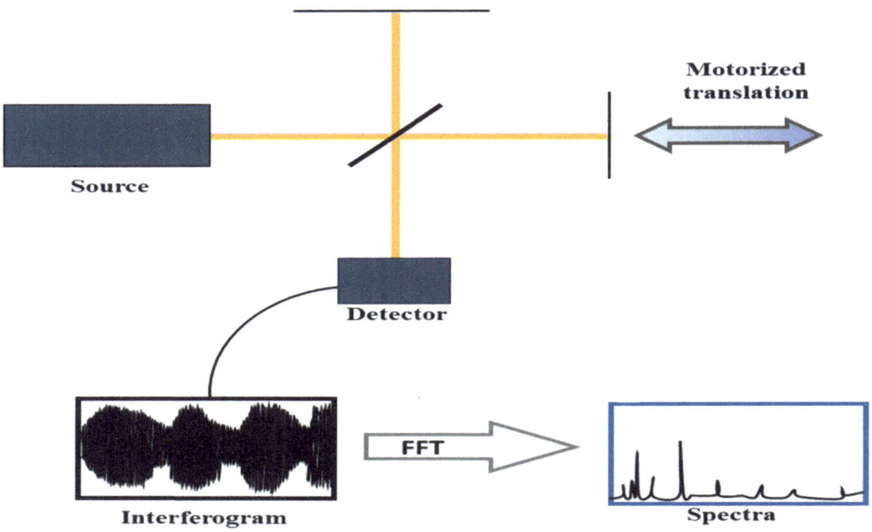

Fig. 8 Schematic of working principle of Fourier Transform infrared spectrometer

Different types of sources for tunable and laser-based methods are Broadband thermal, synchrotron, supercontinuum, and QCL. Provides broadband illumination for simultaneous multiwavelength detection; source brightness and coherence affect possible SNR and the choice of interferometer/design [128, 130].

Interferometers, including Michelson, lamellar-grating, single-mirror/static, wavefront-division, and birefringent types, encode spectral content into an interferogram through variations in optical path difference (OPD). Michelson interferometer is prevalent, static or spatially encoded interferometers which offer advantages in imaging and stability by eliminating moving components [128, 131, 132].

The optics and beam splitters are used to divides and recombines beams; the material or coating used determines the useful spectrum range (for example, diamond beam splitters can cover very broad ranges) and might cause channelling artefacts that need to be fixed [133].

The detector like thermal detectors, photoconductive/photovoltaic detectors, microbolometer arrays, PbSe, InGaAs, etc. converts the intensity of an interferogram into an electrical signal [134]. The detector's responsivity and noise determine its sensitivity and limit its spectral range. Detector arrays allow for static imaging FTIR and high acquisition rate [135–137].

Auxiliary lasers or control interferometers are commonly employed to accurately calibrate optical path difference (OPD) and wavenumber scale for precise spectral measurements [135].

The FTIR spectrum of GO in Fig. 9 as observed by Pooja et al. confirmed the presence of functional groups and the formation of GO from graphite. A broad band around 3400 cm^{-1} is attributed to O–H stretching which is involved in hydrogen bonding and a band at 1710 cm^{-1} confirms the presence of a carboxylic acid functional group in GO. The band between 1000 cm^{-1} to 1250 cm^{-1} corresponds to C–O–C epoxy bonds and the vibrational mode of C-O groups [138]. The band between 1300–1400 cm^{-1} is characteristic of C=C aromatic bonds.

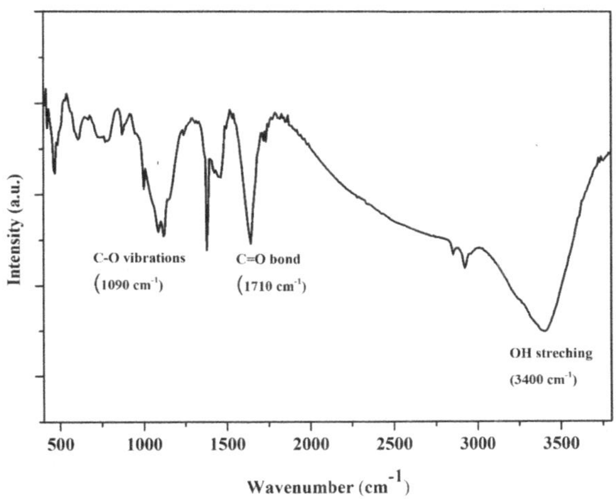

Fig. 9 IR Spectrum of Graphene oxide nanoparticles

2.2.4 NMR Spectroscopy

Near atomic detail comes into view when using nuclear magnetic resonance spectroscopy (NMR) during nanoparticle synthesis, whether materials are dissolved or rigid. Thanks to such capability, tracking how noble metal particles form and what shape they take becomes far easier to understand and depict [139]. While solids and fluids differ greatly in behavior, both allow on-the-spot observation at a molecular scale through this method. As a result, researchers gain clearer insight into assembly pathways and finished arrangements of these tiny structures [140]. The atoms carrying nuclear spin respond under examination and the atoms without spin simply vanish from the signal, leaving no trace behind [140].

Nuclear magnetic resonance is related to spin quantum number S. Because of this, when nuclei carry spin above zero, they also show a detectable magnetic response—this comes from Eq. 5 [141], linking γ (never zero), spin value, and resulting magnetic moment μ. If the spin happens to be zero, no signal appears in NMR; without spin, there's no magnetism to measure. Usually, isotopes made of even counts of both protons and neutrons fall silent in these scans—their lack of magnetic character gives no response in NMR [142].

$$\mu = \gamma \cdot S \tag{5}$$

A nucleus with a spin I = 1/2 exhibits two spin states, +1/2 and −1/2, in the presence of an external magnetic field (B). The energy difference between the two states at a specific external magnetic field (Bx) is expressed in Eq. 6 and Fig. 10, where E is the energy, I is the nuclei's spin, and M is their magnetic moment [142]. Since the energy difference that is consistently observed is small, nuclear magnetic resonance (NMR) requires large magnetic fields to improve the separation of the two energy levels. In case of NMR, the magnetic resonance frequencies at the magnetic fields falls within the radio frequency spectrum [141].

$$E = \frac{\mu B_x}{I} \tag{6}$$

Nuclear magnetic resonance picks up signals tuned to specific nuclei, since each type responds at its own frequency. As a result, just one isotope shows up clearly during any given measurement. That limitation shapes how the method splits into varieties—like ^{1}H, ^{13}C, or ^{31}P detection—each focused on a distinct atomic target [143].

Figure 11 shows an NMR machine built around three key parts. Surrounding the sample, a superconducting magnet generates a steady magnetic field. Transmission and detection of radiofrequency signals happen within the spectrometer unit. Operation of the system and interpretation of results rely on a connected computer.

Usually, nuclear magnetic resonance picks up signals attached to the atomic nuclei that responds to the magnet, allowing them to be identified separately. In this process,

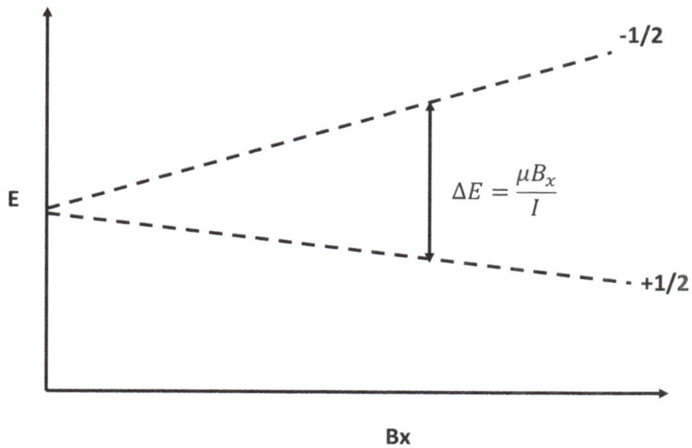

Fig. 10 The potential difference between two spin states as a function of a changing magnetic field flux B

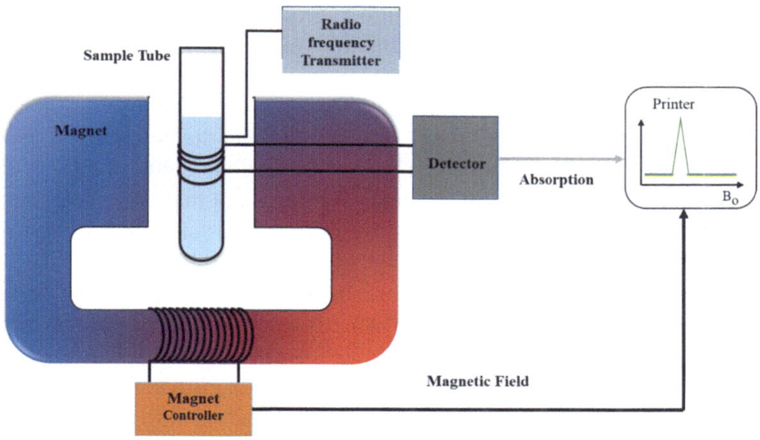

Fig. 11 Working principle of NMR Spectrophotometer

2–50 mg of the substance are mixed into a liquid medium. When placed inside the machine, those reactive nuclei line up under its strong magnetic influence. The frequency at which radiation from a nucleus returns to its ground (lower-energy) state is the same as the frequency at which radio frequencies are released during the energy transition from the ground state to an excited state. The arrangement and the presence of active components reveal how internal magnetic forces contribute meaningfully. With this method, details about molecular motion, shape, surroundings, and reactivity become clearer [144]. NMR spectroscopy is capable of examining the capping ligands involved in the nanoparticles. Through this approach, the core of the atom, their arrangement, and their electronic environment can be known [139]. Information on surfaces, movement of ligands, folding, and packing around particles are observed [145, 146]. By applying NMR, changes in precursor chemistry become visible, revealing effects on nanoparticle shape. Understanding these molecular shifts helps in tracing the evolution of structure during synthesis [147].

The reaction mechanisms involved in NP synthesis can be better understood with the help of nuclear magnetic resonance (NMR), which efficiently screens the chemical transformation of NP precursors in both solid and solution phases. Under various reaction settings and with diverse metal types, NMR reveals strong spatial and chemical detail [103]. Because initial capping ligands sometimes require substitution, this method helps track the ligand replacement progresses and the final formation. Shifting away from uniform patterns, the 1H NMR signal responds directly to changes in electron arrangement and bond context around the nucleus [6].

Changes in chirality affect chemical shifts, making these alterations visible to neighbouring spin centres. Because of this sensitivity, NMR serves as a valuable method to identify chirality in tiny structures such as nanoclusters [148]. Observations through NMR make it possible to track how gases spread when attached to metal nanoparticle surfaces [149, 150]. While TEM and DLS measure physical dimensions, NMR adds unique insights into hydrodynamic behaviour [151]. Diffusion properties forms the basis of size estimation, much like what occurs in dynamic light scattering. When particles move randomly in solution without interference, NMR captures their diffusion rates accurately. From those measurements, the Stokes–Einstein model gets adapted to estimate effective particle diameters [139]. Notably, Alloul and colleagues reported an effect named the "Knight shift," appearing during NMR scans due to certain metallic influences. It turns out that pure metal nanoparticles must typically range between 1 and 5 nm for clear signals, whereas hybrid systems involving polymers allow detection even above 100 nm [152].

Few techniques offer such clear insight into solid catalysts as solid-state nuclear magnetic resonance spectroscopy does when examining surface-level reactions [153].

This method gives an insight on the interactions of the rigid and flexible substances with the ligands and their behaviour in solvents and attachment to particles [65]. The gyromagnetic ratio stands out strongly for ^{31}P, a common and highly responsive NMR-active nucleus. Because of that feature, even when ligands are present in small amounts, clear ^{31}P NMR signals emerge amid minimal background noise [154]. With support from DFT simulations, J-resolved ^{31}P solid-state NMR becomes a tool capable of revealing details about fixed molecular arrangements and how soluble phosphine-based catalysts anchor onto surfaces.

2.2.5 Mass Spectrometry (MS)

It has been discovered that mass spectrometry (MS) is a trustworthy method for characterizing nanoparticles analytically. In order to better understand, NPs, their

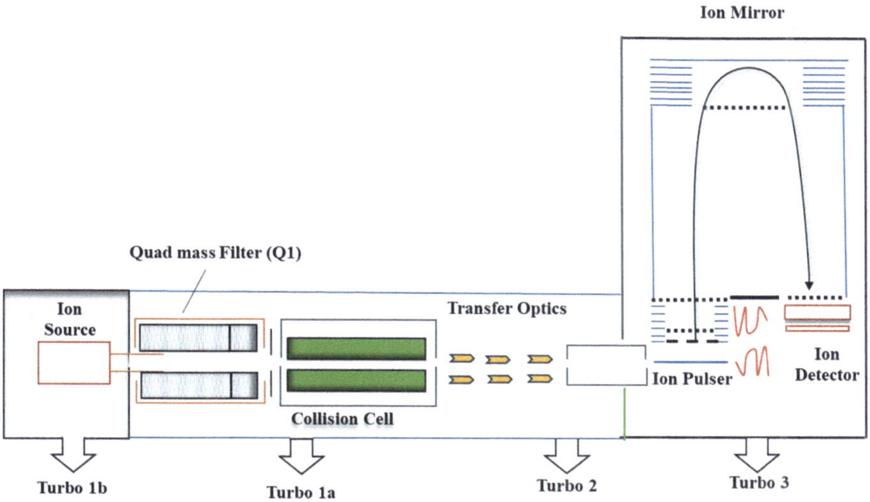

Fig. 12 Working Principle of High resolution mass spectrophotometer

bioconjugation to target biomolecules, and their chemical state, MS provides priceless molecular and elemental data. Mass spectrometry is highly sensitive and works with any kind of sample. Additionally, it may be quickly and readily combined with separation techniques to gather data in real-time [155].

The core principle of mass spectrometry as shown in Fig. 12 encompasses a sequence of procedures to convert a sample into ions, segregate them, detect them, and subsequently analyse the data to produce a mass spectrum [156]. A substance—solid, liquid, or gas is injected into the vacuum chamber of the the mass spectrometer [157, 158]. Some setups use immediate introduction [159]; meanwhile, alternative configurations link with methods like chromatography to separate components beforehand [160, 161].

Ionisation: Gases form when molecules loss or gain electrons during ionisation, a key step because these instruments detect electrically charged fragments [156, 162, 163].

Mass Analysis: A stream of ions enters an electric field, where separation occurs based on how their mass compares to charge [156, 164].

Fourier Transform Ion Cyclotron Resonance Mass Spectrometry: Within a strong magnetic field, ions get trapped during Fourier Transform Ion Cyclotron Resonance Mass Spectrometry. A radiofrequency pulse nudges them into motion. Their movement generates a detectable image current and the measurement is recorded naturally. A Fourier transform is utilised on this signal to ascertain the m/z values with remarkable resolution and mass precision [165–167].

Detection: When the ions are finally separated, they hit a detector, which converts the current of the ions into an electrical signal [168]. The concentration of ions at a specific m/z are reflected by intensity.

Data Processing: The signals from the detector are relayed to a data system for processing by a computer [154]. This procedure entails the transformation of raw data into a mass spectrum, which graphs ion abundance versus m/z [169]. Advanced software solutions are frequently employed for automated data processing, facilitating efficient examination of intricate samples [170].

The properties and potential uses of NPs are better understood with this method. Inductively Coupled Plasma Mass Spectrometry (ICP-MS) is used to analyze the elements contained in nanoparticles (NPs). The system is distinguished by its robust selectivity, wide dynamic range, increased sensitivity, durability, and independence from virtual matrix impacts [171]. It is also easy, typically requiring only basic calibration processes. The technique makes it possible to identify metallic impurities in non-metallic nanoparticles and to precisely quantify and characterize the elemental composition of metallic nanoparticles [172, 173].

Mass spectrometry can be used to better understand the size distribution of tiny clusters. It is also possible to examine combinations of ligands with various stoichiometries [174, 175].

2.3 Characterization of Size

2.3.1 Dynamic Light Scattering (DLS)

A popular method for figuring out the size of nanoparticles in colloidal suspensions in the nanoscale and sub-micrometer ranges is dynamic light scattering (DLS) [176]. Nanoparticles are dispersed within a colloidal medium. Colloidal solutions demonstrate continuous Brownian motion due to the scattering of light by suspended particles [177, 178]. DLS quantifies light scattering over a specified duration. Using this technique in conjunction with the Stokes–Einstein assumption, one may calculate the hydrodynamic diameter of a nanoparticle, which is the combined diameter of the nanoparticle and the solvent molecules diffusing at the same rate as the colloid in solution [179]. Smaller particles have greater velocity, whereas larger particles demonstrate reduced velocity [178].

Light Scattering and Fluctuations: A laser beam illuminating a suspension of these particles results in the scattering of light by the particles. The particles' relative locations are always altering due to Brownian motion. The continuous movement causes variations over time in the intensity of the light that is scattered [180]. The frequency of these fluctuations correlates directly with the velocity of the particles' Brownian motion [178].

Since dynamic light scattering (DLS) uses an autocorrelator to analyze the temporal variations in scattered light intensity, the autocorrelation function is an essential component [177]. The autocorrelation function of the signal is quantified by comparing the scattered light intensity at a specific moment with that at a slightly later moment. The intensity correlation decreases rapidly for smaller, swiftly moving particles, whereas it diminishes more gradually for larger, slowly moving particles [178].

The translational diffusion coefficient (D) of the particles can be found by calculating the decay rate of the autocorrelation function. The Stokes–Einstein in Eq. 7 then links the particles' hydrodynamic radius (Rh) to the diffusion coefficient:

$$D = \frac{kT}{6\pi \eta R_h} \quad (7)$$

D represents the translational diffusion coefficient.

k denotes the Boltzmann constant.

Absolute Temperature is denoted by T.

Viscosity of the solvent η

The effective particle size is represented by the hydrodynamic radius, Rh.

Consequently, by quantifying the variations in scattered light and utilising the Stokes–Einstein equation, Dynamic light scattering (DLS) allows one to find the average and dispersion sizes of particles in a system [181, 182]. It is very efficacious for non-interacting spherical colloids [183]. The different components of DLS are depicted in Fig. 13 and explained below:

Light Source: A monochromatic, coherent laser is utilised as the light source (usually a laser at 632.8 nm). The light needs to be coherent to see the scattered light interference patterns (speckle) [184].

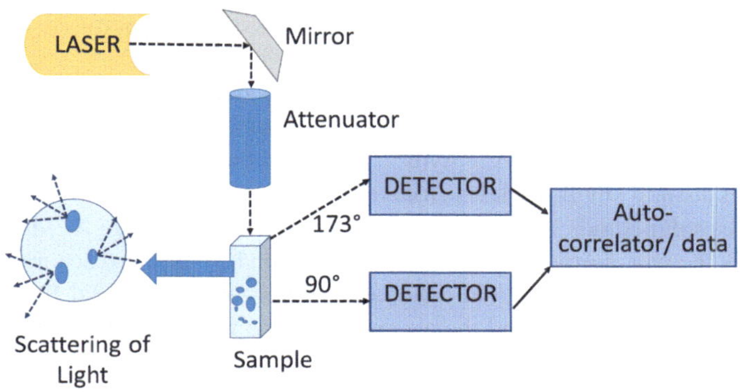

Fig. 13 Working principle of Dynamic light scattering spectrophotometer

Sample Cell (Cuvette): A quartz or glass cuvette holds the sample. To keep viscosity and diffusion rates stable, the cell is generally put in a chamber with a regulated temperature.

Optical System: This sends the laser beam through the sample.

It gathers light that is dispersed at a set angle, which is usually 90° (right-angle scattering) or 173° (backscattering geometry for materials that are very turbid or dense).

Fibre optics are often used to capture data in modern equipment [185].

Detector: A photomultiplier tube (PMT) or avalanche photodiode (APD) picks up scattered photons. This changes the intensity of dispersed light into electrical signals.

Correlator: A digital autocorrelator collects the electrical signal from the detector. This calculates the intensity autocorrelation function, showing the effect of intensity and its connection at different times. Brownian motion of particles causes fluctuations [186].

Computer and Software for Data Analysis: This fits the correlation function to theoretical models, like cumulant analysis and CONTIN algorithms, and helps discover the translational diffusion coefficient (D). Furthermore, the particles' hydrodynamic diameter (Dh) is calculated using the Stokes–Einstein equation [187].

In dynamic light scattering, a low nanoparticle concentration is necessary to avoid multiple scattering effects. Measuring one type of particle at a time is suitable in DLS as- it works fast, gives clear results, yet stays precise. Because it assesses many particles together, the data reflect overall patterns reliably. When samples are uniform, outcomes tend to repeat closely across tests. The movement of the particle and exhibiting Brownian motion is very important because if particles settle or fail to move randomly in the liquid, the analysis is unsuccessful. Under the brightness of light, the particles grows with size; thus, big particles dominate signals easily and hence the presence of large contaminants results in scattering of light thus it becomes difficult to detect the presence of smaller particles [188, 189].

Average size estimates from DLS can misrepresent complex mixtures, since calculations rely on simplifications about particle behavior—particularly when multiple sizes coexist. While the method applies to oddly shaped nanoparticles, interpretations often presume a spherical form [185, 190]. The technique captures the movement of the particles in a liquid, giving hydrodynamic dimensions, but the disadvantage is in distinguishing small particles from aggregates. This method is far better than older dye-based methods, and pairing DLS with imaging tools like AFM or electron microscopy the nanoparticles with polydispersity can be measured accurately [191].

The distribution of size in gold nanoparticles was obtained by DLS as observed in DLS spectrum in Fig. 14 and the average size was found to be 13.95 ± 0.14 nm which was complementing that obtained from TEM and SEM by Basu et al.

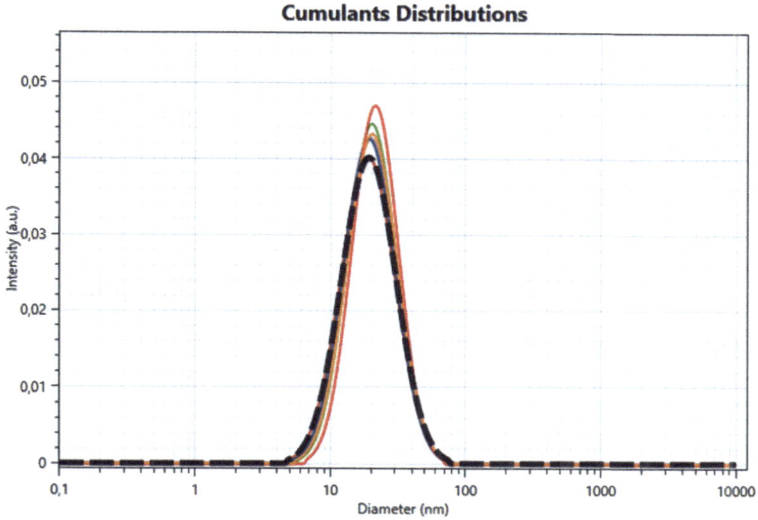

Fig. 14 DLS of gold nanoparticles

2.3.2 Nanoparticle Tracking Analysis

Nanoparticle tracking analysis (NTA), is illustrated in Fig. 15, represents an innovative method that has gained significant acceptance for quantifying nanoparticle size, a concentration detection limit which is lower than dynamic light scattering [192]. Particles in liquid scatter light while moving randomly, allowing size distribution measurement through analysis of their motion [66].

This method stands out because it does not favor larger particles or clumps when measuring the size of the nanoparticles [193]. NTA gives an consistent results with different nanoparticles and can be used for analysing mixtures with two distinct

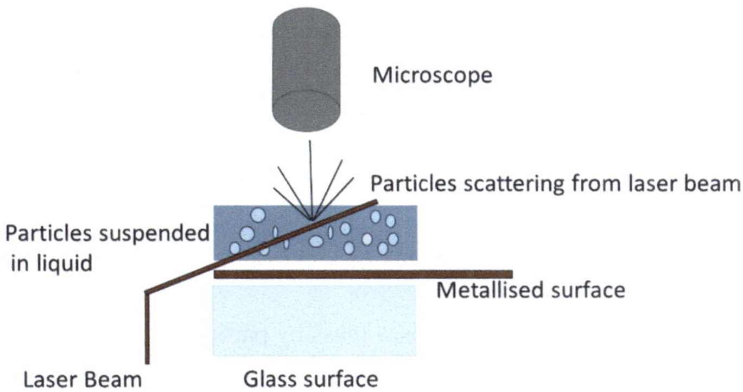

Fig. 15 Diagram of optical set-up of Nanoparticle Tracking Analysis (NTA)

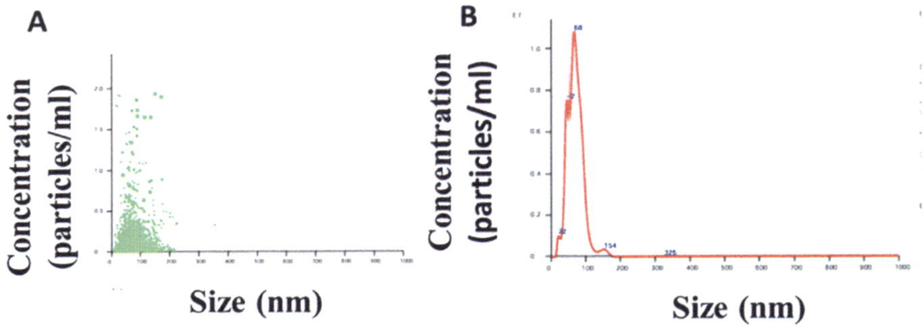

Fig. 16 **a** Distribution of the size of gold nanoparticles **b** Concentration of the nanoparticles with size

particle sizes [66]. A study showed that NTA provided reliable measurements across varied particle sizes, whether uniform or mixed. While DLS results tended toward higher numbers, the averages from NTA aligned closer to true dimensions. Differences emerged clearly when comparing method outputs side by side [66].

This method is considered as slower, and trickier as compared to DLS. Earlier work already showed DLS fails to give clear data on mixed-size samples; still, NTA manages to separate two distinct groups. In this method, one particle is analysed at a time while DLS observes all the particles together and hence it is influenced by the larger particles [66]. The scattering of light by the particles identifies the size of the particles depending on its scattering intensities. The sensitivity of NTA is dependent on the size and composition of the nanomaterials being studied [194]. NTA and image analysis were combined to determine the particle size distribution function of the given samples. The average NP size results generally matched with the values obtained from XRD and TEM examinations [195].

The NTA program detects and monitors individual nanoparticles that experience Brownian motion and correlates their velocity with nanoparticle size [159]. NTA can also determine the refractive index of the sample because of the light-nanoparticles interaction [196].

In Fig. 16, Basu et al. observed the distribution and concentration of gold nanoparticles with size with the instrument NTA. Thus, it can be observed that NTA not only give a very narrow size distribution but also indicate the concentration of their solution and its changes with size distribution as compared to DLS in Fig. 14.

2.4 *Thermogravimetric Analysis (TGA)*

When heated, a nanomaterial sample loses mass by parts and break down at distinct temperatures, which is detected by the method of thermogravimetric analysis [190]. From the initial weight, along with data on temperature shifts and mass reduction, details about organic ligands attached to nanoparticles emerge through TGA measurements [197]. A shift in mass often reveals how substances respond when heated, forming the basis of thermogravimetric methods. As temperature rises under regulated conditions, material transformations occur—sometimes through loss caused by breakdown, air-driven chemical change, or vapor release [198].

In thermogravimetric analysis, a small piece of material is placed inside a machine that tracks its weight while heating up steadily [198]. This process reveals that when substances is heated by raising the temperatures, the mass y break apart, and react with oxygen, or turn into vapor.

The mass of the particles generally reduces during testing because substances such as water, solvents, or decomposition byproducts escape from samples. The different information regarding the material stability, composition, and reaction speed comes from tracking the shifts in weight over time or heat exposure using thermogravimetric analysis (TGA) [199].

Continuous Weighing: A precisely regulated oven applies heat to a tiny amount of substance held in a container. A scale of fine accuracy records the material's weight without interruption. While measuring, it updates constantly. This method captures every small change immediately during testing.

Controlled Atmosphere: A typical setup involves a carefully managed environment—sometimes using nitrogen, sometimes air—to prevent unwanted chemical changes during heating [200]. Depending on the gas used, reactions may be limited or encouraged. Mass, temperature, and timing are captured continuously by the instrument. These measurements come together in a visual format known as a thermogram [201]. Figure 17 depicts the thermogravimetric spectrophotometer with its basic components.

A tiny shift in weight—just one billionth of a gram—can now be detected using micro-thermogravimetric (μ-TGA) analysis, despite its reliance on the same core

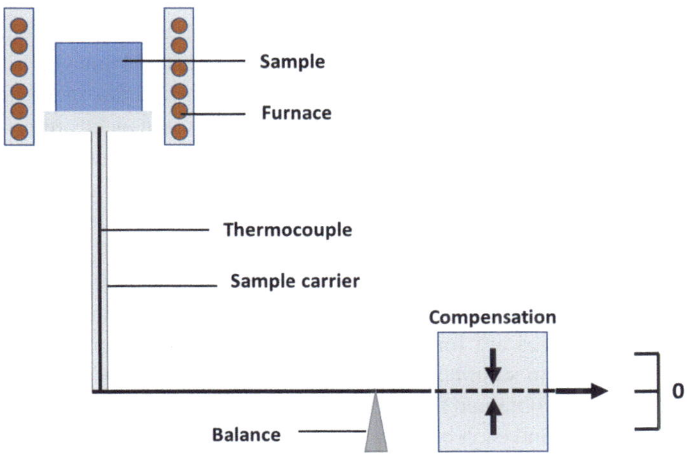

Fig. 17 Working principle of thermogravimetric analysis spectrophotometer

heating method as standard TGA. The advantage of μ-TGA is that it can measure samples weighing barely a millionth of a gram. The older systems often miss slight changes because of the lesser sensitivity in the instrument [202]. Conventional methods inflate costs because of the requirement of more material, Limited by bulk requirements, classic set-ups sometimes hinder small-batch research [203].

A shift in the method of sample are heating allows separation of reactions that happen at the same temperature. For thermogravimetric analysis to accurately track certain chemical groups, their mass changes need to take place at clearly different points on the heat scale. While mass spectrometry detects overlapping signals, interpreting their strength remains a big challenge in characterization of nanoparticles [204].

2.5 Microscopy Based Characterization of Nanoparticles

2.5.1 Transmission Electron Microscopy (TEM)

Electrons moving through ultra-thin samples form images in transmission electron microscopy, where beam energy usually falls between 60 and 150 keV. Although some electrons pass straight through the specimen, others scatter—either preserving energy or losing it during collisions [19]. The amount of scattering depends on thickness, atomic weight, composition, and related physical traits. Images emerge only after analyzing those electrons that make it through without deflection [205]. Earlier discussions highlighted on the influence of the nanoparticle geometry on the behavior—altering reactivity, conductivity, magnetism, biological response, and light interaction [79].

One key concept behind transmission electron microscopy lies in the fact that the particles act like waves. Because high-speed electrons have extremely short wavelengths—due to de Broglie's principle—thus very fine details of the nanoparticles as compared to optical microscopes are observed in the TEM images [206, 207]. In TEM, a specimen must be very thin; only then can a beam of electrons can travel across it [208]. While moving through the sample, some electrons collide with atoms inside the material and not every electron behaves the same: many scatters, yet others continue straight ahead. The magnetic lenses guide these electrons towards the recording device. This setup gives an enlarged view to the interactions of the electrons with the samples [209]. Depending on thickness, density, or atomic structure, regions within the sample alter scattering patterns—resulting in contrast, thus leading to formation of images with intricate features smaller than one nanometre.

Inside a TEM, images are formed when electrons move at great speed. These beams pass through thin samples, revealing tiny details. The resultant image was formed from the energy obtained from the fast-moving particles. The method depends on voltage strength behind each burst of electrons [207, 210]. The electron cannon is located at the top of the instrument and is responsible for creating electrons that travel through the vacuum tube [211]. The electromagnetic lens concentrates the electron

Fig. 18 Schematic representation of Transmission Electron Microscope instrument

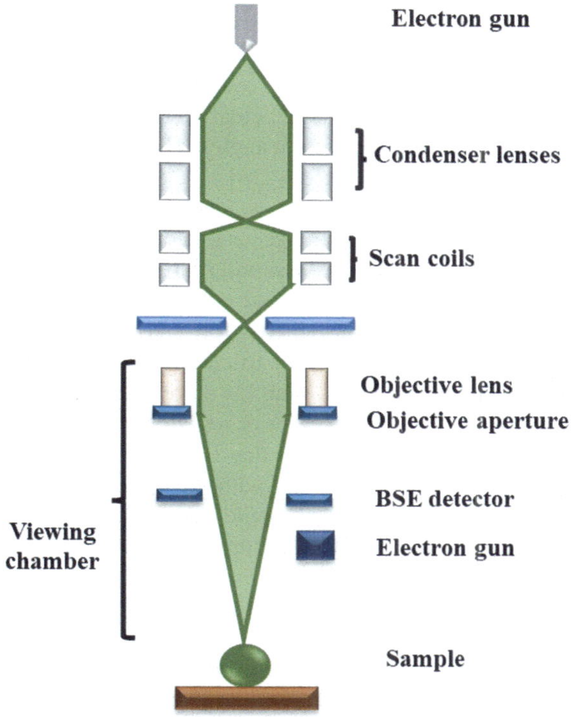

beam that passes through the extremely thin material. The fluorescent screen at the bottom of the microscope is struck by the released electrons. The image of the specimen and all of its parts are shown on the screen in accordance with its density [212]. Figure 18 represents the TEM operating principle schematically.

Equipment of TEM microscope: A glass chamber, sealed tightly, contain nearly all transmission electron microscope components inside. Air stays out of the system so electrons do not bounce off with the gas particles. Inside this space, multiple pieces fit together to form the full system:

Operating the TEM requires a vacuum near 10^{-4} Pa to allow electrons to travel without interference. Initially, vane pumps reduce the pressure so that the operating conditions are established After this the diffusion pumps push further into high-vacuum ranges that is needed needed for imaging [213].

Specimen stage: A small chamber helps guide samples smoothly while keeping pressure stable inside the specimen area. Depending on the use different stages with various shapes and function are used [213].

At the beginning, a stream of electrons are ejected and collided with the sample after being produced by the electron gun. The different parts that are present within the TEM microscope are: the filament, a Wehnelt cap, an extraction anode, along with a biassing setup. The electromagnetic lenses are used for focusing the sample. These lenses prevent random scattering, and guiding the incoming electrons to move in a parallel –path and meet at a set point [214].

A thick block of metal sufficiently block the movement of the unwanted electrons that are scattering at different angles. These barriers form apertures designed to filter out such deflected particles [215]

Despite its widespread use in studying nanoparticle dimensions and shape [137], TEM provides detailed images and accurate evaluation of particles and uniformity. Still, limitations exist as assessing large numbers of particles proves challenging, while misinterpretation may arise during analysis because of the positioning of the samples [216, 217]. The Scherrer formula and Small-Angle X-ray Scattering (SAXS) for larger and spherical nanoparticles are two alternate methods that can characterize extremely homogeneous samples more accurately and investigate larger quantities of nanoparticles. However, prior research is necessary to guarantee that the samples are identical [218]. Methods that investigate larger numbers of nanoparticles, such as X-ray Diffraction (XRD) with the boundaries of the XRD reflections and the Scherrer formula, or Small-Angle X-ray Scattering (SAXS) for bigger and spherical nanoparticles, can provide accurate results for exceedingly homogeneous samples [219].

When two metallic nanoparticles are positioned close to one another, for example, plasmon pairing takes place, resulting in a red shift of the plasmon band and a change in hue. TEM has been utilized to investigate nanoparticle aggregation for several biomedical applications, including sensing and diagnostics [220], in which the presence of analytes or biomarkers is necessary for aggregation to occur; in therapeutic contexts, where aggregation improves the effectiveness of nanoparticles; and imaging, in which the response signal is amplified by aggregation [205, 221]. Careless handling of samples might alter them or cause artifacts like aggregation when the colloidal solution dries, therefore proper sample preparation is essential for reliable results [222]. The production of well-defined nanoparticle superlattices by regulated self-assembly has been made possible in recent years due to substantial control over nanoparticle assembly. Novel multifunctional structures with intriguing and novel attributes are produced by the systematic integration of several nanocrystals. These structures blend the properties of the individual nanocrystals. It is possible that the isostructural analysis of multiple atomic crystal systems by TEM was made possible by the fabrication of super-lattice nanocomposites [223]. Modifying the colloidal surface charge or using directed bonding with DNA can regulate the ultimate shape and composition of the unique three-dimensional arrays, which comprise of diverse nanoparticles (e.g., quantum dots, metals, and magnetic nanoparticles) [113].

Nowadays researchers treat nanoparticles more like evolving entities, which are growing over time and moreover these nanoparticles are used in coalition with living cells. To boost effectiveness across applications in living matter, tracking these shifts are very important [224]. Because agglomeration affects particle behavior, scientists use TEM for examining movement of nanoparticles after entering the cells. This method also reveals damage or modifications inside cell compartments caused by

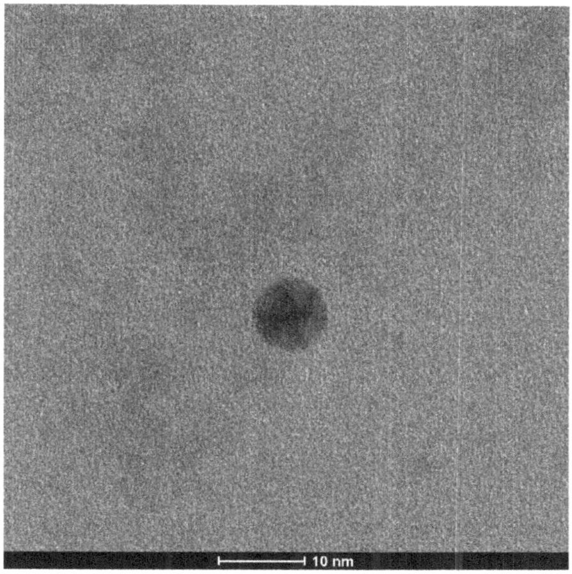

Fig. 19 TEM of gold nanoparticles

nanoparticles [225]. In skin studies, TEM helped to trace the distance travelled by titanium dioxide particles when moving in present of sunblock [110].

The TEM image of gold nanoparticles was depicted in Fig. 19 where the size was observed to be less than 10 nm which is aligned with the DLS study discussed before in Sect. 2.3.1.

High-Resolution Transmission Electron Microscopy (HRTEM)

A technique within TEM called high-resolution transmission electron microscopy relies on both transmitted and scattered beams of electrons to form images through phase contrast. Unlike standard TEM methods, this approach requires a larger objective aperture so it can capture more diffracted signals. At present, no method surpasses phase-contrast imaging when revealing atomic layouts in crystals [226].

Observing the inside layout of tiny particles is is not possible with regular electron microscopes, although such tools handle shape analysis well enough But, high-resolution transmission electron microscopy gives the fine details of the image observable. Because of this clarity, scientists rely on it more than other options for intricate structures present within nanoparticles. Although HRTEM is an effective tool, it's important to remember that it's not always possible to characterize nanoparticles using this technique. The random inclination of the crystals with regard to the source of electron may cause atoms to become misaligned in specific directions, producing complicated pictures that cannot be directly used to infer the structure [227]. Research using HRTEM can also provide light on NP growth and structure-related characteristics. Furthermore, HRTEM has been used to clarify how substrates affect the properties of metal nanoparticles.

Liquid Transmission Electron Microscopy

The vacuum system is an essential part of the TEM since it lessens electron beam dispersion and preserves the filament. The vacuum can be compromised by liquids evaporating, hence traditional transmission electron microscopy has only been used on solids and dried specimens. For many decades, the nanoscale characterization of solid–liquid systems was thus disregarded. In the 1930s, photographing biological material trapped between ultra-thin aluminium layers, was captured by L. Marton and the earliest images of liquid samples were observed. Even so, advancement was stalled because of high maintenance of the vacuum conditions while preventing liquid loss was proved too hard, and this became feasible through modern nanofabrication methods until tiny sealed chambers were formed. A breakthrough arrived in 2003 when Frances M. Ross and her team built a liquid cell for transmission electron microscopy using silicon nitride bonded with epoxy resin [228]. These membranes, are transparent to electrons, and allow the liquid to stay intact under vacuum during imaging. Because polymer-based designs could be reused and assembled quickly within a short spans they are effectively used for this purpose. Today's market favours silicon nitride grids equipped with features like electrochemical control or heating functions, despite their widespread use across labs but the disadvantage is that due to strong electron deflection from both the bulky SiN layer and surrounding fluid volume, the clarity of image is comparatively less. An alternative approach involves observing solid–liquid interactions within ionic liquids, which possess negligible vapor pressure. Since they do not readily turn into gas, these fluids let researchers capture particles suspended in liquid without sealing mechanisms. Without physical barriers like membranes, fewer electrons scatter, boosting image sharpness. On the downside, chemical reactivity limits compatibility, complicating experimental setups and narrowing of the types of materials that can actually be studied.

Observing the change of nanoparticles in real time becomes feasible with liquid transmission electron microscopy, allowing clearer insight into their movement during development and application of particles in living cell. Researchers have applied this imaging method to examine both pathways in synthesis and applications [229].

It was observed that the growth of platinum nanoparticles occurs at the lowest energy level and the attachment is on the surface preferring certain crystal faces by coalescence method [230]. Coalescence drives this development, as seen in experiments tracking particle merging. Hollow structures form when bismuth oxide undergoes the Kirkendall effect, revealing diffusion imbalances at work. Similarly, hollow palladium results from galvanic replacement, where one metal displaces another selectively. Bismuth particles grow in pulses, influenced alternately by Ostwald and reverse Ostwald mechanisms [161, 204]. These patterns emerge clearly under liquid-phase transmission electron microscopy. Real-time imaging captures redox changes while building complex nanostructures like gold–palladium systems. Reaction speeds become visible, layer by layer, without disturbing the process [32].

Cryo-Electron Microscopy (Cryo-TEM)

Using very cold conditions, cryo-TEM captures specimens in a near-natural form by freezing them rapidly and this method also falls under transmission electron microscopy. Samples stay stable because the process prevents ice crystals from forming, and preserving the structure during imaging.

Thus, the sample cooling method allow scientists to examine biological molecules in liquid environments with remarkable clarity through electron imaging. Recognition of Dubochet, Frank, and Henderson came in 2017 when they received chemistry's highest honor—for the advances related to low-temperature techniques for imaging.

Freezing with liquid nitrogen is the most common method. This method sees wide application in colloid chemistry and molecular biology since it limits the distortions caused by strains, and preserving the molecules fold and group them together. The automated commercial plunge-freezers turns the liquids into glass-like solids, and due to to rapid cooling of water-rich mixtures, the freezing was obtained Such speed stops water molecules from organizing into extended crystalline patterns and leads to formation of different forms that. resembles the original fluid state—disordered, without fixed structure [231]. First step is depositing the liquid onto a carbon-coated copper grid. The next step is to surplus the fluid that is drawn off to leave only a fine film. After that, the grid plunges into liquid nitrogen. Lastly, the now-vitrified specimen goes into a vessel bathed in liquid nitrogen.

One of the two main methods used to observe nanoparticle growth before liquid TEM was commercially available was freeze TEM; the other method included stopping nanoparticles at intermediate reaction stages and then characterizing them using conventional TEM [232]. One complex growth mechanism that cryo-TEM has been used to study is the aggregative production of zeolite crystals, which begins with the generation of many amorphous aggregates and continues with their reorganization into the final crystals [233].

It can capture images of the nanoparticles as they grow in their original environment. On the other side, statistical information is usually not yielded by particle concentrations that are very low, in the nanomolar region. Using cryo-TEM in conjunction with NP synthesis, molecular templates like block copolymers and CTAB are observed. These templates allow for the production of Au nanorods and lanthanide-based NPs, respectively [234]. The volumetric and shape variations of thermos-responsive core–shell nanoparticles have been observed using cryo-TEM. The thermosensitive shell's structure remains intact after plunge freezing, making it clearly visible without staining [234].

2.5.2 Scanning Electron Microscopy (SEM)

Instead of light, electrons are used to form images in scanning electron microscopy. This technique reveals fine surface details, reaching down to the nanometer range. Its ability to capture intricate topographies makes it widely used across material studies [235].

A beam of high-energy electrons strikes the sample, initiating signals through kinetic interactions. Due to this interaction various types of electrons emerge that are secondary, backscattered, and diffracted types—each revealing different features of the surface. Imaging relies heavily on secondary and backscattered electrons that actually captures the details. Because they respond differently to atomic number variations, backscattered electrons highlight compositional differences across the regions. Meanwhile, secondary electrons excel at mapping physical contours due to their sensitivity to shape and elevation changes [236].

At the very top of the column are tungsten filament lamps, like those seen in transmission electron microscopes, which produce the electromagnetic lenses and electrons. The release and acceleration of electrons towards the positively charged anode occurs when thermal energy is applied to the electron source [235]. The electron beam causes the specimen surface to release high-energy primary scattered electrons and low-energy secondary electrons. Signals produced by the electron beam's interaction with the item can be used to determine surface topography and material composition. The material can be directly studied under a scanning electron microscope (SEM) even after it has been air-dried. To keep microbes from shriveling up in the microscope's ultra-high vacuum, specimens need to be treated, dehydrated, and dried [237]. The specimens are mounted and covered with a thin layer of heavy metal components to enhance the production of photographs with excellent quality. A thin layer on the sample allows electric charges to spread out in different directions. Moving back and forth rapidly, the electron beam sweeps over a tiny section of the material during imaging. As the incoming electrons strike the surface, they release secondary emissions which are picked up by a dedicated sensor [238]. These emitted electrons travel toward the detection system after passing through a glowing substance that lights up when hit by energetic particles. Light flashes created here are turned into electrical signals by a photomultiplier device. That output feeds into a display unit resembling an old television monitor. Thus, the observed image are obtained directly [236].

When the inclinations is steeper, the electron collection increases. In contrast, shallow angles produce less emission, which limits the capture capacity of the detector. The material traits of the sample play a major role in the behaviour of secondary electron at the sensor [235]. Brighter areas on the screen often come from higher concentrated regions of electrons and darker zones represent the lower-concentrated region.

Figure 20 depicts the scanning electron microscopes with its components. The scanning electron microscope composes of different components. A place where electrons are created by utilizing thermal energy with a voltage range of 1–40 kV is called an electron source. The position near the base, the electrons come together to form a beam for imaging and study the nanoparticles. Field emission guns are placed alongside tungsten filaments and lanthanum hexaboride as main options for releasing those electrons. Moving through the column, the stream of electrons gets tighter because of several condenser lenses shaping it into a precise dot [239].

To move the beam across the sample's surface, scanning coils perform the task. These components steer the beam precisely over the sample and the accuracy depends

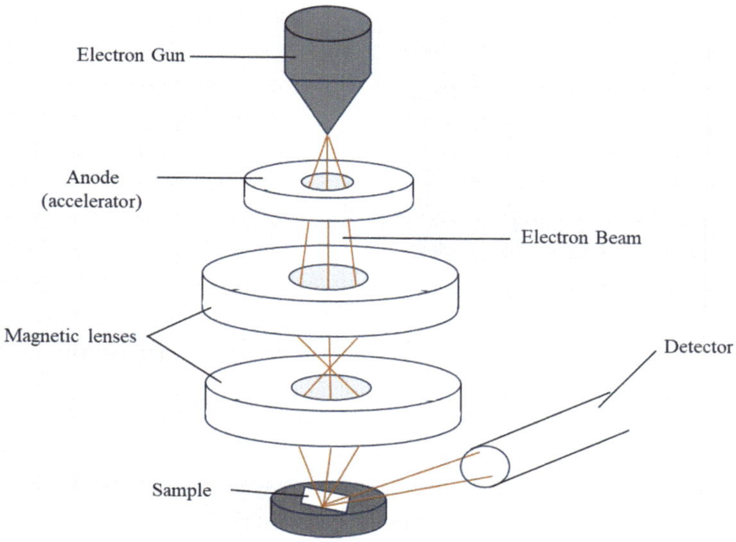

Fig. 20 Schematic Representation of Scanning Electron Microscope

on the positioning of the beam. Their role becomes clear when fine control is required during imaging and the consistent coverage would be difficult to achieve.

A single detector responds the various electrons, like backward scattered, secondary emitted electrons, or bent by diffraction. The fast voltage changes the effects and the performance depending on the thickness of the sample. The various factors that play role in better imaging are low pressure, and sources delivering steady current [240, 241].

In Fig. 21, the SEM image of citrate synthesized gold nanoparticles is observed and the size distribution of the nanoparticles. The average size of the nanoparticles is around 15 nm as obtained from TEM and DLS and discussed in the before sections.

A fresh look at living systems becomes possible through high-resolution scanning electron microscopy when studying nanoparticles. Metallic particles show up clearly using this approach [242]. Biological specimens are particularly vulnerable to radiation damage because metal coating may be required to decrease charging artifacts [243]. In comparison to other imaging methods, HRSEM's strength lies in its scalability, allowing one to study nanometric element combinations in their larger context. It allows for the examination of possible interactions between nanoparticles by facilitating the characterization of their exact spatial configuration. For studying how biological systems interact with metallic nanostructures, it is considered a powerful and versatile tool.

A technique called "transmission in scanning electron microscopy" (T-SEM) allows the scanning electron microscope (SEM) to operate in transmission mode. By collecting extensive data and evaluating NP ensembles, transmission mode enables the execution of complex NP analyses. Using a combination of SEM, T-SEM, electron

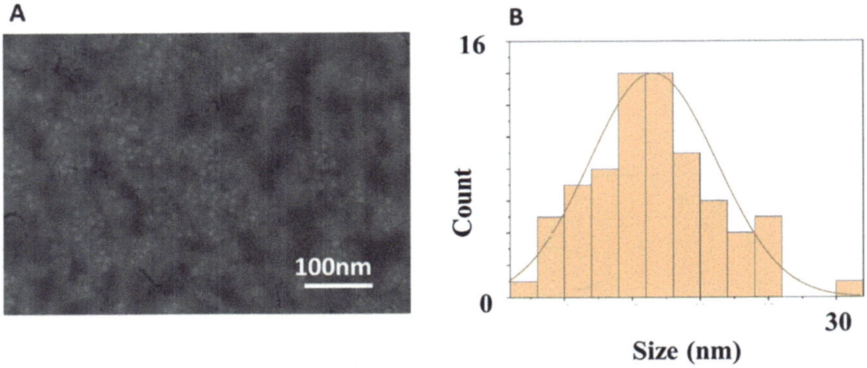

Fig. 21 **a** SEM image of gold nanoparticles **b** Size distribution of gold nanoparticles

density-functional theory (EDX), and scanning Auger microscopy (SAM), Rades et al. proved that a comprehensive method for chemical and morphological analysis of individual titania and silica nanoparticles may be achieved. In contrast, characterization of ensembles of nanoparticles, as opposed to individual particles, is best accomplished using methods like SAXS, DLS, XPS, XRD, and BET. Although T-SEM allows for a quick evaluation of nanoparticle morphology, its lateral resolution is limited to 5–10 nm for nanoparticle size. T-SEM is readily compatible with EDX, enabling a rapid assessment of nanoparticle size and elemental composition even if TEM produces better images [242].

2.5.3 Scanning Transmission Electron Microscopy (STEM)

STEM scans a material at random with a focused electron beam, which is different from traditional TEM [69]. In Fig. 22, we can see the STEM framework in action. After passing via a sophisticated condenser system, the electron beam targeted at a nanoscale probe is directed by a field emission electron cannon [244]. Scanning coils control the raster pattern of this concentrated electron probe as it probes the substance. The material's interaction with the electron probe causes scattered electrons at different angles, which are simultaneously collected by a high-angle annular dark field (HAADF) detector that is placed beneath the sample [245]. Images are generated based on the scattering angles of the dispersed electrons after they are collected, converted into signals, and used [246]. A sequence of repeated scans across a set area leads to the final transmission image. Each location examined on the sample becomes a corresponding dot in the image, step by step [247]. Unlike traditional transmission electron microscopy, where electrons pass through uniformly, here a focused beam moves across the specimen. The method demands precision in mapping, much like drafting fine lines rather than capturing light with a lens [248].

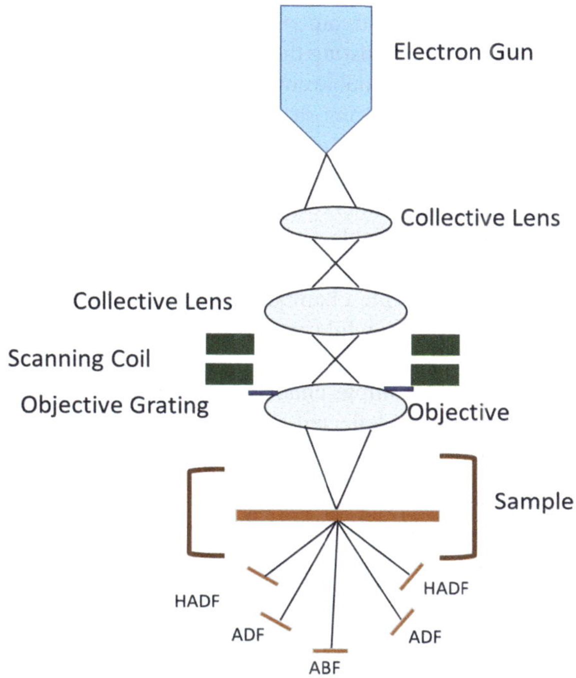

Fig. 22 Working principle of STEM

Even though the beam passes through the sample, making techniques like EDX, EELS, or Z-contrast imaging effective in STEM, interpreting results are very important. Elemental maps revealing atomic-scale details and this emerge clearly when either EDX or EELS integrates into STEM workflows [245]. Because electron sources must handle broad intensity variations, measuring absolute cross sections experimentally becomes necessary for reliable performance. This challenge has limited widespread adoption of STEM methods across diverse particle dimensions [249]. When supported by a verified mass reference, however, mass assessments via STEM deliver rich, detailed outputs.

The 3D shapes of nanoparticles can be mapped effectively using STEM electron tomography, a technique that works well whether the particles are fixed in place or embedded within materials [250]. STEM integrates the characteristics of SEM with TEM, combining scanning imaging with transmission analysis. The instrument's design can be viewed as a fusion of SEM and TEM, chiefly characterized by the integration of scanning components and the placement of electron signal detectors beneath the specimen [251].

Figure 22 demonstrates that the STEM system predominantly consists of the following components: electron gun, electron optical system, sample chamber, annular detector, and image apparatus.

Electron Gun: A stream of charged particles gets shaped carefully by a strong electron emitter, adjusting the intensity, angle, or its movement. In science and engineering work, dependable emitters matter—many setups use field-driven types which are capable of producing sharp, rapid bursts of electrons [244]. Beneath the electron cannon, an arrangement of electromagnetic lenses are placed—this setup narrows the beam down to a minuscule spot, small enough to interact with individual atoms. The electron beam that emerges is not just focused energy but also controls the refinement through magnetic fields arranged in sequence. Each lens contributes incrementally, adjusting trajectory so the final impact lands sharply where needed.

Inside the sample chamber sits the sample stage, making it straightforward to position specimens while enabling field selection through precise adjustments of its location [252].

Beneath the sample chamber, the annular detector uses its central opening to capture electrons deflected across broad angles. By processing these, the electron patterns are detected, resulting in imaging methods form visual outputs.

System for Vacuum: A strong vacuum system is required to avoid any interference from the gas molecules when electron beams are used at high energy levels [253].

Electron microscopes rely on several pumping phases to achieve this environment. Without such setups, both clarity and specimen integrity could suffer. Ensuring clean imaging requires removal of air particles through layered suction mechanisms.

Scanning tools built into transmission electron microscopes supporting atomic-level is critical to fields like nanotech, materials research, and life sciences. As electrons strike atoms within a specimen, they scatter—without or with energy loss, These scattered beams carry details about the sample's structure. Positioned below the specimen, ring-shaped detectors capture this signal from varying locations and the placement allows distinct imaging methods, illustrated in Fig. 22.

Three different kind of fields appear in STEM imaging: bright field, dark field, and high angle annular dark field. Because detectors capture scattered electrons at differing angles, multiple views of a single region emerge at once—each carrying distinct details on the sample's properties [254].

In the centre region of the electron beam used in STEM, a detector set is used to collect signals at small angles—less than about 10 milliradians—and this records mostly unscattered and slightly deflected electrons; this is referred to as annular bright field detection. When placed next to dark field approaches, this mode delivers sharper detail while supporting different image types such as phase-sensitive patterns or crystal lattice views [255]. Often, these outputs serve as additional context when interpreting ABF data Due to its sensitivity to lighter elements, ABF pictures have a contrast that coincides with the cube root of the atomic number Z.

Electrons dispersed by Bragg rays are mostly detected by the Annular Dark Field (ADF) detector, which is located within the angular range of θ_2 (10 mrads < θ_2 < 50 mrads). With the same imaging parameters, ABF images produce better contrast whereas ADF images show less sensitivity to imaging aberrations [256].

The High Angle Annular Dark Field (HAADF) mode mainly detects high-angle incoherent scattered electrons by using a detector to widen the collecting angle to θ_3

(with $\theta_3 > 50$ mrads) inside the annular dark field configuration. The sample's 1 s electrons in the atomic inner shells interact with the input electron beam, resulting in high-angle scattered electrons. Included in this category are features of the atomic nucleus, such as its chemical environment and the configuration of its energy levels [245, 246].

A large number of specimens that are sensitive to electron beams may not be able to undergo the process because of how long it takes to get full tomographs. The special X-ray intensity of the STEM satisfies the needs of a specific sample and detector geometry [257].

High-Angle Annular Dark-Field Scanning Transmission Electron Microscopy (HAADF-STEM)

The annular dark-field imaging method is a STEM mapping strategy. These pictures are made by collecting dispersed electrons with an ADF detector [150]. More sensitive to changes in atomic number inside the sample than Z-contrast images composed of Bragg scattered electrons is an annular dark-field image made entirely of high-angle incoherently scattered electrons (Rutherford scattered from atomic nuclei). Hybrid annular dark-field imaging is another name for this technique [246].

A wide variety of material surfaces have been successfully imaged using HAADF-STEM, making it an effective tool for investigating local atomic structures. The atomic mechanisms responsible for structural alterations and the unique behavior of NPs at the perimeter edge can be readily identified by the sequential HAADF-STEM measurements. In contrast, in order to ascertain atomic positions in conventional HRTEM images—which are plagued by artefacts, especially at surfaces or interfaces—extensive picture simulations are required [258].

Despite HAADF-STEM's obvious interpretability, multiscale simulation is sometimes required to account for the substantial dynamical screening impact when structure-related quantitative analysis data is needed [259]. Tomography, atomic-scale mass/thickness measurement, structural characterization, and composition analysis are some of the fields that can benefit from this technique. An annular dark-field (ADF) detector is the only one that can pick up the scattered electrons while HAADF-STEM scans the material with a highly concentrated beam [260]. TEM with a HAADF was also used to observe gold nanoparticles on titanium dioxide [261]. Direct examination of the atomic structure of heterointerfaces requires this method in combination with HRTEM. Imaging light atoms, like oxygen, presents hurdles, however HAADF-STEM can immediately resolve atomic configurations considering the defocus value and sample thickness, eliminating the requirement for picture simulations [262].

2.5.4 Atomic Force Microscopy (AFM)

One method that can create highly magnified three-dimensional pictures of surfaces is atomic force microscopy (AFM). The original developers were IBM employees Gerard Binns and Heinrich Rohrer in 1986 [263]. Force microscopy (AFM) relies

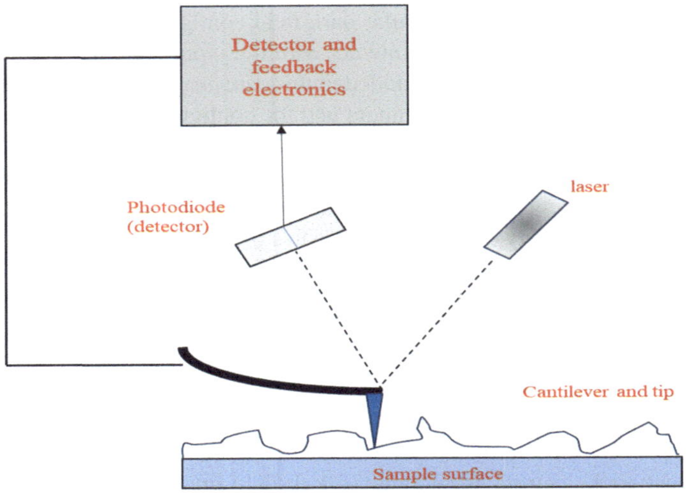

Fig. 23 Instrumentation diagram of atomic force Microscopy

on the measurement of the forces applied by a thin probe to the specimen. As shown in Fig. 23 [126], the probe's pointed end is connected to a cantilever made of silicon or silicon nitride. When an AFM sample is scanned, the cantilever deflects due to attractive and repulsive forces acting on the surface of the sample. By reflecting the laser beam off the back of the cantilever, the deformation may be observed. The forces are calculated by integrating data from the laser's fluctuation with the known cantilever stiffness [27].

The proximity of the probe to the sample determines the AFM mode of operation, which can be either contact, non-contact, or tapping (often called oscillating mode) [260]. The second approach is the most standard for nanoparticle characterization, but the unrestrained amplitude of the pulsating tip makes it extremely responsive. Extra properties, such as the nanoparticle's elasticity, surface energy, and radius of tip curvature, have a greater impact on the ultimate topological values. To mitigate these problems and obtain more reliable results, it is helpful to plot the particle height against the free amplitude of the oscillating probe. On the other hand, when the sample is very delicate and easily affected by the pressures applied by the tip, non-contact mode is preferred [153]. AFM is advantageous since it does not necessitate the application of coatings or surface modifications before imaging [260]. As a result, AFM has been used to investigate the topology of small nanoparticles ($\leq$6 nm), including ion-doped Y_2O_3, without subjecting them to any particular treatment [264]. Inadequately contrasted low-density materials have also been characterized. In order to gain a better understanding of the production process for polymer mixture nanoplatelets, which are both homogeneous and rectangular, AFM was employed [265]. Nanoparticle size analysis using AFM was shown to be on par with that using electron microscopies (SEM and TEM) when compared side by side

[266–268]. AFM provides the benefit of three-dimensional material imaging and enables the characterization of nanoparticle height.

In addition, it saves a lot of money and space in the lab, and it has resolution that is on par with SEM and TEM [267]. The scanning durations of AFM are significantly longer than those of any electron microscope.

Spectroscopic methods including DLS and photon correlation spectroscopy are alternatively used for measuring nanoparticle diameters. When the sample was homogeneous and monodisperse, the DLS and AFM results were comparable [269]. Nanoparticles having a bimodal size distribution could only be adequately characterized by AFM. An investigation into the generation of alumina nano-powder in a solvent by ultrasonic treatment and subsequent sedimentation to a thin layer revealed that the two techniques can produce related results, even if PCS analyzes liquid samples and AFM investigates solid materials [270].

Combining different technological skills, such as the high magnification of HRTEM and the height measurements of AFM, has made it easier to understand long-standing issues in nanoscience, such as how the dendrimer template influences the formation of Pt nanoparticles [271].

It would be challenging to utilize AFM to describe a nanoparticle film with a high density of packing because nearby particles could obscure some of the particles. Consequently, a multitude of techniques have been devised to estimate the size of nanoparticles using only the visible portion of an image. Both spherical and non-spherical particles can be handled using these strategies. The three-dimensional distribution of surface potentials on nanoparticles could be mapped out using atomic force microscopy and Kelvin probe force microscopy. Isolated nanoparticles of iron and stainless steel were subjected to corrosion tests in sulfuric acid to acquire these results [10].

3 Characterization Technique for Evaluating Porosity

3.1 Brunauer–Emmett–Teller (BET) Method

The Brunauer–Emmett–Teller (BET) method is used to describe materials on the nanoscale. The concept of a gas physically sticking to a solid surface is the basis of this theory, which was developed by Brunauer, Emmett, and Teller and is named after their initials [16]. Because it is a simple, fast, and accurate way to determine the surface area of nanostructures, this method is widely used for this purpose [17].

Figure 24 shows the Brunauer–Emmett–Teller (BET) surface area analyzer, a complicated apparatus that estimates the specific surface area and porosity of materials by physical adsorption of gases [272]. The adsorption isotherms produced from the adsorption–desorption of a probe gas at cryogenic temperature, which forms the basis of the fundamental idea and are further interpreted. A single device brings together several parts—each handling different tasks like moving gases, readying

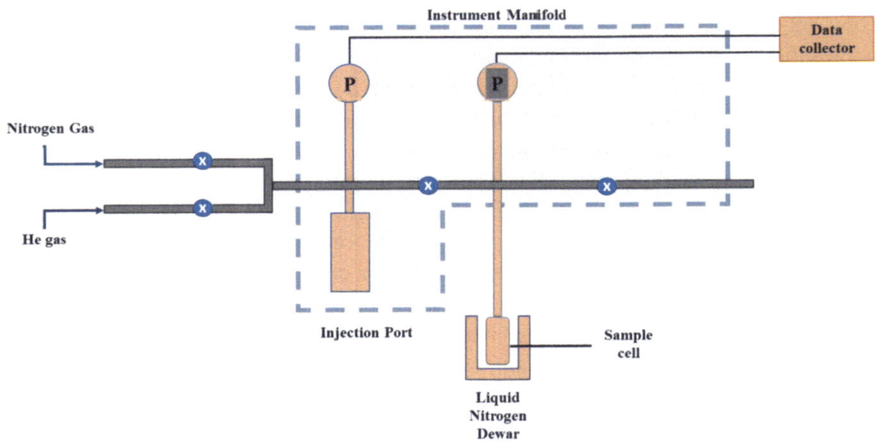

Fig. 24 Instrumentation diagram of BET instrument

samples, measuring the property of the substances adhering to surfaces, or managing information—to achieve precise outcomes [273].

A key element of the setup is the mechanism that manages the movement of gas molecules. Instead of using just one type, two different gases serve distinct roles—one acts as a carrier or purging agent, typically helium or nitrogen, helping prepare the sample by removing contaminants and setting measurement baselines. The second functions as an adsorbate, commonly nitrogen, argon, or krypton, interacting directly with the material's surface area [274]. Each gas comes from pressurized tanks outside the unit, traveling through durable metal tubing before entering the main chamber. Fine regulation happens because of high-accuracy valves that are placed strategically across the network. These components manage the entry and the exit of the gas. Accuracy matters since small shifts in either flow rate or pressure may distort results significantly [275].

The sample enters the degassing unit where unwanted elements begin to break free as the heating is started. Moisture or airborne gases might otherwise block key surface areas needed for accurate readings. Instead of allowing interference, a steady purge happens using either vacuum suction or a stream of neutral gas. Controlled warmth supports this release without altering the structure underneath. Monitoring systems track shifts moment by moment through linked sensors. These feed real-time updates to ensure stable pressure and consistent thermal levels throughout. The shifting of the stage occurs only when the cleanliness of the system reaches its required threshold [275, 276]. The analysis port goes into the sample tube, now dipped into liquid nitrogen after the end of degassing. The critical part is the cold setting that ensures the condensation of gas and adherence to the material, and this is guided by thermal control. Slowly, the tube takes in measured amounts of gas, pausing each time for stability before moving on to the next area [277].

[277] A shift in pressure, is measured by the amount of gas molecules adhering to the surface, and is measured through precise sensors and controlled flow systems. These tools track small variations as molecules gather on solid materials under test conditions. The precise design helps to identify and measure the smallest changes in pressure [273].

The data acquisition unit receives the continuously collected pressure–volume data and compiles it into an adsorption–desorption isotherm. The specific surface area can be calculated using specialized software by applying the BET equation to the isotherm's linear section [273]. A thorough grasp of the porous nature of the material under investigation is obtained by further analyzing the isotherm to obtain data on the distribution of pore sizes and volume. Dual ports are incorporated into the instrument design, as shown, to enable simultaneous analysis and degassing, increasing efficiency. By combining automated valves, digital data recording, and cryogenic cooling, human error is reduced and a high level of reproducibility is achieved [277].

The BET apparatus is a meticulously designed system that balances computer-based data analysis, cryogenic cooling, vacuum technology, and gas handling. The instrument offers a precise and dependable way to measure surface area and porosity by integrating these elements into a single configuration [276]. These characteristics are essential in disciplines like environmental research, materials chemistry, adsorption science, and catalysis [278–280].

In the characterisation of nanoparticles (NPs), different challenges are encountered in identifying a reliable method to assess porosity at the individual NP level. Electron transmission spectroscopy (TEM) and other semi-quantitative porosity measurements have limitations when it comes to studying larger nanoparticles (>100 nm radius) since they make it difficult to determine the relationship between pore width and depth inside the particle [281]. The significant physical feature of pores, referred to as tortuosity, is likewise not obtainable via TEM investigations. A common way to calculate tortuosity is to divide the length of the nanopore by its displacement. This requires monitoring individual pores as a function of depth within the nanoporous structure, as it quantifies the degree of nanopore winding [77]. A nanopore with no tortuosity at all has a value of 1 and can never be less than this, under this definition. Consequently, the ideal tomographic method easily probes and measures the internal porosity and tortuosity of specific nanoparticles with high resolution on a variety of non-specialized substrates [282]. Films, particle aggregates, and microspheres containing hollow cavities have all been cross-sectioned and seen with FIB in the past. In this technique, which is also known as "FIB tomography," a dual-beam scanning electron microscope (SEM) is used to photograph sections sliced using a gallium ion beam [77].

Focused ion beam tomography can also be utilized to find the porosity. The visualization of three-dimensional (3D) images made it possible to measure properties of individual nanoparticles, such as their pore density and nanopore tortuosity, which were previously unavailable [283]. Achieving direct contact at the surface of the substrate above the emulsion by partially immersing the substrate allowed for the electrodeposition of nanoparticles onto transmission electron micrograph grids [284].

A focused electron beam is directed onto a particular surface feature by adjusting the working distance of the stage. In order to bring the sample into alignment with the eucentric height perpendicular to the gallium ion beam, the stage was raised and lowered multiple times. The two beams were able to align to a coincident spot by a blind alignment, which involved patterning within a constrained space, switching to SEM mode, and carefully directing the ion beam towards the electron beam [285]. The nanoparticles were protected from beam damage by using this blind alignment. In order to decrease the likelihood of alignment issues, the approach was refined to within 5 μm of the target particle. Processing for one second each slice, the "cleaning cross section" FIB mode was used to start slicing, which allowed the particle to be patterned at dimensions of $400 \times 10 \times 500$ nm^3 [282]. The scanning electron micrographs (SEMs) taken after each slice revealed the NP's tomography and pores in the cross-section.

The free and open-source ImageJ "Particle Analysis" plugin was utilized to extract data on the size and distribution of pores from SEM slices. The Analysis → Particle Analysis tool was used to quantify the quantity and size of individual nanopores, with a minimum area of 1 nm^2 and a roundness factor of 0.2, in order to reduce the impact of noise and surface flaws caused by FIB milling [286]. Additionally, the slice's overall area was recorded. We transformed the pore regions found in ImageJ into diameter measurements in order to investigate the connection between slice depth and pore size and density. Following this step, analysis of surface-area-to-volume ratio, pore path complexity, and void fraction becomes possible through the softwarwe Avizo. First the regions are segmented, and then the image receive distinct intensity levels via a Watershed method. These divisions then formed a 3D structure when reassembled [263]. Label analysis is used for the measurements of size and occupancy of the space for every separated part. The shift between central points across neighbouring cross-sections helped assess winding paths—each point shaped by how pores contributed within its layer. For better distinction among nearby cavities, visual separation improved with randomly coloured mesh layers. Connections between pores on successive slices emerged naturally through shared voxels, handled inside Avizo's Pore Analysis Wizard [284].

While these technologies come with drawbacks, their potential applications spans through multiple sectors. Still, these methods are used and are expected despite existing constraints. Even so, industry adoption seems difficult at present but likely to occur over the time. So, there are challenges that remain alongside promise of effective method. Though limited now, real-world deployment could grow steadily. Specialized optical equipment is not required to perform the procedures described below; however, a FIB-SEM system with EDX capability must be available [283].

The method produces data on porosity that is dependent on depth and is not dependent on the substrate. By creating a three-dimensional image of these slices, nanopore tortuosity may be seen and measured down to the nanoparticle level. The volume correction provided by the porosity analysis explained the larger deposit size of the nanoparticles. A thorough NP tomographic characterization was possible with three-dimensional visualization of particle surface area, nanopore tortuosity, and surface area to volume ratio [287].

4 Characterization of Nanoparticles (Examples)

This section presents examples from the literature where various methodologies are employed on specific samples to characterise their size, as this attribute is fundamental.

This matter warrants particular consideration for NPsusing TEM, BET, and XRD, and other characterization techniques for analyzing their different properties. Akbari et al. evaluated the size distributions and dimensions of alumina particles. It was discovered that the nanoparticles in question ranged in size from 5 to 95 nm [288]. These particles showed consistency in the size values derived from XRD and TEM. Since spheres tend to show consistent measurements, so BET measurements were similar to that with both TEM and XRD results [288].

Pooja et al. observed the powder X-ray diffraction pattern in Fig. 25 of synthesized GO that showed a sharp peak at $2\Theta = 11.27^\circ$ due to the (002) plane of the GO structure [289]. The interlayer spacing (d) was estimated to be 0.78 nm by well-established Bragg's law:

$$n\bar{\lambda} = 2d\sin\theta \tag{8}$$

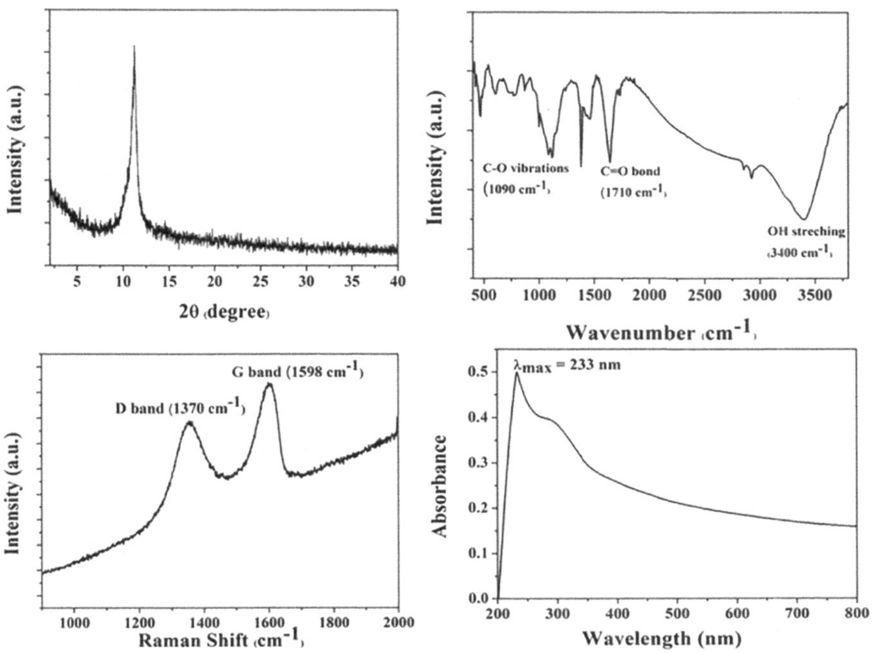

Fig. 25 Characterisation for GO nanoparticles **a** XRD pattern **b** IR spectrum **c** Raman spectrum **d** UV–Vis spectra

where n is the diffraction series and λ is the diffraction wavelength.

According to reports, the interlayer spacing value for graphene oxide (GO) ranges from 0.6 to 1.0, indicating the extent of oxidation and the number of water molecules incorporated into the interlayer spacing [290]. Therefore, the value of 0.78 nm of interlayer spacing can be attributed to the highly oxidized form of graphite, indicating the formation of GO from graphite. The FTIR spectrum of GO in Fig. 25b confirmed the presence of functional groups and the formation of GO from graphite. A broad band around 3400 cm^{-1} is attributed to O–H stretching which is involved in hydrogen bonding and a band at 1710 cm^{-1} confirms the presence of a carboxylic acid functional group in GO. The band between 1000 cm^{-1} to 1250 cm^{-1} corresponds to C–O–C epoxy bonds and the vibrational mode of C-O groups [138]. The band between 1300–1400 cm^{-1} is characteristic of C = C aromatic bonds. The Raman spectrum in Fig. 25c showed two distinct bands: D band (1370 cm^{-1}) and G band (1598 cm^{-1}) which is in line with the literature values for GO [291]. The D band arises due to disordered structure and the intensity of the D band is smaller because of the breakage of stacking order of graphite after oxidation whereas the G band arises due to stretching of the C–C bond in graphitic materials [138]. The UV visible spectrum of GO in Fig. 25d also showed characteristic absorption bands. A well-built absorption peak at 233 nm corresponding to л- $л^*$ transition of aromatic C = C structure and a weak shoulder peak at 305 nm corresponding to n-$л^*$ transition of carbonyl bonds [138] confirms the GO formation in the reaction.

There are various ways to measure silica nanoparticles floating in solution or living cell environments, Gollwitzer's team explored several approaches at once. Among the different methods of characterizing them the motion of individual particles are tracked and the settling of particles are measured by light scattering via DLS, and using X-rays at small angles (SAXS) [292]. Because agglomerates reduced the accuracy of DLS, the findings obtained in the cell culture medium using this method were significantly different from those acquired using other methods [292]. To assess serum silica nanoparticle dimensions, Minelli et al. applied dynamic light scattering, tunable resistive pulse sensing, along with differential centrifugal sedimentation. Aggregates interfered, causing DLS to underperform here [293].

A fresh approach to analyzing gold-based particles emerged through work led by Carney, applying two-dimensional ultracentrifugation. With this technique, details like density, dimensions, and molecular weight can be measured. Researchers rely on measured sedimentation rates alongside diffusion data obtained from scans. These numbers open a path toward precise calculations about each particle's traits without any assumption.

Differential centrifugation Sedimentation (DCS) measured the citrate-coated gold nanoparticles and was found to be smaller as compared via DLS and NTA It is expected that when samples are uniform, Brownian motion analysis in DLS and NTA lead to matching diameter readings. But compared to the other two, DCS proved sharper in resolution and less swayed by measurement distortions. Impurities and agglomerations can interfere in the DLS outcomes, meanwhile NTA faces constraints due to the few number of position and points for one particle that drifts out of sight [294, 295].

5 Conclusion

This chapter aimed to clarify methods used to analyse nanomaterials at the smallest scales. Through diverse techniques are discussed, insight into structure and behaviour emerges for the various characterization techniques. Although each method is different from other but the common goals of observing the shape, size and other factors of nanoparticles tie them together. It is the intention of this comprehensive summary to provide a clear description of the numerous approaches to NP characterization, to provide illustrations of the applications of each approach, to highlight the benefits and limitations of each approach, and to explain how these approaches can be effectively integrated and complement one another. In order to acquire a comprehensive grasp of the various properties associated with a nanomaterial, it is frequently necessary to employ multiple methods. For conducting an exhaustive and precise evaluation of even a single characteristic, it is frequently required to make use of more than one method. With the help of our review, the scientific community will be able to get a more in-depth understanding of the topic at hand. This chapter gives a view on different comparative methods used to characterize a nanoparticle and explains the function of each technique. Researchers will be able to select the approaches that are most suitable for characterisation with the assistance of this approach, which will also make it possible to conduct a more precise evaluation of how these methodologies are used.

There is no doubt that the scientific community is confronted with challenges when it comes to improving the accuracy and precision of a variety of approaches. As a consequence of this, we believe that a careful review of this evaluation will be of use in assessing whether or not effective procedures justify efforts to further improve technical aspects.

References

1. Gupta, R., Chauhan, H., Garg, V.K., Kataria, N.: Chemical and physical properties of nanoparticles and hybrid materials. Sustain. Nanotechnol. Environ. Remediation **199–220**,(2022). https://doi.org/10.1016/B978-0-12-824547-7.00024-2
2. Altammar, K.A.: A review on nanoparticles: characteristics, synthesis, applications, and challenges. Front. Microbiol. **14**, 1155622 (2023). https://doi.org/10.3389/FMICB.2023.1155622/BIBTEX
3. Pan, H., Feng, Y.P.: Semiconductor nanowires and nanotubes: Effects of size and surface-to-volume ratio. ACS Nano **2**, 2410–2414 (2008). https://doi.org/10.1021/NN8004872/ASSET/IMAGES/MEDIUM/NN-2008-004872_0006.GIF
4. Shekhawat, D., Vauth, M., Pezoldt, J.: Size dependent properties of reactive materials. Inorganics **10**, 56 (2022). https://doi.org/10.3390/INORGANICS10040056
5. Baretzky, B., Baró, M.D., Grabovetskaya, G.P., Gubicza, J., Ivanov, M.B., Kolobov, Yu.R., Langdon, T.G., Lendvai, J., Lipnitskii, A.G., Mazilkin, A.A., Nazarov, A.A., Nogués, J., Ovidko, I.A., Protasova, S.G., Raab, G.I., Révész, Á., Skiba, N.V., Starink, M.J., Straumal, B.B., Suriñach, S., Ungár, T., Zhilyaev, A.P.: Fundamentals of interface phenomena in advanced bulk nanoscale materials (2005)

6. Patil, S.P., Burungale, V.V.: Physical and chemical properties of nanomaterials. Nanomedicines for Breast Cancer Theranostics **17–31**,(2020). https://doi.org/10.1016/B978-0-12-820016-2.00002-1
7. Guo, D., Xie, G., physics JL-J of physics D applied, 2013 undefined (2014) Mechanical properties of nanoparticles: basics and applications. iopscience.iop.org 47:25. https://doi.org/10.1088/0022-3727/47/1/013001/META
8. Wu, Q., Miao, W.S., Zhang, Y.Du., Gao, H.J., Hui, D.: Mechanical properties of nanomaterials: a review. degruyterbrill.com **9**, 259–273 (2020). https://doi.org/10.1515/NTREV-2020-0021/HTML
9. Mohajerani, A., Burnett, L., Smith, J.V., Kurmus, H., Milas, J., Arulrajah, A., Horpibulsuk, S., Kadir, A.A.: Nanoparticles in Construction Materials and Other Applications, and Implications of Nanoparticle Use. Materials **12**, 3052 (2019). https://doi.org/10.3390/MA12193052
10. Applied Nanotechnology: The Conversion of Research Results to Products—Jeremy Ramsden—Google Books. https://books.google.co.in/books?hl=en&lr=&id=aB8wDwAAQBAJ&oi=fnd&pg=PP1&dq=academic+appeal+and+potential+technical+applications+of+nanoparticles&ots=5ScjbeqbaX&sig=AUKlzgAtPSclhFrmVU_fLOb6eRs&redir_esc=y#v=onepage&q=academic%20appeal%20and%20potential%20technical%20applications%20of%20nanoparticles&f=false. Accessed 2 Sep 2025
11. Ayuk, E., Ugwu, M.: Journal SA-CR, 2017 undefined A review on synthetic methods of nanostructured materials. eprints.gouni.edu.ng, 97–123 (2017)
12. Jones, M.R., Osberg, K.D., MacFarlane, R.J., Langille, M.R., Mirkin, C.A.: Templated techniques for the synthesis and assembly of plasmonic nanostructures. ACS Publ. **111**, 3736–3827 (2011). https://doi.org/10.1021/CR1004452
13. Nanotechnology, I.H.: Undefined Nanostructure fabrication using block copolymers. iopscience.iop.org (2003). https://doi.org/10.1088/0957-4484/14/10/201/META
14. Technology MR-R research in science and, 2012 undefined (2012) Nanostructures and their applications. core.ac.uk 4:14–19
15. Xia, Y., Rogers, J.A., Paul, K.E., Whitesides, G.M.: Nanostructures & nanomaterials: synthesis, properties & applications (2004)
16. Nanotechnology: An Introduction to Nanostructuring Techniques—Michael Köhler, Wolfgang Fritzsche—Google Books. https://books.google.co.in/books?hl=en&lr=&id=Jm48VBDQuU8C&oi=fnd&pg=PP7&dq=techniques+to+create+nanostructures&ots=HBEXtDSzbt&sig=KJLv_DYrZ-Xng3cRHLaZyViJRXs&redir_esc=y#v=onepage&q=techniques%20to%20create%20nanostructures&f=false. Accessed 2 Sep 2025
17. Surface characterization of nanomaterials and nanoparticles: Important needs and challenging opportunities. pubs.aip.org
18. Dhirendra Prasad, R., Prasad, R.S., Prasad, R.B., Prasad, S.R., Singha, S.B., Singha, A.D., Prasad, R.J., Teli, S.B., Sinha, P., Kumar Vaidya, A., Saxena, S., Saxena, U.R., Harale, A., Deshmukh, M.B., Padvi, M.N., Navathe, G.J.: A review on modern characterization techniques for analysis of nanomaterials and biomaterials. espublisher.com 23, 1087–1088 ((2024)). https://doi.org/10.30919/esee1087
19. Rao, C., Chemistry, KB-AR of A.: Undefined (2009) Characterization of nanomaterials by physical methods. annualreviews.org **2**, 435–462 (2009). https://doi.org/10.1146/ANNUREV-ANCHEM-060908-155236
20. Jayawardena, H.S.N., Liyanage, S.H., Rathnayake, K., Patel, U., Yan, M.: Analytical methods for characterization of nanomaterial surfaces. ACS Publ. **93**, 1889–1911 (2021). https://doi.org/10.1021/ACS.ANALCHEM.0C05208
21. Prasad, R.D., Prasad, R.S., Prasad, R.B., Prasad, S.R., Singha, S.B., Singha, A.D., Prasad, R.J., Teli, S.B., Sinha, P., Vaidya, A.K., Saxena, S., Saxena, U.R., Harale, A., Deshmukh, M.B., Padvi, M.N., Navathe, G.J.: A Review on modern characterization techniques for analysis of nanomaterials and biomaterials. ES Energy Environ. **23**,(2024). https://doi.org/10.30919/ESEE1087

22. Mourdikoudis, S., Pallares, R.M., Thanh, N.T.K.: Characterization techniques for nanoparticles: comparison and complementarity upon studying nanoparticle properties. Nanoscale **10**, 12871–12934 (2018). https://doi.org/10.1039/C8NR02278J
23. Ealias, A.M., Saravanakumar, M.P.: A review on the classification, characterisation, synthesis of nanoparticles and their application. IOP Conf Ser Mater Sci Eng **263**, 032019 (2017). https://doi.org/10.1088/1757-899X/263/3/032019
24. Gupta, P., Rai, N., Verma, A., Gautam, V.: Microscopy based methods for characterization, drug delivery, and understanding the dynamics of nanoparticles. Wiley Online Library **44**, 138–168 (2024). https://doi.org/10.1002/MED.21981
25. Lapresta-Fernández, A., Salinas-Castillo, A., Anderson De La Llana, S., Costa-Fernández, J.M., Domínguez-Meister, S., Cecchini, R., Capitán-Vallvey, L.F., Moreno-Bondi, M.C., Marco, M.P., Sánchez-López, J.C., Anderson, I.S.: A general perspective of the characterization and quantification of nanoparticles: imaging, spectroscopic, and separation techniques. Taylor & Francis **39**, 423–458 (2014). https://doi.org/10.1080/10408436.2014.899890
26. Pettitt, M.E., Lead, J.R.: Minimum physicochemical characterisation requirements for nanomaterial regulation. Environ. Int. **52**, 41–50 (2013). https://doi.org/10.1016/J.ENVINT.2012.11.009
27. Zhang, B., Yan, B.: Analytical strategies for characterizing the surface chemistry of nanoparticles. Anal. Bioanal. Chem. **396**, 973–982 (2010). https://doi.org/10.1007/S00216-009-2996-1/METRICS
28. Dolai, J., Mandal, K., Jana, N.R.: Nanoparticle size effects in biomedical applications. ACS Publ. **4**, 6471–6496 (2021). https://doi.org/10.1021/ACSANM.1C00987
29. Nanoparticles, Y.K.-M.: undefined (2009) Magnetism of nanoparticles: effects of size, shape, and interactions. Wiley Online Library, 1–466 (2009). https://doi.org/10.1002/9783527627561
30. Burda, C., Chen, X., Narayanan, R., El-Sayed, M.A.: Chemistry and properties of nanocrystals of different shapes. Chem. Rev. **105**, 1025–1102 (2005). https://doi.org/10.1021/CR030063A/ASSET/CR030063A.FP.PNG_V03
31. Santos, C.S.C., Gabriel, B., Blanchy, M., Menes, O., García, D., Blanco, M., Arconada, N., Neto, V.: Industrial applications of nanoparticles—a prospective overview. Mater. Today Proc. **2**, 456–465 (2015). https://doi.org/10.1016/J.MATPR.2015.04.056
32. Stark, W.J., Stoessel, P.R., Wohlleben, W., Hafner, A.: Industrial applications of nanoparticles. Chem. Soc. Rev. **44**, 5793–5805 (2015). https://doi.org/10.1039/C4CS00362D
33. Dejene, B.K.: Reviewing the manufacturing challenges and scientific debates: insights into the antibacterial capabilities and potential applications of PLA/ZnO nanocomposites. J. Thermoplast. Compos. Mater. (2024). https://doi.org/10.1177/08927057241292298/ASSET/FE07D439-4E39-4D23-B5CE-0F8C678829B1/ASSETS/IMAGES/LARGE/10.1177_08927057241292298-IMG01.JPG
34. MacCuspie, R.I., Rogers, K., Patra, M., Suo, Z., Allen, A.J., Martin, M.N., Hackley, V.A.: Challenges for physical characterization of silver nanoparticles under pristine and environmentally relevant conditions. J. Environ. Monit. **13**, 1212–1226 (2011). https://doi.org/10.1039/C1EM10024F
35. Fernandes, C., Jathar, M., Sawant, B.K.S., Warde, T.: Scale-up of nanoparticle manufacturing process, 173–203 (2023). https://doi.org/10.1007/978-3-031-31380-6_12
36. Cho, E.J., Holback, H., Liu, K.C., Abouelmagd, S.A., Park, J., Yeo, Y.: Nanoparticle characterization: state of the art, challenges, and emerging technologies. ACS Publ. **10**, 2093–2110 (2013). https://doi.org/10.1021/MP300697H
37. Modena, M.M., Rühle, B., Burg, T.P., Wuttke, S.: Nanoparticle characterization: what to measure? Wiley Online Library **31**,(2019). https://doi.org/10.1002/ADMA.201901556
38. Nam, J.M., Owen, J.S., Talapin, D.V.: The ligand–surface interface and its influence on nanoparticle properties. ACS Publ. **56**, 2265–2266 (2023). https://doi.org/10.1021/ACS.ACCOUNTS.3C00416
39. Bergmann, A., Roldan Cuenya, B.: Operando insights into nanoparticle transformations during catalysis. ACS Publ. **9**, 10020–10043 (2019). https://doi.org/10.1021/ACSCATAL.9B01831

40. Ali, A., Chiang, Y., Minerals, R.S.: Undefined X-ray diffraction techniques for mineral characterization: a review for engineers of the fundamentals, applications, and research directions. mdpi.com (2022)
41. Giannini, C., Ladisa, M., Altamura, D., Siliqi, D., Crystals, T.S.: Undefined X-ray diffraction: a powerful technique for the multiple-length-scale structural analysis of nanomaterials. mdpi.com (2016)
42. Muniz, F.T.L., Miranda, M.A.R., Morilla Dos Santos, C., Sasaki, J.M.: The Scherrer equation and the dynamical theory of X-ray diffraction. urn:issn, 2053–2733 **72**, 385–390 (2016). https://doi.org/10.1107/S205327331600365X
43. Epp, J.: X-ray diffraction (XRD) techniques for materials characterization. Materials Characterization Using Nondestructive Evaluation (NDE) Methods 81–124 (2016). https://doi.org/10.1016/B978-0-08-100040-3.00004-3
44. Blanton, T., Huang, T., Toraya, H, … CH-P.: Undefined JCPDS—International Centre for Diffraction Data round robin study of silver behenate. A possible low-angle X-ray diffraction calibration standard. cambridge.org (1995)
45. Browning, N., Applied SP-J of PD: Undefined Direct experimental determination of the atomic structure at internal interfaces. iopscience.iop.org (1996). https://doi.org/10.1088/0022-3727/29/7/013/META
46. Petkov, V., Prasai, B., Shastri, S., Chen, T.Y.: 3D Atomic arrangement at functional interfaces inside nanoparticles by resonant high-energy X-ray diffraction. ACS Publ. **7**, 23265–23277 (2015). https://doi.org/10.1021/ACSAMI.5B07391
47. Reviews, B.I.-C.: Undefined (2015) X-ray scattering characterisation of nanoparticles. Taylor & Francis **21**, 229–303 (2015). https://doi.org/10.1080/0889311X.2015.1024114
48. Holder, C.F., Schaak, R.E.: Tutorial on powder X-ray diffraction for characterizing nanoscale materials. ACS Publ. **13**, 7359–7365 (2019). https://doi.org/10.1021/ACSNANO.9B05157
49. Zhang, J., Liao, P., Zhou, H., … RL-CS.: Undefined Single-crystal X-ray diffraction studies on structural transformations of porous coordination polymers. pubs.rsc.org (2014)
50. Newman, J.A., Iuzzolino, L., Tan, M., Orth, P., Bruhn, J., Lee, A.Y.: From powders to single crystals: a crystallographer's toolbox for small-molecule structure determination. ACS Publ. **19**, 2133–2141 (2022). https://doi.org/10.1021/ACS.MOLPHARMACEUT.2C00020
51. Bunaciu, A.A., Udriştioiu E gabriela, Aboul-Enein, H.Y.: X-ray diffraction: instrumentation and applications. Taylor & Francis 45, 289–299 ((2015)). https://doi.org/10.1080/10408347.2014.949616
52. Principles and Practice of X-Ray Spectrometric Analysis—E.P. Bertin—Google Books. https://books.google.co.in/books?hl=en&lr=&id=2Dj2BwAAQBAJ&oi=fnd&pg=PA3&dq=xrd+instrumentation++principle&ots=mArW-V69Jr&sig=66yPr0wQRImYKcmwd2CIxPlre88&redir_esc=y#v=onepage&q&f=false. Accessed 7 Sep 2025
53. Education PF-P: Undefined Electron diffraction and the Bragg equation. iopscience.iop.org (1971). https://doi.org/10.1088/0031-9120/6/1/002/META
54. Zolotoyabko Emil: Basic concepts of X-ray diffraction (2014)
55. Dinnebier, R., Billinge, S.: The Bragg equation derived (2019)
56. Sun, S., Zhang, X., Cui, J., Liang, S.: Identification of the Miller indices of a crystallographic plane: a tutorial and a comprehensive review on fundamental theory, universal methods based on different case studies and matters needing attention. Nanoscale **12**, 16657–16677 (2020). https://doi.org/10.1039/D0NR03637D
57. Mustapha, S., Ndamitso, M.M., Abdulkareem, A.S., Tijani, J.O., Shuaib, D.T., Mohammed, A.K., Sumaila, A.: Comparative study of crystallite size using Williamson-Hall and Debye-Scherrer plots for ZnO nanoparticles. iopscience.iop.org **10**, 45013 (2019). https://doi.org/10.1088/2043-6254/AB52F7/META
58. Chérif, I., Dkhil, Y.O., Smaoui, S., Elhadef, K., Ferhi, M., Ammar, S.: X-ray diffraction analysis by modified scherrer, williamson–hall and size–strain plot methods of ZnO nanocrystals synthesized by oxalate route: a potential. Springer **34**, 623–638 (2023). https://doi.org/10.1007/S10876-022-02248-Z

59. Ciupină, V., Zamfirescu, S, ... GP--Toxicological I and: Undefined Evaluation of mean diameter values using Scherrer equation applied to electron diffraction images. Springer (2007). https://doi.org/10.1007/978-1-4020-6076-2.PDF
60. Miranda, M., Crystallography JS-F of.: undefined The limit of application of the Scherrer equation. journals.iucr.org (2018)
61. физика AV-H, химия undefined, математика undefined, 2018 undefined (2018) Scherrer formula: estimation of error in determining small nanoparticle size. cyberleninka.ru 9:364–369. https://doi.org/10.17586/2220-8054-2018-9-3-364-369
62. Hajalilou, A., Hashim, M., Mohamed Kamari, H.: Structure and magnetic properties of Ni0.64Zn0.36Fe2O4 nanoparticles synthesized by high-energy milling and subsequent heat treatment. Springer **26**, 1709–1718 (2015). https://doi.org/10.1007/S10854-014-2597-4
63. Materials, J.G.-A.E.: Undefined (2020) Annealing-induced hardening in ultrafine-grained and nanocrystalline materials. Wiley Online Library **22**,(2020). https://doi.org/10.1002/ADEM.201900507
64. In MM-R research advancements exploring advanced materials: in-depth analysis using x-ray diffraction methodology. swamivivekanandauniversity.ac.in
65. Pooja, R.P., Kumari, P., Sharma, R.K.: Selective adsorption of azine group containing pesticides on carbon-based materials. J. Integr. Sci. Technol. **13**,(2025). https://doi.org/10.62110/SCIENCEIN.JIST.2025.V13.1061
66. Filipe, V., Hawe, A., Jiskoot, W.: Critical evaluation of nanoparticle tracking analysis (NTA) by NanoSight for the measurement of nanoparticles and protein aggregates. Springer **27**, 796–810 (2010). https://doi.org/10.1007/S11095-010-0073-2
67. Shekunov, B., Pharmaceutical BS-E of.: undefined (2020) Crystal Form Analysis of Drug Substances. api.taylorfrancis.com 723–741 (2013). https://doi.org/10.1081/E-EPT4-120050229
68. Upadhyay, S., Parekh, K., Pandey, B.: Influence of crystallite size on the magnetic properties of Fe3O4 nanoparticles. J. Alloys Compd. **678**, 478–485 (2016). https://doi.org/10.1016/J.JALLCOM.2016.03.279
69. Turtelli, R., Pascarelli, S., Ruffoni, M., ... RG-J of M.: undefined Extended X-ray absorption fine structure (EXAFS) and X-ray absorption near-edge structure (XANES) study of melt-spun Fe85Ga15 ribbons. ElsevierRS Turtelli, S Pascarelli, M Ruffoni, R Grössinger, C Eisenmenger-SittnerJournal of Magnetism and Magnetic Materials, 2008•Elsevier (2008)
70. Sharma, A., Varshney, M., Park, J., Ha, T.K., Chae, K.H., Shin, H.J.: XANES, EXAFS and photocatalytic investigations on copper oxide nanoparticles and nanocomposites. RSC Adv. **5**, 21762–21771 (2015). https://doi.org/10.1039/C4RA16217J
71. Jabeen, S., Veg, E., Ahmad, M.I., Bala, S., Khan, T.: A comprehensive Review on metal oxide-based nanomaterials: synthesis, characterization, environmental, and therapeutic applications. Wiley Online Library **10**,(2025). https://doi.org/10.1002/SLCT.202500080
72. Krishnan, V., Selvan, R.K., Augustin, C.O., Gedanken, A., Bertagnolli, H.: EXAFS and XANES investigations of $CuFe_2O_4$ nanoparticles and $CuFe_2O_4-MO2$ (M = Sn, Ce) Nanocomposites. ACS Publications **111**, 16724–16733 (2007). https://doi.org/10.1021/JP073746T
73. Reviews EM-RC, 2011 undefined (2011) X-Ray diffraction structure diagnostics of nanomaterials. iopscience.iop.orgEM MorozRussian Chemical Reviews, 2011•iopscience.iop.org **80**, 293–312. https://doi.org/10.1070/RC2011V080N04ABEH004163/META
74. Lee, P., B GB-PR, 1977 undefined (1977) New method for the calculation of atomic phase shifts: application to extended X-ray absorption fine structure (EXAFS) in molecules and crystals. APS **15**, 2862–2883 (1977). https://doi.org/10.1103/PHYSREVB.15.2862
75. Ghirardello, M., Candeo, A., Ardini, B., Valentini, G., Manzoni, C., Calligaro, T., Pichon, L., Bai, X., Lenz, R., Alberti, R., Gironda, M., Comelli, D.: Time-resolved photoluminescence imaging for the mapping of weakly luminescent pigments in paintings. Springer **138**, 906 (2023). https://doi.org/10.1140/EPJP/S13360-023-04485-1

76. Giovannantonio, M. Di, Contini, G., Lim, H., Dewaraja, Y.K., Maccallini, E., Manini, P., Conte, A., Siviero, F., Bonucci SAES, A.:, SpA G New Approach to meet vacuum requirements in UHV/XHV systems by Non Evaporable Getter Technology. iopscience.iop.org. https://doi.org/10.1088/1742-6596/390/1/012006/META
77. Glasscott, M.W., Pendergast, A.D., Choudhury, M.H., Dick, J.E.: Advanced characterization techniques for evaluating porosity, nanopore tortuosity, and electrical connectivity at the single-nanoparticle level. ACS Publ. **2**, 819–830 (2019). https://doi.org/10.1021/ACSANM.8B02051
78. Wang, Y.C., Engelhard, M.H., Baer, D.R., Castner, D.G.: Quantifying the impact of nanoparticle coatings and nonuniformities on XPS analysis: gold/silver core–shell nanoparticles. ACS Publ. **88**, 3917–3925 (2016). https://doi.org/10.1021/ACS.ANALCHEM.6B00100
79. Research CF-NI and M in P.: Undefined X-ray photoelectron spectroscopy: From origins to future directions. Elsevier (2009)
80. Liu, Y., Su, X., Ding, J., Zhou, J., Liu, Z., … XW-CS.: Undefined Progress and challenges in structural, in situ and operando characterization of single-atom catalysts by X-ray based synchrotron radiation techniques. pubs.rsc.org (2024)
81. Stevie, F., A CD-J of VS& T, 2020 undefined (2020) Introduction to x-ray photoelectron spectroscopy. pubs.aip.org **38**, 63204. https://doi.org/10.1116/6.0000412
82. Wang, X., Nature, L.W.: Undefined (1999) Observation of negative electron-binding energy in a molecule. nature.com (1999)
83. Chemical PN-J of MCA, 2005 undefined Materials characterization by X-ray photoelectron spectroscopy. Elsevier
84. Hawn, D.D., Cieslinski, R.C., Klassen, H.E.: Application of surface analysis techniques in the study of catalyst systems. American Chem. Soc. Div. Petrol. Chem. Preprints **29**, 705 (1985). https://doi.org/10.1021/BK-1985-0288.CH004
85. Chemistry US-A and bioanalytical.: undefined (2013) Sputter-induced cross-contamination in analytical AES and XPS instrumentation: utilization of the effect for the in situ deposition of ultrathin functional layers. Springer 405, 7145–7151 (2013). https://doi.org/10.1007/S00216-013-6840-2
86. Polaschegg, H.D., Jungel, A., Schirk, E.: A high pressure preparation lock for surface analysis. Wiley Online Library **1**, 100–103 (1979). https://doi.org/10.1002/SIA.740010307
87. Rutkowski, M., McNicholas, K., … ZZ-R of S.: Undefined Design of an ultrahigh vacuum transfer mechanism to interconnect an oxide molecular beam epitaxy growth chamber and an x-ray photoemission spectroscopy. pubs.aip.org (2013)
88. Tauber, G., Instruments HV-R of S.: Undefined Ultrahigh vacuum lock. ui.adsabs.harvard.edu (1976)
89. Arndt, J., Stockert, O., Borth, R.: Soft x-ray angle-resolved and resonance photoemission study of CeCu2Ge2 and LaCu2Ge2. iopscience.iop.org 592:12003 (2015). https://doi.org/10.1088/1742-6596/592/1/012003/META
90. Felser, C., Zhang, S., Yan, B., Wang, Y., Gedik, N.: Circular dichroism in angle-resolved photoemission spectroscopy of topological insulators. Wiley Online Library **7**, 64–71 (2013). https://doi.org/10.1002/PSSR.201206458
91. Park, S.R., Song, D.J., Leem, C.S., Kim, C., Kim, C., Kim, B.J., Eisaki, H.: Angle-resolved photoemission spectroscopy of electron-doped cuprate superconductors: isotropic electron-phonon coupling. APSSR Park, DJ Song, CS Leem, C Kim, C Kim, BJ Kim, H EisakiPhysical review letters, 2008•APS 101 (2008). https://doi.org/10.1103/PHYSREVLETT.101.117006
92. Vinai, G., Motti, F., Petrov, A., … VP-R of S.: Undefined An integrated ultra-high vacuum apparatus for growth and in situ characterization of complex materials. pubs.aip.org (2020)
93. Brundle, C.R.: Ultra-high vacuum techniques of surface characterization. ACS Symp. Ser. **13–32**,(1982). https://doi.org/10.1021/BK-1982-0199.CH002
94. Scheithauer, U.: Vacuum Quality Monitoring of Analytical Sputter Depth Profile Equipment by Recontamination Measurements of a Sputtered Ti Surface (2017)
95. Vacuum, R.R.: undefined Ultra high vacuum aspects of the investigation of surface phenomena. Elsevier (1971)

96. Comini, N., Huthwelker, T., ... JD-J of V: Undefined Factors influencing surface carbon contamination in ambient-pressure x-ray photoelectron spectroscopy experiments. pubs.aip.org (2021)
97. Joyner, R., Roberts, M., Science KY-S.: Undefined A "high-pressure" electron spectrometer for surface studies. Elsevier (1979)
98. Pantförder, J., Pöllmann, S., ... JZ-R of scientific: Undefined New setup for in situ x-ray photoelectron spectroscopy from ultrahigh vacuum to 1mbar. pubs.aip.org (2005)
99. Offi, F., Fondacaro, A., Paolicelli, G., ... ADL-NI and: Undefined Design and test of a lens system for a high energy and high resolution electron spectrometer. Elsevier (2005)
100. Shiraki, S., Ishii, H., Related YN-J of ES and: undefined Design of a new high angle-resolving electron energy analyzer. Elsevier (2001)
101. Gelius, U., Asplund, L., Basilier, E., ... SH-NI and: Undefined A high resolution multipurpose ESCA instrument with X-ray monochromator. Elsevier (1984)
102. Anwar, M., Abdul, M.S, Khan, U., Materials, M.H.: Undefined A review of X-ray photoelectron spectroscopy technique to analyze the stability and degradation mechanism of solid oxide fuel cell cathode materials. mdpi.com (2022)
103. Passos, M., Measurement MS: undefined Detection in UV-visible spectrophotometry: Detectors, detection systems, and detection strategies. Elsevier (2019)
104. Conceição, F. Da, ... PB-J of P: Undefined Fundamental studies to develop certified reference material to calibrate spectrophotometer in the ultraviolet region. iopscience.iop.org (2016). https://doi.org/10.1088/1742-6596/733/1/012016/META
105. Vigata, M., Meinert, C., Hutmacher, D., Pharmaceutics, N.B.: Undefined Hydrogels as drug delivery systems: A review of current characterization and evaluation techniques. mdpi.com (2020)
106. Sciences JØ-AT in the P.: Undefined (2016) UV/VIS spectrophotometry and UV imaging. Springer, 3–27 (2016). https://doi.org/10.1007/978-1-4939-4029-5_1
107. Dadi, M., Colorimetry, M.Y.: Undefined Spectroscopy and Spectrophotometry: Principles and Applications for Colorimetric. books.google.com (2022)
108. Zhang, Y., Kubicki, J., Platz, M.S.: Ultrafast UV−visible and infrared spectroscopic observation of a singlet vinylcarbene and the intramolecular cyclopropenation reaction. ACS Publ. **131**, 13602–13603 (2009). https://doi.org/10.1021/JA905992P
109. Ahmad, I., Sheraz, M, ... SA-P of D: Undefined Multicomponent spectrometric analysis of drugs and their preparations. books.google.com (2018)
110. Jia, J., Zhou, X., Granjeiro, J.M., Rosa, A., Ribeiro, L., Sanches, P.L., De Oliveira, R., Geaquinto, L., Cruz, R., Schuck, D.C., Lorencini, M.: Titanium dioxide and zinc oxide nanoparticles in sunscreens: focus on their safety and effectiveness. Taylor & Francis **4**, 95–112 (2011). https://doi.org/10.2147/NSA.S19419
111. Khurana, K., Jaggi, N.: Localized surface plasmonic properties of Au and Ag nanoparticles for sensors: a review. Springer **16**, 981–999 (2021). https://doi.org/10.1007/S11468-021-01381-1
112. Behzadi, S., Ghasemi, F., Ghalkhani, M., Ashkarran, A.A., Akbari, S.M., Pakpour, S., Hormozi-Nezhad, M.R., Jamshidi, Z., Mirsadeghi, S., Dinarvand, R., Atyabi, F., Mahmoudi, M.: Determination of nanoparticles using UV-Vis spectra. Nanoscale **7**, 5134–5139 (2015). https://doi.org/10.1039/C4NR00580E
113. Joshi, R., Reviews JS-CS.: undefined Assembly of one dimensional inorganic nanostructures into functional 2D and 3D architectures. Synthesis, arrangement and functionality. pubs.rsc.org (2012)
114. Ashoka, A.H., Aparin, I.O., Reisch, A., Klymchenko, A.S.: Brightness of fluorescent organic nanomaterials. pubs.rsc.org 52:4525 (2023). https://doi.org/10.1039/d2cs00464j
115. Cao, L., Meziani, M.J., Sahu, S., Sun, Y.P.: Photoluminescence properties of graphene versus other carbon nanomaterials. ACS Publ. **46**, 171–182 (2013). https://doi.org/10.1021/AR300128J
116. Zacharioudaki, D., Fitilis, I., Molecules, M.K.: undefined Review of fluorescence spectroscopy in environmental quality applications. mdpi.com (2022)

117. Gutiérrez, L., Cueva, L.D. la, Moros, M, … EM.: Undefined (2019) Aggregation effects on the magnetic properties of iron oxide colloids. iopscience.iop.org. https://doi.org/10.1088/1361-6528/AAFBFF/META
118. Siraj, N., El-Zahab, B., Hamdan, S., Karam, T.E., Haber, L.H., Li, M., Fakayode, S.O., Das, S., Valle, B., Strongin, R.M., Patonay, G., Sintim, H.O., Baker, G.A., Powe, A., Lowry, M., Karolin, J.O., Geddes, C.D., Warner, I.M.: Fluorescence, phosphorescence, and chemiluminescence. ACS Publ. **88**, 170–202 (2016). https://doi.org/10.1021/ACS.ANALCHEM.5B04109
119. (a) MR status solidi, 2023 undefined (2023) Photoluminescence from vacancy-containing defects in GaN. Wiley Online Library 220. https://doi.org/10.1002/PSSA.202200402
120. Moffatt, J.E., Tsiminis, G., Klantsataya, E., de Prinse, T.J., Ottaway, D., Spooner, N.A.: A practical review of shorter than excitation wavelength light emission processes. Taylor & Francis **55**, 327–349 (2020). https://doi.org/10.1080/05704928.2019.1672712
121. Comedi, D., Zalloum, O.H.Y., Irving, E.A., Wojcik, J., Roschuk, T., Flynn, M.J., Mascher, P.: X-ray-diffraction study of crystalline Si nanocluster formation in annealed silicon-rich silicon oxides. pubs.aip.org (2006). https://doi.org/10.1063/1.2162989
122. Skurlov, I.D., Onishchuk, D.A., Parfenov, P.S., Litvin, A.P.: An experimental setup for analysis of weak photoluminescence in the near-infrared spectral region. Springer **125**, 756–759 (2018). https://doi.org/10.1134/S0030400X18110279
123. Lee, H.-N., Chang, C.-C.: On the estimation of target spectrum for filter-array based spectrometers. Optics Express **16**(2), 1056–1061 16:1056–1061 (2008). https://doi.org/10.1364/OE.16.001056
124. Nevin, A., Cesaratto, A., Bellei, S., D'Andrea, C., Sensors, L.T.: undefined Time-resolved photoluminescence spectroscopy and imaging: new approaches to the analysis of cultural heritage and its degradation. mdpi.com (2014)
125. Jung, H., Kang, S., Lee, H., Yu, Y.J., Jeong, J.H., Song, J., Jeon, Y., Park, J.: High efficiency and long lifetime of a fluorescent blue-light emitter made of a Pyrene core and optimized side groups. ACS Appl. Mater. Interfaces **10**, 30022–30028 (2018). https://doi.org/10.1021/ACSAMI.8B09013
126. Zalloum, O.H.Y., Flynn, M., Roschuk, T., Wojcik, J., Irving, E., Mascher, P.: Laser photoluminescence spectrometer based on charge-coupled device detection for silicon-based photonics. Rev. Sci. Instrum. **77**,(2006). https://doi.org/10.1063/1.2173030
127. And PG-AA of F-I to M, 1980 undefined (1980) Accessories and Sampling Techniques for FT-IR Spectrometry. SpringerPR GriffithsAnalytical Applications of FT-IR to Molecular and Biological Systems, 141–148. Springer (1980). https://doi.org/10.1007/978-94-009-9070-8_9
128. Lagos, M., Paredes, R., Retamal, C.: Fourier Transform Spectrometry with Fourier Analysis of the Interferogram as Just an Optional Tool. ACS PublicationsM Lagos, R Paredes, C RetamalACS omega, 2018•ACS Publications 3:18258–18262 (2018). https://doi.org/10.1021/ACSOMEGA.8B02914
129. Sheikh-Sofla, S., Neshat, M.: Theory and Instrumentation of Fourier Transform Spectroscopy (2023)
130. Childs, D.T.D., Hogg, R.A., Revin, D.G., Rehman, I.U., Cockburn, J.W., Matcher, S.J.: Sensitivity advantage of QCL tunable-laser mid-infrared spectroscopy over FTIR spectroscopy. Taylor & FrancisDTD Childs, RA Hogg, DG Revin, IU Rehman, JW Cockburn, SJ Matcher-Applied Spectroscopy Reviews, 2015•Taylor & Francis 50:822–839 (2015). https://doi.org/10.1080/05704928.2015.1075208
131. Köhler, M., Nguyen, T., Kienle, P, … XD-O: Undefined Hyperspectral imager for the mid-infrared spectral range using a single-mirror interferometer and a windowing method. opg.optica.org MH Köhler, TT Nguyen, P Kienle, X Dong, M Schardt, M Jakobi, AW KochOSA Continuum, 2019•opg.optica.org (2019)
132. Zorin, I., Gattinger, P., Ricchiuti, G., Lendl, B., Heise, B., Brandstetter, M.: All-mirror wavefront division interferometer for mid-infrared spectrometry (2024). https://doi.org/10.1364/OE.545267

133. Strelnikov, D., Kern, B., … CS-R of S.: undefined (2016) 15 cm− 1 to 12 000 cm− 1 spectral coverage without changing optics: Diamond beam splitter adaptation of an FTIR spectrometer. pubs.aip.org (2017)
134. King, B., Ellis, T., Spectroscopic TO-N-G.: Undefined Performance and applications of a hypertemporal hyperspectral Fourier-transform infrared spectroradiometer. spiedigitallibrary.orgBH King, T Ellis, TE OldNext-Generation Spectroscopic Technologies II, 2009•spiedigitallibrary.org (2009). https://doi.org/10.1117/12.818654.SHORT
135. Fathy, A., Sabry, Y.M., Nazeer, S., Bourouina, T., Khalil, D.A.: On-chip parallel Fourier transform spectrometer for broadband selective infrared spectral sensing. nature.comA Fathy, YM Sabry, S Nazeer, T Bourouina, DA KhalilMicrosystems & Nanoengineering. nature.com, 1234567890 (2020). https://doi.org/10.1038/s41378-019-0111-0
136. Köhler, M.H., Schardt, M., Ghazala, H.B., Colicchia, E., Kienle, P., Dong, X., Wang, K., Koch, A.W.: Static Fourier transform mid-infrared spectrometer with continuous background correction. spiedigitallibrary.orgMH Köhler, M Schardt, HB Ghazala, E Colicchia, P Kienle, X Dong, K Wang, AW KochApplied Optical Metrology III, 2019•spiedigitallibrary.org 43 (2019). https://doi.org/10.1117/12.2528404
137. Sabry, Y.M., Hassan, K., Anwar, M., Alharon, M.H., Medhat, M., Adib, G.A., Dumont, R., Saadany, B., Khalil, D.: Ultra-compact MEMS FTIR spectrometer. spiedigitallibrary.org YM Sabry, K Hassan, M Anwar, MH Alharon, M Medhat, GA Adib, R Dumont, B SaadanyNext-generation spectroscopic technologies X, 2017•spiedigitallibrary.org 10210:102100H (2017). https://doi.org/10.1117/12.2268078
138. Khalili, D.: Graphene oxide: A promising carbocatalyst for the regioselective thiocyanation of aromatic amines, phenols, anisols and enolizable ketones by hydrogen peroxide/KSCN in water. New J. Chem. **40**, 2547–2553 (2016). https://doi.org/10.1039/c5nj02314a
139. Marbella, L.E., Millstone, J.E.: NMR techniques for noble metal nanoparticles. ACS Publ. **27**, 2721–2739 (2015). https://doi.org/10.1021/CM504809C
140. Marbella, L. NMR characterization of metal nanoparticle formation, structure, and performance (2016)
141. Khare, E.: Nuclear magnetic resonance (2015)
142. Harris, R.K., Becker, E.D., Cabral De Menezes, S.M., Granger, P., Hoffman, R.E., Zilm, K.W.: Further conventions for NMR shielding and chemical shifts (IUPAC Recommendations 2008). degruyterbrill.com **80**, 59–84 (2008). https://doi.org/10.1351/PAC200880010059/HTML
143. Takeda, K., Ichijo, N., Noda, Y., resonance KT-J of magnetic: Undefined Elemental analysis by NMR. Elsevier (2012)
144. Hall, D.G., Manku, S., Wang, F.: Solution-and solid-phase strategies for the design, synthesis, and screening of libraries based on natural product templates: a comprehensive survey. ACS Publ. **3**, 125–150 (2001). https://doi.org/10.1021/CC0001001
145. biomolecular API-A review of biophysics and, 2001 undefined (2001) NMR probes of molecular dynamics: overview and comparison with other techniques. annualreviews.org 30, 129–155. https://doi.org/10.1146/ANNUREV.BIOPHYS.30.1.129
146. Maity, S., Gundampati, R.K., Kumar, T.K.S.: NMR methods to characterize protein-ligand interactions. journals.sagepub.com 14 (2019). https://doi.org/10.1177/1934578X19849296
147. Fan, T., spectroscopy AL-P in nuclear magnetic resonance: Undefined Applications of NMR spectroscopy to systems biochemistry. Elsevier (2016)
148. Günther, H.: NMR spectroscopy: basic principles, concepts and applications in chemistry (2013)
149. Babu, P.K., Chung, J.H., Kuk, S.T., Kobayashi, T., Oldfield, E., Wieckowski, A.: Metallic nature and surface diffusion of CO adsorbed on Ru nanoparticles in aqueous media: A 13C NMR study. ACS Publ. **109**, 2474–2477 (2005). https://doi.org/10.1021/JP040729K
150. Witherspoon, V.J., Xu, J., Reimer, J.A.: Solid-state NMR investigations of carbon dioxide gas in metal–organic frameworks: insights into molecular motion and adsorptive behavior. ACS Publ. **118**, 10033–10048 (2018). https://doi.org/10.1021/ACS.CHEMREV.7B00695
151. Dumas, C., Meledandri, C.J.: Insights into the partitioning behavior of secondary surfactants in a microemulsion-based synthesis of metal nanoparticles: a DLS and 2D NMR spectroscopic. ACS Publ. **31**, 7193–7203 (2015). https://doi.org/10.1021/ACS.LANGMUIR.5B00417

152. Kovtunov, K.V., Pokochueva, E.V., Salnikov, O.G., Cousin, S.F., Kurzbach, D., Vuichoud, B., Jannin, S., Chekmenev, E.Y., Goodson, B.M., Barskiy, D.A., Koptyug, I.V.: Hyperpolarized NMR spectroscopy: d-DNP, PHIP, and SABRE techniques. Wiley Online Library **13**, 1857–1871 (2018). https://doi.org/10.1002/ASIA.201800551
153. Zhang, W., Xu, S., Han, X.: Reviews XB-CS, 2012 undefined In situ solid-state NMR for heterogeneous catalysis: a joint experimental and theoretical approach. pubs.rsc.org
154. Porter, C.J., Beynon, J.H., Ast, T.: The modern mass spectrometer—a complete chemical laboratory. Wiley Online Library **16**, 101–114 (1981). https://doi.org/10.1002/OMS.1210160302
155. Li, C., Chu, S., Tan, S., Yin, X., Jiang, Y., Dai, X., Gong, X., Fang, X., Tian, D.: Towards higher sensitivity of mass spectrometry: a perspective from the mass analyzers. frontiersin.org 9 (2021). https://doi.org/10.3389/FCHEM.2021.813359/FULL
156. Paré, J.: Analytical VY-T and instrumentation in, 1997 undefined Mass spectrometry: principles and applications. Elsevier
157. Arigò, A., Famiglini, G., Marittimo, N., Agostini, M., Renzoni, C., Palma, P., Cappiello, A.: (123AD) Extractive-liquid sampling electron ionization-mass spectrometry (E-LEI-MS): a new powerful combination for direct analysis. nature.com. https://doi.org/10.1038/s41598-023-33647-5
158. Shi, X., Xi, B., Jasbi, P., Turner, C., Jin, Y., Gu, H.: Comprehensive isotopic targeted mass spectrometry: reliable metabolic flux analysis with broad coverage. ACS Publ. **92**, 11728–11738 (2020). https://doi.org/10.1021/ACS.ANALCHEM.0C01767
159. Medicine WZ-N impacts on cell biology and, 2014 undefined (2014) Nanoparticle aggregation: principles and modeling. Springer 811, 20–43. https://doi.org/10.1007/978-94-017-8739-0_2
160. Baba, T., Campbell, J.L., Le Blanc, J.C.Y., Baker, P.R.S.: Distinguishing cis and trans isomers in intact complex lipids using electron impact excitation of ions from organics mass spectrometry. ACS Publ. **89**, 7307–7315 (2017). https://doi.org/10.1021/ACS.ANALCHEM.6B04734
161. Spectrometry JB-BM, 1981 undefined (1981) The history of mass spectrometry and the search for zero. Wiley Online Library 8:380–383. https://doi.org/10.1002/BMS.1200080906
162. Laboratory GS-JJ of the A for.: Undefined (2004) An Introduction to Mass Spectrometry Ionization: An Excerpt from The Expanding Role of Mass Spectrometry in Biotechnology, 2nd ed.; MCC Press: San Diego, 2005. journals.sagepub.com **9**, 50–63 (2004). https://doi.org/10.1016/J.JALA.2004.01.004
163. Ribeiro, C.: Environmental MT-IJ of, 2015 undefined (2015) Development and validation of a gas chromatography mass spectrometry method for the analysis of phytoestrogens, phytosterols and mycotoxins in estuarine water. Taylor & Francis **95**, 187–202. https://doi.org/10.1080/03067319.2014.1002492
164. Ribeiro, H., Klassen, A., Pedrini, M., … MC-P.: undefined A preliminary study of bipolar disorder type I by mass spectrometry-based serum lipidomics. Elsevier (2017)
165. Barrow, M., Headley, J., Peru, K., Chromatography, P.D.-J.: Undefined Fourier transform ion cyclotron resonance mass spectrometry of principal components in oilsands naphthenic acids. Elsevier (2004)
166. Marshall, A., … CH-M spectrometry: Undefined Fourier transform ion cyclotron resonance mass spectrometry: a primer. Wiley Online Library (1998). https://doi.org/10.1002/(SICI)1098-2787(1998)17:1
167. Zhang, Z., Pan, H., Chen, X.: Mass spectrometry for structural characterization of therapeutic antibodies. Wiley Online Library **28**, 147–176 (2009). https://doi.org/10.1002/MAS.20190
168. Preparation AH-M proteomics–Sample, and analysis, 2016 undefined (2016) Mass analyzers and mass spectrometers. Springer 919, 157–169. https://doi.org/10.1007/978-3-319-41448-5_7
169. Cannataro, M., Guzzi, P.H., Mazza, T., Tradigo, G., Veltri, P.: On the preprocessing of mass spectrometry proteomics data. Springer 3931 LNCS:127–131 (2006). https://doi.org/10.1007/11731177_19

170. Liu, C.: Discovery HZ-E opinion on drug, 2024 undefined (2024) Data processing for high-throughput mass spectrometry in drug discovery. Taylor & Francis **19**, 815–825. https://doi.org/10.1080/17460441.2024.2354871
171. Meermann, B., Spectrometry VN-J of AA.: undefined (2016) ICP-MS for the analysis at the nanoscale–a tutorial review. pubs.rsc.org **31**, 1–346 (2018). https://doi.org/10.1039/C8JA00037A
172. Bolea, E., Jimenez, M., Perez-Arantegui, J, ... JV-A.: Undefined Analytical applications of single particle inductively coupled plasma mass spectrometry: a comprehensive and critical review. pubs.rsc.org (2021). https://doi.org/10.1039/D1AY00761K
173. Foisner, E.: Analysis of nanoparticle composition using laser ablation inductively coupled plasma quadrupole mass spectrometry (2025).https://doi.org/10.34726/hss.2025.131408
174. Montaño, M.D., Olesik, J.W., Barber, A.G., Challis, K., Ranville, J.F.: Single Particle ICP-MS: advances toward routine analysis of nanomaterials. Springer **408**, 5053–5074 (2016). https://doi.org/10.1007/S00216-016-9676-8
175. Pan, L., Dai, F., Li, G., Energy, S.L.: Undefined a TGA/DTA-MS investigation to the influence of process conditions on the pyrolysis of Jimsar oil shale. Elsevier (2015)
176. Nanoparticles FB-C of, 2020 undefined Dynamic light scattering (DLS). Elsevier
177. Kaszuba, M., McKnight, D., Connah, M.T., McNeil-Watson, F.K., Nobbmann, U.: Measuring sub nanometre sizes using dynamic light scattering. Springer **10**, 823–829 (2008). https://doi.org/10.1007/S11051-007-9317-4
178. Stetefeld, J., McKenna, S.A., Patel, T.R.: Dynamic light scattering: a practical guide and applications in biomedical sciences. Springer **8**, 409–427 (2016). https://doi.org/10.1007/S12551-016-0218-6
179. Technology VU-J of dispersion science and, 2012 undefined (2012) Dynamic light scattering based microelectrophoresis: main prospects and limitations. Taylor & Francis **33**, 1762–1786. https://doi.org/10.1080/01932691.2011.625523
180. Human JW-DLSS of the: Undefined (2022) Dynamic Light Scattering (DLS) Spectroscopy, 13–17. Springer (2022). https://doi.org/10.1007/978-3-031-06624-5_3
181. Bremer, L.G.B., Deriemaeker, L., Finsy, R., Geladé, E., Joosten, J.G.H.: Fiber optic dynamic light scattering, neither homodyne nor heterodyne. ACS Publ. **9**, 2008–2014 (1993). https://doi.org/10.1021/LA00032A019
182. International FH-F research, 1994 undefined Particle size analysis by dynamic light scattering. Elsevier
183. Hassan, Z.U., Abbas, Z., Bakht, K., Ayoub, M.H., Ahmad, S., Khan, A.M., Farooq, U., Khan, M.S., Shaikh, A.J.: Dynamic light scattering and zeta-potential as a tool for understanding the mechanism of pesticides binding toward individual components of transition metal. Taylor & Francis **57**, 932–947 (2022). https://doi.org/10.1080/03601234.2022.2147348
184. Alexander, M., Hallett, F.R., Dalgleish, D.G.: Design of a small angle spetrometer: application to milk casein. ACS Symp. Ser. **693**, 2–13 (1998). https://doi.org/10.1021/BK-1998-0693.CH001
185. Jin, L., Jarand, C., Technology, M.B. ... S and: Undefined Angle-dependent effects in DLS measurements of polydisperse particles. iopscience.iop.org (2022). https://doi.org/10.1088/1361-6501/AC42B2/META
186. Chastek, T., Beers, K., Instruments EA-R of S.: Undefined (2007) Miniaturized dynamic light scattering instrumentation for use in microfluidic applications. pubs.aip.org (2007). https://doi.org/10.1063/1.2755569
187. Charsooghi, M., Akhlaghi, E, ... ST-CP: Undefined A MATLAB program to calculate translational and rotational diffusion coefficients of a single particle. Elsevier (2011)
188. Marucco, A., Aldieri, E., Leinardi, R., Materials, E.B.: Undefined Applicability and limitations in the characterization of poly-dispersed engineered nanomaterials in cell media by dynamic light scattering (DLS). mdpi.com (2019)
189. Shukla, V., Niveria, K, ... PS-AT.: Undefined Dynamic light scattering (DLS) particle size analysis for biomedical nanotechnology. iopscience.iop.org (2023)

190. Prague BM-UUDCTU, 2017 undefined Characterization of size of polydisperse nanoparticles of various shape: Comparative study of AFM and DLS methods. core.ac.uk
191. Tomaszewska, E., Soliwoda, K., Kadziola, K., Tkacz-Szczesna, B., Celichowski, G., Cichomski, M., Szmaja, W., Grobelny, J.: Detection limits of Dls and Uv-Vis spectroscopy in characterization of polydisperse nanoparticles colloids. Wiley Online Library **2013**,(2013). https://doi.org/10.1155/2013/313081
192. James, A., Analyst JD: Undefined Monitoring gold nanoparticle conjugation and analysis of biomolecular binding with nanoparticle tracking analysis (NTA) and dynamic light scattering (DLS). pubs.rsc.org (2013)
193. Libralato, G., Galdiero, E., Falanga, A., Molecules, R.C.: Undefined Toxicity effects of functionalized quantum dots, gold and polystyrene nanoparticles on target aquatic biological models: a review. mdpi.com (2017)
194. Gallego-Urrea, J., Tuoriniemi, J., Analytical MH-TT in: Undefined Applications of particle-tracking analysis to the determination of size distributions and concentrations of nanoparticles in environmental, biological and food samples. Elsevier (2011)
195. Ryu, J.H., Park, G.S., Kim, K.M., Lim, C.S., Yoon, J.W., Shim, K.B.: Synthesis of CaWO4 nanocolloidal suspension via pulsed laser ablation and its optical properties. Springer **88**, 731–736 (2007). https://doi.org/10.1007/S00339-007-4051-4
196. Van Der Pol, E., Coumans, F.A.W., Sturk, A., Nieuwland, R., Van Leeuwen, T.G.: Refractive index determination of nanoparticles in suspension using nanoparticle tracking analysis. ACS Publ. **14**, 6195–6201 (2014). https://doi.org/10.1021/NL503371P
197. Saadatkhah, N., Carillo Garcia, A., Ackermann, S., Leclerc, P., Latifi, M., Samih, S., Patience, G.S., Chaouki, J.: Experimental methods in chemical engineering: thermogravimetric analysis—TGA. Wiley Online Library **98**, 34–43 (2020). https://doi.org/10.1002/CJCE.23673
198. Faroque, F., Garimella, A., Science SN-J of C: undefined Analysis and Modeling of Thermogravimetric Curves of Chemically Modified Wheat Straw Filler-Based Biocomposites Using Machine Learning Techniques. mdpi.com (2025)
199. Yiga, V.A., Katamba, M., Lubwama, M., Adolfsson, K.H., Hakkarainen, M., Kamalha, E.: Combustion, kinetics and thermodynamic characteristics of rice husks and rice husk-biocomposites using thermogravimetric analysis. Springer **148**, 11435–11454 (2023). https://doi.org/10.1007/S10973-023-12458-W
200. Devaraju, S., Vengatesan, M.R., Ashok Kumar, A., Alagar, M.: Polybenzoxazine–silica (PBZ–SiO2) hybrid nanocomposites through in situ sol–gel method. Springer **60**, 33–40 (2011). https://doi.org/10.1007/S10971-011-2547-Z
201. Subramanian, N., Paramashivan, S.S., Mahadevan, S., Renganathan, V.: Thermal analysis of kerosene with 1, 2-methyl ethyl 1-methyl pyrrolidinium bis (trifluoromethanesulfonyl) amide as an additive. Springer **141**, 2347–2354 (2020). https://doi.org/10.1007/S10973-020-09792-8
202. Mourdikoudis, S., Pallares, R., Nanoscale, N.T.: Undefined (2018) Characterization techniques for nanoparticles: comparison and complementarity upon studying nanoparticle properties. pubs.rsc.org **10**, 12871 (2018). https://doi.org/10.1039/c8nr02278j
203. Nickle, S., Meyers, K., Conference LN-SAT: undefined Shortcomings in the use of TGA/DSC techniques to evaluate in-situ combustion. onepetro.org (1987)
204. Smith, R.D., Shen, Y., Tang, K.: Ultrasensitive and quantitative analyses from combined separations– mass spectrometry for the characterization of proteomes. ACS Publ. **37**, 269–278 (2004). https://doi.org/10.1021/AR0301330
205. Rashid, M.H., Sen, P.: Recent Advancements in Biosensors for the Detection and Characterization of Amyloids: A Review. Springer 43, 656–674 (2024). https://doi.org/10.1007/S10930-024-10205-0
206. Kisielowski, C., Specht, P., Freitag, B, … EK-M and: Undefined Coherence and Inelastic Scattering in Electron Microscopy. cambridge.org (2022)
207. Winey, M., Meehl, J.B., O'Toole, E.T., Giddings, T.H.: Conventional transmission electron microscopy. molbiolcell.org **25**, 319–323 (2014). https://doi.org/10.1091/MBC.E12-12-0863

208. Williams, D., for CC-T electron microscopy: a textbook, 2009 undefined (2009) High-resolution TEM, 483–509. Springer. https://doi.org/10.1007/978-0-387-76501-3_28
209. Williams, D., for CC-TEMAT, 2009 undefined (2009) Diffraction in TEM. Springer 197–209. https://doi.org/10.1007/978-0-387-76501-3_11
210. Franken, R., Tromp, P., Kringlen Ervik, T., Staff, J., Alstrup Jensen, K., Eypert-Blaison, C., Brostrøm, A., Cannizzaro, A., Teresa Sanchez Cabo, M., Rosaria Bruno, M., Sofia Fonseca, A., Davies, L., Graff, P., Spaan, S.: European harmonization of asbestos exposure assessment: comparing PCM, SEM, and TEM to derive conversion factors. academic.oup.com 69, 575–591 (2025). https://doi.org/10.1093/ANNWEH/WXAF023/8148733
211. Zuo, J., Spence, J.: Advanced transmission electron microscopy (2017)
212. Preparation RS-TEMS, 2025 undefined (2024) Introduction to Transmission Electron Microscopy (TEM). Springer, 1–20. https://doi.org/10.1007/978-3-031-82967-3_1
213. Erts, D., Lõhmus, A., Lõhmus, R., Olin, H.: Instrumentation of STM and AFM combined with transmission electron microscope. Springer **72**,(2001). https://doi.org/10.1007/S003390100636
214. Preparation RS-TEMS, 2025 undefined (2024) Unique Case Studies in TEM Instrumentation, Analysis and Interpretation. Springer, 289–317. https://doi.org/10.1007/978-3-031-82967-3_11
215. Javed, Y., Ali, K., Akhtar, K., Hussain, J.M.I., Ahmad, G., Arif, T.: TEM for atomic-scale study: fundamental, instrumentation, and applications in nanotechnology. Springer **147–216**,(2018). https://doi.org/10.1007/978-3-319-92955-2_5
216. Hodoroaba, V., Motzkus, C, … TM-M.: Undefined of high-resolution SEM/EDX systems equipped with transmission mode (TSEM) for imaging and measurement of size and size distribution of spherical nanoparticles. academic.oup.com (2014)
217. Technology SZ-CR and, 2011 undefined (2011) A critical review of orientation microscopy in SEM and TEM. Wiley Online Library **46**, 607–628. https://doi.org/10.1002/CRAT.201100125
218. Li, T., Senesi, A.J., Lee, B.: Small angle X-ray scattering for nanoparticle research. ACS Publ. **116**, 11128–11180 (2016). https://doi.org/10.1021/ACS.CHEMREV.5B00690
219. Peddis, D., Cannas, C., Musinu, A., Ardu, A., Orruì, F., Fiorani, D., Laureti, S., Rinaldi, D., Muscas, G., Concas, G., Piccaluga, G.: Beyond the effect of particle size: influence of $CoFe_2O_4$ nanoparticle arrangements on magnetic properties. ACS Publ. **25**, 2005–2013 (2013). https://doi.org/10.1021/CM303352R
220. Kairdolf, B.A., Qian, X., Nie, S.: Bioconjugated nanoparticles for biosensing, in vivo imaging, and medical diagnostics. ACS Publ. **89**, 1015–1031 (2017). https://doi.org/10.1021/ACS.ANALCHEM.6B04873
221. Chen, G., Roy, I., Yang, C., Prasad, P.N.: Nanochemistry and nanomedicine for nanoparticle-based diagnostics and therapy. ACS Publ. **116**, 2826–2885 (2016). https://doi.org/10.1021/ACS.CHEMREV.5B00148
222. Preparation RS-TEMS, 2025 undefined (2024) TEM Sample Preparation for Powder and Liquid Formulations. Springer, 61–95.https://doi.org/10.1007/978-3-031-82967-3_4
223. Si, K.J., Chen, Y., Shi, Q., Cheng, W.: Nanoparticle superlattices: the roles of soft ligands. Wiley Online Library **5**,(2018). https://doi.org/10.1002/ADVS.201700179
224. Mu, Q., Jiang, G., Chen, L., Zhou, H., Fourches, D., Tropsha, A., Yan, B.: Chemical basis of interactions between engineered nanoparticles and biological systems. ACS Publ. **114**, 7740–7781 (2014). https://doi.org/10.1021/CR400295A
225. Sciences MM-I journal of molecular: Undefined Transmission electron microscopy as a powerful tool to investigate the interaction of nanoparticles with subcellular structures. mdpi.com (2021)
226. Li, G., Zhang, H., Han, Y.: Applications of transmission electron microscopy in phase engineering of nanomaterials. ACS Publ. **123**, 10728–10749 (2023). https://doi.org/10.1021/ACS.CHEMREV.3C00364
227. Mayeen, A., Shaji, L., Nair, A., of NK-C.: undefined Morphological characterization of nanomaterials. Elsevier (2018)

228. Science, F.R.: Undefined (2015) Opportunities and challenges in liquid cell electron microscopy. science.org 350 (2015). https://doi.org/10.1126/SCIENCE.AAA9886
229. Dachraoui, W., Henninen, T., Keller, D., Reports RE-S: Undefined (123AD) Multi-step atomic mechanism of platinum nanocrystals nucleation and growth revealed by in-situ liquid cell STEM. nature.com 11, 23965 (2021). https://doi.org/10.1038/s41598-021-03455-w
230. Grieshaber, R.V.: Nanoparticle Synthesis and Characterization: A Mechanistic Study of Nanoparticle Morphological Evolution. BS (2018)
231. Nanomedicine, P.S.-W.I.R.: Undefined (2017) Cryo-electron microscopy and cryo-electron tomography of nanoparticles. Wiley Online Library **9**,(2017). https://doi.org/10.1002/WNAN.1417
232. Pu, S., Gong, C., Robertson, A.W.: Liquid cell transmission electron microscopy and its applications. R Soc Open Sci **7**,(2020). https://doi.org/10.1098/RSOS.191204
233. Mallette, A.J., Shilpa, K., Rimer, J.D.: The current understanding of mechanistic pathways in zeolite crystallization. ACS Publ. **124**, 3416–3493 (2024). https://doi.org/10.1021/ACS.CHEMREV.3C00801
234. Kabius, B., Browning, N., Thevuthasan, S., Diehl, B.: Dynamic Processes in Biology, Chemistry, and Materials Science: Opportunities for UltraFast Transmission Electron Microscopy (Workshop Summary Report) (2012)
235. Zhou, W., Apkarian, R., Wang, Z., applications DJ-: techniques and, 2006 undefined Fundamentals of scanning electron microscopy (SEM). Springer
236. Akhtar, K., Khan, S.A., Khan, S.B., Asiri, A.M.: Scanning electron microscopy: principle and applications in nanomaterials characterization. Springer **113–145**,(2018). https://doi.org/10.1007/978-3-319-92955-2_4
237. of MK-A textbook on fundamentals and applications, 2018 undefined Scanning electron microscopy: Principle, components and applications. researchgate.net
238. Datye, A., Catalyst AD-SH of A: Undefined (2023) Scanning electron microscopy (SEM), 359–380. Springer (2023). https://doi.org/10.1007/978-3-031-07125-6_18
239. Goldstein, J.I., Newbury, D.E., Michael, J.R., Ritchie, N.W.M., Scott, J.H.J., Joy, D.C.: Scanning electron microscope (SEM) instrumentation. Springer **65–91**,(2018). https://doi.org/10.1007/978-1-4939-6676-9_5
240. Goldstein, J.I., Newbury, D.E., Echlin, P., Joy, D.C., Lyman, C.E., Lifshin, E., Sawyer, L., Michael, J.R.: The SEM and its modes of operation. Springer **21–60**,(2003). https://doi.org/10.1007/978-1-4615-0215-9_2
241. Mohammed, A., … AA the 2018 international conference on, 2018 undefined Scanning electron microscopy (SEM): A review. fluidas.ro
242. Goldstein, A., Soroka, Y., Frušić-Zlotkin, M., Popov, I., Kohen, R.: High resolution SEM imaging of gold nanoparticles in cells and tissues. Wiley Online Library **256**, 237–247 (2014). https://doi.org/10.1111/JMI.12179
243. Mehdizadeh Kashi, A., Tahermanesh, K., Chaichian, S., Taghi Joghataei, M., Moradi, F., Mohammad Tavangar, S., Sadat Mousavi Najafabadi, A., Lotfibakhshaiesh, N., Pour Beyranvand, S., Fazel Anvari-Yazdi, A., Mehr Abed, S.: How to prepare biological samples and live tissues for scanning electron microscopy (SEM). GMJ **3**, 63–80 (2014)
244. Microscopy PN-SH of, 2019 undefined (2019) Scanning transmission electron microscopy. Springer, 49–99. https://doi.org/10.1007/978-3-030-00069-1_2
245. He, Q., Li, A., Yao, B., Zhou, W., Kiely, C.J.: STEM high angle annular dark-field imaging. Springer **409–448**,(2023). https://doi.org/10.1007/978-3-031-07125-6_20
246. Nukala, H.: Quantitative thickness mapping in high-angle annular dark-field (HAADF) scanning transmission electron microscopy (STEM) (2008)
247. Technique MO-J of electron microscopy, 1991 undefined (1991) High-Angle annular dark-field imaging on a tem/stem system. Wiley Online Library **17**, 221–230. https://doi.org/10.1002/JEMT.1060170209
248. Wille, G., Hellal, J., Ollivier, P., Richard, A., Burel, A., Jolly, L., Crampon, M., Michel, C., Burel, A.: Cryo-scanning electron microscopy (SEM) and scanning transmission electron microscopy (STEM)-in-SEM for bio-and organo-mineral interface characterization in the. academic.oup.com 23. 10.1017/S143192761701265Xï (2017)

249. Mata, M. de la, Nanomaterials, S.M.: undefined STEM tools for semiconductor characterization: beyond high-resolution imaging. mdpi.com (2022)
250. House, S., Chen, Y., Jin, R., Ultramicroscopy, J.Y.: undefined High-throughput, semi-automated quantitative STEM mass measurement of supported metal nanoparticles using a conventional TEM/STEM. Elsevier (2017)
251. Nanotechnology JC-H of M for, Undefined Scanning transmission electron microscopy. Springer, 455–491 (2005). https://doi.org/10.1007/1-4020-8006-9_15
252. Nanotechnology JC-H of M for, Undefined Scanning transmission electron microscopy. SpringerJM CowleyHandbook of Microscopy for Nanotechnology, 2005•Springer 455–491 (2005). https://doi.org/10.1007/1-4020-8006-9_15
253. Fang, K., Yuan, W., Wagner, J.B., Zhang, Z., Wang, Y.: In-situ gas transmission electron microscopy. Springer **251–325**,(2023). https://doi.org/10.1007/978-981-19-6845-7_9
254. Okunishi, E., Ishikawa, I., ... HS-M and: Undefined Visualization of light elements at ultrahigh resolution by STEM annular bright field microscopy. cambridge.org (2009)
255. Okunishi, E., Sawada, H., Micron, Y.K.: Undefined Experimental study of annular bright field (ABF) imaging using aberration-corrected scanning transmission electron microscopy (STEM). Elsevier (2012)
256. Tong, Y., Zhang, Q., B LG-CP.: Undefined Scanning transmission electron microscopy: A review of high angle annular dark field and annular bright field imaging and applications in lithium-ion batteries. iopscience.iop.org (2018). https://doi.org/10.1088/1674-1056/27/6/066107/META
257. STEM-EDX tomography of bimetallic nanoparticles: A methodological investigation. Elsevier
258. Akita, T., Tanaka, S., Tanaka, K., Haruta, M., Kohyama, M.: Sequential HAADF-STEM observation of structural changes in Au nanoparticles supported on CeO_2. Springer **46**, 4384–4391 (2011). https://doi.org/10.1007/S10853-011-5370-6
259. Mitchson, G., Ditto, J., ... KW-S.: undefined Application of HAADF STEM image analysis to structure determination in rotationally disordered and amorphous multilayered films. iopscience.iop.org (2016). https://doi.org/10.1088/0268-1242/31/8/084003/META
260. Xiang, W., Tian, Y., Engineering XL-P: Undefined Dynamic analysis of tapping mode atomic force microscope (AFM) for critical dimension measurement. Elsevier (2020)
261. Findlay, S., Shibata, N., Ultramicroscopy, Y.I.: Undefined What atomic resolution annular dark field imaging can tell us about gold nanoparticles on TiO2 (1 1 0). Elsevier (2009)
262. Research CO-AR of M.: undefined (2023) Quantitative scanning transmission electron microscopy for materials science: Imaging, diffraction, spectroscopy, and tomography. annualreviews.org **53**, 105–141 (2023). https://doi.org/10.1146/ANNUREV-MATSCI-080921-092646
263. Zaefferer, S., Wright, S.I., Raabe, D.: Three-dimensional orientation microscopy in a focused ion beam–scanning electron microscope: a new dimension of microstructure characterization. Springer **39**, 374–389 (2008). https://doi.org/10.1007/S11661-007-9418-9
264. Qamar, H., Iqbal, A., Ahmad, A.S., Kiran, S., Ahmad, S., Usman, M., Muneer, M.H., Hussain, K., Shahbakht, R.M., Rehman, A., Haider, A.: Nanomaterials Characterization. uniquescientificpublishers.com. https://doi.org/10.47278/book.CAM/2024.233
265. Qiu, H., Gao, Y., Boott, C.E., Gould, O.E.C., Harniman, R.L., Miles, M.J., Webb, S.E.D., Winnik, M.A., Manners, I.: Uniform patchy and hollow rectangular platelet micelles from crystallizable polymer blends. science.org **352**, 697–701 (2016). https://doi.org/10.1126/SCIENCE.AAD9521
266. Delvallée, A., Feltin, N., ... SD-M: Undefined Direct comparison of AFM and SEM measurements on the same set of nanoparticles. iopscience.iop.org (2015). https://doi.org/10.1088/0957-0233/26/8/085601/META
267. Gwaze, P., Annegarn, H., Huth, J., Research GH-A.: undefined Comparison of particle sizes determined with impactor, AFM and SEM. Elsevier (2007)
268. Hubenthal, F., Sánchez, D., Sciences FT-A.: undefined Morphological Parameters of Supported Gold Nanoparticles: Comparison of AFM Combined with Optical Spectroscopy and Theoretical Modeling versus TEM. mdpi.com (2012)

269. Leong, S.S., Ng, W.M., Lim, J.K., Yeap, S.P.: Dynamic light scattering: effective sizing technique for characterization of magnetic nanoparticles. Springer **77–111**,(2018). https://doi.org/10.1007/978-3-319-92955-2_3
270. Viguié, J., Sukmanowski, J., … BN-C and SA.: Undefined Study of agglomeration of alumina nanoparticles by atomic force microscopy (AFM) and photon correlation spectroscopy (PCS). Elsevier (2007)
271. Gu, Y., Xie, H., Gao, J., Liu, D., Williams, C.T., Murphy, C.J., Ploehn, H.J.: AFM characterization of dendrimer-stabilized platinum nanoparticles. ACS Publ. **21**, 3122–3131 (2005). https://doi.org/10.1021/LA047843E
272. Shaji, A.: Measurement AZ-T and R, 2017 undefined Surface area analysis of nanomaterials. Elsevier
273. Irwansyah, F.S., Amal, A.I., Diyanthi, E.W., Hadisantoso, E.P., Noviyanti, A.R., Rakhmawaty Eddy, D., Risdiana, R.: How to read and determine the specific surface area of inorganic materials using the Brunauer-Emmett-Teller (BET) method. ejournal.kjpupi.id **4**, 61–70 (2024). https://doi.org/10.17509/xxxx.xxxx
274. Lowell, S., Shields, J., Thomas, M., Thommes, M.: Characterization of porous solids and powders: surface area, pore size and density (2012)
275. Jayakantha, D.N.P.R., Gunawardhana, N., Bandara, H.M.N., Comini, E., Gunawardana, N.M., Karunarathne, S.M.M.L.: Low cost automated instrumentation for the measurement of the specific surface area of powders by the Brunauer–Emmett–Teller (BET) method. Taylor & Francis **50**, 47–56 (2021). https://doi.org/10.1080/10739149.2021.1953522
276. Bosch, H., Instruments APPES: Undefined Automatic and low-cost determination of BET surface areas. iopscience.iop.org (1977). https://doi.org/10.1088/0022-3735/10/6/012/META
277. Brame, J., Griggs, C.: Surface area analysis using the Brunauer-Emmett-Teller (BET) method: scientific operation procedure series: SOP-C (2016)
278. Management EK-J of E: Undefined Optimum BET surface areas for activated carbon produced from textile sewage sludges and its application as dye removal. Elsevier (2016)
279. Walton, K., Society, RS-J of the AC.: Undefined (2007) Applicability of the BET method for determining surface areas of microporous metal– organic frameworks. ACS Publ. **129**, 8552–8556 2007.https://doi.org/10.1021/JA071174K
280. Wang, T.C., Bury, W., Gómez-Gualdrón, D.A., Vermeulen, N.A., Mondloch, J.E., Deria, P., Zhang, K., Moghadam, P.Z., Sarjeant, A.A., Snurr, R.Q., Stoddart, J.F., Hupp, J.T., Farha, O.K.: Ultrahigh surface area zirconium MOFs and insights into the applicability of the BET theory. ACS Publ. **137**, 3585–3591 (2015). https://doi.org/10.1021/JA512973B
281. Liu, F., Luo, X., Zhou, X., Shen, Y., Chen, J., … XL-MT: Undefined A new strategy for semi quantitative analysis of pore structure for lithium-ion battery electrodes. Elsevier (2024)
282. Aluru, N.R., Aydin, F., Bazant, M.Z., Blankschtein, D., Brozena, A.H., de Souza, J.P., Elimelech, M., Faucher, S., Fourkas, J.T., Koman, V.B., Kuehne, M., Kulik, H.J., Li, H.K., Li, Y., Li, Z., Majumdar, A., Martis, J., Misra, R.P., Noy, A., Pham, T.A., Qu, H., Rayabharam, A., Reed, M.A., Ritt, C.L., Schwegler, E., Siwy, Z., Strano, M.S., Wang, Y.H., Yao, Y.C., Zhan, C., Zhang, Z.: Fluids and electrolytes under confinement in single-digit nanopores. ACS Publ. **123**, 2737–2831 (2023). https://doi.org/10.1021/ACS.CHEMREV.2C00155
283. Kubis, A.J., Shiflet, G.J., Dunn, D.N., Hull, R.: Focused ion-beam tomography. Springer 35 A, 1935–1943 (2004). https://doi.org/10.1007/S11661-004-0142-4
284. Uchic, M.D., Holzer, L., Inkson, B.J., Principe, E.L., Munroe, P.: Three-dimensional microstructural characterization using focused ion beam tomography. Springer **32**, 408–416 (2007). https://doi.org/10.1557/MRS2007.64
285. Raffa, V., Castrataro, P., Menciassi, A., Probe, P.D.-A.S.: Undefined (2006) Focused ion beam as a scanning probe: methods and applications. Springer **361–412**,(2006). https://doi.org/10.1007/3-540-27453-7_11
286. Lawrence, T., Carr, S., Manning, A., Methods X JW.: Undefined A novel 3D volumetric method for directly quantifying porosity and pore space morphology in flocculated suspended sediments. Elsevier (2023)

287. Garum, M., Glover, P.W.J., Lorinczi, P., Scott, G., Hassanpour, A.: Ultrahigh-resolution 3D imaging for quantifying the pore nanostructure of shale and predicting gas transport. ACS Publ. **35**, 702–717 (2021). https://doi.org/10.1021/ACS.ENERGYFUELS.0C03225
288. Akbari, B, ... MT-IranJMater, S.: Undefined Particle size characterization of nanoparticles–a practical approach. scholar.archive.org (2011)
289. Hidayah, N.M.S., Liu, W.W., Lai, C.W., Noriman, N.Z., Khe, C.S., Hashim, U., Lee, H.C.: Comparison on graphite, graphene oxide and reduced graphene oxide: Synthesis and characterization. In: AIP Conference Proceedings. American Institute of Physics Inc. (2017)
290. Marcano, D.C., Kosynkin, D.V., Berlin, J.M., Sinitskii, A., Sun, Z., Slesarev, A., Alemany, L.B., Lu, W., Tour, J.M.: Improved synthesis of graphene oxide. ACS Nano **4**, 4806–4814 (2010). https://doi.org/10.1021/nn1006368
291. Bîru, E.I., Iovu, H.: Graphene Nanocomposites Studied by Raman Spectroscopy. Raman Spectroscopy (2018). https://doi.org/10.5772/INTECHOPEN.73487
292. Gollwitzer, C., Bartczak, D., Goenaga-Infante, H., Kestens, V., Krumrey, M., Minelli, C., Almai, M.P., Ramaye, Y., Roebben, G., Sikora, A., Zolt, Z., Varga, Z.: A comparison of techniques for size measurement of nanoparticles in cell culture medium. pubs.rsc.org (2016). https://doi.org/10.1039/c6ay00419a
293. Sikora, A., Shard, A.G., Minelli, C.: Size and ζ-potential measurement of silica nanoparticles in serum using tunable resistive pulse sensing. ACS Publ. **32**, 2216–2224 (2016). https://doi.org/10.1021/ACS.LANGMUIR.5B04160
294. Han, Q., Candiloro, Z.P.J., Cai, X., El Mohamad, M., Dyett, B.P., Rosado, C.J., Zhai, J., Bryant, G., Drummond, C.J., Greaves, T.L.: Silica nanoparticle-protein aggregation and protein corona formation investigated with scattering techniques. ACS Appl. Mater. Interfaces **17**, 8574–8587 (2025). https://doi.org/10.1021/ACSAMI.4C19591
295. Vetter, V.C., Yazdi, M., Gialdini, I., Pöhmerer, J., Seidl, J., Höhn, M., Lamb, D.C., Wagner, E.: Ionic Coating of siRNA Polyplexes with cRGD-PEG-hyaluronic acid to modulate serum stability and in vivo performance. Biochemistry **64**, 1509–1529 (2025). https://doi.org/10.1021/ACS.BIOCHEM.4C00650

Application of Polymeric Nanomaterials as Therapeutic Agents

Sunil Luhar and Prashant Kumar

1 Introduction

In recent decades, the landscape of global healthcare has been challenged by the increasing complexity and diversity of diseases, ranging from infectious pathogens and multidrug-resistant (MDR) bacteria to cancer and chronic inflammatory disorders [1]. According to the WHO, antimicrobial resistance (AMR) ranks among the top threats to public health, posing serious risks to the efficacy of existing drugs. Simultaneously, cancer remains a leading cause of mortality worldwide, with millions of new cases diagnosed each year [2]. In addition, chronic conditions such as diabetes, cardiovascular diseases, and neurodegenerative disorders are rising steadily, fueled by aging populations and lifestyle-related risk factors [3]. Despite progress in diagnostic and therapeutic technologies, conventional drug delivery approaches often suffer from significant limitations, including poor bioavailability, nonspecific distribution, systemic toxicity, and the inability to cross physiological barriers [4]. Traditional pharmaceuticals typically operate under a one-size-fits-all paradigm, which fails to address the complexity of disease microenvironments or patient-specific variability. Furthermore, many therapeutic agents, particularly those used in oncology and infectious disease treatment, are either highly toxic or rapidly metabolized, leading to suboptimal therapeutic outcomes [5]. These challenges highlight the urgent need for innovative drug delivery systems capable of improving therapeutic efficacy while minimizing side effects. Amidst this pressing need, nanotechnology has

S. Luhar (✉)
Nalecz Institute of Biocybernetics and Biomedical Engineering, Polish Academy of Sciences, Warsaw, Poland
e-mail: scientistksv99@gmail.com

Bhavira Technologies Pty Ltd, RMIT Activator, Melbourne, VIC 3030, Australia

P. Kumar
Next-Generation Magnet Development Collaboration Unit, RIKEN-Center for Integrative Medical Sciences, Yokohama, Japan

G. Tailor et al. (eds.), *Nanomaterials*, Engineering Materials,
https://doi.org/10.1007/978-981-92-0678-0_5

emerged as a powerful tool for revolutionizing disease treatment [6, 7]. Among the various nanocarrier systems investigated, polymeric nanomaterials (PNPs) have gained considerable attention due to their favorable physicochemical characteristics and high degree of design flexibility [8]. These nanoscale structures, engineered from natural or synthetic polymers, can encapsulate, protect, and selectively deliver a broad range of therapeutic cargos, including small-molecule drugs, nucleic acids, proteins, and vaccines [9]. Their ability to overcome biological barriers, control drug release profiles, and target specific tissues has positioned PNPs at the forefront of modern therapeutic strategies. Polymeric nanomaterials are typically engineered within the size range of 10–500 nm and exhibit a high surface-area-to-volume ratio, making them ideal for interaction with biological systems at the cellular and molecular levels [10]. One of the major advantages of PNPs is their ability to encapsulate therapeutic agents, thereby shielding them from premature degradation and clearance in the bloodstream. This protective encapsulation improves the pharmacokinetic and pharmacodynamic profiles of drugs, extending their circulation time and allowing for more effective dosing schedules [11]. Moreover, PNPs can be designed to offer controlled and sustained drug release, reducing the frequency of administration and enhancing patient compliance. Through careful selection of polymer types and crosslinking strategies, properties can tune the release kinetics of encapsulated agents in response to specific physiological stimuli such as pH, temperature, enzymes, or redox conditions. This capability is particularly valuable in cancer therapy, where the tumor microenvironment often exhibits distinct pH or enzymatic activity that can be exploited for stimuli-responsive drug release [12, 13]. The surface functionalization of PNPs with ligands, antibodies, or antimicrobial peptides can facilitate targeted bacterial recognition and internalization, minimizing off-target effects and reducing the emergence of resistance [14, 15].

Beyond antibacterial and anticancer applications, polymeric nanomaterials are increasingly being explored in gene therapy, vaccine development, and regenerative medicine. As non-viral vectors, PNPs offer a safer alternative to viral carriers for delivering nucleic acids such as siRNA, mRNA, or plasmid DNA. Their tunable properties enable efficient cellular uptake and endosomal escape, crucial steps in achieving effective gene expression modulation. In vaccine delivery, PNPs can protect antigens from degradation, enhance immune cell activation, and promote sustained antigen release, contributing to stronger and longer-lasting immune responses. In tissue engineering, PNP-based scaffolds and delivery vehicles facilitate the localized release of growth factors, supporting cell proliferation, differentiation, and tissue regeneration. An additional advantage of PNPs is their biocompatibility and biodegradability, particularly when constructed from FDA-approved polymers like poly (lactic-co-glycolic acid) (PLGA), polyethylene glycol (PEG), chitosan, or polylactic acid (PLA). These polymers break down into nontoxic metabolites within the body, reducing the risk of long-term accumulation or immune response [16, 17].

2 Fundamentals of Polymeric Nanomaterials (PNPs)

Polymers are large macromolecules composed of repeating structural units known as monomers, which are covalently bonded to form long chains or networks. These monomers can be identical (homopolymers) or different (copolymers), and their chemical structure largely determines the physical and chemical properties of the resulting polymer [18]. Polymers can be derived from natural sources-such as proteins, polysaccharides, or DNA-or synthesized from chemical monomers to produce materials with desirable features such as durability, flexibility, and biodegradability. Because of their versatility and compatibility with living systems, polymers have become essential in various scientific and industrial fields, particularly in the development of materials for biomedical applications. Nanoparticles, on the other hand, are ultrafine particles typically ranging from 1 to 100 nanometers in size. Due to their small size and large surface-area-to-volume ratio, nanoparticles exhibit unique physicochemical properties, including enhanced solubility, reactivity, and the ability to cross biological barriers [19–21]. When these two concepts-polymers and nanoparticles-are combined, they give rise to PNPs: nanoscale drug carriers made from polymeric materials. These carriers are engineered to encapsulate, protect, and deliver a wide range of therapeutic agents such as drugs, genes, and proteins to targeted areas within the body [22]. Polymeric nanoparticles formulated using natural polymers like chitosan, alginate, or gelatin, known for their biocompatibility and biological activity, or synthetic polymers like polylactic acid (PLA), polyethylene glycol (PEG), polycaprolactone (PCL), and poly (lactic-co-glycolic acid) (PLGA), which offer precise control over molecular weight, degradation rates, and mechanical strength [23, 24]. PNPs may exist as nanospheres (matrix systems where the drug is uniformly dispersed) or nanocapsules (reservoir systems with a core-shell structure) shown in Fig. 1. Their properties—such as size, shape, surface charge, and hydrophobicity-can be tuned by modifying the polymer composition or processing conditions. This structural flexibility makes them ideal for overcoming biological barriers, improving drug solubility, enabling controlled and sustained release, and minimizing side effects by targeting specific cells or tissues [25].

2.1 Structure and Morphology of PNPs

PNPs can be designed with various internal architectures. The two most common types are nanospheres and Nano capsules. Nanospheres are solid matrix systems in which the polymer forms a continuous structure throughout the particle. All incorporated materials are distributed uniformly or trapped within the matrix. Nano capsules consist of a core-shell structure where the core can hold fluids or other agents, surrounded by a polymeric wall. This architecture allows for compartmentalization and the possibility of surface modifications without affecting the contents. The specific structure of a PNP impacts its behavior in biological systems, including how

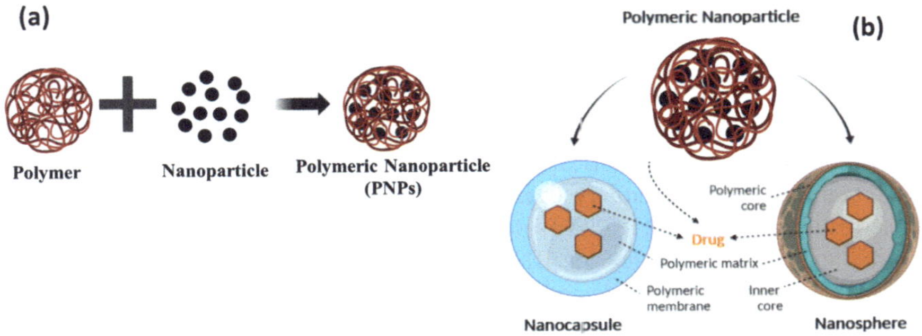

Fig. 1 **a** Schematic representation of a polymeric nanoparticle showing core-shell structure for drug encapsulation and delivery. **b** Types of polymeric nanoparticles, including nanospheres with uniformly dispersed drugs and nanocapsules with drug enclosed in a polymeric shell [26]

it interacts with cells, how long it circulates in the bloodstream, and how it degrades over time. Other structural forms include polymeric micelles, dendrimers, and polymersomes. Polymeric micelles, for example, self-assemble from amphiphilic block copolymers in aqueous environments, forming a hydrophobic core and hydrophilic shell. Dendrimers are highly branched, tree-like structures with a high density of surface functional groups. Polymersomes, resembling liposomes, are vesicular systems with polymer bilayers [27, 28]. Each morphology offers unique advantages depending on the intended biological environment or therapeutic challenge, and their performance is governed by key physicochemical properties such as size, surface charge, hydrophilicity, and colloidal stability, as summarized shown in the Table 1.

3 Methods of Synthesis and Functionalization

PNPs can be synthesized using a range of techniques that allow precise control over particle size, morphology, surface characteristics, and payload encapsulation. The choice of synthesis method largely depends on the nature of the polymer, the intended application, and the desired physicochemical properties. Among the widely used methods are nanoprecipitation, emulsion techniques, and self-assembly. Post-synthesis, surface functionalization further tailors PNPs for specific biological interactions, including ligand attachment and stimuli-responsive modifications [30–32].

Table 1 Key physicochemical properties of polymeric nanoparticles (PNPs) influencing their biological performance [29]

Property	Description	Implications
Size	Typically, 10–1000 nm; often optimized below 200 nm	Affects cellular uptake, biodistribution, and ability to cross biological barriers
Surface area	High surface-area-to-volume ratio	Enables efficient drug loading and surface functionalization
Surface charge (zeta potential)	Typically, −30 mV to + 30 mV; tunable via polymer or ligand choice	Influences colloidal stability, cellular interaction, and immune recognition
Morphology	Spherical, rod-like, or core-shell structures	Impacts drug release profile, cellular uptake, and circulation behavior
Hydrophilicity/ hydrophobicity	Determined by polymer composition (e.g., PEG = hydrophilic, PLA = hydrophobic)	Controls solubility, drug compatibility, and dispersion in biological fluids
Biocompatibility	Non-toxic, non-immunogenic materials	Essential for safety and long-term clinical use
Biodegradability	Ability to degrade into non-toxic byproducts	Allows for clearance from the body post-delivery
Thermal stability	Resistance to heat degradation during storage and processing	Important for shelf life and stability in body temperature conditions
Mechanical stability	Structural integrity under stress	Prevents premature drug release or particle breakdown
Polydispersity index (PDI)	Measurement of size distribution (ideal <0.2)	Low PDI = uniform behavior, higher predictability and reproducibility
Crystallinity/amorphous nature	Refers to the internal arrangement of polymer chains	Affects drug loading, release kinetics, and degradation behavior
Colloidal stability	Ability to resist aggregation in suspension	Ensures consistent dosing and delivery in physiological conditions
Protein corona formation	Adsorption of biomolecules upon exposure to biological fluids	Alters biological identity, uptake, circulation, and immune system interactions
Barrier penetration ability	Capability to cross physiological barriers (e.g., BBB, mucosa)	Essential for targeting hard-to-reach tissues or systemic delivery
PEGylation or surface coating	Modification with polyethylene glycol or other molecules	Enhances "stealth" properties, reduces immune clearance, prolongs circulation time

3.1 Solvent Emulsion Evaporation Method

The solvent emulsion evaporation method is a well-established and extensively employed technique for the synthesis of PNPs, particularly suited for encapsulating hydrophobic drugs and bioactive molecules. This method relies on the formation of an oil-in-water (O/W) emulsion, where the polymer and the hydrophobic drug are first dissolved in a volatile organic solvent that is immiscible with water. The resulting organic phase is then emulsified into an aqueous phase containing surfactants or stabilizers to generate nanoscale droplets of the polymer solution dispersed in water. Upon evaporation of the organic solvent, the polymer precipitates, forming solid nanoparticles encapsulating the drug within their matrix. The process begins with dissolving the chosen polymer-commonly biodegradable and biocompatible materials such as PLGA, PLA, or PCL-in a suitable organic solvent like DCM, chloroform, or ethyl acetate. Hydrophobic drugs are either dissolved or uniformly dispersed within this organic solution. Subsequently, this organic phase is added dropwise or rapidly injected into an aqueous solution containing a surfactant such as PVA, SDS, or Tween 80. The role of the surfactant is critical as it reduces the interfacial tension between the organic and aqueous phases, stabilizing the emulsion droplets and preventing coalescence. Emulsification is generally achieved by mechanical agitation, such as high-speed homogenization or ultrasonic sonication, which helps to produce droplets of uniform size in the nanometer range. The size and uniformity of the resulting nanoparticles can be tuned by controlling parameters including polymer concentration, surfactant type and concentration, homogenization speed, and duration of sonication. Typically, smaller particle sizes are obtained with higher energy input during emulsification. Once a stable emulsion is formed, the organic solvent is gradually evaporated either by stirring the emulsion under ambient conditions or by applying reduced pressure using a rotary evaporator. As the solvent evaporates, the polymer solidifies, encapsulating the drug within the nanoparticle matrix. This phase transition from liquid droplets to solid nanoparticles is central to the formation of stable PNPs. After complete solvent removal, the nanoparticle suspension is subjected to purification processes such as centrifugation, ultrafiltration, or dialysis to remove excess surfactant, free drug, and residual solvents. The purified nanoparticles are then resuspended in a suitable aqueous medium. To enhance the long-term stability and ease of storage, nanoparticles can be lyophilized (freeze-dried) in the presence of cryoprotectants like trehalose or mannitol, which prevent aggregation and preserve the particle structure during drying. The resulting dry powder can be reconstituted in aqueous media when needed. The solvent emulsion evaporation method offers several advantages: it is a relatively simple, reproducible, and scalable process compatible with a wide range of polymers and hydrophobic drugs. It generally produces nanoparticles with a narrow size distribution (commonly 100–300 nm), which is critical for consistent biological performance. Moreover, the encapsulation efficiency and drug loading capacity can be optimized by adjusting the formulation parameters. This method is widely used in pharmaceutical and biomedical fields for the controlled delivery of anticancer drugs, vaccines, and other therapeutic agents. However, there are some limitations to consider. The use of organic solvents poses potential toxicity risks, necessitating thorough solvent removal and verification before in vivo applications. Additionally, this method is less effective for hydrophilic drugs, which have low solubility in organic solvents, often leading to poor encapsulation efficiencies. Alternative techniques like double emulsion or nanoprecipitation are preferred for hydrophilic molecules. Furthermore, the process requires careful control of parameters to ensure reproducibility and uniform particle size [33, 34].

3.2 *Double Emulsification Solvent Evaporation Method (W/O/W)*

The double emulsification solvent evaporation method, also known as the water-in-oil-in-water (W/O/W) double emulsion technique, is a widely used approach for preparing PNPs, particularly designed for encapsulating hydrophilic drugs, proteins, peptides, and nucleic acids. This method extends the classical solvent emulsion evaporation by incorporating two emulsification steps, allowing effective encapsulation of water-soluble bioactive molecules that otherwise exhibit poor solubility in organic solvents. The process starts with the preparation of the primary emulsion (W/O), where an aqueous solution containing the hydrophilic drug or biomolecule is emulsified into an organic phase comprising a polymer dissolved in a volatile solvent such as DCM, chloroform, or ethyl acetate. The polymer is typically a biodegradable and biocompatible polyester such as PLGA, PLA, or PCL. The primary W/O emulsion is formed using high-energy methods like probe sonication or high-speed homogenization to generate fine droplets of the aqueous phase dispersed in the organic solvent containing the polymer. Once the primary W/O emulsion is prepared, it is immediately added to a larger volume of an external aqueous phase containing a surfactant or stabilizer such as PVA, sodium cholate, or Tween 80. This secondary aqueous phase stabilizes the emulsion and prevents coalescence of droplets during emulsification. The addition is followed by further emulsification under mechanical agitation, forming the secondary emulsion (W/O/W). The water droplets containing the drug are thus encapsulated within the polymer solution droplets, which themselves are dispersed in the continuous aqueous phase. This two-step emulsification creates a more complex structure in which hydrophilic drugs can be effectively entrapped inside the nanoparticle core or matrix. Subsequently, the organic solvent is removed by evaporation under reduced pressure or continuous stirring at ambient conditions. As the solvent evaporates, the polymer precipitates and solidifies around the inner aqueous droplets, yielding solid polymeric nanoparticles encapsulating the hydrophilic drug. The resulting nanoparticle suspension is then purified by centrifugation, ultrafiltration, or dialysis to remove residual solvent, free drug, and excess surfactant. The double emulsification technique offers advantages in improving encapsulation efficiency for hydrophilic drugs, which typically exhibit poor retention in single emulsion methods. It also allows for the controlled release of therapeutic molecules by modulating polymer composition, particle size, and surfactant concentration. Additionally, the size distribution and surface characteristics of the nanoparticles can be tailored by adjusting process parameters such as sonication time, homogenization speed, polymer concentration, and the ratio of inner aqueous phase to organic phase. Despite its versatility, the W/O/W method has challenges including potential leakage of hydrophilic drugs into the external aqueous phase during emulsification or solvent evaporation, leading to reduced encapsulation efficiency. To mitigate this, additives such as osmotic agents (e.g., salts, sugars) can be incorporated into the internal aqueous phase to maintain osmotic balance and stabilize the inner droplets. Moreover, the method requires optimization of formulation parameters to balance particle size, stability, and drug loading. This technique has been extensively applied in the delivery of proteins, peptides, nucleic acids, and vaccines where preservation of biological activity is crucial. The use of biodegradable polymers like PLGA allows for sustained drug release and biodegradation into non-toxic metabolites. Furthermore, the method supports surface modification of nanoparticles for targeting and improved biodistribution [35].

3.3 Emulsion Diffusion Method

The emulsion diffusion method is an effective and widely used technique for the preparation of PNPs particularly suited for encapsulating both hydrophobic and hydrophilic drugs with good control over particle size and drug loading. This method is considered a modification of the solvent evaporation technique, designed to improve nanoparticle stability and reduce the use of toxic solvents. The emulsion diffusion process begins with the dissolution of the polymer, such as PLGA, PLA, or PCL, in a partially water-miscible organic solvent like ethyl acetate, benzyl alcohol, or methyl ethyl ketone. The choice of solvent is critical as it must be miscible with water to allow diffusion but should not mix completely to maintain an organic phase during emulsification. After polymer dissolution, the organic phase is emulsified into an aqueous phase containing a stabilizer or surfactant such as PVA, poloxamers, or SDS. The emulsification is typically achieved by high-shear homogenization, ultrasonication, or mechanical stirring, forming an oil-in-water (O/W) emulsion with fine droplets. Once the emulsion is formed, the process exploits the diffusion of the organic solvent from the dispersed phase into the continuous aqueous phase. This solvent diffusion leads to a gradual reduction in solvent concentration within the oil droplets, causing polymer precipitation and nanoparticle solidification. Because the solvent diffuses out rather than evaporating, the process typically occurs under moderate stirring at room temperature or slightly elevated temperature without the need for vacuum or high-temperature evaporation. This helps preserve the structural integrity and bioactivity of sensitive drug molecules. An essential advantage of the emulsion diffusion method is that the diffusion-driven precipitation often results in nanoparticles with narrow size distribution and improved colloidal stability. The diffusion mechanism also minimizes nanoparticle aggregation by enabling controlled and gradual polymer solidification. Particle size can be finely tuned by adjusting parameters such as polymer concentration, surfactant type and concentration, emulsification energy (sonication or homogenization time), and organic-to-aqueous phase ratio. The method is particularly attractive for preparing nanoparticles loaded with poorly water-soluble drugs, as the polymer and drug co-precipitate during solvent diffusion. It also allows for encapsulation of hydrophilic drugs by including them in the aqueous phase or by modifying the polymer matrix. The resultant nanoparticles exhibit controlled drug release profiles that depend on the polymer degradation rate and particle size. However, the method has certain limitations, such as the need to carefully select organic solvents that are biocompatible and approved for pharmaceutical use. Residual solvents must be removed efficiently to meet safety standards. Also, the diffusion process can be slower than solvent evaporation, potentially affecting production throughput [36].

3.4 Nanoprecipitation (Solvent Displacement)

Nanoprecipitation, also known as solvent displacement, is a straightforward, rapid, and highly reproducible technique for fabricating polymeric nanoparticles. It was first introduced by Fessi et al. in 1989 and has since become one of the most popular methods for preparing nanoscale drug delivery systems due to its simplicity, mild operating conditions, and ability to produce nanoparticles with a narrow size distribution. This method is especially advantageous for hydrophobic drug encapsulation and polymer systems soluble in water-miscible solvents. The process begins with dissolving the polymer, such as PLGA, PCL, or PLA, along with the hydrophobic drug in a water-miscible organic solvent like acetone, ethanol, or acetonitrile. The organic phase, containing both polymer and drug, is then rapidly injected or poured into an aqueous phase containing a surfactant or stabilizer, commonly PVA, Tween 80, or poloxamers. The aqueous phase is typically under moderate stirring or gentle agitation to facilitate homogeneous mixing. Nanoparticles form instantaneously due to the diffusion of the solvent into the aqueous phase, which leads to supersaturation and precipitation of the polymer and encapsulated drug. This diffusion-driven process causes the polymer chains to collapse and form stable nanosized particles. Since nanoprecipitation relies on solvent displacement rather than energy-intensive emulsification, it avoids harsh conditions like high shear or sonication, preserving sensitive molecules' integrity, such as proteins and peptides. Key parameters influencing particle size and morphology include the polymer concentration, type and volume ratio of organic to aqueous phases, surfactant concentration, and mixing rate. Generally, lower polymer concentrations and higher surfactant amounts lead to smaller particle sizes, often ranging from 50 to 300 nm. The choice of solvent is also critical: solvents with higher miscibility in water promote faster diffusion and finer particle sizes, whereas solvents with limited miscibility yield larger particles and may cause aggregation. Nanoprecipitation is well-suited for encapsulating hydrophobic drugs due to their solubility in the organic phase and their entrapment within the polymer matrix upon precipitation. However, the method is less effective for highly hydrophilic drugs unless modified, such as by forming conjugates or using double emulsion techniques. Encapsulation efficiency typically ranges between 50% and 90%, depending on the drug-polymer compatibility and process parameters. An advantage of the nanoprecipitation method is its scalability and ease of adaptation to continuous production processes, such as microfluidics or impinging jet mixers, facilitating industrial applications. Moreover, it produces nanoparticles with high colloidal stability due to the presence of surfactants and the avoidance of residual solvents, which are easily removed via evaporation or dialysis. Despite its benefits, nanoprecipitation has limitations related to solvent choice, since only water-miscible solvents can be used, restricting the range of polymers and drugs that can be processed. Additionally, residual organic solvent traces must be carefully eliminated to meet regulatory standards for biomedical applications [37].

3.5 Ionotropic Gelation Method

The ionotropic gelation method is a widely used, mild, and straightforward technique for preparing polymeric nanoparticles, especially suited for biopolymers such as chitosan, alginate, and other polyelectrolytes. This method exploits the electrostatic interaction between oppositely charged polymers or between charged polymers and multivalent ions to induce gelation, resulting in nanoparticle formation without the need for harsh chemical crosslinkers or organic solvents. Its simplicity, biocompatibility, and ability to produce nanoparticles under aqueous and mild conditions make it highly attractive for pharmaceutical and biomedical applications, such as drug delivery and gene therapy. Typically, the ionic gelation process begins with dissolving a cationic polymer like chitosan in an acidic aqueous medium, where it acquires a positive charge due to protonation of its amino groups. Then, a polyanion such as sodium tripolyphosphate (TPP) or alginate solution is slowly added under constant stirring. The negatively charged groups of TPP interact electrostatically with the positively charged amino groups of chitosan, leading to crosslinking and formation of a three-dimensional gel network. This interaction causes the spontaneous formation of nanoparticles, with size and stability depending on factors such as polymer concentration, the ratio of polymer to crosslinker, pH, ionic strength, and mixing conditions. One key advantage of ionic gelation is its ability to encapsulate sensitive bioactive molecules, including proteins, nucleic acids, and drugs, under gentle conditions that preserve their bioactivity. Nanoparticles produced by this method typically exhibit good stability, controlled size distribution, and surface charge, which can be tuned by modifying the formulation parameters. Particle sizes generally range from 50 to 300 nm, which is ideal for cellular uptake and enhanced permeability in biological systems. Moreover, the surface of the nanoparticles can be functionalized easily through further chemical modifications or by incorporating other polymers during the gelation process, enabling targeted delivery or stimuli-responsive behaviour. The ionic gelation method has been extensively applied in fabricating chitosan nanoparticles for diverse applications, including antimicrobial agents, vaccine delivery, and gene carriers. Alginate-based nanoparticles prepared through ionic gelation are also prominent in encapsulating therapeutic proteins and enzymes due to their gentle gelation mechanism with divalent cations like calcium ions. Furthermore, the method's aqueous and solvent-free nature aligns with green chemistry principles and regulatory requirements for biocompatible materials. Despite its advantages, ionic gelation has some limitations. The mechanical strength of nanoparticles formed solely by ionic interactions can be lower than those crosslinked chemically, which may affect stability under certain physiological conditions. To address this, dual crosslinking strategies combining ionic gelation with covalent crosslinking have been developed to enhance nanoparticle robustness. Additionally, reproducibility can be influenced by the precise control of process parameters, demanding careful optimization for scale-up and commercial manufacturing shown in Fig. 2 [38].

4 Functionalization: Ligand Attachment and Stimuli-Responsive Modifications

PNPs have gained extensive attention as versatile carriers for drug delivery, diagnostics, and biomedical applications due to their tunable size, biocompatibility, and ability to encapsulate diverse therapeutic agents. Beyond their basic structure, functionalization of these nanoparticles plays a critical role in enhancing their performance, targeting capability, and responsiveness to physiological conditions. Two major strategies in PNP functionalization include ligand attachment for targeted delivery and stimuli-responsive modifications for controlled release. Ligand Attachment for Targeted Delivery: Ligand attachment involves chemically or physically conjugating specific molecules onto the nanoparticle surface to impart targeting ability toward desired cells or tissues. Common ligands include antibodies, peptides, aptamers, small molecules, or polysaccharides, which recognize and bind to receptors overexpressed on target cells, such as cancer cells or inflamed tissues. This active targeting approach improves nanoparticle accumulation at the site of interest while minimizing off-target effects and systemic toxicity. The functionalization process typically involves surface activation of polymeric nanoparticles, often via carboxyl, amine, or hydroxyl groups inherent to polymers like PLGA, chitosan, or PEGylated polymers. These reactive groups enable covalent coupling through carbodiimide chemistry (e.g., EDC/NHS coupling), click chemistry, or maleimide-thiol linkages, ensuring stable ligand immobilization. The density, orientation, and stability of ligands on the nanoparticle surface critically influence targeting efficacy and biological interactions. Ligand-functionalized PNPs have demonstrated improved cellular uptake and therapeutic outcomes in cancer therapy, gene delivery, and imaging applications. For example, folate-conjugated nanoparticles exploit the overexpression of folate receptors in many tumors, enhancing selective uptake by malignant

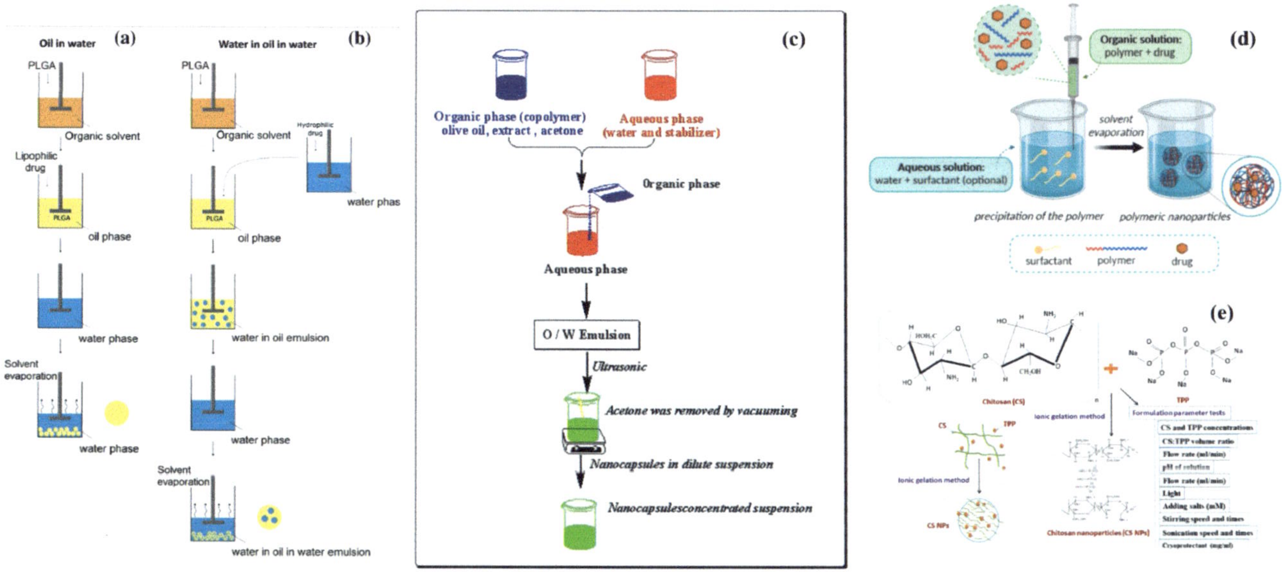

Fig. 2 **a** Solvent Emulsion Evaporation Method [39], **b** Double Emulsification Solvent Evaporation Method (W/O/W), **c** Emulsion Diffusion Method [40] **d** Nanoprecipitation (Solvent Displacement) [41], and **e** Ionotropic Gelation Method—schematic representations of common techniques used for the preparation of polymeric nanoparticles [38]

cells. Similarly, antibody-conjugated PNPs provide specificity toward cell-surface antigens, enabling precision medicine approaches. Stimuli-Responsive Modifications for Controlled Release: Stimuli-responsive functionalization endows polymeric nanoparticles with the ability to release their payloads in response to specific internal or external triggers, offering spatiotemporal control over drug delivery. Common stimuli include pH changes, temperature, redox potential, enzymes, light, magnetic fields, and ultrasound. This functionalization is achieved by incorporating stimuli-sensitive moieties either in the polymer backbone or as surface grafts. For instance, pH-sensitive linkers or polymers undergo conformational changes or bond cleavage in acidic tumor microenvironments or endosomal compartments, triggering drug release selectively in diseased tissues shown in Fig. 3. Temperature-sensitive polymers like poly (N-isopropylacrylamide) (PNIPAM) exhibit phase transitions near physiological temperature, facilitating temperature-triggered release. Redox-responsive nanoparticles often contain disulfide bonds that cleave in the reductive intracellular environment (high glutathione concentration), promoting intracellular payload release. Enzyme-responsive systems are designed to degrade or alter in the presence of overexpressed enzymes such as matrix metalloproteinases (MMPs) in cancerous tissues. These modifications improve therapeutic index by minimizing premature drug leakage during circulation and enhancing drug concentration at the target site. Combining stimuli-responsiveness with ligand targeting offers synergistic benefits, enabling nanoparticles to navigate biological barriers and release cargo precisely where needed [42].

5 Applications of Polymeric Nanoparticles

PNPs have emerged as a versatile and promising platform in various scientific and technological fields due to their tunable physicochemical properties, biocompatibility, and ability to encapsulate a wide range of therapeutic and diagnostic agents. The use of biodegradable carbohydrate-and protein-based nanoparticles is economically, socially, and environmentally sustainable, leading to their application across a wide range of fields shown in Fig. 4.

5.1 Drug Delivery

Drug delivery is a cornerstone of modern therapeutics, aimed at enhancing the efficacy and safety of medicines. Traditional drug administration methods, such as oral ingestion or intravenous injection, often encounter several significant challenges, including poor solubility of drugs, rapid clearance, low bioavailability, and nonspecific distribution throughout the body. These limitations can reduce the therapeutic effect of drugs and increase the likelihood of side effects [45]. To address these issues, PNPs have emerged as an innovative and highly versatile platform for drug delivery

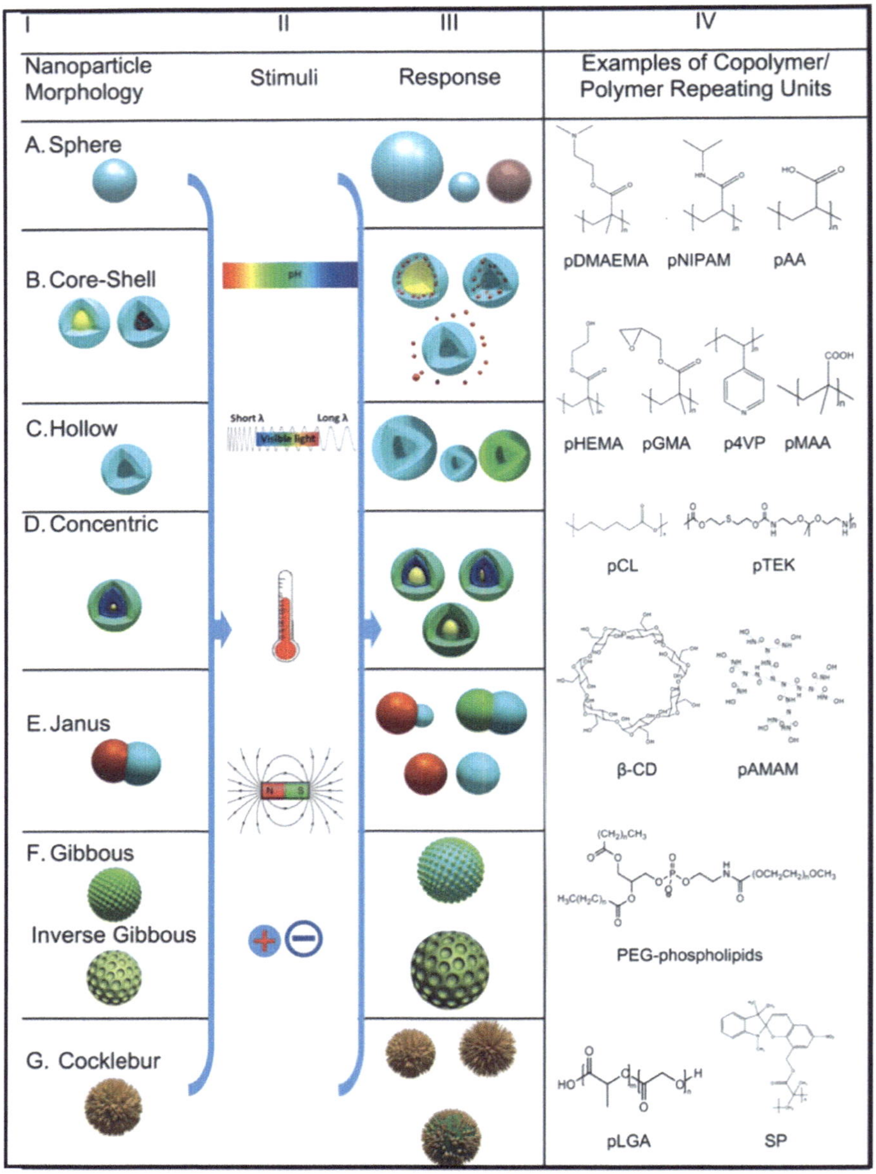

Fig. 3 Nanoparticle morphologies, stimuli, responsive behaviors, and commonly used polymers/ copolymers [101]

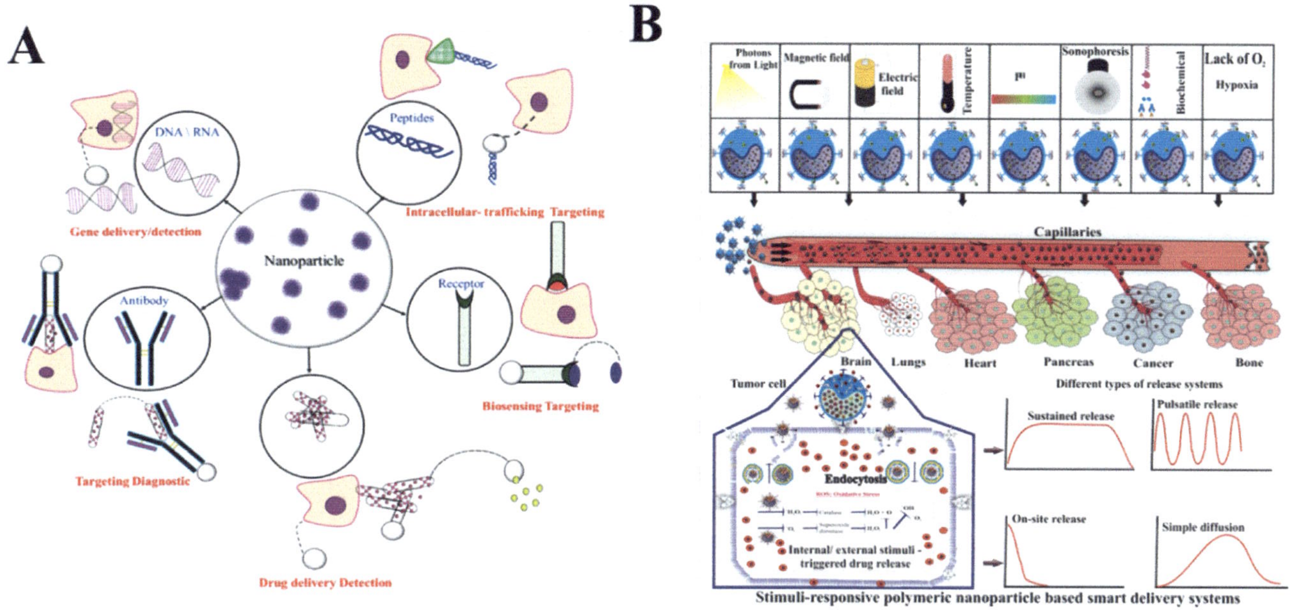

Fig. 4 **a** Applications of biodegradable carbohydrate-and protein-based nanoparticles [43]. **b** Environmentally responsive smart polymeric nanoparticle delivery systems, showing stimulus-induced phase transitions for controlled drug release [44]

systems. Polymeric nanoparticles are nano-sized carriers typically ranging from 10 to 500 nm, composed of biodegradable and biocompatible polymers such as PLGA, PCL, chitosan, PLA, and PEG. These carriers can encapsulate a wide variety of therapeutic agents, from hydrophobic drugs to proteins and nucleic acids, protecting them from premature degradation and improving their solubility and bioavailability. One of the primary advantages of polymeric nanoparticles in drug delivery is their ability to improve drug stability and solubility. Many therapeutic agents, particularly hydrophobic drugs, have limited water solubility, leading to poor absorption in the body. When these drugs are encapsulated within polymeric nanoparticles, their effective solubility is increased because the nanoparticles can disperse the drug in an aqueous environment more efficiently. Furthermore, the nanoparticle matrix protects sensitive drugs from enzymatic degradation, pH changes, and other destabilizing factors encountered in the bloodstream or gastrointestinal tract. This protection ensures that a higher proportion of the active drug reaches its intended site of action. Controlled and sustained release is another significant benefit offered by polymeric nanoparticles. Unlike conventional drug formulations that release the drug rapidly, resulting in spikes and troughs in drug concentration, polymeric nanoparticles can be engineered to release the drug slowly over extended periods. This controlled release is typically achieved by tailoring the polymer composition, molecular weight, and degradation rate of the nanoparticles. Such sustained release improves therapeutic outcomes by maintaining drug concentrations within the therapeutic window, reduces dosing frequency, and enhances patient compliance. For example, PLGA-based nanoparticles degrade hydrolytically over days or weeks, allowing drugs to be released gradually at the target site [46]. Figure 4b illustrates the environmentally responsive smart polymeric nanoparticle-based delivery systems discussed in this chapter, highlighting how changes in local stimuli induce phase transitions that trigger controlled drug release. Targeted drug delivery is one of the most transformative applications of polymeric nanoparticles. By modifying the surface of nanoparticles with ligands such as antibodies, peptides, or small molecules that recognize specific receptors overexpressed on diseased cells, these nanosystems can selectively deliver drugs to target tissues or cells. This active targeting enhances drug accumulation at the site of disease-such as tumors or inflamed tissues-while minimizing exposure to healthy organs, thereby reducing systemic toxicity and adverse effects. In cancer therapy, for instance, nanoparticles conjugated with folic acid or antibodies can preferentially bind to cancer cells and facilitate internalization of chemotherapeutic agents, improving treatment efficacy and safety. The versatility of polymeric nanoparticles extends to the range of therapeutic molecules they can carry. Small molecule drugs, proteins, peptides, and even genetic material like DNA or siRNA can be encapsulated or conjugated to the polymer matrix. This flexibility allows the development of multifunctional drug delivery systems that can combine therapy with diagnostics (theranostics) or co-delivery of multiple drugs for synergistic effects. Different nanoparticle synthesis methods, including solvent evaporation, nanoprecipitation, ionic gelation, and double emulsification, enable optimization of size, surface charge, drug loading capacity, and release kinetics tailored to the specific therapeutic application [47]. Cellular uptake and intracellular delivery are critical factors influencing the success of drug delivery systems. Due to their nanoscale size, polymeric nanoparticles are readily internalized by cells through endocytosis pathways. This property allows for efficient intracellular delivery of drugs, which is especially important for therapies targeting intracellular pathogens, cancer cells, or genetic material. Moreover, surface modifications with cell-penetrating peptides or other functional groups can further enhance cellular uptake and enable escape from endosomes, improving the bioavailability of drugs within cells. Polymeric nanoparticles have found extensive applications in various therapeutic fields. In oncology, they are used to deliver chemotherapeutic agents such as doxorubicin, paclitaxel, and cisplatin. These nanoparticle formulations exploit the enhanced permeability and retention (EPR) effect in tumors, where leaky vasculature allows nanoparticles to accumulate preferentially in tumor tissues. This targeted accumulation combined with controlled drug release significantly improves the therapeutic index of anticancer drugs while reducing their systemic toxicity. In gene therapy, polymeric nanoparticles protect nucleic acids from enzymatic degradation and enhance cellular uptake, improving transfection efficiency for the treatment of genetic disorders, cancers, and viral infections. Additionally, polymeric nanoparticles serve as adjuvants and delivery vehicles in vaccine formulations, improving antigen stability and immune responses. They also have applications in ocular drug delivery, where sustained release from nanoparticles reduces the need for frequent eye drops or injections, enhancing patient compliance. The mechanisms governing drug release from polymeric nanoparticles generally involve diffusion of the drug through the polymer matrix and degradation of the polymer itself. Biodegradable polymers such as PLGA undergo hydrolysis in physiological conditions, gradually breaking down into biocompatible byproducts, thereby releasing the encapsulated drug in a controlled manner. The release rate can be finely tuned by adjusting factors such as polymer molecular weight, composition ratio of lactide to glycolide in PLGA, nanoparticle size, and the method of drug incorporation. This tunability enables the design of nanoparticle systems tailored for immediate release, delayed release, or long-term sustained release depending on clinical needs [48, 49].

5.2 Cancer Therapy

PNPs have gained significant attention as advanced drug delivery systems in cancer therapy due to their unique physicochemical properties and ability to improve the therapeutic index of anticancer drugs [50]. These nanoparticles are typically composed of biodegradable polymers such as PLGA, PLA, chitosan, and PEG-based polymers. One of the primary advantages of PNPs in cancer treatment is their capacity to deliver chemotherapeutic agents directly to tumor sites, minimizing systemic toxicity and enhancing drug efficacy. This targeting is achieved via two main mechanisms: passive and active targeting. Passive targeting exploits the Enhanced Permeability and Retention (EPR) effect, where nanoparticles accumulate preferentially in tumor tissues due to their leaky vasculature and poor lymphatic drainage. For example, PLGA nanoparticles loaded with paclitaxel have demonstrated increased accumulation in tumors in vivo by leveraging the EPR effect, resulting in improved antitumor activity and reduced side effects compared to free drug administration. Active targeting involves surface functionalization of PNPs with ligands such as folic acid, antibodies, or peptides that bind specifically to receptors overexpressed on cancer cells. For instance, folate-conjugated chitosan nanoparticles have been shown to target folate receptors overexpressed on certain cancer cells, enhancing cellular uptake through receptor-mediated endocytosis. This specificity improves drug delivery efficiency and reduces off-target toxicity. Additionally, stimuli-responsive polymeric nanoparticles have been developed to respond to tumor microenvironment characteristics such as acidic pH, redox potential, or enzymatic activity. For example, pH-sensitive polymers can release their drug payload preferentially in the acidic tumor microenvironment, providing localized therapy while sparing healthy tissues. Mechanistically, these nanoparticles encapsulate hydrophobic anticancer drugs within their polymeric matrix or adsorb them onto their surface, protecting the drugs from premature degradation and improving solubility. PEGylation—a common surface modification—provides a hydrophilic "stealth" layer that prolongs circulation time by evading the mononuclear phagocyte system. Moreover, dual-functional nanoparticles combining targeting ligands with stimuli-responsiveness have been reported; for instance, PLGA nanoparticles functionalized with tumor-targeting peptides and loaded with doxorubicin showed enhanced uptake and pH-triggered release in breast cancer models, leading to improved therapeutic outcomes [51–53]. Despite the promising results, challenges remain in clinical translation, including tumor heterogeneity that affects nanoparticle penetration and variable responses to treatment. Nonetheless, several polymeric nanoparticle formulations, such as Abraxane (albumin-bound paclitaxel) and PEGylated liposomal doxorubicin, have successfully reached the market or clinical trials, validating the potential of polymeric nanomedicines in cancer therapy [54].

5.3 Vaccine Delivery and Immunotherapy

PNPs have emerged as a versatile platform for vaccine delivery and immunotherapy, offering significant advantages in enhancing immune responses and enabling targeted delivery of antigens and immunomodulatory agents. Traditional vaccines often suffer from limitations such as poor stability of antigens, insufficient immune activation, and the need for multiple booster doses [55]. PNPs can address these challenges by protecting antigens from degradation, enabling controlled release, and facilitating targeted delivery to antigen-presenting cells (APCs), such as dendritic cells and macrophages, which are critical for initiating robust adaptive immune responses. In vaccine delivery, PNPs are typically engineered from biodegradable polymers like PLGA, chitosan, or PCL, which are biocompatible and safely degrade into non-toxic byproducts. These nanoparticles can encapsulate protein antigens, peptides, DNA, or RNA vaccines, effectively shielding them from enzymatic degradation in biological environments. Moreover, their nano-size facilitates efficient uptake by APCs through endocytosis or phagocytosis, promoting antigen processing and presentation via major histocompatibility complexes (MHC) I and II pathways, thereby stimulating both humoral and cellular immunity. For example, PLGA nanoparticles encapsulating influenza antigens have been shown to induce strong antibody and T-cell responses in preclinical models, demonstrating enhanced immunogenicity compared to soluble antigens. Immunotherapy applications using PNPs extend to cancer vaccines and immune checkpoint modulation, where nanoparticles can deliver tumor-associated antigens or adjuvants to stimulate the immune system to recognize and attack cancer cells [56]. The surface of PNPs can be modified with targeting ligands such as mannose to specifically bind to dendritic cells, optimizing antigen presentation and T-cell activation. Additionally, PNPs can be engineered to co-deliver antigens along with immune adjuvants like CpG oligonucleotides or cytokines, further boosting the immune response. A notable example includes chitosan-based nanoparticles loaded with tumor antigens and Toll-like receptor agonists that have shown promise in preclinical melanoma models by activating cytotoxic T lymphocytes and reducing tumor growth. Another critical advantage of polymeric nanoparticles in immunotherapy is their ability to enable controlled and stimuli-responsive release of cargo. Tumor microenvironments or infected tissues often exhibit specific physiological triggers such as acidic pH or oxidative stress. Designing pH-sensitive or redox-responsive PNPs allows for site-specific release of immunomodulators, minimizing systemic side effects and improving therapeutic efficacy. For instance, acid-degradable PNPs have been used to deliver antigens that release their payload within the acidic endosomal compartments of dendritic cells, enhancing antigen cross-presentation and subsequent cytotoxic T-cell responses [57, 58].

PNPs also facilitate the development of next-generation nucleic acid vaccines, including mRNA vaccines. Encapsulation of mRNA within PNPs protects the fragile nucleic acids from nuclease degradation and assists in cellular uptake, making polymeric nanocarriers a promising alternative to lipid nanoparticles. For example, recent studies have explored biodegradable polymers such as poly (beta-amino esters) for mRNA delivery, demonstrating efficient transfection and potent immune responses in animal models [59].

5.4 Diagnostic Imaging

PNPs have increasingly become crucial agents in the field of diagnostic imaging, providing enhanced sensitivity, specificity, and multifunctionality compared to conventional imaging probes. These nanoparticles serve as versatile carriers that can be engineered to incorporate or conjugate imaging agents such as fluorescent dyes, magnetic resonance imaging (MRI) contrast agents, radionuclides for positron emission tomography (PET), and computed tomography (CT) contrast materials

[60]. Their small size and modifiable surface chemistry allow for prolonged circulation time, targeted delivery to specific tissues or cells, and reduced nonspecific background signal, which together improve image clarity and diagnostic accuracy. One key advantage of PNPs in diagnostic imaging is their ability to be functionalized with targeting ligands—such as antibodies, peptides, or small molecules—that direct the nanoparticles selectively to disease biomarkers or cellular receptors. This active targeting enhances accumulation of imaging agents at pathological sites, such as tumors or sites of inflammation, enabling early and precise detection. For instance, PLGA nanoparticles conjugated with folate have been used to target folate receptors that are overexpressed on various cancer cells, facilitating enhanced MRI contrast specifically in tumor tissues. In addition to MRI, PNPs have been widely applied in optical imaging techniques such as fluorescence imaging. Quantum dots or near-infrared (NIR) dyes encapsulated within biodegradable polymeric matrices provide bright and stable fluorescence signals with reduced photobleaching compared to free dyes. For example, chitosan-based nanoparticles loaded with NIR dyes have demonstrated effective in vivo imaging of lymph nodes and tumors, aiding surgical guidance and disease monitoring. Furthermore, polymeric nanoparticles allow the development of multimodal imaging probes by co-loading different imaging agents, enabling simultaneous detection by multiple imaging modalities. For example, PLGA nanoparticles co-encapsulating superparamagnetic iron oxide nanoparticles (SPIONs) for MRI and fluorescent dyes for optical imaging have been reported to provide complementary information, improving diagnostic confidence [61–63]. This multifunctionality is critical for complex disease diagnosis and therapy monitoring. Polymeric nanocarriers also enhance the biocompatibility and pharmacokinetic profile of contrast agents, often reducing toxicity and improving clearance from the body. Additionally, stimuli-responsive PNPs have been developed to release imaging agents selectively in response to environmental triggers like pH or enzymatic activity, further increasing imaging specificity [64].

5.5 Gene Therapy

Gene therapy offers a transformative approach to treat genetic disorders, cancers, and infectious diseases by delivering nucleic acids such as DNA, RNA, or small interfering RNA (siRNA) into target cells to modulate gene expression. However, the safe and efficient delivery of these genetic materials remains a major challenge due to their instability, susceptibility to enzymatic degradation, and difficulties crossing cellular membranes. PNPs have emerged as promising non-viral vectors that can overcome many limitations associated with conventional viral vectors, including immunogenicity, limited cargo capacity, and safety concerns. PNPs provide a protective nano-scale environment that shields nucleic acids from degradation by nucleases in the bloodstream, enhancing their stability and circulation time. Polymers such as PLGA, PEI, chitosan, and poly (β-amino esters) (PBAEs) have been widely used to fabricate nanoparticles that condense and encapsulate DNA or RNA molecules [65,

66]. These polymers facilitate cellular uptake via endocytosis and enable endosomal escape mechanisms, crucial for releasing the genetic cargo into the cytoplasm or nucleus to achieve efficient gene expression or silencing. A key feature of polymeric nanoparticles is their surface modifiability, which allows conjugation of targeting ligands such as antibodies, peptides, or aptamers, enabling selective delivery to specific cell types or tissues. For example, folate-modified chitosan nanoparticles have been employed to target cancer cells that overexpress folate receptors, enhancing the delivery efficiency of therapeutic siRNA to silence oncogenes. Similarly, PEGylation of nanoparticles can improve biocompatibility and reduce immune recognition, prolonging systemic circulation and improving gene transfer efficacy. Mechanistically, once polymeric nanoparticles carrying nucleic acids are internalized by cells, the acidic environment of endosomes triggers protonation of amine groups in cationic polymers like PEI, leading to a "proton sponge effect." This effect causes osmotic swelling and rupture of the endosomal membrane, releasing the genetic material into the cytoplasm. From here, plasmid DNA can enter the nucleus to enable gene expression, while siRNA remains in the cytoplasm to mediate RNA interference pathways and gene silencing. Several preclinical and clinical studies have demonstrated the therapeutic potential of polymeric nanoparticles in gene therapy. For instance, PEI-based nanoparticles have been used for delivery of tumor suppressor genes such as p53 in cancer models, showing significant tumor growth inhibition. PLGA nanoparticles loaded with siRNA targeting viral genes have shown promise in treating viral infections such as hepatitis B by reducing viral replication. Moreover, polymeric nanoparticles have been explored for CRISPR-Cas9 delivery systems to enable precise gene editing, further expanding their utility [67, 68].

The advantages of polymeric nanoparticles over viral vectors include lower immunogenicity, easier large-scale production, and the ability to carry large or multiple nucleic acid cargos. Furthermore, polymeric nanoparticles can be engineered for stimuli-responsive release, such as pH or redox-sensitive linkages, to enable controlled and site-specific gene delivery, minimizing off-target effects and toxicity [69, 70].

5.6 *Antimicrobial Applications*

The rise of antimicrobial resistance has become a critical global health concern, necessitating innovative strategies to combat bacterial, fungal, and viral infections. PNPs offer a versatile and effective approach for antimicrobial applications by enhancing drug delivery, improving the stability and bioavailability of antimicrobial agents, and enabling targeted and controlled release. Their nanoscale size allows PNPs to interact intimately with microbial membranes, enhancing penetration and efficacy while reducing systemic side effects. One of the major challenges in treating infections is the presence of biofilms-complex communities of microorganisms embedded within a self-produced extracellular polymeric matrix that protects them from antibiotics and host immune responses. Biofilms are implicated in chronic infections and are highly resistant to conventional antimicrobial therapies. Polymeric nanoparticles can penetrate biofilms more effectively than free drugs due to their small size and surface functionalization, facilitating targeted delivery and enhanced antimicrobial action within these protective barriers [71]. PNPs can be engineered to encapsulate a wide range of antimicrobial agents, including antibiotics, antifungals, antiviral drugs, and natural antimicrobial peptides. Polymers such as chitosan, PLGA, and poly (ε-caprolactone) (PCL) are commonly used due to their biocompatibility and biodegradability. Chitosan-based nanoparticles, in particular, possess intrinsic antimicrobial properties because of their positive charge, which interacts electrostatically with negatively charged bacterial cell walls, leading to membrane disruption and leakage of intracellular contents [72]. This inherent activity, combined with loaded antibiotics, results in a synergistic effect that can overcome resistant strains and disrupt biofilms. Mechanistically, PNPs improve antimicrobial efficacy by enabling sustained and controlled release of drugs, maintaining therapeutic concentrations at the infection site over extended periods. This controlled release reduces the frequency of dosing and limits the emergence of resistant microbes due to sub-therapeutic exposure [73]. Targeting moieties can also be attached to PNP surfaces to direct nanoparticles toward specific pathogens or biofilms, which are notoriously difficult to eradicate. For example, nanoparticles functionalized with antibodies or peptides that bind biofilm components of Staphylococcus aureus or Pseudomonas aeruginosa have shown enhanced targeting, penetration, and biofilm disruption capabilities. In practical applications, polymeric nanoparticles have been utilized for topical treatment of wound infections, oral delivery of antifungal agents, and inhalation therapies against respiratory pathogens. For instance, PLGA nanoparticles loaded with the antibiotic vancomycin have demonstrated improved treatment outcomes against methicillin-resistant Staphylococcus aureus (MRSA) biofilm-associated infections. Similarly, chitosan nanoparticles encapsulating antifungal drugs like fluconazole have enhanced mucosal adhesion and drug retention, increasing therapeutic effectiveness against Candida biofilms. In addition to direct antimicrobial activity, polymeric nanoparticles can be used to deliver gene-editing tools or small interfering RNA (siRNA) that target microbial resistance genes, providing a novel approach to combat multidrug-resistant strains. Furthermore, stimuli-responsive polymeric nanoparticles that release antimicrobial agents upon environmental triggers such as pH changes or bacterial enzymes offer site-specific therapy, minimizing damage to beneficial microbiota and maximizing action within biofilms [74].

Table 2 Common synthesis methods of polymeric nanoparticles, the polymers used, their applications, representative examples, and references

Synthesis method	Polymers used	Application	Example	Reference
Solvent emulsion evaporation	PEG PLGA	Cancer drug delivery	PLGA nanoparticles for doxorubicin delivery	[75]
Solvent emulsion evaporation	PLA-PEG	Targeted drug delivery	PEGylated PLA nanoparticles for paclitaxel	[76]
Double emulsification (W/O/W)	PLGA, PEG-PLA	Protein and vaccine delivery	PLGA nanoparticles loaded with insulin	[77]
Double emulsification (W/O/W)	Chitosan	Gene delivery	Chitosan nanoparticles for siRNA	[78]
Emulsion diffusion	PLGA, PCL	Sustained drug release	PLGA nanoparticles encapsulating anticancer agents	[79]
Emulsion diffusion	PEG-PCL	Hydrophobic drug delivery	PEG-PCL nanoparticles for curcumin delivery	[80]
Nanoprecipitation (solvent displacement)	PCL, PEG-PLA, PLGA	Gene and drug delivery	PEG-PLA nanoparticles for siRNA	[81]
Ionic gelation	Chitosan, alginate	Gene delivery, antimicrobial	Chitosan nanoparticles for DNA delivery	[82]
Ionic gelation	Hyaluronic acid	Targeted drug delivery	HA-chitosan nanoparticles targeting CD44 receptors	[83]

6 Biocompatibility and Toxicity Assessment of Polymeric Nanoparticles

The biocompatibility of polymeric nanoparticles used as drug delivery vehicles is a critical factor in the development of effective and safe therapeutic systems. Comprehensive evaluation through in vitro and in vivo assays, including cell viability, hemocompatibility, and histological analyses, is essential to ensure their suitability for clinical applications [84].

6.1 In Vitro Cytotoxicity Assessments

A widely employed method for evaluating the in vitro cytotoxicity of polymeric nanoparticles is the MTT assay, which measures cellular metabolic activity via mitochondrial dehydrogenase enzyme function. In this assay, the reduction of 3-(4,5-dimethylthiazol-2-yl)-2,5-diphenyltetrazolium bromide (MTT) to blue formazan crystals correlates with the number of metabolically active cells, where higher formazan production indicates greater cell viability [85]. For instance, Dionísio et al. investigated the cytotoxicity of pullulan-based nanoparticles using the MTT assay on Calu-3 respiratory cells, an immortalized lung adenocarcinoma cell line. After 24 hours of exposure, cell viability ranged from 74% to 94%, independent of polymer modification, concentration, or incubation duration, suggesting that these nanoparticles maintain acceptable biocompatibility without compromising cellular metabolic functions [86]. Similarly, the MTT assay was used to assess the cytotoxicity of poly (vanillin oxalate) (PVO) nanoparticles on RAW 264.7 cells. The results indicated dose-dependent cytotoxicity after 24 hours, with negligible toxic effects observed at concentrations below 500 µg/mL, highlighting the satisfactory biocompatibility of PVO nanoparticles at lower doses. Another study evaluated core/shell poly (methyl methacrylate)/biosurfactant nanoparticles using the MTT assay on peripheral blood mononuclear cells (PBMCs). Cell viability decreased with increasing nanoparticle concentrations, while a hemolysis assay demonstrated that surfactin-coated nanoparticles exhibited lower hemolytic activity compared to those coated with rhamnolipid or trehalose lipid biosurfactants. An alternative to the MTT assay, the CellTiter96® Aqueous Non-Radioactive Cell Proliferation Assay (MTS), was employed by Papa et al. to evaluate the viability of microglial cells exposed to poly (methyl methacrylate) nanoparticles. No significant reduction in cell viability was observed after up to 6 days of incubation, indicating excellent biocompatibility [87, 88].

6.2 In Vivo Biocompatibility Studies

In vivo assessments further elucidate the biocompatibility of polymeric nanoparticles. For example, Yoo et al. examined lung biocompatibility in Balb/c mice following intratracheal administration of hydroxybenzyl alcohol-incorporated polyoxalate nanoparticle suspensions. Histological analysis revealed minimal inflammatory responses and no structural changes in the airway, suggesting high tissue compatibility [89]. Similarly, intramuscular injection of PVO nanoparticles (1 mg/mL in phosphate-buffered saline) in Balb/c mice resulted in minimal inflammatory cell infiltration after 7 days, as confirmed by histological analysis. Katiyar et al. investigated the ocular tolerability of dorzolamide-loaded chitosan nanoparticles for potential glaucoma treatment using the hen's egg test-chorioallantoic membrane (HET-CAM) assay. This assay monitors adverse effects such as hemorrhage, hyperemia, or coagulation on the chorioallantoic membrane. The results indicated non-irritant properties, supporting the suitability of these nanoparticles for ocular applications [90]. Narayanan et al. assessed the hemocompatibility of ibuprofen-loaded gelatin nanoparticles, both with and without poly (ethylene glycol) (PEG) modification. Neither formulation induced hemolysis at concentrations up to 1 mg/mL, nor did they prolong coagulation times within the 0.1–1 mg/mL range. In vitro studies with PBMCs showed minimal immune activation, as evidenced by low cytokine induction, limited lymphocyte proliferation, and negligible toxicity. In vivo studies in Sprague Dawley rats following intravenous administration of nanoparticles (2.5 mg/kg) demonstrated prolonged drug release and enhanced bioavailability from PEGylated nanoparticles, attributed to the hydrophilic PEG coating, which reduces protein adsorption and extends circulation time. Histopathological analysis revealed no significant tissue damage in the liver or kidneys, and cytokine profiling showed minimal activation of pro-inflammatory (INF-γ, TNF-α) and anti-inflammatory (IL-4, IL-10) markers [91].

6.3 Targeted Toxicity in Cancer Therapy

In certain therapeutic contexts, such as cancer treatment, controlled toxicity of nanoparticle carriers can be advantageous. Kulhari et al. utilized the MTT assay to evaluate PLGA nanoparticles loaded with docetaxel and conjugated with bombesin peptide for targeting gastrin-releasing peptide (GRP) receptor-overexpressing MDA-MB-231 breast cancer cells. These nanoparticles exhibited 12-fold higher cytotoxicity compared to free docetaxel or Taxotere, with enhanced toxicity in a dose-dependent manner (6.25–100 ng/mL), underscoring their potential for targeted anticancer therapy [92]. Similarly, Zhao and Feng employed the MTT assay to assess Herceptin-functionalized poly (lactide)-D-α-tocopheryl polyethylene glycol 1000 succinate (PLA-TPGS) nanoparticles loaded with doxorubicin in SK-BR-3 breast cancer cells. The functionalized nanoparticles demonstrated dose-dependent cytotoxicity and reduced cell viability compared to non-functionalized carriers, likely due to the intrinsic toxicity of Herceptin. The biocompatibility and toxicity of polymeric nanoparticles are rigorously evaluated using a combination of in vitro assays (e.g., MTT, MTS, hemolysis) and in vivo studies (e.g., histological and pharmacokinetic analyses). These assessments confirm the safety and efficacy of polymeric nanoparticles for drug delivery, with tailored toxicity profiles being advantageous in applications such as targeted cancer therapy. The choice of evaluation method depends on the nanoparticle composition, intended application, and target tissue, ensuring a comprehensive understanding of their biological interactions [84].

7 Future Directions

As the field continues to evolve, future research should prioritize the clinical translation of PNPs through comprehensive in vivo studies, toxicity profiling, and robust clinical trials. One key emerging area is the integration of Artificial Intelligence (AI) and machine learning (ML) into nanoparticle research and design [93–95]. AI can assist in predicting nanoparticle behavior, drug loading efficiency, and pharmacokinetics, as well as optimizing formulation parameters to enhance therapeutic outcomes. Predictive algorithms can analyze large datasets to identify the best polymer–drug combinations or surface modifications for specific diseases, accelerating the development of targeted therapies. Additionally, AI can play a pivotal role in customizing PNPs for personalized medicine by integrating patient-specific data such as genetic profiles and disease biomarkers. Furthermore, the development of theranostic PNPs, which combine therapeutic and diagnostic functions, offers exciting potential for real-time disease monitoring and precision treatment. Designing smart, stimuli-responsive PNPs that respond to pH, temperature, redox conditions, or enzymatic activity could further improve targeted drug delivery. Advancements in green chemistry and scalable manufacturing methods will be essential for the widespread commercial adoption of these systems. Finally, establishing clear regulatory pathways and safety guidelines will be crucial in ensuring the safe and ethical use of AI-guided PNP platforms in clinical settings [96–100].

8 Conclusion

This book chapter provides a comprehensive overview of the fundamental concepts, synthesis techniques, surface functionalization approaches, and diverse biomedical applications of PNPs. The versatility of synthesis methods-ranging from solvent evaporation, nanoprecipitation, and emulsification to ionic gelation, polymerization, and supercritical fluid techniques-enables the precise design of nanoparticles with tailored physicochemical properties. Additionally, surface modifications, such as ligand conjugation and stimuli-responsive coatings, significantly enhance targeting efficiency and controlled release in biological systems. PNPs have shown exceptional promise in multiple biomedical applications, including drug delivery, cancer treatment, antimicrobial therapies, vaccine and gene delivery, immunotherapy, diagnostic imaging, and tissue engineering. Their capacity to encapsulate various therapeutic agents, improve bioavailability, extend systemic circulation time, and facilitate both passive and active targeting mechanisms addresses critical clinical challenges like multidrug resistance, tumor heterogeneity, and poor solubility of drugs. Furthermore, the integration of Artificial Intelligence (AI) and machine learning is revolutionizing the development of PNPs by accelerating nanoparticle optimization, predicting biological interactions, and enabling data-driven decisions, thus steering the field toward personalized nanomedicine. The synergy of AI, nanotechnology, and healthcare promises to deliver precision therapeutics that are more effective and patient-centric. Looking ahead, the future of PNPs lies in interdisciplinary collaboration, development of scalable and sustainable production methods, and thorough evaluation of safety and efficacy to facilitate clinical translation. Navigating regulatory pathways and addressing ethical considerations will also be crucial for the widespread adoption of these nanomaterials. In summary, polymeric nanoparticles have evolved from experimental entities to robust platforms with transformative potential in diagnostics and therapeutics. This book chapter encapsulates the essence and flavor of this dynamic field, highlighting how ongoing innovations and technological integrations position PNPs at the forefront of next-generation healthcare solutions, offering hope for safer, more targeted, and efficient treatment modalities.

References

1. Mahmoud, A.: Achieving a "grand convergence" in global health by 2035. Vaccine. **35**, A2–A5 (2017). https://doi.org/10.1016/J.VACCINE.2016.10.085
2. Leal, H.F., Mamani, C., Quach, C., Bédard, E.: Survey on antimicrobial resistance knowledge and perceptions in university students reveals concerning trends on antibiotic use and procurement. J. Assoc. Med. Microbiol. Infect. Dis. Can. **7** (2022). https://doi.org/10.3138/jammi-2022-0008
3. Bertagnolio, S., Suthar, A.B., Tosas, O., Van Weezenbeek, K.: Antimicrobial resistance: strengthening surveillance for public health action. PLoS Med. **20** (2023). https://doi.org/10.1371/journal.pmed.1004265
4. Ezike, T.C., Okpala, U.S., Onoja, U.L., Nwike, C.P., Ezeako, E.C., Okpara, O.J., Okoroafor, C.C., Eze, S.C., Kalu, O.L., Odoh, E.C., Nwadike, U.G., Ogbodo, J.O., Umeh, B.U., Ossai, E.C., Nwanguma, B.C.: Advances in drug delivery systems, challenges and future directions. Heliyon. **9** (2023). https://doi.org/10.1016/j.heliyon.2023.e17488
5. Mitchell, M.J., Billingsley, M.M., Haley, R.M., Wechsler, M.E., Peppas, N.A., Langer, R.: Engineering precision nanoparticles for drug delivery. Nat. Rev. Drug Discov. **20** (2021). https://doi.org/10.1038/s41573-020-0090-8
6. Garbayo, E., Pascual-Gil, S., Rodríguez-Nogales, C., Saludas, L., Estella-Hermoso de Mendoza, A., Blanco-Prieto, M.J.: Nanomedicine and drug delivery systems in cancer and regenerative medicine. Wiley Interdiscip. Rev. Nanomed. Nanobiotechnol. **12** (2020). https://doi.org/10.1002/wnan.1637
7. Simos, Y.V., Spyrou, K., Patila, M., Karouta, N., Stamatis, H., Gournis, D., Dounousi, E., Peschos, D.: Trends of nanotechnology in type 2 diabetes mellitus treatment. Asian J. Pharm. Sci. **16** (2021). https://doi.org/10.1016/j.ajps.2020.05.001
8. Bhardwaj, H., Jangde, R.K.: Current updated review on preparation of polymeric nanoparticles for drug delivery and biomedical applications. Next Nanotechnol. **2**, 100013 (2023). https://doi.org/10.1016/J.NXNANO.2023.100013
9. De, R., Mahata, M.K., Kim, K.T.: Structure-based varieties of polymeric Nanocarriers and influences of their physicochemical properties on drug delivery profiles. Adv. Sci. **9** (2022). https://doi.org/10.1002/advs.202105373
10. Herdiana, Y., Wathoni, N., Shamsuddin, S., Muchtaridi, M.: Scale-up polymeric-based nanoparticles drug delivery systems: development and challenges. Open Nano. **7**, 100048 (2022). https://doi.org/10.1016/J.ONANO.2022.100048
11. A. Zielinska, F. Carreiró, A.M. Oliveira, A. Neves, B. Pires, D. Nagasamy Venkatesh, A. Durazzo, M. Lucarini, P. Eder, A.M. Silva, A. Santini, E.B. Souto, Polymeric Nanoparticles: Production, Characterization, Toxicology and Ecotoxicology, Molecules 25 (2020). doi:https://doi.org/10.3390/molecules25163731
12. Singh, A., Kumar, P., Pathak, S., Jain, K., Garg, P., Pant, M., Mahapatro, A.K., Rath, D., Wang, L., Kim, S.K., Maurya, K.K., Pant, R.P.: A threefold increase in SAR performance for magnetic hyperthermia by compositional tuning in zinc-substituted iron oxide superparamagnetic nanoparticles with superior biocompatibility. J. Alloys Compd. **968**, 171868 (2023). https://doi.org/10.1016/J.JALLCOM.2023.171868
13. Kumar, P., Pathak, S., Singh, A., Khanduri, H., Basheed, G.A., Wang, L., Pant, R.P.: Microwave spin resonance investigation on the effect of the post-processing annealing of CoFe2O4 nanoparticles. Nanoscale Adv. **2**, 1939–1948 (2020). https://doi.org/10.1039/D0NA00156B

14. Aloke, C., Egwu, C.O., Onisuru, O.O., Otun, S., Achilonu, I.: Exploiting the therapeutic efficacy of nanoparticles in the treatment of multidrug-resistant bacteria: excitements and pitfalls. J. Drug Deliv. Sci. Technol. **104**, 106501 (2025). https://doi.org/10.1016/J.JDDST.2024.106501
15. Xuan, J., Feng, W., Wang, J., Wang, R., Zhang, B., Bo, L., Chen, Z.S., Yang, H., Sun, L.: Antimicrobial peptides for combating drug-resistant bacterial infections. Drug Resist. Updat. **68**, 100954 (2023). https://doi.org/10.1016/J.DRUP.2023.100954
16. Lu, Y., Qi, J., Wu, W.: Lipid nanoparticles: in vitro and in vivo approaches in drug delivery and targeting. Drug Target. Stimuli Sensitive Drug Deliv. Syst. (2018). https://doi.org/10.1016/B978-0-12-813689-8.00020-3
17. Chai, Q.Q., Li, D., Zhang, M., Gu, Y.W., Li, A.X., Wu, X., Liu, X.Y., Liu, J.Y.: Engineering nanoplatforms of bacterial outer membrane vesicles to overcome cancer therapy resistance. Drug Resist. Updat. **83**, 101277 (2025). https://doi.org/10.1016/J.DRUP.2025.101277
18. Handbook of polymers. (2022). https://doi.org/10.1016/c2021-0-00291-1
19. Kumar, P., Pathak, S., Singh, A., Verma, R., Khanduri, H., Jain, K., Tawale, J., Wang, L., Pant, R.P.: Augmented magnetic nanoparticle assimilation in rGO sheets for tailored static and dynamic magnetic properties in surface functionalized Co0.8Zn0.2Fe_2O_4 nanoferrite–rGO hybrid structures. J. Mater. Chem. C Mater. **12**, 18036–18047 (2024). https://doi.org/10.1039/D4TC03498H
20. Kumar, P., Pathak, S., Singh, A., Jain, K., Khanduri, H., Wang, L., Kim, S.-K., Pant, R.P.: Observation of intrinsic fluorescence in cobalt ferrite magnetic nanoparticles by Mn^{2+} substitution and tuning the spin dynamics by cation distribution. J. Mater. Chem. C Mater. **10**, 12652–12679 (2022). https://doi.org/10.1039/D2TC02605H
21. Kumar, P., Pathak, S., Singh, A., Khanduri, H., Jain, K., Tawale, J., Wang, L., Basheed, G.A., Pant, R.P.: Enhanced static and dynamic magnetic properties of PEG-400 coated $CoFe_{2-x}Er_xO_4$ ($0.7 \leq x \leq 0$) nanoferrites. J. Alloys Compd. **887**, 161418 (2021)
22. Jiang, X., Abedi, K., Shi, J.: Polymeric nanoparticles for RNA delivery. Encycl. Nanomater. (2023). https://doi.org/10.1016/B978-0-12-822425-0.00017-8
23. Kumar, P., Khanduri, H., Pathak, S., Singh, A., Basheed, G.A., Pant, R.P.: Temperature selectivity for single phase hydrothermal synthesis of PEG-400 coated magnetite nanoparticles. Dalton Trans. **49**, 8672–8683 (2020). https://doi.org/10.1039/D0DT01318H
24. Kumar, P., Pathak, S., Jain, K., Singh, A., Kuldeep, G.A., Basheed, R.P.P.: Low-temperature large-scale hydrothermal synthesis of optically active PEG-200 capped single domain MnFe2O4 nanoparticles. J. Alloys Compd. **904**, 163992 (2022). https://doi.org/10.1016/J.JALLCOM.2022.163992
25. Esteruelas, G., Souto, E.B., Cano, A., Ettcheto, M., Espina, M., Camins, A., García, M.L., Sánchez-López, E.: Polymeric nanoparticles as drug delivery systems for dementia. Nanomedicine-Based Approaches Treat. Dement. (2023). https://doi.org/10.1016/B978-0-12-824331-2.00006-6
26. Aleksandra, Z.'n., Filipa, C., Oliveira Ana, M., Neves Andreia, P., Bárbara, Nagasamy, V.D., Alessandra, D., Lucarini, M., Piotr, E., M, S.A., Antonello, S., B, S.E.: Polymeric nanoparticles: production, characterization, toxicology and ecotoxicology. Molecules. **25** (2020)
27. Pathak, S., Verma, R., Singhal, S., Chaturvedi, R., Kumar, P., Sharma, P., Pant, R.P., Wang, X.: Spin dynamics investigations of multifunctional ambient scalable Fe3O4 surface decorated ZnO magnetic nanocomposite using FMR. Sci. Rep. **11**, 3799 (2021)
28. Kumar, P., Khanduri, H., Singh, A., Jain, K., Pathak, S., Gautam, A., Wang, L., Pant, R.P.: Design and development of ferrofluids materials. In: Material Aspects of Ferrofluids, pp. 1–36. CRC Press (2023)
29. Haider, A., Ikram, M., Shahzadi, I., Asif Raza, M.: Properties and characterization of polymeric nanomaterials. In: Springer Series in Biomaterials Science and Engineering (2023). https://doi.org/10.1007/978-3-031-39947-3_3
30. Haider, A., Ikram, M., Shahzadi, I., Asif Raza, M.: Polymeric nanoparticles for bovine Mastitogens. In: Springer Series in Biomaterials Science and Engineering (2023). https://doi.org/10.1007/978-3-031-39947-3_6

31. Kumar, P., Pathak, S., Singh, A., Khanduri, H., Wang, X., Basheed, G.A., Pant, R.P.: Optimization of cobalt concentration for improved magnetic characteristics and stability of cox Fe3-xO4 mixed ferrite nanomagnetic fluids. Mater. Chem. Phys. **265**, 124476 (2021)
32. Singh, A., Kumar, P., Pathak, S., Jain, K., Garg, P., Pant, M., Mahapatro, A.K., Singh, R.K., Rajput, P., Kim, S.-K.: Tailored nanoparticles for magnetic hyperthermia: highly stable aqueous dispersion of Mn-substituted magnetite superparamagnetic nanoparticles by double surfactant coating for improved heating efficiency. J. Alloys Compd. **976**, 172999 (2024)
33. Sousa, S., Costa, A., Silva, A., Simões, R.: Poly (lactic acid)/cellulose films produced from composite spheres prepared by emulsion-solvent evaporation method. Polym. (Basel). **11** (2019). https://doi.org/10.3390/polym11010066
34. Alshamsan, A.: Nanoprecipitation is more efficient than emulsion solvent evaporation method to encapsulate cucurbitacin I in PLGA nanoparticles. Saudi Pharm. J. **22** (2014). https://doi.org/10.1016/j.jsps.2013.12.002
35. Iqbal, M., Zafar, N., Fessi, H., Elaissari, A.: Double emulsion solvent evaporation techniques used for drug encapsulation. Int. J. Pharm. **496**, 173–190 (2015). https://doi.org/10.1016/J.IJPHARM.2015.10.057
36. Piñón-Segundo, E., Llera-Rojas, V.G., Leyva-Gómez, G., Urbán-Morlán, Z., Mendoza-Muñoz, N., Quintanar-Guerrero, D.: The emulsification-diffusion method to obtain polymeric nanoparticles: two decades of research, nanoscale fabrication, optimization. Scale-Up Biol. Asp. Pharm. Nanotechnol., 51–83 (2018). https://doi.org/10.1016/B978-0-12-813629-4.00002-4
37. Martínez Rivas, C.J., Tarhini, M., Badri, W., Miladi, K., Greige-Gerges, H., Nazari, Q.A., Galindo Rodríguez, S.A., Román, R.Á., Fessi, H., Elaissari, A.: Nanoprecipitation process: from encapsulation to drug delivery. Int. J. Pharm. **532**, 66–81 (2017). https://doi.org/10.1016/J.IJPHARM.2017.08.064
38. Algharib, S.A., Dawood, A., Zhou, K., Chen, D., Li, C., Meng, K., Zhang, A., Luo, W., Ahmed, S., Huang, L., Xie, S.: Preparation of chitosan nanoparticles by ionotropic gelation technique: effects of formulation parameters and in vitro characterization. J. Mol. Struct. **1252**, 132129 (2022). https://doi.org/10.1016/J.MOLSTRUC.2021.132129
39. Lagreca, E., Onesto, V., Di Natale, C., La Manna, S., Netti, P.A., Vecchione, R.: Recent advances in the formulation of PLGA microparticles for controlled drug delivery. Prog. Biomater. **9** (2020). https://doi.org/10.1007/s40204-020-00139-y
40. Esmaeili, A., Rahnamoun, S., Sharifnia, F.: Effect of O/W process parameters on Crataegus azarolus L nanocapsule properties. J. Nanobiotechnology. **11** (2013). https://doi.org/10.1186/1477-3155-11-16
41. Mahajan, K., Thakur, N.: Simran, review on polymer based nanoparticles for increase the Bioavalibilty of poorly water soluble drug. J. Pharm. Res. Int. (2021). https://doi.org/10.9734/jpri/2021/v33i55b33847
42. Lee, D.S., Choi, H.G., Chung, K.H., Lee, B.Y., Yoon, H.C.: Fabrication of an integrated microfluidic device based on a heat-sensitive poly (N-isopropylacrylamide) polymer and micromachining protocols for programmed bio-molecular patterning. Sens. Actuators B Chem. **130** (2008). https://doi.org/10.1016/j.snb.2007.07.134
43. Kuperkar, K., Atanase, L.I., Bahadur, A., Crivei, I.C., Bahadur, P.: Degradable polymeric bio (nano) materials and their biomedical applications: a comprehensive overview and recent updates. Polymers (Basel). **16** (2024). https://doi.org/10.3390/polym16020206
44. Bennet, D., Kim, S.: Polymer nanoparticles for smart drug delivery. Appl. Nanotechnol. Drug Deliv. (2014). https://doi.org/10.5772/58422
45. Begines, B., Ortiz, T., Pérez-Aranda, M., Martínez, G., Merinero, M., Argüelles-Arias, F., Alcudia, A.: Polymeric nanoparticles for drug delivery: recent developments and future prospects. Nano. **10** (2020). https://doi.org/10.3390/nano10071403
46. Elmowafy, M., Shalaby, K., Elkomy, M.H., Alsaidan, O.A., Gomaa, H.A.M., Abdelgawad, M.A., Mostafa, E.M.: Polymeric nanoparticles for delivery of natural bioactive agents: recent advances and challenges. Polymers (Basel). **15** (2023). https://doi.org/10.3390/polym15051123

47. Lee, H., Lytton-Jean, A.K.R., Chen, Y., Love, K.T., Park, A.I., Karagiannis, E.D., Sehgal, A., Querbes, W., Zurenko, C.S., Jayaraman, M., Peng, C.G., Charisse, K., Borodovsky, A., Manoharan, M., Donahoe, J.S., Truelove, J., Nahrendorf, M., Langer, R., Anderson, D.G.: Molecularly self-assembled nucleic acid nanoparticles for targeted in vivo si RNA delivery. Nat. Nanotechnol. **7** (2012). https://doi.org/10.1038/nnano.2012.73
48. Kozielski, K.L., Tzeng, S.Y., Hurtado De Mendoza, B.A., Green, J.J.: Bioreducible cationic polymer-based nanoparticles for efficient and environmentally triggered cytoplasmic si RNA delivery to primary human brain cancer cells. ACS Nano. **8** (2014). https://doi.org/10.1021/nn500704t
49. Weng, J., Tong, H.H.Y., Chow, S.F.: In vitro release study of the polymeric drug nanoparticles: development and validation of a novel method. Pharmaceutics. **12** (2020). https://doi.org/10.3390/pharmaceutics12080732
50. Caraway, C.A., Gaitsch, H., Wicks, E.E., Kalluri, A., Kunadi, N., Tyler, B.M.: Polymeric nanoparticles in brain cancer therapy: a review of current approaches. Polymers (Basel). **14** (2022). https://doi.org/10.3390/polym14142963
51. Borzęcka, W., Domiński, A., Kowalczuk, M.: Recent progress in phthalocyanine-polymeric nanoparticle delivery systems for cancer photodynamic therapy. Nano. **11** (2021). https://doi.org/10.3390/nano11092426
52. Madej, M., Kurowska, N., Strzalka-Mrozik, B.: Polymeric nanoparticles—tools in a drug delivery system in selected cancer therapies. Appl. Sci. (Switz.). **12** (2022). https://doi.org/10.3390/app12199479
53. Hosseini, S.M., Mohammadnejad, J., Salamat, S., Beiram Zadeh, Z., Tanhaei, M., Ramakrishna, S.: Theranostic polymeric nanoparticles as a new approach in cancer therapy and diagnosis: a review. Mater. Today Chem. **29** (2023). https://doi.org/10.1016/j.mtchem.2023.101400
54. Chapter 6: Electrospun Bioscaffolds That Mimic the Topology of Extracellular Matrix, in: Nanomedicine in Cancer, 2017. doi:10.1201/9781315114361-7
55. De Souza Rebouças, J., Esparza, I., Ferrer, M., Sanz, M.L., Irache, J.M., Gamazo, C.: Nanoparticulate adjuvants and delivery systems for allergen immunotherapy. J. Biomed. Biotechnol. **2012** (2012). https://doi.org/10.1155/2012/474605
56. Danaeifar, M., Negahdari, B., Eslam, H.M., Zare, H., Ghanaat, M., Koushali, S.S., Malekshahi, Z.V.: Polymeric nanoparticles for DNA vaccine-based cancer immunotherapy: a review. Biotechnol. Lett. **45** (2023). https://doi.org/10.1007/s10529-023-03383-x
57. Bolhassani, A., Safaiyan, S., Rafati, S.: Improvement of different vaccine delivery systems for cancer therapy. Mol. Cancer. **10** (2011). https://doi.org/10.1186/1476-4598-10-3
58. Priyanka, M.A.H., Abusalah, H., Chopra, A., Sharma, S.A., Mustafa, O.P., Choudhary, M., Sharma, M., Dhawan, R., Khosla, A., Loshali, A., Sundriyal, J.: Saini, Nanovaccines: a game changing approach in the fight against infectious diseases. Biomed. Pharmacother. **167** (2023). https://doi.org/10.1016/j.biopha.2023.115597
59. Kim, H., Griffith, T.S., Panyam, J.: Poly (D, L-lactide-co-glycolide) nanoparticles as delivery platforms for TLR7/8 agonist-based cancer vaccine. J. Pharmacol. Exp. Ther. **370** (2019). https://doi.org/10.1124/jpet.118.254953
60. Perumal, V., Sivakumar, P.M., Zarrabi, A., Muthupandian, S., Vijayaraghavalu, S., Sahoo, K., Das, A., Das, S., Payyappilly, S.S., Das, S.: Near infra-red polymeric nanoparticle based optical imaging in cancer diagnosis. J. Photochem. Photobiol. B. **199** (2019). https://doi.org/10.1016/j.jphotobiol.2019.111630
61. Tiwari, H., Rai, N., Singh, S., Gupta, P., Verma, A., Singh, A.K., Kajal, P., Salvi, S.K., Singh, V.G.: Recent advances in nanomaterials-based targeted drug delivery for preclinical cancer diagnosis and therapeutics. Bioengineering. **10** (2023). https://doi.org/10.3390/bioengineering10070760
62. Choi, K.Y., Liu, G., Lee, S., Chen, X.: Theranostic nanoplatforms for simultaneous cancer imaging and therapy: current approaches and future perspectives. Nanoscale. **4** (2012). https://doi.org/10.1039/c1nr11277e

63. Barmin, R.A., Moosavifar, M.J., Dasgupta, A., Herrmann, A., Kiessling, F., Pallares, R.M., Lammers, T.: Polymeric materials for ultrasound imaging and therapy. Chem. Sci. **14** (2023). https://doi.org/10.1039/d3sc04339h
64. Miedema, I.H.C., Zwezerijnen, G.J.C., Huisman, M.C., Doeleman, E., Mathijssen, R.H.J., Lammers, T., Hu, Q., van Dongen, G.A.M.S., Rijcken, C.J.F., Vugts, D.J., C.W.: Menke-van der Houven van Oordt, PET-CT imaging of polymeric nanoparticle tumor accumulation in patients. Adv. Mater. **34** (2022). https://doi.org/10.1002/adma.202201043
65. Kim, J., Kang, Y., Tzeng, S.Y., Green, J.J.: Synthesis and application of poly (ethylene glycol)-co-poly (β-amino ester) copolymers for small cell lung cancer gene therapy. Acta Biomater. **41** (2016). https://doi.org/10.1016/j.actbio.2016.05.040
66. Lin, G., Zhang, H., Huang, L.: Smart polymeric nanoparticles for cancer gene delivery. Mol. Pharm. **12** (2015). https://doi.org/10.1021/mp500656v
67. Bhandari, R., Sharma, A., Kuhad, A.: Novel Nanotechnological approaches for targeting dorsal root ganglion (DRG) in mitigating diabetic neuropathic pain (DNP). Front. Endocrinol. (Lausanne). **12** (2022). https://doi.org/10.3389/fendo.2021.790747
68. Lee, J., Jeong, E.J., Lee, Y.K., Kim, K., Kwon, I.C., Lee, K.Y.: Optical imaging and gene therapy with neuroblastoma-targeting polymeric nanoparticles for potential Theranostic applications. Small. **12** (2016). https://doi.org/10.1002/smll.201501913
69. Zhou, X., Lian, H., Li, H., Fan, M., Xu, W., Jin, Y.: Nanotechnology in cervical cancer immunotherapy: therapeutic vaccines and adoptive cell therapy. Front. Pharmacol. **13** (2022). https://doi.org/10.3389/fphar.2022.1065793
70. Wan, Z., Zheng, R., Moharil, P., Liu, Y., Chen, J., Sun, R., Song, X., Ao, Q.: Polymeric micelles in cancer immunotherapy. Molecules. **26** (2021). https://doi.org/10.3390/molecules26051220
71. Mishra, A., Goswami, R., Bhatt, N.: Polymeric nanomaterials as broad-spectrum antimicrobial compounds. Adv. Polym. Nanomater. Biomed. Appl., 225–266 (2021). https://doi.org/10.1016/B978-0-12-814657-6.00008-2
72. Ul-Islam, M., Alabbosh, K.F., Manan, S., Khan, S., Ahmad, F., Ullah, M.W.: Chitosan-based nanostructured biomaterials: synthesis, properties, and biomedical applications. Adv. Ind. Eng. Polym. Res. **7**, 79–99 (2024). https://doi.org/10.1016/J.AIEPR.2023.07.002
73. Sánchez, A., Mejía, S.P., Orozco, J.: Recent advances in polymeric nanoparticle-encapsulated drugs against intracellular infections. Molecules. **25** (2020). https://doi.org/10.3390/molecules25163760
74. Werfel, T., Duvall, C.: Polymeric nanoparticles for gene delivery, polymers and nanomaterials for. Gene Ther., 147–188 (2016). https://doi.org/10.1016/B978-0-08-100520-0.00007-2
75. Zhang, D., Liu, L., Wang, J., Zhang, H., Zhang, Z., Xing, G., Wang, X., Liu, M.: Drug-loaded PEG-PLGA nanoparticles for cancer treatment. Front. Pharmacol. **13** (2022). https://doi.org/10.3389/fphar.2022.990505
76. Ruan, G., Feng, S.S.: Preparation and characterization of poly (lactic acid)–poly (ethylene glycol)–poly (lactic acid) (PLA–PEG–PLA) microspheres for controlled release of paclitaxel. Biomaterials. **24**, 5037–5044 (2003). https://doi.org/10.1016/S0142-9612(03)00419-8
77. Xu, B., Jiang, G., Yu, W., Liu, D., Liu, Y., Kong, X., Yao, J.: Preparation of poly (lactic-co-glycolic acid) and chitosan composite nanocarriers via electrostatic self assembly for oral delivery of insulin. Mater. Sci. Eng. C. **78** (2017). https://doi.org/10.1016/j.msec.2017.04.113
78. Wu, D., Zhu, L., Li, Y., Zhang, X., Xu, S., Yang, G., Delair, T.: Chitosan-based colloidal polyelectrolyte complexes for drug delivery: a review. Carbohydr. Polym. **238** (2020). https://doi.org/10.1016/j.carbpol.2020.116126
79. Hernández-Giottonini, K.Y., Rodríguez-Córdova, R.J., Gutiérrez-Valenzuela, C.A., Peñuñuri-Miranda, O., Zavala-Rivera, P., Guerrero-Germán, P., Lucero-Acuña, A.: PLGA nanoparticle preparations by emulsification and nanoprecipitation techniques: effects of formulation parameters. RSC Adv. **10** (2020). https://doi.org/10.1039/c9ra10857b
80. Grossen, P., Witzigmann, D., Sieber, S., Huwyler, J.: PEG-PCL-based nanomedicines: a biodegradable drug delivery system and its application. J. Control. Release. **260**, 46–60 (2017). https://doi.org/10.1016/J.JCONREL.2017.05.028

81. Almoustafa, H.A., Alshawsh, M.A., Chik, Z.: Technical aspects of preparing PEG-PLGA nanoparticles as carrier for chemotherapeutic agents by nanoprecipitation method. Int. J. Pharm. **533**, 275–284 (2017). https://doi.org/10.1016/J.IJPHARM.2017.09.054
82. Masalova, O., Kulikouskaya, V., Shutava, T., Agabekov, V.: Alginate and chitosan gel nanoparticles for efficient protein entrapment. Phys. Procedia. **40**, 69–75 (2013). https://doi.org/10.1016/J.PHPRO.2012.12.010
83. Wang, T., Hou, J., Su, C., Zhao, L., Shi, Y.: Hyaluronic acid-coated chitosan nanoparticles induce ROS-mediated tumor cell apoptosis and enhance antitumor efficiency by targeted drug delivery via CD44. J. Nanobiotechnology. **15** (2017). https://doi.org/10.1186/s12951-016-0245-2
84. Zhao, J., Feng, S.-S.: Effects of PEG tethering chain length of vitamin E TPGS with a Herceptin-functionalized nanoparticle formulation for targeted delivery of anticancer drugs. Biomaterials. **35**, 3340–3347 (2014)
85. Jana, S., Maji, N., Nayak, A.K., Sen, K.K., Basu, S.K.: Development of chitosan-based nanoparticles through inter-polymeric complexation for oral drug delivery. Carbohydr. Polym. **98**, 870–876 (2013)
86. Dionísio, M., Cordeiro, C., Remuñán-López, C., Seijo, B., da Costa, A.M.R., Grenha, A.: Pullulan-based nanoparticles as carriers for transmucosal protein delivery. Eur. J. Pharm. Sci. **50**, 102–113 (2013)
87. Hazra, C., Kundu, D., Chatterjee, A., Chaudhari, A., Mishra, S.: Poly (methyl methacrylate)(core)–biosurfactant (shell) nanoparticles: size controlled sub-100 nm synthesis, characterization, antibacterial activity, cytotoxicity and sustained drug release behavior. Colloids Surf. A Physicochem. Eng. Asp. **449**, 96–113 (2014)
88. Kwon, J., Kim, J., Park, S., Khang, G., Kang, P.M., Lee, D.: Inflammation-responsive antioxidant nanoparticles based on a polymeric prodrug of vanillin. Biomacromolecules. **14**, 1618–1626 (2013)
89. Yoo, D., Guk, K., Kim, H., Khang, G., Wu, D., Lee, D.: Antioxidant polymeric nanoparticles as novel therapeutics for airway inflammatory diseases. Int. J. Pharm. **450**, 87–94 (2013)
90. Katiyar, S., Pandit, J., Mondal, R.S., Mishra, A.K., Chuttani, K., Aqil, M., Ali, A., Sultana, Y.: In situ gelling dorzolamide loaded chitosan nanoparticles for the treatment of glaucoma. Carbohydr. Polym. **102**, 117–124 (2014)
91. Narayanan, D., Geena, M.G., Lakshmi, H., Koyakutty, M., Nair, S., Menon, D.: Poly-(ethylene glycol) modified gelatin nanoparticles for sustained delivery of the anti-inflammatory drug ibuprofen-sodium: an in vitro and in vivo analysis. Nanomedicine. **9**, 818–828 (2013)
92. Kulhari, H., Pooja, D., Shrivastava, S., VGM, N., Sistla, R.: Peptide conjugated polymeric nanoparticles as a carrier for targeted delivery of docetaxel. Colloids Surf. B Biointerfaces. **117**, 166–173 (2014)
93. López-Rios de Castro, R., Ziolek, R.M., Ulmschneider, M.B., Lorenz, C.D.: Therapeutic peptides are preferentially solubilized in specific microenvironments within PEG-PLGA polymer nanoparticles. Nano Lett. **24** (2024). https://doi.org/10.1021/acs.nanolett.3c04558
94. Alshawwa, S.Z., Kassem, A.A., Farid, R.M., Mostafa, S.K., Labib, G.S.: Nanocarrier drug delivery systems: characterization, limitations, future perspectives and implementation of artificial intelligence. Pharmaceutics. **14** (2022). https://doi.org/10.3390/pharmaceutics14040883
95. Youshia, J., Ali, M.E., Lamprecht, A.: Artificial neural network based particle size prediction of polymeric nanoparticles. Eur. J. Pharm. Biopharm. **119** (2017). https://doi.org/10.1016/j.ejpb.2017.06.030
96. Lashari, N., Ganat, T., Otchere, D., Kalam, S., Ali, I.: Navigating viscosity of GO-SiO2/HPAM composite using response surface methodology and supervised machine learning models. J. Pet. Sci. Eng. **205** (2021). https://doi.org/10.1016/j.petrol.2021.108800
97. Gong, D., Ben-Akiva, E., Singh, A., Yamagata, H., Est-Witte, S., Shade, J.K., Trayanova, N.A., Green, J.J.: Machine learning guided structure function predictions enable in silico nanoparticle screening for polymeric gene delivery. Acta Biomater. **154** (2022). https://doi.org/10.1016/j.actbio.2022.09.072

98. Tao, H., Wu, T., Aldeghi, M., Wu, T.C., Aspuru-Guzik, A., Kumacheva, E.: Nanoparticle synthesis assisted by machine learning. Nat. Rev. Mater. **6** (2021). https://doi.org/10.1038/s41578-021-00337-5
99. Singh, A., Kumar, P., Jain, K., Singh, V.N., Maurya, K.K., Pant, R.P.: Insight on efficient biocompatible magnetic nanoparticles for cancer treatment using magnetic hyperthermia technique. In: Material Aspects of Ferrofluids, pp. 168–183. CRC Press (2023)
100. Singh, A., Kumar, P., Rajput, P., Jain, K., Pathak, S., Puri, S., Maurya, K.K., Pant, R.P.: Investigating the impact of zinc ion substitution on the rheological properties and hyperthermia potential of cobalt ferrite-based ferrofluids. Mater. Today Proc. (2023)
101. Liu, X., Yang, Y., Urban, M.W.: Stimuli-Responsive Polymeric Nanoparticles. Macromol. Rapid Commun. **38** (2017). https://doi.org/10.1002/marc.201700030

Recent Approaches of Metal Oxide-Based Nanomaterials in Drug Delivery Systems (DDS)

Sudipto Debnath, Deepak Kumar, Susmita Mondal, Manosi Das, Ch. V. Narasimajhi, and Gajji Babu

1 Introduction

The Oxford English Dictionary traces the term 'nano' to the Latin word *nanus* and its Greek root *nanos*, both of which mean 'dwarf' [1]. In 1947, at the 14th IUPAC (International Union of Pure and Applied Chemistry) conference, the prefix *nano* was formally adopted to indicate one-billionth (10^{-9}) of a unit [2]. Since then, *nano* has gained wide usage in contemporary science, giving rise to numerous terms now found in dictionaries, such as nanometre, nanoscale, nanoscience, nanotechnology, nanostructure, nanotube, nanowire, nanorobot, etc. [3]. In these usages, the prefix signifies entities or phenomena that are 'extremely small', typically measured on the nanometre scale. The term 'nanotechnology' was first coined by Nobel laureate Richard P. Feynman during his famous lecture "*There's Plenty of Room at the Bottom*" in 1959; since then, various ground-breaking developments have taken place in the field of nanotechnology [4]. With the aid of nanotechnology, numerous materials of various types are produced at the nanoscale level. Nanoparticles are generally referred to as any particulate matter for which at least one dimension lies in the range of 1–100 nm [5]. The various chemical and physical processes involved in the synthesis of materials on the nanometre scale can be categorised as Nano-chemistry, which plays a crucial part in tailoring the physical and chemical properties of the nanoparticles. When nanoparticles are used in medicine, it is called nanomedicine. Nanomedicine is a blooming field of research with enormous prospects for the improvement of the diagnosis and treatment of human diseases. The fields of advanced medicine seek to address various medical challenges and shortcomings faced by conventional medicine, which include poor bioavailability,

S. Debnath (✉) · D. Kumar · S. Mondal · M. Das · Ch. V. Narasimajhi · G. Babu
Central Ayurveda Research Institute, Salt Lake, Sector-V, Bidhannagar, Kolkata, India
e-mail: debnath1304@gmail.com; sudpta.debnath@ccras.nic.in

S. Debnath
Central Ayurveda Research Institute, Gwalior Road, Jhansi, India

G. Tailor et al. (eds.), *Nanomaterials*, Engineering Materials,
https://doi.org/10.1007/978-981-92-0678-0_6

impaired target specificity, systemic and organ toxicity, etc. The enhanced therapeutic performance of nanoparticles is largely linked to their minute dimensions and high surface-to-volume ratio, which facilitate efficient cellular uptake and enable penetration into the nucleus, specific molecular targets, and even deeper tissue layers. In recent years, nanoparticle-based drug delivery systems have emerged as an adjunct and enabling technology, offering transformative potential for modern medical applications [6]. Nanoparticles can be synthesized using a wide spectrum of strategies, including eco-friendly green routes as well as established chemical and physical methods [7, 8]. Products obtained through these techniques demonstrate notable advantages such as pronounced antimicrobial and anti-inflammatory activities, excellent stability, simplicity of preparation, ease of integration into formulations, and suitability for surface modification with diverse functional moieties [9, 10]. Metal oxide nanoparticles (MONPs) have gained substantial attention in biomedicine and are increasingly utilized as antibacterial surface coatings, agents for wound management, biosensing platforms, anticancer therapeutics, and imaging contrast agents [11]. Widely investigated MONPs include zinc oxide (ZnO), cerium oxide (CeO_2), iron oxide (Fe_2O_3), silver oxide (AgO), magnesium oxide (MgO), titanium dioxide (TiO_2), nickel oxide (NiO), zirconium oxide (ZrO), and cadmium oxide (CdO). These nanomaterials exhibit a broad range of biological activities, supported by extensive in vitro and in vivo investigations. Among them, ZnO nanoparticles are regarded as relatively biocompatible, with their biological performance being strongly influenced by physicochemical attributes such as particle size, shape, and aspect ratio [12]. Notably, ZnO nanoparticles demonstrate preferential cytotoxic effects against cancer cells, primarily through reactive oxygen species (ROS) generation and interference with zinc-dependent cellular proteins, a mechanism that has contributed to their FDA recognition for antitumor applications [13, 14]. Cerium oxide (CeO_2) nanoparticles are distinguished by their reversible redox cycling between Ce^{3+} and Ce^{4+} states [15], coupled with their capacity for oxygen storage and release [16]. This dynamic surface chemistry underlies their dual antioxidant and anticancer properties. Nevertheless, in acidic microenvironments or at elevated doses, CeO_2 nanoparticles may shift toward pro-oxidant behaviour, leading to cytotoxic effects that are strongly influenced by concentration and synthesis methodology [17]. Titanium dioxide (TiO_2) nanoparticles are extensively applied in tissue and bone regeneration owing to their ability to enhance osseointegration, facilitate wound healing, and support cellular adhesion [18]. In addition, they exhibit pronounced antibacterial activity through reactive oxygen species (ROS) generation under ultraviolet irradiation [19], a mechanism that also contributes to their reported anticancer efficacy [20]. Iron oxide (Fe_2O_3) nanoparticles serve as a highly adaptable system, either alone or in hybrid formulations. Their magnetic properties enable targeted drug delivery strategies, particularly in regenerative medicine and oncology, where payload release can be guided by external magnetic fields or functional surface ligands [21]. Furthermore, their use in magnetic cell labelling allows for non-invasive tracking and monitoring of therapeutic outcomes [22, 23]. Magnesium oxide (MgO) nanoparticles are generally regarded as biocompatible due to their aqueous solubility, which minimizes long-term tissue accumulation. Consequently, they have been investigated for use in antibacterial wound dressings and as carriers for anticancer therapeutics [24]. Nickel oxide (NiO) nanoparticles display a wide spectrum of antimicrobial, antifungal, and anticancer activities, highlighting their potential in biomedical applications [25]. Zirconium oxide (ZrO) nanoparticles offer favorable mechanical stability and biodegradability; however, comprehensive evaluations of their antibacterial, regenerative, and anticancer properties remain scarce [26]. Cadmium oxide (CdO) nanoparticles demonstrate potent antibacterial and antifungal effects mediated by metal ion release and ROS generation [27]. At carefully controlled low doses, they have also shown comparatively strong antitumor activity among heavy-metal oxide nanoparticles, although concerns regarding toxicity significantly restrict their clinical applicability [28]. This review primarily emphasizes recent advances in metal oxide–based nanomaterials for drug delivery systems (DDS), while also discussing the underlying chemistry of their synthesis, biological activities, therapeutic applications, and the critical parameters governing their functional performance.

2 Why are Nanoparticles So Different?

As the definition goes, any particulate matter created with at least one dimension in the range of 1–100 nm is referred to as nanoparticles [29–31]. The properties of Nanoparticles are widely distinct from their bulk counterpart [32]. This is because of one phenomenon, which is known as the 'quantum effect', which actually governs the characteristics and properties of the Nanoparticles [33, 34]. Properties of the particles depend on the size in this nano range [35]. So, when the dimension of the particles is in the nano range, various properties like melting point, hardness, optical phenomena like fluorescence, colour, transparency, electrical conductivity, magnetic permeability, chemical reactivity, etc., change as the size of the particle changes. This concept can be clearly explained through a familiar and illustrative example. Gold (Au) is commonly known as a yellow metal in its bulk form; however, when its particle size is reduced to the nanometre scale, it exhibits strikingly different colours, such as red or purple. This colour variation arises because, at the nanoscale, the movement of electrons becomes spatially confined, leading to a modified interaction with incident light compared to bulk gold. Such size-dependent optical behaviour [36] has significant practical relevance. For instance, gold nanoparticles can preferentially localize within tumour tissues, enabling enhanced imaging and allowing targeted laser-mediated destruction of cancer cells while sparing surrounding healthy tissue. By exploiting this so-called "quantum effect," the physicochemical properties of nanoparticles can be deliberately adjusted to meet specific functional requirements. Another important characteristic of nanoparticles is their enhanced reactivity, which is largely a consequence of their exceptionally high surface area [37]. When compared to bulk materials of the same mass, nanomaterials possess a much larger surface-to-volume ratio, allowing a greater proportion of atoms to interact with the surrounding environment. This increased surface exposure markedly improves chemical reactivity and plays a crucial role in the development of highly efficient catalytic systems.

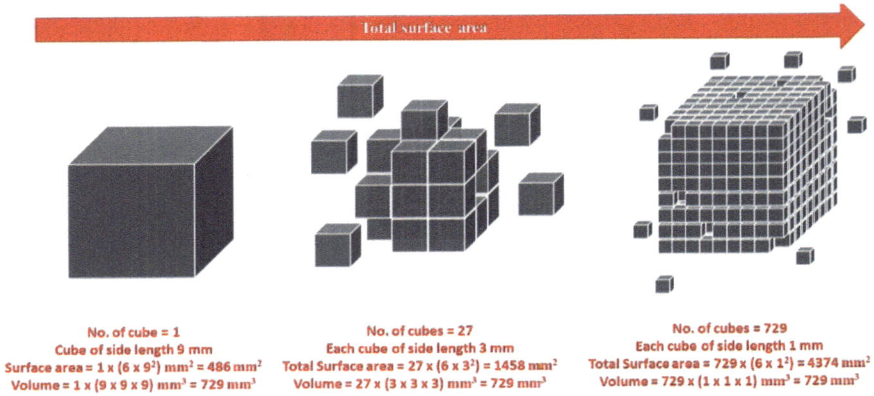

Fig. 1 An illustration to show the increased surface area when bulk material is disintegrated without changing the total volume

The relationship between particle size and surface area can be easily demonstrated through a simple example. Consider a cube with a side length of 9 mm, which has a total surface area of 486 mm^2. If this cube is divided into 27 smaller cubes, the combined surface area increases to 1458 mm^2. Further fragmentation of the original cube into 729 even smaller cubes results in a total surface area of 4374 mm^2 (see Fig. 1). This straightforward illustration clearly shows that, while the overall volume remains constant, reducing the size of a bulk material dramatically increases its total surface area. This principle underlies many of the unique properties and applications of nanomaterials.

Besides these two factors, i.e., quantum effect and increased surface area of nanoparticles, there is another factor, namely 'surface area to volume ratio' [38], which also contributes to the enhanced activity of the nanoparticles. With the increase in the surface area (SA) to volume (V) ratio, the percentage of atoms at the surface increases and surface forces dominate. Let's understand this concept with an example.

Let us consider a spherical nanoparticle having a diameter of 100 nm. We now calculate the SA: V ratio for this nanoparticle.

$$V = \frac{4}{3}\pi r^3 = \frac{1}{6}\pi R^3 = \frac{\pi \times (100 \times 10^{-9})^3}{6} = 5.24 \times 10^{-22}\ m^3$$

$$SA = 4\pi r^2 = \pi R^2 = \pi \times (100 \times 10^{-9})^2 = 3.14 \times 10^{-14}\ m^2$$

$$\text{Radius} = r = \frac{R}{2};\ R = \text{Diameter} \quad (1\ nm = 10^{-9}\ m)$$

$$\text{So, } SA{:}V = \frac{3.14 \times 10^{-14}\ m^2}{5.24 \times 10^{-22}\ m^3} = 0.599 \times 10^{8}\ m^{-1}$$

So, the ratio becomes approximately $>10^8{:}1$, which is substantially higher than a macro-sized particle.

Below (see Table 1) is the comparison of the SA: V ratio with the increased diameter of the nanoparticles, and it has also been represented graphically (see Fig. 2).

Table 1 Comparison of the SA: V ratio with the increased diameter of the nanoparticle

Diameter (nm)	Volume (nm^3)	Surface area (nm^2)	SA:V ratio (nm^{-1})
1	0.52375	3.1425	6
10	523.75	314.25	0.6
100	523,750	31,425	0.06
1000	523,750,000	3,142,500	0.006

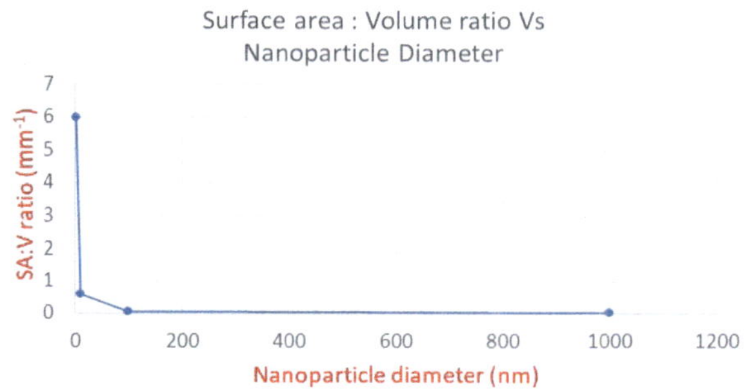

Fig. 2 Graphical representation of SA:V ratio versus nanoparticle diameter

From the comparison table and the graphical representation, it is quite clear that when the nanoparticle diameter is below 100 nm, the SA: V ratio increases drastically.

3 Synthesis of Nanoparticles

Nanoparticles synthesised utilising green techniques and biological organisms are known as biogenic Nanoparticles [39]. Biological organisms include bacteria, actinomycetes, fungi, plants, viruses, yeasts/fungi, enzymes, algae, waste materials, etc. [40]. In particular, there are two general pathways for the synthesis of nanoparticles: the top-down approach and the bottom-up approach. In the top-down approach, the bulk material is disintegrated into smaller pieces (i.e., nanoparticles), whereas in the bottom-up approach, a larger structure (i.e., nanoparticles) is built starting from atoms, molecules or clusters (see Fig. 3). So, it can be said that technique-wise, one process is just reversed from that of another method, but in the end, both processes give more or less the same result [41]. In practice, it has been seen that the bottom-up approach is superior to the top-down approach because the former process has the advantage of producing nanoparticles with fewer defects, homogeneous chemical composition, better short- and long-range ordering, etc. [42].

While discussing the synthesis of Nanoparticles, the physical and chemical methods are in use at large. Synthesis of Nanoparticles using biological means is not that popular compared to the other two techniques. Please see Fig. 4 which represents various techniques involved in these three processes.

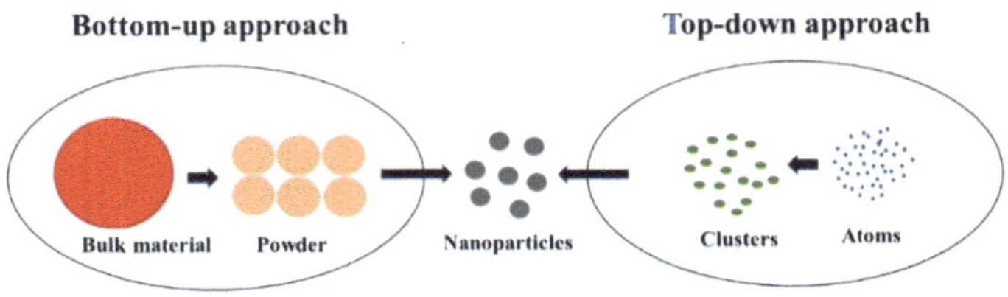

Fig. 3 Schematic representation of bottom-up and top-down approaches for nanoparticle synthesis

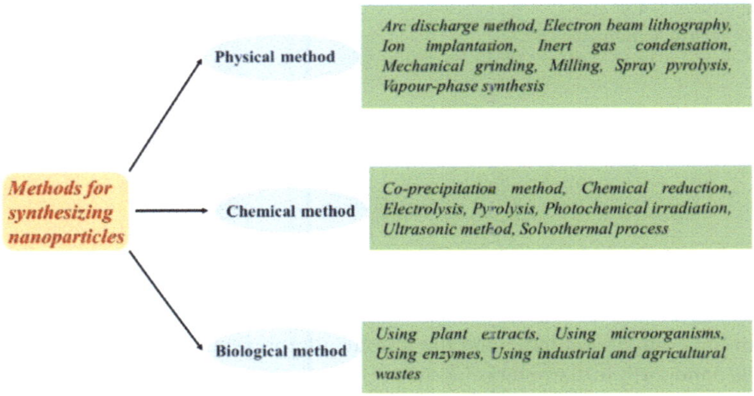

Fig. 4 Various methods for synthesising nanoparticles

But the production of nanoparticles through conventional physical and chemical methods has shortcomings, such as the use of toxic chemicals that could exert potential hazards like carcinogenicity, toxicity, and environmental toxicity, along with the generation of toxic by-products at the end of the process, which is potentially hazardous for the environment. Besides that, hazardous substances like reducing agents, organic solvents, and stabilisers/capping agents make the toxicity problems more pronounced [43]. Chemicals are known for preventing the coagulation of nanoparticles in colloidal synthesis. But the fact is that the use of toxic solvents and chemical contaminants restricts the use of Nanoparticles in various clinical and biomedical applications. Conventional strategies like chemical and physical methods for Nanoparticle synthesis can be used for their production in large quantities with definite morphologies in a short time span; although, these procedures are complicated, costly, inefficient, and outmoded [44].

4 Characterisation of Nanoparticles

After the synthesis of nanoparticles, the crucial step is the characterisation of the synthesised Nanoparticles. Why is characterisation so important? A simple answer is that, at first, one has to ascertain whether Nanoparticles have formed or not. Yes, one has to ensure that the material that he or she has synthesised is actually in the nano range. Once it is confirmed that the synthesised materials are on the nanoscale, then one focuses on some other parameters. As this review intends to shed light on the recent advances, approaches, applications, and prospects of metal oxide-based nanomaterials in drug delivery systems, we will limit ourselves to preliminary discussions regarding the techniques used for the characterisation of Nanoparticles. Please see Table 2 for some basic parameters and the characterisation technique [45] required for their measurement.

Table 2 Characterisation techniques for nanoparticles

Parameters to be studied [a]	Characterisation techniques [b]
Detection of nanoparticles	FESEM[c], TEM[d]
Size	FESEM, TEM, DLS[e], AFM[f], MALDI[g], NMR[h]
Shape	TEM, HRTEM[i], AFM, FMR[j]
Crystal structure	XRD[k], STEM[l]
Elemental-chemical composition	ICP-OES[m], SEM-EDX[n]
Size distribution	DLS, FMR
Chemical state-oxidation state	XPS[o], EELS[p]
Surface charge	ζ-Potential
Surface characterization	FT-IR[q]
Density	DCS[r], RMM-MEMS[s]
Surface area	BET[t]
3D visualization	AFM, 3D-tomography
Optical properties	UV-Vis, PL[u], EELS-STEM
Magnetic properties	SQUID[v], VSM[w]

[a]Some common parameters have been included, [b]This list is not exclusive (most common techniques have been listed), [c]Field emission scanning electron microscopy, [d]Transmission electron microscopy, [e]Dynamic light scattering, [f]Atomic force microscopy, [g]Matrix assisted laser desorption/ionization, [h]Nuclear magnetic resonance spectroscopy, [i]High-resolution transmission electron microscopy, [j]Ferromagnetic resonance spectrometer, [k]X-ray diffraction, [l]Scanning transmission electron microscope, [m]Inductively coupled plasma- optical electron microscopy, [n]Scanning electron microscopy with energy dispersive X-ray spectroscopy, [o]X-ray photoelectron spectroscopy, [p]Electron energy-loss spectroscopy, [q]Fourier-transform infrared spectroscopy, [r]Differential centrifugal sedimentation, [s]Resonant mass measurement microelectron-mechanical system, [t]Brunauer-Emmett-Teller [u]Photoluminescence spectroscopy, [v]Superconducting quantum interface device, [w]Vibrating sample magnetometer.

5 Metal Oxide Nanoparticles: Syntheses and Key Characteristics

Metal oxide nanoparticles (MONPs) are most often produced through bottom-up physical and chemical methodologies. Although these conventional approaches typically provide high yields, they are associated with several limitations, such as elevated operational costs, the use of hazardous chemicals, and the necessity for stabilising or capping agents. Residual toxic reagents may remain adhered to the nanoparticle surface, which raises serious concerns regarding their suitability for biomedical applications [46]. Widely employed fabrication techniques include chemical precipitation, wet-chemical synthesis, hydrothermal and solvothermal processes, sol–gel methods, solid-state pyrolysis, thermal decomposition, and microwave-assisted synthesis. While these methods are technically efficient, the presence of chemical impurities often renders the resulting nanoparticles less appropriate for direct biological use. In response to these challenges, biogenic or green synthesis strategies have emerged as promising alternatives. Such approaches utilise eco-friendly and non-toxic biological resources including plant extracts, biopolymers, algae, and microorganisms to mediate nanoparticle formation. Nanoparticles produced via these routes generally exhibit enhanced biocompatibility, greater colloidal stability, and improved biodegradability, making them more suitable for biomedical and pharmaceutical applications [47]. Please see Fig. 5 which represents a schematic overview of these different synthetic strategies.

Chemical Precipitation relies on a precipitating agent (e.g., NaOH, NH_4OH, or urea) added to a precursor solution, followed by high-temperature annealing to

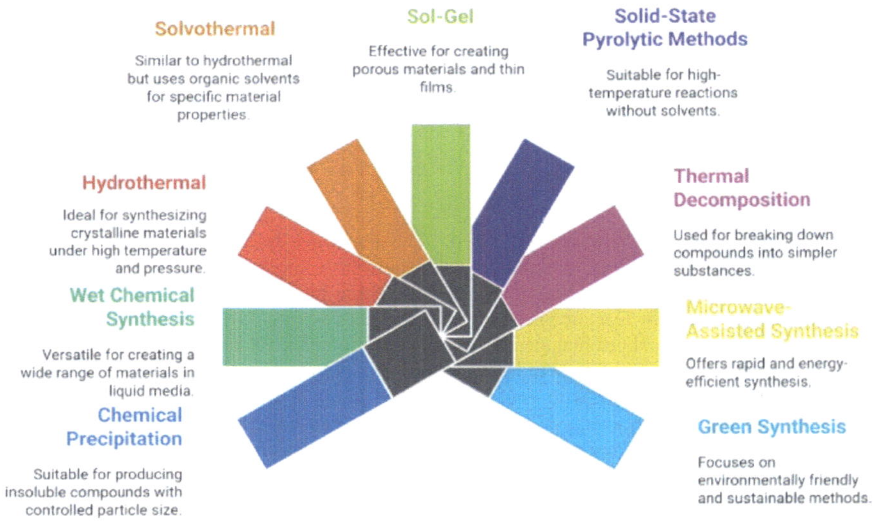

Fig. 5 Some Bottom-up approaches to metal oxide nanoparticle synthesis methods

generate MONPs. Although this technique is capable of producing nanoparticles with high purity, uniform size distribution, and reduced particle dimensions, it is not without limitations. Issues such as inadequate crystallinity and the possibility of contamination may arise, and precipitation may sometimes proceed spontaneously even in the absence of an external inducing agent. Despite these constraints, the method has been successfully applied to the synthesis of several metal oxide nanoparticles, including ZnO, CeO_2, Fe_2O_3, TiO_2, MgO, and NiO [48]. Wet-chemical synthesis can be regarded as a modified form of chemical precipitation; wherein stabilising agents are deliberately introduced to prevent agglomeration and enhance particle stability. For example, starch has been effectively utilised as a stabiliser during the preparation of ZnO nanoparticles [49], while a range of other stabilising additives has been employed for oxides such as CeO_2, Fe_2O_3, TiO_2, MgO, NiO, ZrO, and CdO [48]. Among the available approaches, hydrothermal synthesis is particularly favoured due to its capacity to precisely control particle morphology, size, and degree of aggregation. The method is also amenable to scale-up and typically yields nanoparticles of high purity. Nevertheless, it involves extended reaction durations and often necessitates multiple downstream processing steps. In a standard procedure, the precursor solution is enclosed within a stainless-steel autoclave along with a precipitating agent and heated at temperatures ranging from 80 to 200 °C for several hours, followed by washing and thermal treatment [50]. Using this approach, metal oxide nanoparticles such as ZnO, CeO_2, Fe_2O_3, TiO_2, MgO, NiO, and CdO have been synthesised [48]. Solvothermal Synthesis is similar to the hydrothermal synthesis, but using organic solvents instead of water; this method operates across a wide temperature (100–1000 °C) and pressure range (1 bar–10 kbar) [51]. Commonly employed solvents include diethanolamine, ethanol, methanol, butanediol, toluene, and ethylene glycol, while stabilising agents such as trisodium citrate are occasionally incorporated, particularly for biomedical applications [52, 53]. Among synthesis routes, the sol–gel technique is widely preferred at the industrial scale due to its ability to produce nanoparticles with controlled size distribution, high chemical purity, and uniform composition under comparatively mild thermal conditions. Nevertheless, drawbacks such as the use of organic solvents, restricted availability of suitable precursors, and prolonged processing times remain notable challenges. In this approach, metal alkoxide precursors undergo hydrolysis and condensation to form a sol, which subsequently transitions into a gel that is dried and calcined. This method has been extensively utilised for the synthesis of metal oxide nanoparticles including ZnO, Fe_2O_3, MgO, NiO, and CdO [48]. Solid-state pyrolysis involves thermal decomposition of metal-containing precursors at elevated temperatures, where the applied temperature plays a decisive role in determining particle size and the extent of agglomeration [54]. For example, ZnO nanoparticles with sizes ranging from approximately 8–35 nm have been obtained by adjusting the pyrolysis temperature [55]. In thermal decomposition techniques, organometallic precursors dissolved in organic media are decomposed at high temperatures under inert atmospheres. This route typically produces highly crystalline nanoparticles without requiring subsequent annealing, although the overall material yield is relatively low. Particle dimensions can be precisely tailored by modifying reaction conditions such as precursor selection and temperature [56, 57]. Microwave-assisted synthesis utilises microwave irradiation to achieve rapid and uniform heating, thereby enhancing reaction kinetics and significantly reducing synthesis time. This method often results in higher product yields and minimised particle aggregation. Despite challenges associated with large-scale production, microwave-assisted approaches have been successfully employed for the fabrication of various metal oxide nanoparticles, including ZnO, CeO_2, Fe_2O_3, TiO_2, NiO, and CdO [48]. Besides these, mechanochemical synthesis (e.g. Fe_2O_3), flow-chemistry precipitation (e.g. Fe_2O_3), continuous-flow synthesis (e.g. TiO_2, Fe_2O_3), successive ionic layer adsorption and reaction (e.g. NiO), and plasma-based methods (e.g. NiO) are the less conventional routes. In contrast to chemical methods, biosynthetic approaches use non-toxic reducing and capping agents naturally present in plants, microbes, or biological products. These approaches are environmentally benign, economically viable, and capable of producing nanoparticles with superior biocompatibility [58]. Biological entities such as bacteria, algae, and fungi can withstand synthesis conditions and function as effective bio-factories for nanoparticle generation [59]. In a similar manner, plant-derived extracts serve dual roles as reducing as well as stabilising agents, allowing for straightforward and relatively fast nanoparticle synthesis. Nonetheless, drawbacks such as slower kinetics and restricted control over particle size and morphology remain. Green synthesis strategies have been successfully employed for the production of ZnO, Fe_2O_3, CeO_2, TiO_2, MgO, CuO, NiO, ZrO/ZrO_2, and CdO nanoparticles [48]. The biological performance of metal oxide nanoparticles (MONPs) is strongly governed by their physicochemical properties. Parameters such as particle size, shape, and degree of crystallinity play crucial roles in cellular uptake, dissolution behaviour, and toxicity profiles [60, 61]. Surface-related attributes, including charge and dispersion stability, further influence aggregation tendencies and interactions with biological systems [62, 63]. Moreover, intrinsic photocatalytic activity and chemical composition regulate the generation of reactive species, such as hydroxyl radicals, peroxide ions, and metal ions, which may elicit either therapeutic benefits or adverse biological responses [64, 65].

6 Need for Functionalization of Metal Oxide Nanoparticles

Certain metal oxide nanoparticles possess inherent biological activity even in their native, unmodified form; however, for many others, their functional performance can be markedly improved through deliberate surface tailoring. Consequently, surface functionalization has emerged as a pivotal strategy to adapt nanoparticles for biomedical use. Nanoparticles synthesized via conventional chemical routes frequently exhibit hydrophobic surface characteristics, largely attributable to reaction environments and the presence of stabilizing surfactants, which restrict their dispersion in aqueous and physiological media [66]. To address this limitation, a range of surface-engineering approaches including surface coating, molecular conjugation, self-assembled architectures, encapsulation strategies, and core–shell designs have been developed. These modifications generally impart hydrophilic properties to the nanoparticle surface, thereby enhancing aqueous solubility and biological compatibility [67, 68]. In biomedical contexts, metal oxide nanoparticles can function as delivery platforms that facilitate site-specific accumulation of therapeutic agents, resulting in improved therapeutic outcomes while reducing off-target toxicity. The introduction of a protective surface layer not only improves colloidal stability but also plays a critical role in regulating pharmacokinetic behaviour, tissue distribution, and overall safety profiles [69]. A broad spectrum of pharmacologically active compounds including anticancer, antibacterial, antiviral, antifungal, anticonvulsant, and immunomodulatory drugs has been successfully immobilized onto metal oxide nanoparticles via covalent bonding, electrostatic attraction, or multistep surface modification processes. The release characteristics of these agents can be precisely controlled by adjusting the nature of the surface interaction and the physicochemical conditions of the surrounding microenvironment [70]. In addition to small-molecule drugs, both synthetic polymers and naturally derived biopolymers are widely employed as surface modifiers to further enhance functionality and biocompatibility. Synthetic polymers such as polyethylene glycol, poly(lactic-co-glycolic acid), poly(vinyl alcohol), poly(vinylpyrrolidone), poly(methyl methacrylate), polycaprolactone, poly-cyanoacrylates, poly(ethyleneimine), and polydopamine are frequently applied [71]. Such surface modifications enhance colloidal stability, prolong systemic circulation, and allow the integration of multiple functionalities. However, apprehensions regarding the potential toxicity of synthetic polymers particularly at elevated doses or upon long-term exposure—have encouraged the use of naturally derived biopolymers. Materials such as chitosan, dextran, cellulose, lignin, heparin, peptides, and proteins are frequently favoured owing to their intrinsic biocompatibility and biodegradability. In parallel, inorganic surface coatings have gained considerable attention. A variety of agents, including surfactants (e.g., sodium dodecyl sulphate and sodium oleate), silanes, carboxylates, phosphates, and notably silica, have been extensively utilised. Among these, silica is especially valued for its ability to improve particle stability and biological compatibility while providing a flexible surface for subsequent functional modifications, particularly in the case of magnetic nanoparticles. Another important approach is bioconjugation, wherein nanoparticles are chemically coupled with biomolecules engineered to trigger desirable biological interactions. This process commonly involves the use of linker molecules to achieve robust attachment and favourable spatial orientation of the conjugated biomolecules [72, 73]. Such functional assemblies promote targeted delivery and enhanced interactions with specific cell types or tissues [74]. It must be noted that the selection of an appropriate functionalization strategy is strongly influenced by the intrinsic composition and surface characteristics of the nanoparticles, and therefore, a universally applicable method does not exist. Please see Fig. 6 which provides a schematic overview of different surface modifications of MONPs usually applied for improving the bioactivities.

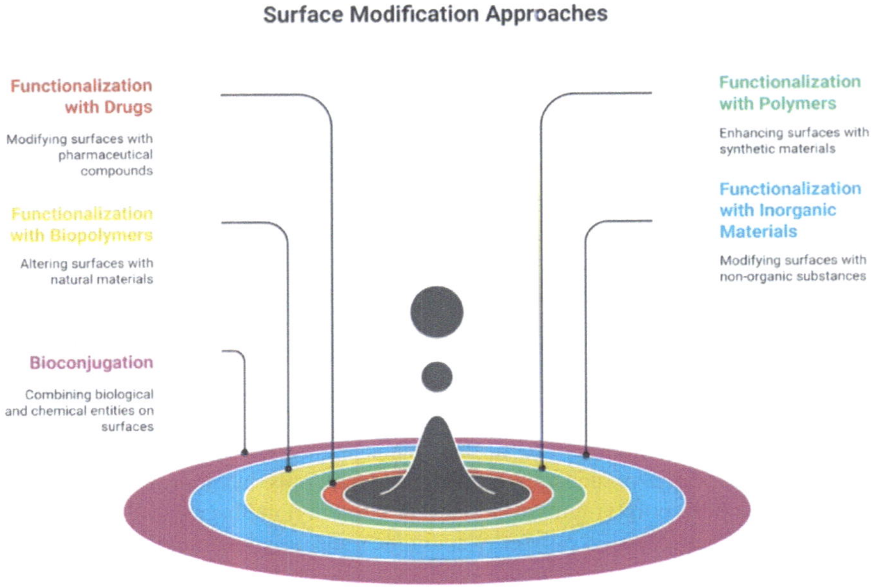

Fig. 6 Schematic overview of different surface modifications of MONPs usually applied for improving the bioactivities

7 Applications of Metal Oxide Nanoparticles as Drug Delivery Agents

In recent years, nanotechnology has emerged as a pivotal contributor to regenerative medicine, accelerating its translation from bench research to clinical use. Metal oxide nanoparticles (MONPs) have garnered significant interest because of their unique physicochemical properties and extensive biomedical potential. One prominent area of research is wound healing, as injuries from trauma, burns, dermatological disorders, or surgical procedures often lead to tissue damage that is susceptible to infection and delayed recovery [11]. This has created a demand for wound dressings that can simultaneously provide antimicrobial effects and support tissue repair [75]. Among MONPs, zinc oxide (ZnO) nanoparticles have shown strong antibacterial activity along with the capacity to promote epithelial regeneration [76]. For example, Manuja et al. [77] incorporated ZnO nanoparticles into hydrogels composed of sodium alginate and gum acacia, demonstrating wound-healing activity in sheep fibroblasts at low concentrations, though higher doses were associated with cytotoxicity. Importantly, the hydrogel scaffold reduced the inherent toxicity of ZnO while maintaining its antimicrobial function. Since excessive reactive oxygen species (ROS) and oxidative stress can hinder the healing process [78], several studies have explored the antioxidant capabilities of MONPs in wound repair. Raju and colleagues [79] reported that spherical cerium oxide nanoparticles (~160 nm) accelerated wound closure and enhanced collagen deposition in rats without causing scarring. Similarly, incorporating CeO_2 nanoparticles into polycaprolactone/gelatin films promoted fibroblast proliferation most effectively at a 1.5% concentration [80], while Wu et al. [81] demonstrated that CeO_2-based tissue adhesives reduced ROS-driven tissue injury and inflammation, promoting scar-free regeneration both *in vitro* and *in vivo*. Fe_2O_3 nanoparticles have also shown promise as wound-healing agents. Jain and co-workers [82] reported that polycaprolactone-Fe_2O_3 thin films exhibited antibacterial effects and promoted fibroblast proliferation. Biocompatible wound dressings incorporating Fe_2O_3 nanoparticles within alginate [83], silk fibroin [84], or nanofluid coatings with natural antimicrobials [85] all revealed enhanced antibacterial and anti-biofilm properties. Beyond their antimicrobial roles, the magnetic responsiveness of Fe_2O_3 nanoparticles offers unique therapeutic opportunities. Wu et al. [86] functionalized these nanoparticles with basic fibroblast growth factor (bFGF), finding that they enhanced pro-healing macrophage polarisation and cell proliferation. Moreover, magnetically guided Fe_2O_3-loaded mesenchymal stem cells significantly improved skin regeneration in rat models [87]. Beyond cutaneous applications, MONPs are also being investigated for bone regeneration. ZnO, CeO_2, and Fe_2O_3 nanoparticles have demonstrated osteoinductive potential by modulating osteoblast and osteoclast activity. For example, *Scutellaria baicalensis*-ZnO nanoconjugates enhanced osteoblast proliferation while inhibiting osteoclastogenesis [88]. Poly(e-caprolactone)-ZnO scaffolds supported both osteogenic and chondrogenic differentiation of stem cells [89], while functionalized ZnO nanocrystals encouraged bone proliferation even at elevated doses [90]. CeO_2 nanoparticles displayed size and concentration-dependent effects on osteoblast mineralisation [91], promoted osteogenesis in MSCs [92], and enhanced new bone formation on titanium and glass foam scaffolds even without exogenous osteogenic cues [93]. Similarly, Fe_2O_3 nanoparticles integrated into polymeric scaffolds [94, 95] improved osteogenic differentiation, cell adhesion, and bone proliferation, with additional benefits under applied magnetic fields. Despite encouraging preclinical findings, translating MONPs into clinical bone and wound therapies requires further clarification of their physicochemical profiles and long-term biosafety. Parallel to regenerative applications, MONPs are under intensive evaluation in oncology, as cancer remains a leading cause of global mortality, with cases projected to reach 2.84 crore annually by 2040 [96]. Current therapeutic approaches, such as surgery, chemotherapy, radiation, targeted therapy, and immunotherapy, are limited by systemic toxicity, poor selectivity, or incomplete tumour removal [97]. Thus, nanotechnology offers new avenues for precision cancer therapy, with MONPs serving as both intrinsic anticancer agents and drug-delivery vehicles [98]. ZnO nanoparticles have been widely studied for their selective cytotoxicity, largely mediated by ROS generation and electrostatic interactions. They induce apoptosis preferentially in cancer cells, as demonstrated in myoblastoma, glioma, cervical, ovarian, and colon cancer models [99–101]. Surface modification with PEG, hyaluronic acid, or Triton-X has improved their biocompatibility without diminishing antitumor potency [102, 103]. Moreover, ZnO nanoparticles have been engineered as drug carriers for doxorubicin, paclitaxel, curcumin, ferulic acid, and methotrexate, yielding synergistic cytotoxicity and enhanced tumour suppression both *in vitro* and *in vivo* [104–106]. Similarly, CeO_2 nanoparticles, initially investigated for radioprotective properties [107, 108], have demonstrated direct cytotoxic effects against diverse cancer cell types, including colon, pancreatic, ovarian, and lung cancers [109, 110]. Their activity depends on valence state, particle size, and surface chemistry, with smaller or + 4 valence forms often showing greater uptake and cytotoxicity. *In vivo*, CeO_2 nanoparticles reduced tumour growth, angiogenesis, and metastasis in xenograft models. Furthermore, CeO_2-based delivery systems have enabled controlled release of doxorubicin, camptothecin, photosensitizers, and curcumin, often achieving synergistic antitumor responses with minimal toxicity to normal tissue [111, 112]. Fe_2O_3 nanoparticles, valued for their magnetic properties, are being exploited in hyperthermia-based therapies. Alternating magnetic fields induce localised tumour heating (>40 °C), selectively killing cancer cells while sparing normal tissue [113, 114]. Fe_2O_3 nanoparticles have also been used to deliver chemotherapeutics such as doxorubicin, enhancing efficacy when combined with magnetic-field-induced drug release [115, 116]. Other MONPs, including MgO and CuO NPs, have shown cytotoxicity against breast, lung, liver, and melanoma cells *via* ROS-mediated apoptosis [117–119]. Their biocompatibility, biodegradability, and cost-effectiveness further support their potential for drug delivery and therapeutic use. Overall, MONPs represent a multifaceted platform in regenerative medicine and oncology, offering both intrinsic therapeutic benefits and advanced drug-delivery capabilities. However, systematic studies on long-term toxicity, biodistribution, and clinical translation remain essential before widespread adoption. TiO_2 nanoparticles are among the most studied metal oxides in oncology research. Their photocatalytic properties make them attractive for tumour suppression. For instance, animal experiments with glioma-bearing mice demonstrated that light-activated TiO_2 nanoparticles not only reduced tumour progression but also prolonged survival [120]. Modifications such as nitrogen doping further improve their performance; anatase *N*-doped TiO_2 exhibited stronger visible-light absorption and induced nearly 93% melanoma cell death under UV exposure [121]. Comparable outcomes were noted when ruthenium-complex-loaded TiO_2 NPs were tested on melanoma cells, though the efficiency under UV illumination surpassed that of visible light [122]. Since UV penetration in tissues is limited and harmful, surface engineering of TiO_2 NPs has been pursued to enable safer and deeper activation. Recent advances include core-shell architectures responsive to NIR light, which achieved significant antitumor effects against HeLa xenografts in nude mice [123]. Likewise, Zhang and co-workers [124] showed that TiO_2-coated up-conversion nanoparticles suppressed oral squamous carcinoma growth *in vivo*. Drug delivery applications have also been widely explored. Paclitaxel (PTX)-loaded hyaluronic acid (HA)/TiO_2 nanocarriers displayed superior therapeutic efficacy in hepatocarcinoma models compared with free PTX [125]. Similarly, cisplatin-loaded HA/TiO_2 improved intracellular drug uptake in ovarian cancer cells relative to unformulated cisplatin [126]. Evidence further indicates that the geometry of TiO_2 nanostructures affects drug deposition at tumour sites, with elongated carriers showing greater retention and payload delivery following intravenous injection [127, 128]. In line with this, Schmuki and co-workers

[129] fabricated uniform TiO_2 nanotubes and nanocylinders *via* anodization, conjugated them with doxorubicin, and demonstrated pH-sensitive release inside acidic compartments of HeLa cells. Hybrid systems have also been developed, for example, TiO_2 nanotubes loaded with Fe_2O_3 nanoparticles enabled magnetic field-guided targeting and site-specific release, achieving up to 90% HeLa cell killing with camptothecin and complete uptake with oligonucleotide cargo [130]. Other metal oxides exhibit promising anticancer effects as well. Nickel oxide (NiO) nanoparticles reduced HepG2 hepatocellular carcinoma survival rates by up to 84% in a dose-dependent manner [131]. Green-synthesised NiO showed cytotoxicity against A549 lung and HepG2 liver cells, attributed to nanoparticle accumulation and apoptosis induction [129]. Zhang et al. [132] reported growth inhibition in oesophageal cancer cells, with higher sensitivity compared to colon carcinoma. Zirconium oxide (ZrO_2) NPs also demonstrated cytotoxicity: exposure decreased viability of breast cancer (MCF-7, MDA-MB-231) and liver cancer cells, particularly when doped with iron–manganese sulphates [133, 134]. Similarly, cadmium oxide (CdO) nanoparticles, though less frequently studied, exhibited strong cytotoxicity against colon and cervical cancer lines, even at low doses. Beyond oncology, MONPs have been intensively investigated for antimicrobial applications. Their relevance arises from the growing resistance to conventional antibiotics and the demand for antibacterial coatings in implants, stents, catheters, and wound dressings. MONPs such as Ag_2O, TiO_2, CuO, ZnO, and MgO consistently display bactericidal activity [135]. Their mechanisms generally involve reactive oxygen species (ROS) mediated oxidative stress or the release of toxic ions [136], with efficacy strongly influenced by particle size, morphology, and surface chemistry [137, 138]. Smaller and positively charged nanoparticles interact more efficiently with negatively charged bacterial membranes, thereby amplifying antimicrobial potency [139]. Zinc oxide nanoparticles exemplify these principles: they dissolve to release Zn^{2+} ions, which disrupt bacterial respiration by binding thiol groups, simultaneously triggering ROS generation and damaging membranes, mitochondria, and DNA [140]. ZnO has shown activity against both Gram-positive and Gram-negative bacteria, as well as spore-forming strains [141]. Numerous studies confirmed its size- and dose-dependence [142, 143]. Moreover, ZnO enhances antibiotic efficacy; for example, ciprofloxacin and ceftazidime exhibited stronger action against *Acinetobacter baumannii* in the presence of ZnO NPs [144]. ZnO also inhibited *Vibrio cholerae* [145], *Campylobacter jejuni* [146], and even *Mycobacterium tuberculosis* [147]. TiO_2 nanoparticle's antibacterial activity is linked to ROS generation under photocatalysis. Studies showed that TiO_2 can enhance the effects of various antibiotic classes, including cephalosporins and macrolides, against resistant *Staphylococcus aureus* [148]. Lipid peroxidation-driven membrane destabilisation has been reported as a contributing mechanism [149]. However, practical use is limited under UV light due to potential genotoxicity [150]. Copper oxide (CuO) NPs are also broad-spectrum bactericidals, disrupting bacterial membranes and enzymes. They are effective against both Gram-positive and Gram-negative strains, with potency scaling with dose and particle size. Magnesium oxide (MgO) nanoparticles, on the other hand, exert antibacterial effects by generating superoxide and raising local pH, which destabilises membranes

[151]. Composite ZnO-MgO nanostructures demonstrated particularly strong inhibition against *Bacillus subtilis* [152]. Silver oxide (Ag_2O) nanoparticles, long known for antimicrobial properties, exhibit strong activity even against resistant bacteria. Their mechanism involves ROS generation and interference with cell cycle progression [153]. Cerium oxide (CeO_2) nanoparticles operate differently, as they cannot penetrate bacterial membranes. Instead, they disturb redox balance at the membrane interface and impair respiration, selectively inhibiting Gram-negative strains [154, 155]. Nickel oxide and zirconium oxide nanoparticles have also displayed antibacterial effects. NiO has been shown to act against *E. coli*, *S. aureus*, *B. subtilis*, and *P. aeruginosa* [131]. ZrO_2 nanoparticles inhibited *E. coli*, *S. typhi*, and *B. subtilis*, with improved performance when synthesised via biological routes [156, 157]. Cadmium oxide (CdO) nanoparticles demonstrated selective bactericidal effects, with greater inhibition zones against *Proteus vulgaris* compared to *S. aureus* [158]. Altogether, MONPs represent powerful antimicrobial and anticancer agents, with potential in combination therapies to counteract drug resistance and improve biomedical device safety. Nevertheless, clinical translation requires further work to mitigate toxicity while preserving therapeutic efficacy.

8 Toxicity Associated with Metal Oxide Nanoparticles

The biomedical potential of metal oxide nanoparticles (MONPs) is undeniable; however, their safety remains a major concern [159]. Reported toxicity is inconsistent, with mechanisms proposed including ROS overproduction, lipid peroxidation, ion release, and altered nanoparticle-cell interactions. Yet, no single pathway has been universally validated. Importantly, toxicity is influenced by both intrinsic properties and exposure duration, emphasising the need for standardised risk assessment protocols [160–163]. Among widely used MONPs, TiO_2 nanoparticles are generally regarded as safe, but in vitro studies have shown variable results ranging from DNA damage and micronuclei formation to negligible cytotoxicity, largely dependent on dispersion methods and cell type [164, 165]. ZnO nanoparticles, despite their broad biomedical applications, demonstrate marked cytotoxicity due to Zn^{2+} ion release, disrupting intracellular homeostasis and inducing oxidative stress. Their toxicity is strongly dose-, size-, and time-dependent, with smaller particles exerting greater effects [166, 167]. CeO_2 nanoparticles exhibit dual antioxidant/pro-oxidant behaviour. While often cytotoxic to tumour cells, they show inconsistent effects in non-cancerous cells, reflecting dependence on particle characteristics and cellular context [168, 169]. Similarly, Fe_2O_3 nanoparticles induce ROS-driven damage, with toxicity varying across cell lines and surface modifications, typically showing a dose-dependent response. In vivo studies confirm organ-level toxicities [170, 171]. CeO_2 nanoparticles accumulate due to poor clearance, leading to lung inflammation, fibrosis, and systemic effects [172]. TiO_2 nanoparticles have induced pulmonary inflammation and even carcinogenesis in rodent models following inhalation [173].

ZnO nanoparticles distribute widely across organs, causing neurotoxicity, hepatotoxicity, and pulmonary fibrosis in a size- and concentration-dependent manner [174, 175]. Fe_2O_3 nanoparticles accumulate in reticuloendothelial organs, with outcomes differing according to exposure route and animal model. Overall, MONP toxicity remains controversial, primarily due to heterogeneity in synthesis methods, experimental conditions, and biological systems. Standardised protocols for nanoparticle characterisation and toxicological testing are urgently needed to resolve these discrepancies and ensure safe biomedical translation.

9 Future Perspectives

If challenges related to toxicity assessment and methodological standardisation are systematically resolved, metal oxide nanoparticles (MONPs) are poised to become powerful platforms for advanced biomedical interventions. Their potential extends beyond therapeutic roles, with strong prospects in non-invasive diagnostics and theranostics, enabling simultaneous treatment and imaging of diseases at the cellular and tissue levels. Ongoing research is increasingly focusing on functionalized MONPs designed for site-specific drug delivery, a strategy particularly relevant to oncology, where precision and minimal side effects are critical. In the future, MONPs could contribute substantially to the development of multimodal therapies and even patient-tailored treatment regimens. This trend aligns with the growing emphasis on personalised medicine, which aims to enhance safety, efficacy, and patient outcomes. Consequently, the coming years are likely to witness not only an expansion in the number of clinical trials but also tangible improvements in therapeutic success rates, especially in cancer care.

10 Conclusion

The nanoscale dimensions and distinct physicochemical features of metal oxide nanoparticles (MONPs) give them properties markedly different from their bulk forms, which explains their expanding role in biomedical applications. Their ability to stimulate reactive oxygen species (ROS) production, leading to cellular damage and death, underpins their utility as anticancer, antibacterial, and tissue-regenerative agents. MONPs can act either independently or as carriers for therapeutic molecules, offering versatile biomedical tools. However, the main limitation hindering their wider adoption is the inherent toxicity associated with many formulations, compounded by the absence of universally accepted standards for assessing their safety in both *in vitro* and *in vivo* models. Variability in synthesis routes further complicates the field, as differences in morphology, size, chemical composition, and surface modifications often yield divergent biological outcomes. Among the commonly employed synthesis strategies, precipitation, sol-gel, hydrothermal, and green/biogenic routes are predominant. Notably, green synthesis methods are gaining importance due to their cost-effectiveness, sustainability, and favourable safety profile. Addressing these methodological inconsistencies through systematic documentation and standardised protocols would significantly reduce current limitations and accelerate clinical translation.

References

1. Boholm, M.: The use and meaning of nano in American English: Towards a systematic description. Ampersand. ***3***, 163–173 (2016)
2. Joudeh, N., Linke, D.: Nanoparticle classification, physicochemical properties, characterization, and applications: A comprehensive review for biologists. J. Nanobiotechnol. ***20***(1), 262 (2022)
3. Buzea, C., Pacheco, I.I., Robbie, K.: Nanomaterials and nanoparticles: sources and toxicity. Biointerphases. ***2***(4), 17–71 (2007). https://doi.org/10.1116/1.2815690
4. Feynman, R.P.: There's plenty of room at the bottom. Eng. Sci. ***22***, 22–36 (1960)
5. Vert, M., Doi, Y., Hellwich, K.H., Hess, M., Hodge, P., Kubisa, P., Rinaudo, M., Schue, F.O.: Terminology for biorelated polymers and applications (IUPAC Recommendations 2012). Pure Appl. Chem. ***84***(2), 377–410 (2012). https://doi.org/10.1351/PAC-REC-10-12-04
6. Chen, G., Roy, I., Yang, C., Prasad, P.N.: Nanochemistry and nanomedicine for nanoparticle-based diagnostics and therapy. Chem. Rev. ***116***, 2826–2885 (2016). https://doi.org/10.1021/acs.chemrev.5b00148
7. Devesa, S.: Comparative overview of metal oxide nanoparticle synthesis methods: conventional sol-gel versus green approaches. IntechOpen. (2025). https://doi.org/10.5772/intechopen.1008457
8. Ying, S., Guan, Z., Ofoegbu, P.C., Clubb, P., Rico, C., He, F., Hong, J.: Green synthesis of nanoparticles: current developments and limitations. Environ. Technol. Innov. ***26***, 102336 (2022). https://doi.org/10.1016/j.eti.2022.102336
9. Chaudhary, R.G., Bhusari, G.S., Tiple, A.D., Rai, A.R., Somkuvar, S.R., Potbhare, A.K., Lambat, T.L., Ingle, P.P., Abdala, A.A.: Metal/metal oxide nanoparticles: toxicity, applications, and future prospects. Curr. Pharm. Des. ***25***(37), 4013–4029 (2019)
10. Yadavalli, T., Shukla, D.: Role of metal and metal oxide nanoparticles as diagnostic and therapeutic tools for highly prevalent viral infections. Nanomedicine. ***13***(1), 219–230 (2017). https://doi.org/10.1016/j.nano.2016.08.016
11. Augustine, R., Hasan, A.: Emerging applications of biocompatible phytosynthesized metal/metal oxide nanoparticles in healthcare. J. Drug. Deliv. Sci. Technol. **56**, 101516 (2020)
12. Mirzaei, H., Darroudi, M.: Zinc oxide nanoparticles: Biological synthesis and biomedical applications. Ceram Int. **43**, 907–914 (2017)
13. Shen, C., James, S.A., de Jonge, M.D., Turney, T.W., Wright, P.F.A., Feltis, B.N.: Relating cytotoxicity, zinc ions, and reactive oxygen in ZnO nanoparticle-exposed human immune cells. Toxicol. Sci. ***136***, 120–130 (2013)
14. Elshama, S.S., Abdallah, M.E., Abdel-Karim, R.I.: Zinc oxide nanoparticles: therapeutic benefits and toxicological hazards. Open Nanomedicine J. ***5***(1), 16–22 (2018)
15. Korsvik, C., Patil, S., Seal, S., Self, W.T.: Superoxide dismutase mimetic properties exhibited by vacancy engineered ceria nanoparticles. Chem. Commun. ***10***, 1056–1058 (2007)
16. Deshpande, S., Patil, S., Kuchibhatla, S.V., Seal, S.: Size dependency variation in lattice parameter and valency states in nanocrystalline cerium oxide. Appl. Phys. Lett. ***87***, 133113 (2005)
17. Asati, A., Santra, S., Kaittanis, C., Nath, S., Perez, J.M.: Oxidase-like activity of polymer-coated cerium oxide nanoparticles. Angew. Chem. Int. Ed. ***48***, 2308–2312 (2009)

18. Haugen, H.J., Monjo, M., Rubert, M., Verket, A., Lyngstadaas, S.P., Ellingsen, J.E., Rønold, H.J., Wohlfahrt, J.C.: Porous ceramic titanium dioxide scaffolds promote bone formation in rabbit peri-implant cortical defect model. Acta Biomaterialia. ***9***, 5390–5399 (2013)
19. Fu, G., Vary, P.S., Lin, C.T.: Anatase TiO_2 Nanocomposites for Antimicrobial Coatings. Int. J. Nanomedicine. ***109***, 8889–8898 (2005)
20. Jafari, S., Mahyad, B., Hashemzadeh, H., Janfaza, S., Gholikhani, T., Tayebi, L.: Biomedical Applications of TiO_2 Nanostructures: Recent Advances. Int. J. Nanomedicine. ***15***, 3447–3470 (2020)
21. Friedrich, R.P., Cicha, I., Alexiou, C.: Iron oxide nanoparticles in regenerative medicine and tissue engineering. Nanomaterials. ***11***, 2337 (2021)
22. Mathiasen, A.B., Hansen, L., Friis, T., Thomsen, C., Bhakoo, K., Kastrup, J.: Optimal labeling dose, labeling time, and magnetic resonance imaging detection limits of ultrasmall superparamagnetic iron-oxide nanoparticle labeled mesenchymal stromal cells. Stem. Cells Int. **2013**, 353105 (2013). https://doi.org/10.1155/2013/353105
23. Parashurama, N., Ahn, B.C., Ziv, K., Ito, K., Paulmurugan, R., Willmann, JK., Chung, J., Ikeno, F., Swanson, JC., Merk, DR., et al.: Multimodality molecular imaging of cardiac cell transplantation: part ii. In vivo imaging of bone marrow stromal cells in Swine with PET/CT and MR imaging. Radiology. Molecular imaging, *280*, 826–836 (2016)
24. Nejati, M., Rostami, M., Mirzaei, H., Rahimi-Nasrabadi, M., Vosoughifar, M., Nasab, A.S., Ganjali, M.R.: Green methods for the preparation of MgO nanomaterials and their drug delivery, anti-cancer and anti-bacterial potentials: a review. Inorg. Chem. Commun. ***136***, 109107 (2022)
25. Narender, S.S., Varma, V.V.S., Srikar, C.S., Ruchitha, J., Varma, P.A., Praveen, B.V.S.: Nickel oxide nanoparticles: a brief review of their synthesis, characterization, and applications. Chem. Eng. Technol. ***45***, 397–409 (2022)
26. Alsharari, S.S., Alenezi, M.A., Al Tami, M.S., Soliman, M.: Recent advances in the biosynthesis of zirconium oxide nanoparticles and their biological applications. Baghdad. Sci. J., 41–57 (2022) https://doi.org/10.21123/bsj.2022.7055
27. Skheel, A.Z., Jaduaa, M.H., Abd, A.N.: Green synthesis of cadmium oxide nanoparticles for biomedical applications (antibacterial, and anticancer activities). Mater. Today: Proc. ***45***, 5793–5799 (2021)
28. Gowri, S., Gopinath, K., Arumugam, A.: Experimental and computational assessment of mycosynthesized CdO nanoparticles towards biomedical applications. J. Photochem. Photobiol. B: Biology. **180**, 166–174 (2018). https://doi.org/10.1016/j.jphotobiol.2018.02.009
29. Prasad, P.N.: Introduction to nanomedicine and nanobioengineering. Wiley-Interscience, Hoboken, NJ (2012)
30. Prasad, P.N.: Introduction to biophotonics. Wiley-Interscience, Hoboken, NJ (2003)
31. Prasad, P.N.: Nanophotonics. Wiley-Interscience, Hoboken, NJ (2004)
32. Burda, C., Chen, X.B., Narayanan, R., El-Sayed, M.A.: Chemistry and properties of nanocrystals of different shapes. Chem. Rev. ***105***(4), 1025–1102 (2005)
33. Alivisatos, A.P.: Semiconductor clusters, nanocrystals, and quantum dots. Science. **271**, 933–937 (1996). https://doi.org/10.1126/science.271.5251.933
34. Taylor-Smith, K.: An Introduction to the Quantum Mechanics of Nanoparticles. AZoQuantum. (2024) https://www.azoquantum.com/Article.aspx?ArticleID=179
35. Roduner, E.: Size matters: why nanomaterials are different. Chem. Soc. Rev. ***35***, 583–592 (2006)
36. Kreibig, U., Volmer, M.: Optical Properties of Metal Clusters, Springer Series in Materials Science, vol. 25. Springer, Berlin (1995)
37. Roduner, E.: Nanoscopic materials: size-dependent phenomena. The Royal Society of Chemistry, Cambridge (2006)
38. Poole, C., Owens, F.: Introduction to nanotechnology. Wiley, New Jersey (2003)
39. Debnath, S., Kumar, D., Mondal, S., Basak, S., Dixit, A.K., Dey, R.K., Das, M.: Green synthesis of gold nanomaterials: recent studies and scopes. Lett. Org. Chem. ***22***(12), e15701786395012 (2025). https://doi.org/10.2174/0115701786395012250721014326

40. Sajjad, S., Leghari, S.A.K., Ryma, N.U.A., Farooqi, S.A.: Green synthesis of metal-based nanoparticles and their applications. In: Kanchi, S., Ahmed, S. (eds.) Green metal nanoparticles, pp. 23–77. Scrivener Publishing LLC. (2018) https://doi.org/10.1002/9781119418900.ch2
41. Kumari, S., Raturi, S., Kulshrestha, S., Chauhan, K., Dhingra, S., András, K., Thu, K., Khargotra, R., Singh, T.A.: Comprehensive review on various techniques used for synthesizing nanoparticles. J. Mater. Res. Technol. **27**, 1739–1763 (2023). https://doi.org/10.1016/j.jmrt.2023.09.291
42. Tripathy, S., Rodrigues, J., Shimpi, N.G.: Top-down and bottom-up approaches for synthesis of nanoparticles. Mater. Res. Proc. **145**, 92–130 (2023) https://doi.org/10.21741/9781644902370-4
43. Nagime, P.V., Chandak, V.S.: A comprehensive review of nanomaterials synthesis: physical, chemical, and biological approaches and emerging challenges. Biocatal. Agric. Biotechnol. ***2024***, 103458 (2024). https://doi.org/10.1016/j.bcab.2024.103458
44. Jamkhande, P.G., Ghule, N.W., Bamer, A.H., Kalaskar, M.G.: Metal nanoparticles synthesis: An overview on methods of preparation, advantages and disadvantages, and applications. J. Drug. Deliv. Sci. Technol. ***53***, 101174 (2019). https://doi.org/10.1016/j.jddst.2019.101174
45. Mourdikoudis, S., Pallares, R.M., Thanh, N.T.K.: Characterisation techniques for nanoparticles: comparison and complementarity upon studying nanoparticle properties. Nanoscale. ***10***, 12871–12934 (2018). https://doi.org/10.1039/C8NR02278J
46. Jain, K., Takuli, A., Gupta, T.K., Gupta, D.: Rethinking nanoparticle synthesis: a sustainable approach vs. traditional methods. Chem.-An Asian J. ***19***(21), e202400701 (2024). https://doi.org/10.1002/asia.202400701
47. Saxena, R., Kotnala, S., Bhatt, S.C., Uniyal, M., Rawat, B.S., Negi, P., Riyal, M.K.: A review on green synthesis of nanoparticles toward a sustainable environment. Sustain. Chem. Clim. Action. **6**, 100071 (2025). https://doi.org/10.1016/j.scca.2025.100071
48. Negrescu, A.M., Killian, M.S., Raghu, S.N.V., Schmuki, P., Mazare, A., Cimpean, A.: Metal oxide nanoparticles: review of synthesis, characterisation and biological effects. J. Funct. Biomater. ***13***, 274 (2022). https://doi.org/10.3390/jfb13040274
49. Sharma, H., Kumar, K., Choudhary, C., Mishra, P.K., Vaidya, B.: Development and characterisation of metal oxide nanoparticles for the delivery of anticancer drug. Artif. Cells, Nanomedicine, Biotechnol. **44**, 672–679 (2016)
50. Tang, Z.X., Lv, B.F.: MgO nanoparticles as antibacterial agent: Preparation and activity. Braz. J. Chem. Eng. **31**, 591–601 (2014)
51. Walton, R.I.: Subcritical solvothermal synthesis of condensed inorganic materials. Chem. Soc. Rev. **31**, 230–238 (2002)
52. Razali, R., Zak, A.K., Majid, W.A., Darroudi, M.: Solvothermal synthesis of microsphere ZnO nanostructures in DEA media. Ceram. Int. **37**, 3657–3663 (2011)
53. Rashid, A.N.M., Haw, C., Chiu, W., Khanis, N.H., Rohaizad, A., Khiew, P., Rahman, A.S.: Structural- and optical properties analysis of single crystalline hematite (α-Fe2O3) nanocubes prepared by one-pot hydrothermal approach. CrystEngComm. **18**, 4720–4732 (2016)
54. Mishra, P.K., Mishra, H., Ekielski, A., Talegaonkar, S., Vaidya, B.: Zinc oxide nanoparticles: a promising nanomaterial for biomedical applications. Drug Discov. Today. ***22***, 1825–1834 (2017)
55. Wang, Z., Zhang, H., Zhang, L., Yuan, J., Yan, S., Wang, C.: Low-temperature synthesis of ZnO nanoparticles by solid-state pyrolytic reaction. Nanotechnology. **14**, 11–15 (2003)
56. Nikam, A.V., Prasad, B.L.V., Kulkarni, A.A.: Wet chemical synthesis of metal oxide nanoparticles: a review. CrystEngComm. **20**, 5091–5107 (2018)
57. Mehra, S., Bergerud, A., Milliron, D.J., Chan, E.M., Salleo, A.: Core/shell approach to dopant incorporation and shape control in colloidal zinc oxide nanorods. Chem. Mater. J. **28**, 3454–3461 (2016)
58. Chouke, P.B., Shrirame, T., Potbhare, A.K., Mondal, A., Chaudhary, A.R., Mondal, S., Thakare, S.R., Nepovimova, E., Valis, M., Kuca, K., et al.: Bioinspired metal/metal oxide nanoparticles: a road map to potential applications. Mater. Today Adv. **16**, 100314 (2022)

59. Gade, A.K., Bonde, P., Ingle, A.P., Marcato, P.D., Durán, N., Rai, M.K.: Exploitation of aspergillus niger for synthesis of silver nanoparticles. J. Biobased. Mater. Bioenergy. **2**, 243–247 (2008)
60. Elsabahy, M., Wooley, K.L.: Design of polymeric nanoparticles for biomedical delivery applications. Chem. Soc. Rev. **41**, 2545–2561 (2012)
61. Ankamwar, B.: Size and Shape Effect on Biomedical Applications of Nanomaterials. InTech, London, UK (2012). https://doi.org/10.5772/46121
62. Utembe, W., Potgieter, K., Stefaniak, A.B., Gulumian, M.: Dissolution and biodurability: Important parameters needed for risk assessment of nanomaterials. Part. Fibre Toxicol. ***12***, 11 (2015)
63. Michaelis, M., Fischer, C., Ciacchi, L.C., Luttge, A.: Variability of zinc oxide dissolution rates. Environ. Sci. Technol. ***51***, 4297–4305 (2017)
64. Augustine, R., Mathew, A.P., Sosnik, A.: Metal oxide nanoparticles as versatile therapeutic agents modulating cell signaling pathways: linking nanotechnology with molecular medicine. Appl. Mater. Today. ***7***, 91–103 (2017)
65. Schanen, B.C., Das, S., Reilly, C.M., Warren, W.L., Self, W.T., Seal, S., Drake, D.R.: Immunomodulation and T Helper TH1/TH2 response polarization by CeO2 and TiO2 nanoparticles. PLoS ONE. ***8***, e62816 (2013)
66. Kumar, R., Pulikanti, G.R., Shankar, K.R., Rambabu, D., Mangili, V., Kumbam, L.R., Sagara, P.S., Nakka, N., Yogesh, M.: Surface coating and functionalization of metal and metal oxide nanoparticles for biomedical applications. In: Mondal, K. (ed.) Metal oxides for biomedical and biosensor applications, pp. 205–231. Elsevier, Amsterdam, The Netherlands (2022)
67. Nikolova, M.P., Chavali, M.S.: Metal Oxide Nanoparticles as Biomedical Materials. *Biomimetics*. **5**, 27 (2020)
68. Sun, S.N., Wei, C., Zhu, Z.Z., Hou, Y.L., Venkatraman, S.S., Xu, Z.C.: Magnetic iron oxide nanoparticles: synthesis and surface coating techniques for biomedical applications. Chin. Phys. B. ***23***, 037503 (2014)
69. Arias, L.S., Pessan, J.P., Vieira, A.P.M., de Lima, T.M.T., Delbem, A.C.B., Monteiro, D.R.: Iron oxide nanoparticles for biomedical applications: a perspective on synthesis, drugs, antimicrobial activity, and toxicity. Antibiotics. ***7***, 46 (2018)
70. Raghu, S.N.V., Onyenso, G., Mohajernia, S., Killian, M.S.: Functionalization strategies to facilitate multi-depth, multi-molecule modifications of nanostructured oxides for triggered release applications. Surf. Sci. ***719***, 122024 (2022)
71. Aktan, M.K., Coppola, G.A., Van der Gucht, M., Yoshioka, T., Killian, M.S., Lavigne, R., Van der Eycken, E., Steenackers, H.P., Braem, A.: Influence of polydopamine functionalization on the rapid protein immobilization by alternating current electrophoretic deposition. Surf. Interfaces. ***34***, 102347 (2022)
72. Sperling, R.A., Parak, W.J.: Surface modification, functionalization and bioconjugation of colloidal inorganic nanoparticles. Philos. Trans. A. ***368***, 1333–1383 (2010)
73. Killian, M.S., Schmuki, P.: Influence of bioactive linker molecules on protein adsorption. Surf. Interface Anal. ***46***, 193–197 (2014)
74. Farokhzad, O.C., Karp, J.M., Langer, R.: Nanoparticle- aptamer bioconjugates for cancer targeting. Expert Opin. Drug. Deliv. ***3***, 311–324 (2006)
75. Augustine, R.: Skin bioprinting: A novel approach for creating artificial skin from synthetic and natural building blocks. Prog. Biomater. ***7***, 77–92 (2018)
76. Lansdown, A.B.G., Mirastschijski, U., Stubbs, N., Scanlon, E., Ågren, M.S.: Zinc in wound healing: theoretical, experimental, and clinical aspects. Wound Repair Regen. **15**, 2–16 (2007)
77. Raguvaran, R., Manuja, B.K., Chopra, M., Thakur, R., Anand, T., Kalia, A., Manuja, A.: Sodium alginate and gum acacia hydrogels of ZnO nanoparticles show wound healing effect on fibroblast cells. Int. J. Biol. Macromol. ***96***, 185–191 (2017)
78. Chen, J., Patil, S., Seal, S., McGinnis, J.F.: Rare earth nanoparticles prevent retinal degeneration induced by intracellular peroxides. Nat. Nanotechnol. ***1***, 142–150 (2006)
79. Davan, R., Prasad, R.G.S.V., Jakka, V.S., Aparna, R.S.L., Phani, A.R., Jacob, B., Salins, P.C., Raju, D.B.: Cerium oxide nanoparticles promotes wound healing activity in in-vivo animal model. J. Bionanoscience. ***6***, 78–83 (2012)

80. Naseri-Nosar, M., Farzamfar, S., Sahrapeyma, H., Ghorbani, S., Bastami, F., Vaez, A., Salehi, M.: Cerium oxide nanoparticlecontaining poly (-caprolactone)/gelatin electrospun film as a potential wound dressing material: In vitro and in vivo evaluation. Mater. Sci. Eng.: C. **81**, 366–372 (2017)
81. Wu, H., Li, F., Wang, S., Lu, J., Li, J., Du, Y., Sun, X., Chen, X., Gao, J., Ling, D.: Ceria nanocrystals decorated mesoporous silica nanoparticle-based ROS-scavenging tissue adhesive for highly efficient regenerative wound healing. Biomaterials. ***151***, 66–77 (2018)
82. Pai, B.G., Kulkarni, A.V., Jain, S.: Study of smart antibacterial PCL-xFe3O4 thin films using mouse NIH-3T3 fibroblast cells in vitro. J. Biomed. Mater. Res. Part B: Appl. Biomater. ***105***, 795–804 (2019)
83. Grumezescu, A.M., Holban, A.M., Andronescu, E., Mogosanu, G.D., Vasile, B.S., Chifiriuc, M.C., Lazar, V., Andrei, E., Constantinescu, A., Maniu, H.: Anionic polymers and 10nm Fe3O4@UA wound dressings support human foetal stem cells normal development and exhibit great antimicrobial properties. Int. J. Pharm. **463**(2), 146–154 (2014). https://doi.org/10.1016/j.ijpharm.2013.08.026
84. Bunea, M.C., Vasile, E., Galateanu, B., Hudita, A., Serban, M., Zaharia, C.: Silk fibroin films decorated with magnetic nanoparticles for wound healing applications. Mater. Plast. ***54***, 83–87 (2017)
85. Anghel, I., Holban, A.M., Grumezescu, A.M., Andronescu, E., Ficai, A., Anghel, A.G., Maganu, M., Lazˇar, V., Chifiriuc, M.C.: Modified wound dressing with phyto-nanostructured coating to prevent staphylococcal and pseudomonal biofilm development. Nanoscale Res. Lett. ***7***, 690 (2012)
86. Wu, J., Zhu, J., Wu, Q., An, Y., Wang, K., Xuan, T., Zhang, J., Song, W., He, H., Song, L., et al.: Mussel-inspired surface immobilisation of heparin on magnetic nanoparticles for enhanced wound repair via sustained release of a growth factor and M2 macrophage polarisation. ACS Appl. Mater. Interfaces. ***13***, 2230–2244 (2021)
87. Li, X., Wei, Z., Zhang, W., Lv, H., Li, J., Wu, L., Zhang, H., Yang, B., Zhu, M., Jiang, J.: Anti-inflammatory effects of magnetically targeted mesenchymal stem cells on laser-induced skin injuries in rats. Int. J. Nanomedicine. ***15***, 5645–5659 (2020)
88. Tang, Y., Rajendran, P., Veeraraghavan, V.P., Hussain, S., Balakrishna, J.P., Chinnathambi, A., Alharbi, S.A., Alahmadi, T.A., Rengarajan, T., Mohan, S.K.: Osteogenic differentiation and mineralization potential of zinc oxide nanoparticles from Scutellaria baicalensis on human osteoblast-like MG-63 cells. Mater. Sci. Eng.: C. ***119***, 111656 (2021)
89. Khader, A., Arinzeh, T.L.: Biodegradable zinc oxide composite scaffolds promote osteochondral differentiation of mesenchymal stem cells. Biotechnol. Bioeng. ***117***, 194–209 (2020)
90. Garino, N., Sanvitale, P., Dumontel, B., Laurenti, M., Colilla, M., Izquierdo-Barba, I., Cauda, V., Vallet-Regì, M.: Zinc oxide nanocrystals as a nanoantibiotic and osteoinductive agent. RSC Adv. ***9***, 11312–11321 (2019)
91. Zhou, G., Gu, G., Li, Y., Zhang, Q., Wang, W., Wang, S., Zhang, J.: Effects of cerium oxide nanoparticles on the proliferation, differentiation, and mineralization function of primary osteoblasts in vitro. Biol. Trace Elem. Res. **153**, 411–418 (2013)
92. Wei, F., Neal, C.J., Sakthivel, T.S., Kean, T., Seal, S., Coathup, M.J.: Multi-functional cerium oxide nanoparticles regulate inflammation and enhance osteogenesis. Mater. Sci. Eng.: C. **124**, 112041 (2021)
93. Li, J., Wen, J., Li, B., Li, W., Qiao, W., Shen, J., Jin, W., Jiang, X., Yeung, K.W.K., Chu, P.K.: Valence state manipulation of cerium oxide nanoparticles on a titanium surface for modulating cell fate and bone formation. Adv. Sci. ***5***, 1700678 (2018)
94. Singh, R.K., Patel, K.D., Lee, J.H., Lee, E.J., Kim, J.H., Kim, T.H., Kim, H.W.: Potential of magnetic nanofiber scaffolds with mechanical and biological properties applicable for bone regeneration. PLoS ONE. ***9***, e91584 (2014)
95. Cojocaru, F.D., Balan, V., Popa, M.I., Lobiuc, A., Antoniac, A., Antoniac, I.V., Verestiuc, L.: Biopolymers-calcium phosphates composites with inclusions of magnetic nanoparticles for bone tissue engineering. Int. J. Biol. Macromol. ***125***, 612–620 (2019)

96. Sung, H., Ferlay, J., Siegel, R.L., Laversanne, M., Soerjomataram, I., Jemal, A., Bray, F.: Global cancer statistics 2020: GLOBOCAN estimates of incidence and mortality worldwide for 36 cancers in 185 countries. CA: Cancer J. Clin. **71**, 209–249 (2021)
97. Arruebo, M., Vilaboa, N., Sáez-Gutierrez, B., Lambea, J., Tres, A., Valladares, M., González-Fernández, Á.: Assessment of the evolution of cancer treatment therapies. Cancers. ***3***, 3279–3330 (2011). https://doi.org/10.3390/cancers3033279
98. Vinardell, M.P., Mitjans, M.: Antitumor activities of metal oxide nanoparticles. Nanomaterials. ***5***, 1004–1021 (2015)
99. Chandrasekaran, M., Pandurangan, M.: In Vitro selective anti-proliferative effect of zinc oxide nanoparticles against Co-cultured C2C12 myoblastoma cancer and 3T3-L1 normal cells. Biol. Trace Elem. Res. ***172***, 148–154 (2016)
100. Wahab, R., Kaushik, N.K., Kaushik, N., Choi, E.H., Umar, A., Dwivedi, S., Musarrat, J., Al-Khedhairy, A.A.: ZnO nanoparticles induces cell death in malignant human T98G gliomas, KB and non-malignant HEK cells. J. Biomed. Nanotechnol. ***9***, 1181–1189 (2013)
101. Premanathan, M., Karthikeyan, K., Jeyasubramanian, K., Manivannan, G.: Selective toxicity of ZnO nanoparticles toward Grampositive bacteria and cancer cells by apoptosis through lipid peroxidation. Nanomedicine: Nanotechnol, Biol. Med. *7*, 184–192 (2011)
102. Chakraborti, S., Chakraborty, S., Saha, S., Manna, A., Banerjee, S., Adhikary, A., Sarwar, S., Hazra, T.K., Das, T., Chakrabarti, P.: PEG-functionalized zinc oxide nanoparticles induce apoptosis in breast cancer cells through reactive oxygen species-dependent impairment of DNA damage repair enzyme NEIL2. Free Radic. Biol. Med. **103**, 35–47 (2017)
103. Rahman, H.S., Azizi, S., Namvar, F., Mohamad, R., Rasedee, A., Soltani, M., Rahim, R.A.: Green synthesis, characterization, and anticancer activity of hyaluronan/zinc oxide nanocomposites. OncoTargets. Ther. ***9***, 4549–4559 (2016)
104. Wu, H., Zhang, J.: Chitosan-based zinc oxide nanoparticle for enhanced anticancer effect in cervical cancer: a physicochemical and biological perspective. Saudi Pharm J. **26**, 205–210 (2018)
105. Singh, T.A., Das, J., Sil, P.C.: Zinc oxide nanoparticles: a comprehensive review on its synthesis, anticancer and drug delivery applications as well as health risks. Adv. Colloid Interface Sci. **286**, 102317 (2020)
106. Zhang, J., Wu, D., Li, M.F., Feng, J.: Multifunctional mesoporous silica nanoparticles based on charge-reversal plug-gate nanovalves and acid-decomposable ZnO quantum dots for intracellular drug delivery. ACS Appl. Mater. Interfaces. **7**, 26666–26673
107. Colon, J., Hsieh, N., Ferguson, A., Kupelian, P., Seal, S., Jenkins, D.W., Baker, C.H.: Cerium oxide nanoparticles protect gastrointestinal epithelium from radiation-induced damage by reduction of reactive oxygen species and upregulation of superoxide dismutase 2. Nanomedicine: Nanotechnol, Biol. Med. ***6***, 698–705 (2010)
108. Madero-Visbal, R.A., Alvarado, B.E., Colon, J.F., Baker, C.H., Wason, M.S., Isley, B., Seal, S., Lee, C.M., Das, S., Mañon, R.: Harnessing nanoparticles to improve toxicity after head and neck radiation. Nanomedicine: Nanotechnol, Biol. Med. **8**, 1223–1231 (2012)
109. Wason, M.S., Colon, J., Das, S., Seal, S., Turkson, J., Zhao, J., Baker, C.H.: Sensitization of pancreatic cancer cells to radiation by cerium oxide nanoparticle-induced ROS production. Nanomedicine: Nanotechnol, Biol. Med. **9**, 558–569 (2013)
110. Cheng, G., Guo, W., Han, L., Chen, E., Kong, L., Wang, L., Ai, W., Song, N., Li, H., Chen, H.: Cerium oxide nanoparticles induce cytotoxicity in human hepatoma SMMC-7721 cells via oxidative stress and the activation of MAPK signalling pathways. Toxicol. Vitro. ***27***, 1082–1088 (2013)
111. Muhammad, F., Wang, A., Qi, W., Zhang, S., Zhu, G.: Intracellular antioxidants dissolve man-made antioxidant nanoparticles: using redox vulnerability of nanoceria to develop a responsive drug delivery system. ACS Appl. Mater. Interfaces. ***6***, 19424–19433 (2014)
112. Das, J., Choi, Y.J., Han, J.W., Reza, A.M.M.T., Kim, J.H.: Nanoceria-mediated delivery of doxorubicin enhances the anti-tumour efficiency in ovarian cancer cells via apoptosis. Sci. Rep. **7**, 9513 (2017)

113. Giustini, AJ., Petryk, AA., Cassim, SM., Tate, JA., Baker, I., & Hoopes, PJ. (2010). Magnetic nanoparticle hyperthermia in cancer treatment. Nano Life, 1, 17–32
114. Hilger, I., Hiergeist, R., Hergt, R., Winnefeld, K., Schubert, H., Kaiser, W.A.: Thermal ablation of tumours using magnetic nanoparticles. Investig. Radiol. **37**, 580–586 (2002)
115. Plichta, Z., Horák, D., Mareková, D., Turnovcová, K., Kaiser, R., Jendelová, P.: Poly [N-(2-hydroxypropyl)methacrylamide]- modified magnetic-F2O3 nanoparticles conjugated with doxorubicin for glioblastoma treatment. ChemMedChem. ***15***, 96–104 (2020)
116. Li, S., Zhang, R., Wang, D., Feng, L., Cui, K.: Synthesis of hollow maghemite (<gamma>-Fe2O3) particles for magnetic field and pH-responsive drug delivery and lung cancer treatment. Ceram. Int. **47**, 7457–7464 (2021)
117. MubarakAli, D., Manzoor, M.A., Sabarinathan, A., Devi, C.A., Rekha, P., Thajuddin, N., Lee, S.Y.: An investigation of antibiofilm and cytotoxic property of MgO nanoparticles. Biocatal. Agric. Biotechnol. ***18***, 101069 (2019)
118. Karthik, K., Dhanuskodi, S., Gobinath, C., Prabukumar, S., Sivaramakrishnan, S.: Fabrication of MgO nanostructures and its efficient photocatalytic, antibacterial and anticancer performance. J. Photochem. Photobiol. B: Biol. ***190***, 8–20 (2019)
119. Mangalampalli, B., Dumala, N., Grover, P.: Acute oral toxicity study of magnesium oxide nanoparticles and microparticles in female albino Wistar rats. Regul. Toxicol. Pharmacol. ***90***, 170–184 (2017)
120. Wang, C., Cao, S., Tie, X., Qiu, B., Wu, A., Zheng, Z.: Induction of cytotoxicity by photoexcitation of TiO2 can prolong survival in glioma-bearing mice. Mol. Biol. Rep. **38**, 523–530 (2011)
121. Zeni, P.F., Dos Santos, D.P., Canevarolo, R.R., Yunes, J.A., Padilha, F.F., Júnior, R.L.C.D.A., Egues, S.M., Hernández-Macedo, M.L.: Photocatalytic and cytotoxic effects of nitrogen-doped TiO2 nanoparticles on melanoma cells. J. Nanosci. Nanotechnol. ***18***, 3722–3728 (2018)
122. Nešic, M., Žakula, J., Koricanac, L., Stepic, M., Radoicic, M., Popovic, I., Šaponjic, Z., Petkovic, M.: Light controlled metallo-drug delivery system based on the TiO2-nanoparticles and Ru-complex. J. Photochem. Photobiol. A: Chem. ***347***, 55–66 (2017)
123. Hou, Z., Zhang, Y., Deng, K., Chen, Y., Li, X., Deng, X., Cheng, Z., Lian, H., Li, C., Lin, J.: UV-emitting Up-conversion-based TiO2 photosensitizing nanoplatform: near-infrared light mediated in-vivo photodynamic therapy via mitochondria-involved apoptosis pathway. ACS Nano. **9**, 2584–2599 (2015)
124. Lucky, S.S., Idris, N.M., Li, Z., Huang, K., Soo, K.C., Zhang, Y.: Titania Coated Upconversion Nanoparticles for Near-Infrared Light Triggered Photodynamic Therapy. ACS Nano. ***9***, 191–205 (2015)
125. Venkatasubbu, G.D., Ramasamy, S., Reddy, G.P., Kumar, J.: In vitro and In-vivo anticancer activity of surface modified paclitaxel attached hydroxyapatite and titanium dioxide nanoparticles. Biomed. Microdevices. **15**, 711–726 (2013)
126. Liu, E., Zhou, Y., Liu, Z., Li, J., Zhang, D., Chen, J., Cai, Z.: Cisplatin-loaded hyaluronic acid-modified TiO2 nanoparticles for neoadjuvant chemotherapy of ovarian cancer. J. Nanomater. ***2015***, 390358 (2015)
127. Parodi, A., Quattrocchi, N., van de Ven, A.L., Chiappini, C., Evangelopoulos, M., Martinez, J.O., Brown, B.S., Khaled, S.Z., Yazdi, I.K., Enzo, M.V., et al.: Synthetic nanoparticles functionalized with biomimetic leukocyte membranes possess cell-like functions. Nat. Nanotechnol. ***8***, 61–68 (2013)
128. Delalat, B., Sheppard, V.C., Ghaemi, S.R., Rao, S., Prestidge, C.A., McPhee, G., Rogers, M.L., Donoghue, J.F., Pillay, V., Johns, T.G., et al.: Targeted drug delivery using genetically engineered diatom bio-silica. Nat. Commun. ***6***, 8791 (2015)
129. Kafshgari, M.H., Mazare, A., Distaso, M., Goldmann, W.H., Peukert, W., Fabry, B., Schmuki, P.: Intracellular drug delivery with anodic titanium dioxide nanotubes and nanocylinders. ACS Appl. Mater. Interfaces. **11**, 14980–14985 (2019)
130. Kafshgari, M.H., Kah, D., Mazare, A., Nguyen, N.T., Distaso, M., Peukert, W., Goldmann, W.H., Schmuki, P., Fabry, B.: Anodic titanium dioxide nanotubes for magnetically guided therapeutic delivery. Sci. Rep. ***9***, 14336 (2019)

131. Abbasi, B.A., Iqbal, J., Mahmood, T., Ahmad, R., Kanwal, S., Afridi, S.: Plant-mediated synthesis of nickel oxide nanoparticles (NiO) via geranium wallichianum: characterization and different biological applications. Mater. Res. Express. ***6***, 0850a7 (2019)
132. Zhang, Y., Mahdavi, B., Mohammadhosseini, M., Rezaei-Seresht, E., Paydarfard, S., Qorbani, M., Karimian, M., Abbasi, N., Ghaneialvar, H., Karimi, E.: Green synthesis of NiO nanoparticles using Calendula officinalis extract: chemical charactrization, antioxidant, cytotoxicity, and anti-esophageal carcinoma properties. Arab. J. Chem. ***14***, 103105 (2021)
133. Tabassum, N., Kumar, D., Verma, D., Bohara, R.A., Singh, M.P.: Zirconium oxide (ZrO2) nanoparticles from antibacterial activity to cytotoxicity: a next-generation of multifunctional nanoparticles. Mater. Today Commun. **26**, 102156 (2021)
134. El Zowalaty, M., Al-Fahdawi, M.Q., Al-Qubaisi, M.S., Alhassan, F., Rosli, R., Webster, T.J., Naadja, S.E., Taufiq-Yap, Y.H., Abdullah, R.: Cytotoxicity and physicochemical characterization of iron-manganese-doped sulfated zirconia nanoparticles. Int. J. Nanomedicine. **10**, 5739–5750 (2015)
135. Dizaj, S.M., Lotfipour, F., Barzegar-Jalali, M., Zarrintan, M.H., Adibkia, K.: Antimicrobial activity of the metals and metal oxide nanoparticles. Mater. Sci. Eng: C. **44**, 278–284 (2014)
136. Besinis, A., De Peralta, T., Handy, R.D.: The antibacterial effects of silver, titanium dioxide and silica dioxide nanoparticles compared to the dental disinfectant chlorhexidine on Streptococcus mutans using a suite of bioassays. Nanotoxicology. ***8***, 1–16 (2014)
137. Webster, T.J., Seil, J.T.: Antimicrobial applications of nanotechnology: methods and literature. Int. J. Nanomedicine. **7**, 2767–2781 (2012)
138. Khosro, A., Mahmood, A.B., Mohammad, B.J., Gobad, M., Mehdi, S.A.: Evaluation and optimization of factors affecting novel diclofenac sodium-eudragit RS100 nanoparticles. Afr. J. Pharm. Pharmacol. **6**, 941–947 (2012)
139. Buzea, C., Pacheco, I.I., Robbie, K.: Nanomaterials and nanoparticles: Sources and toxicity. Biointerphases. ***2***, MR17–MR71 (2007)
140. Dwivedi, S., Siddiqui, M.A., Farshori, N.N., Ahamed, M., Musarrat, J., Al-Khedhairy, A.A.: Synthesis, characterization and toxicological evaluation of iron oxide nanoparticles in human lung alveolar epithelial cells. Colloids Surf. B: Biointerfaces. ***122***, 209–215 (2014)
141. Azam, A., Ahmed, A.S., Oves, M., Khan, M.S., Habib, S.S., Memic, A.: Antimicrobial activity of metal oxide nanoparticles against Gram-positive and Gram-negative bacteria: a comparative study. Int. J. Nanomedicine. **7**, 6003–6009 (2012)
142. Padmavathy, N., Vijayaraghavan, R.: Enhanced bioactivity of ZnO nanoparticles—An antimicrobial study. Sci. Technol. Adv. Mater. **9**, 035004
143. Hosseinkhani, P., Zand, A.M., Imani, S., Rezayi, M., Rezaei, Z.S.: Determining the antibacterial effect of ZnO nanoparticle against the pathogenic bacterium, Shigella dysenteriae (type 1). Int. J. Nano Dimens. ***1***, 279–285 (2011)
144. Ghasemi, F., Jalal, R.: Antimicrobial action of zinc oxide nanoparticles in combination with ciprofloxacin and ceftazidime against multidrug-resistant Acinetobacter baumannii. J.. Glob Antimicrob. Resist. ***6***, 118–122 (2016)
145. Sarwar, S., Chakraborti, S., Bera, S., Sheikh, I.A., Hoque, K.M., Chakrabarti, P.: The antimicrobial activity of ZnO nanoparticles against Vibrio cholerae: variation in response depends on biotype. Nanomedicine: Nanotechnol. Biol. Med. ***12***, 1499–1509 (2016)
146. Manzoor, U., Siddique, S., Ahmed, R., Noreen, Z., Bokhari, H., Ahmad, I.: Antibacterial, structural and optical characterization of mechano-chemically prepared ZnO nanoparticles. PLoS ONE. ***11***, e0154704 (2016)
147. Jafari, A.R., Mosavi, T., Mosavari, N., Majid, A., Movahedzade, F., Tebyaniyan, M., Kamalzadeh, M., Dehgan, M., Jafari, S., Arastoo, S.: Mixed metal oxide nanoparticles inhibit growth of mycobacterium tuberculosis into THP-1 cells. Int. J. Mycobacteriol. **5**, S181–S183 (2016)
148. Roy, A.S., Parveen, A., Koppalkar, A.R., Prasad, M.V.N.A.: Effect of nano-titanium dioxide with different antibiotics against methicillin-resistant staphylococcus aureus. J. Biomater. Nanobiotechnol. ***1***, 37–41 (2010)

149. Carré, G., Hamon, E., Ennahar, S., Estner, M., Lett, M.C., Horvatovich, P., Gies, J.P., Keller, V., Keller, N., Andre, P.: TiO2 Photocatalysis Damages Lipids and Proteins in Escherichia coli. Appl. Environ. Microbiol. **80**, 2573–2581 (2014)
150. Allahverdiyev, A.M., Abamor, E.S., Bagirova, M., Rafailovich, M.: Antimicrobial effects of TiO2 and Ag2O nanoparticles against drug-resistant bacteria and leishmania parasites. Future Microbiol. ***6***, 933–940 (2011)
151. Jin, T., He, Y.: Antibacterial activities of magnesium oxide (MgO) nanoparticles against foodborne pathogens. J. Nanoparticle Res. ***13***, 6877–6885 (2011)
152. Vidic, J., Stankic, S., Haque, F., Ciric, D., Le Goffic, R., Vidy, A., Jupille, J., Delmas, B.: Selective antibacterial effects of mixed ZnMgO nanoparticles. J. Nanoparticle. Res. ***15***, 1595 (2013)
153. Sondi, I., Salopek-Sondi, B.: Silver nanoparticles as antimicrobial agent: a case study on E. Coli as a model for Gram-negative bacteria. J. Colloid. Interface Sci. **275**, 177–182 (2004)
154. Pelletier, DA., Suresh, AK., Holton, GA., McKeown, CK., Wang, W., Gu, B., Mortensen, NP., Allison, D.P., Joy, DC., Allison, MR., et al. (2010). Effects of engineered cerium oxide nanoparticles on bacterial growth and viability. Appl. Environ. Microbiol., 76, 7981–7989
155. Ravishankar, T.N., Ramakrishnappa, T., Nagaraju, G., Rajanaika, H.: Synthesis and characterization of CeO2 nanoparticles via solution combustion method for photocatalytic and antibacterial activity studies. ChemistryOpen. **4**, 146–154 (2015)
156. Gowri, S., Gandhi, R.R., Sundrarajan, M.: Structural, optical, antibacterial and antifungal properties of zirconia nanoparticles by biobased protocol. J. Mater. Sci. Technol. ***30***, 782–790 (2014)
157. Kumaresan, M., Anand, K.V., Govindaraju, K., Tamilselvan, S., Kumar, V.G.: Seaweed Sargassum wightii mediated preparation of zirconia (ZrO2) nanoparticles and their antibacterial activity against gram-positive and gram-negative bacteria. Microb. Pathog. **124**, 311–315 (2018)
158. Skheel, A.Z., Jaduaa, M.H., Abd, A.N.: Biosynthesis of Cadmium Oxide Nanoparticles (CdO NPS) Using Aqueous Rhizome Extract of Curcuma for Biological Applications. AIP Conf. Proc. **2398**, 020030 (2022). https://doi.org/10.1063/5.0097587
159. Manuja, A., Kumar, B., Kumar, R., Chhabra, D., Ghosh, M., Manuja, M., Brar, B., Pal, Y., Tripathi, B.N., Prasad, M.: Metal/metal oxide nanoparticles: toxicity concerns associated with their physical state and remediation for biomedical applications. Toxicol. Rep. **8**, 1970–1978 (2021). https://doi.org/10.1016/j.toxrep.2021.11.020
160. Fu, P.P., Xia, Q., Hwang, H.M., Ray, P.C., Yu, H.: Mechanisms of nanotoxicity: generation of reactive oxygen species. J. Food Drug Analysis. ***22***(1), 64–75 (2014). https://doi.org/10.1016/j.jfda.2014.01.005
161. Gonzalez, L., Lison, D., Kirsch-Volders, M.: Genotoxicity of engineered nanomaterials: A critical review. Nanotoxicology. *2*(4), 252–273 (2008). https://doi.org/10.1080/17435390802464986
162. Thomas, J., Kumar, V., Sharma, N., John, N., Umesh, M., Huligowda, L.K.D., Kaur, K., Utreja, D.: Recent approaches in nanotoxicity assessment for drug delivery applications: challenges and prospects. Med. Drug. Discov. **25**, 100204 (2025). https://doi.org/10.1016/j.medidd.2025.100204
163. Kumar, R., Kumar, A., Bhardwaj, S., Sikarwar, M., Sriwastaw, S., Sharma, G., Gupta, M.: Nanotoxicity unveiled: Evaluating exposure risks and assessing the impact of nanoparticles on human health. J. Trace. Elem. Miner. **13**, 100252 (2025). https://doi.org/10.1016/j.jtemin.2025.100252
164. Rashid, M.M., Forte Tavčer, P., Tomšič, B.: Influence of titanium dioxide nanoparticles on human health and the environment. Nanomaterials. ***11***(9), 2354 (2021). https://doi.org/10.3390/nano11092354
165. Zhang, X., Li, W., Yang, Z.: Toxicology of nanosized titanium dioxide: An update. Arch. Toxicol. **89**, 2207–2217 (2015). https://doi.org/10.1007/s00204-015-1594-6
166. Ragu Prasath, A., Selvam, K.: A review of the zinc oxide nanoparticles synthesis and their emerging biomedical potential. Biomed. Mater. Devices. (2025). https://doi.org/10.1007/s44174-025-00422-1

167. Singh, S.: Zinc oxide nanoparticles impacts: cytotoxicity, genotoxicity, developmental toxicity, and neurotoxicity. Toxicol. Mech. Methods. ***29***(4), 300–311 (2019). https://doi.org/10.1080/15376516.2018.1553221
168. Gallucci, N., Vitiello, G., Di Girolamo, R., Imbimbo, P., Monti, D.M., Tarallo, O., Vergara, A., Krauss, R.I., Paduano, L.: Towards the development of antioxidant cerium oxide nanoparticles for biomedical applications: controlling the properties by tuning synthesis conditions. Nanomaterials (Basel). ***11***(2), 542 (2021). https://doi.org/10.3390/nano11020542
169. Liu, B., Pan, Y., Han, Z., Shu, Y., Liu, Z., Zhang, M., Wan, A., Wang, M., Tan, Y., Wang, Z.: Interplay of structural properties and redox behavior in CeO2 nanoparticles: impact on reactivity and bioavailability. Environ. Sci. Technol. ***59***(9), 4641–4651 (2025). https://doi.org/10.1021/acs.est.4c10490
170. Yadav, R.K., Verma, S., Ramakant, A.K., Pandey, V.: Iron oxide nanoparticle genotoxicity, ROS production, and epigenetic alterations: a comprehensive review. Int. J. For, Anim. Fish. Res. (IJFAF). ***9***(2), 1–16 (2025) https://doi.org/10.22161/ijfaf.9.2.1
171. Luo, C., Li, X., Yan, H., Guo, Q., Liu, J., Li, Y.: Iron oxide nanoparticles induce ferroptosis under mild oxidative stress in vitro. Sci. Rep. **14**, 31383 (2024). https://doi.org/10.1038/s41598-024-82917-3
172. Mohajeri, M., Momenai, R., Karami-Mohajeri, S., Ohadi, M., Estabragh, M.A.R.: Cerium oxide nanoparticles, physical and chemical properties, applications and toxicological implications: A review. Results Chem. ***15***, 102302 (2025). https://doi.org/10.1016/j.rechem.2025.102302
173. Yamano, S., Umeda, Y.: Fibrotic pulmonary dust foci is an advanced pneumoconiosis lesion in rats induced by titanium dioxide nanoparticles in a 2-year inhalation study. Part Fibre. Toxicol. **22**, 7 (2025). https://doi.org/10.1186/s12989-025-00623-y
174. Almansour, M.I., Alferah, M.A., Shraideh, Z.A., Jarrar, B.M.: Zinc oxide nanoparticles hepatotoxicity: Histological and histochemical study. Environ. Toxicol. Pharmacol. **51**, 124–130 (2017). https://doi.org/10.1016/j.etap.2017.02.015
175. Chen, C., Pei, X., Yu, Y., Gao, C., Wang, J., Zhu, R., Liu, S., Tang, S., Li, D.: Zinc oxide nanoparticle-induced neurotoxicity: underlying molecular mechanisms and future perspectives. Toxics. ***14***(1), 11 (2026). https://doi.org/10.3390/toxics14010011

Drug Delivery Agents: Necessary Modifications and Condition in Nano Materials

Prakash K. Mandal, Deepak Kumar, Sreya Dutta, Kalyan Hazra, Sudipto Debnath, Giriraj Tailor, and Venkata Narasimhaji

1 Introduction

1.1 Emergence of Nanotechnology in Drug Delivery

Drug delivery being an expansive as well as innovative field for mankind, so many researchers from different fields, such as organic synthesis, material science, drug discovery, pharmacology, medicinal inventions, get involved themselves with their relentless effort for improvement in drug design, discovery and most importantly specific and safe delivery [1]. On the other hand, nanotechnology is a rapidly expanding arena in research and in actual, we are surrounded by nanomaterials inextricably from morning to night [2]. The **emergence of nanotechnology in drug delivery** has revolutionized modern medicine by enabling precise, proficient, and targeted therapeutic interventions [3]. Traditional drug delivery methods often face challenges such as poor solubility, limited bioavailability, rapid metabolism, and non-specific distribution, leading to reduced efficacy and adverse side effects [4]. Nanotechnology addresses these limitations through the development of nanoscale carriers like liposomes, dendrimers, polymeric nanoparticles, micelles, and metallic nanostructures that can encapsulate, protect, and deliver drugs directly to diseased sites with improved **targeted delivery and controlled release** [5]. By functionalizing nanoparticle surfaces with ligands, antibodies, or peptides, prodrugs can selectively bind to specific receptors overexpressed on diseased cells, such as cancerous tissues

P. K. Mandal · D. Kumar (✉) · S. Dutta · K. Hazra (✉) · S. Debnath · V. Narasimhaji
Central Ayurveda Research Institute, Salt Lake Sector-V, Kolkata, India
e-mail: dpkjangid89@gmail.com

K. Hazra
e-mail: kalyan987@gmail.com

G. Tailor
Department of Chemistry, Mewar University, Gangrar, Chittorgarh, Rajasthan, India

G. Tailor et al. (eds.), *Nanomaterials*, Engineering Materials,
https://doi.org/10.1007/978-981-92-0678-0_7

or inflamed sites, thereby minimizing harm to healthy cells. Additionally, stimuli-responsive nanoparticles activated by pH, temperature, or enzymes allow for site-specific and time-controlled drug release, further enhancing therapeutic precision. Moreover, advanced nanotechnologies have unveiled the futuristic possibilities for **gene delivery, vaccine formulation, and crossing biological barriers**; including the integration of diagnostic and therapeutic functions through simultaneous imaging with personalized medicinal treatment evolving in a single nano-platform, known as theranosticsw [6]. It can be evident that the emergence of nanotechnology in drug delivery marks a paradigm shift from traditional pharmacotherapy toward **smart, efficient, and patient-specific treatments**, holding immense promise for combating complex diseases such as cancer, neurological disorders, and infectious diseases with enhanced safety and efficacy (Fig. 1).

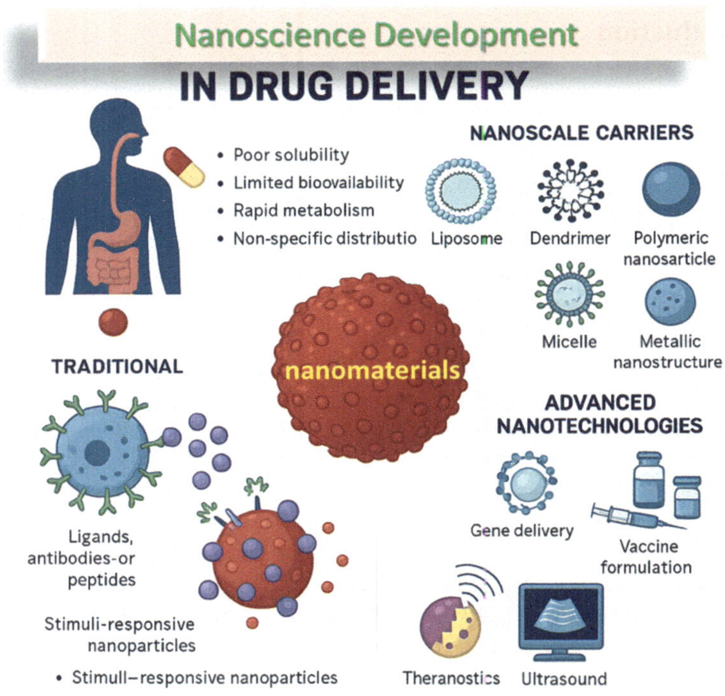

Fig. 1 Features of nanoscience evolved drug delivery systems

2 Scope and Objectives of the Chapter

Proficient modifications and precise optimized conditions of nanomaterials are crucial to ensure safety, stability, and therapeutic efficiency in nanotechnology-based drug delivery programs. To achieve that enhancement in biocompatibility, target specificity, and circulation time would be the primary lookout, which can be executed by surface functionalization through polymers, ligands, or antibodies [7]. In this regard, adjusting particle size, charge, and morphology of a nanoparticle can influence the cellular uptake and bio-distribution, while maintaining uniformity prevents aggregation and premature clearance. Furthermore, stimuli-responsive designs enable controlled drug release under specific physiological triggers such as pH, temperature, or enzymes [8]. In medicinal approach background, the chemical composition and surface coating must also minimize toxicity and immune response, which is an essential criteria for effective drug delivery in the disease site. In this chapter, aspects of nanomaterial-based drug delivery agents, diverse surface functionalization techniques for upgrading the nanomaterials towards better performance, different physiological parameters for stability and biocompatibility along with challenges to address the shortcomings are discussed elaborately in a vivid and lucid manner to ensure the better interpretation of this important contemporary area of research.

3 Nanomaterials as Drug Delivery Agent

3.1 *Types of Nanomaterials-Based Drug Delivery Agents*

As mentioned earlier that nanomaterials have revolutionized drug delivery by allowing therapies to be administered with precision, control, and targeted accuracy and their unique physicochemical properties as well, such as nanoscale size, high surface area-to-volume ratio, and tunable surface chemistry, which make them ideal for improving drug solubility, stability, bioavailability, and cellular uptake [9]. A wide range of nanomaterials has been engineered, each characterized by unique structural features, compositional attributes, and modes of action.

The choice of an appropriate nanocarrier is governed by the physicochemical nature of the therapeutic agent, the required release kinetics, and the intended biological target. With continuous progress in nanotechnology, hybrid and multifunctional nanocarriers are increasingly being developed to combine diagnostic, therapeutic, and monitoring functions, thereby advancing personalized and precision medicine [10]. Consequently, nanomaterials used as drug delivery platforms possess tremendous potential to transform modern pharmacotherapy by improving safety, efficacy, and patient specificity. This chapter therefore presents an overview of the principal classes of nanomaterials currently employed as drug delivery agents.

3.2 Liposomes

Liposomes are spherical vesicular structures consisting of one or more phospholipid bilayers enclosing an aqueous core (Fig. 2) [11]. Owing to their amphiphilic nature, they can encapsulate hydrophilic drugs within the aqueous core and hydrophobic drugs within the lipid bilayer. Their inherent biocompatibility and biodegradability make liposomes well suited for clinical use. Surface modification with polyethylene glycol (PEG), known as PEGylation, prolongs systemic circulation, while ligand conjugation enables site-specific or tumor-targeted delivery [12]. Notable examples include Doxil® (liposomal doxorubicin) for cancer treatment and liposomal vaccines for infectious diseases [13].

3.2.1 Polymeric Nanoparticles

Polymeric nanoparticles, synthesized from biodegradable polymers such as poly-lactic acid (PLA), polyglycolic acid (PGA), and their copolymer PLGA, provide controlled and sustained release of drugs [14]. The polymer matrix protects encapsulated drugs from premature degradation, allowing gradual release over time. Surface

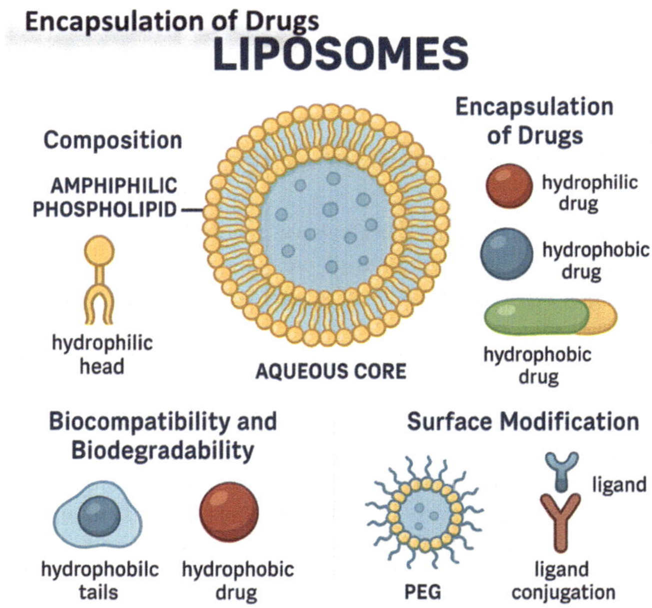

Fig. 2 Structural insight and drug delivery application of liposomes

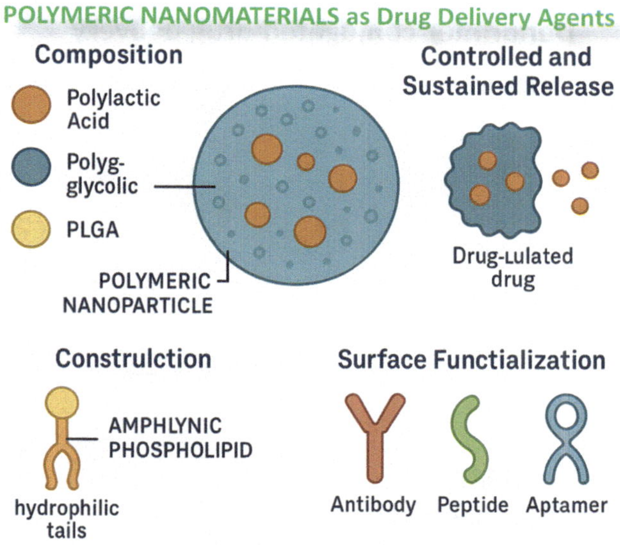

Fig. 3 Surface functionalized polymeric materials in drug delivery processes

functionalization with targeting ligands (antibodies, peptides, or aptamers) further improves specificity toward diseased cells (Fig. 3). Polymeric nanoparticles are widely studied for cancer, cardiovascular, and neurological drug delivery due to their high versatility and tenability. The **solution dynamics of 5-fluorouracil entrapped in PLGA microsphere, a study with 1D selective NMR methods** to analyze molecular motion and environment of 5-fluorouracil within polymeric microspheres, advancing drug-delivery characterization [15].

3.2.2 Dendrimers

Dendrimers are monodisperse, tree-like branched macromolecules with a well-defined architecture consisting of a central core, internal branches, and multiple surface functional groups. This highly ordered structure offers abundant sites for drug conjugation, targeting ligands, or imaging agents (Fig. 4). Drugs can be encapsulated in their internal cavities or attached covalently to surface groups. Poly(amidoamine) (PAMAM) dendrimers are among the most studied for drug and gene delivery, providing enhanced solubility, bioavailability, and reduced toxicity when properly modified [16].

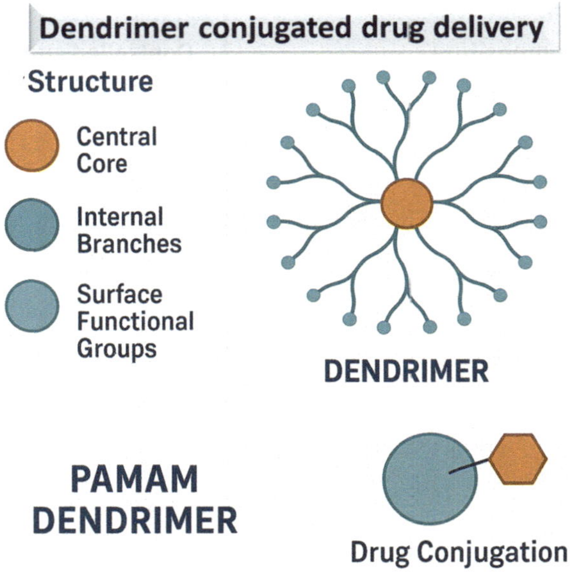

Fig. 4 Drug conjugation with dendrimer for site-selective drug delivery

3.2.3 Metal and Metal Oxide Nanoparticles

Metallic nanomaterials such as gold (Au), silver (Ag), chromium (Cr), iron oxide (Fe_3O_4) and zinc oxide (ZnO) nanoparticles (Fig. 5) have gained prominence in drug delivery and theranostics [17–20]. The Au nanoparticles are inert, easily functionalized, and suitable for targeted delivery and photothermal therapy [21]. Ag nanoparticles possess strong antimicrobial activity [22], while iron oxide nanoparticles are superparamagnetic and widely used in magnetic drug targeting and magnetic resonance imaging (MRI) [23]. These nanoparticles can be combined with polymers or ligands for enhanced biocompatibility and reduced toxicity.

3.2.4 Carbon-Based Nanomaterials

Carbon nanomaterials, including carbon nanotubes (CNTs), graphene oxide (GO), and fullerenes, offer exceptional surface area and mechanical strength (Fig. 6) [24]. Their ability to penetrate cell membranes facilitates intracellular drug and gene delivery. Functionalization of their surfaces with hydrophilic and biocompatible moieties improves solubility and reduces toxicity. Graphene oxide, for instance, can adsorb a variety of drugs through π–π stacking and hydrogen bonding, allowing controlled release in response to pH or temperature [25].

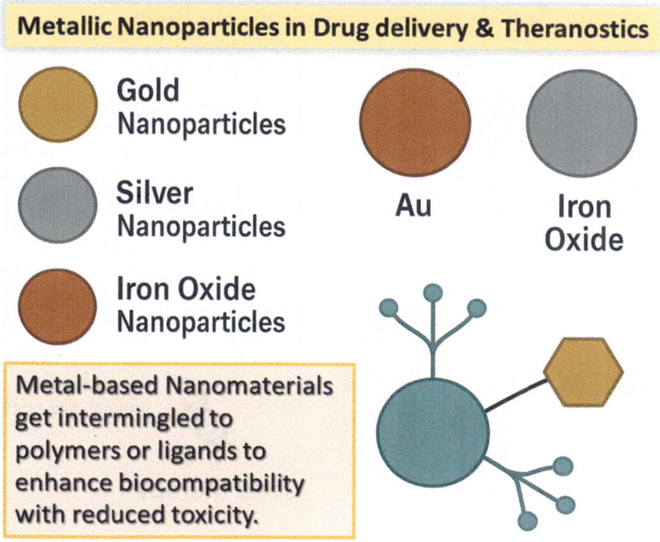

Fig. 5 Functionalized metal-based nanoparticles show enhanced biocompatibility

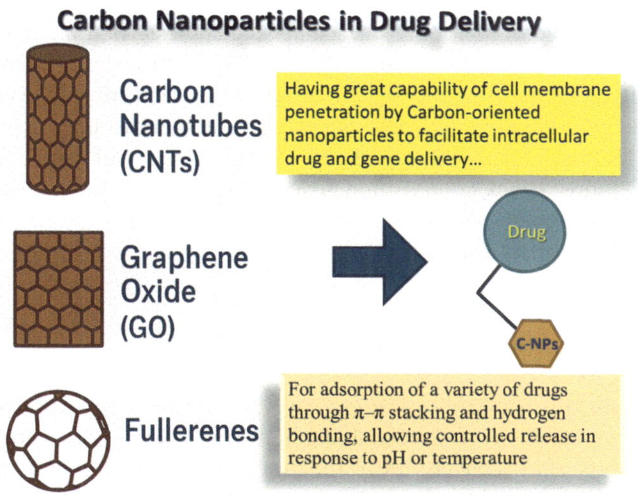

Fig. 6 Carbon nanoparticles are efficient drug delivery agents via cell membrane penetration

3.2.5 Micelles and Niosomes

Polymeric micelles are formed by the self-assembly of amphiphilic block copolymers in aqueous environments, resulting in a hydrophobic core and hydrophilic shell

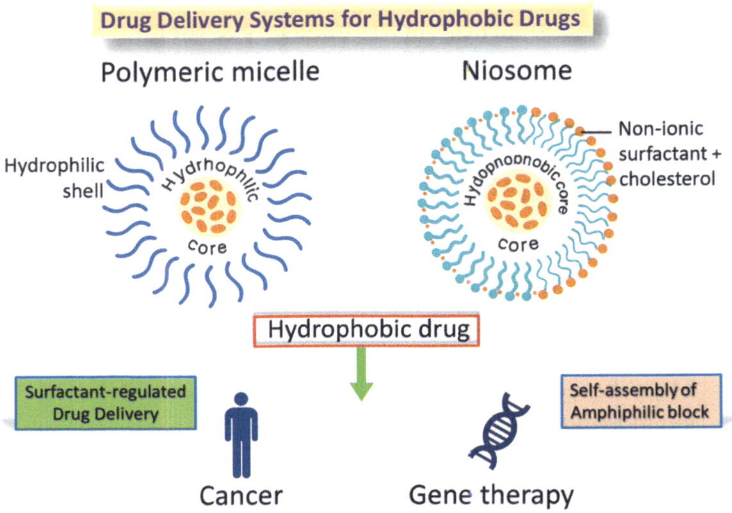

Fig. 7 Polymeric micelle and niosome as drug delivery agents for hydrophobic drugs

(Fig. 7) [26]. They are especially effective for delivering poorly soluble drugs [27]. Niosomes, composed of non-ionic surfactants and cholesterol, function similarly to liposomes but are more stable and cost-effective [28]. Both systems enhance the solubility, stability, and bioavailability of hydrophobic drugs and are extensively explored in cancer and gene therapy. Wei et al. reported pH- and ultrasound-responsive polymeric micelles formed from amphiphilic block copolymers with varied hydrophilic–hydrophobic ratios. Block composition governed self-assembly, stability, and drug-loading efficiency, enabling tunable, stimuli-triggered drug release for improved controlled delivery performance [29].

3.2.6 Mesoporous Silica Nanoparticles (MSNs)

Mesoporous silica nanoparticles possess a well-defined porous structure with large surface area and tunable pore size, enabling high drug loading and controlled release (Fig. 8). Their surface can be easily functionalized with targeting molecules or stimuli-responsive gates that release drugs upon specific triggers such as pH change or enzyme activity [30]. MSNs are chemically stable, biocompatible, and suitable for co-delivery of multiple drugs or imaging agents.

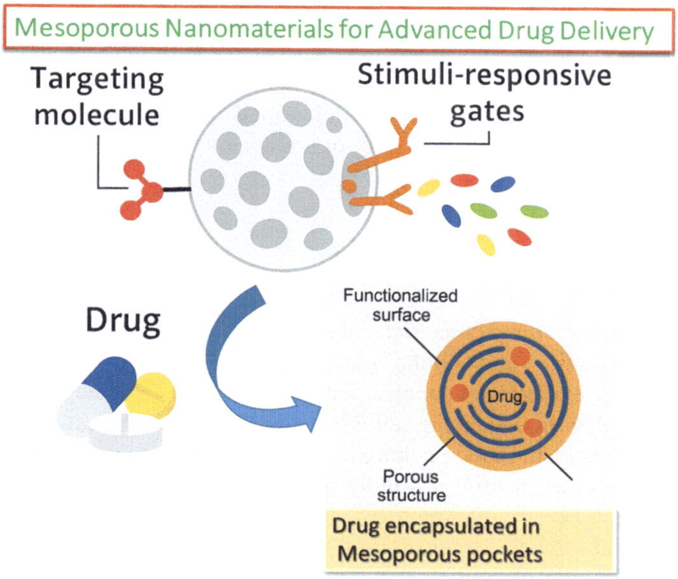

Fig. 8 Mesoporous silica NPs as advanced drug vehicles via stimuli-responsive holes

4 Properties of Nanomaterial-Based Drug Delivery Agents

Nanomaterial-based drug delivery systems have revolutionized modern medicine by enabling controlled, targeted, and efficient therapeutic delivery. Their distinctive physicochemical and biological characteristics help address major limitations of conventional drugs, including poor solubility, limited stability, and rapid systemic clearance [31]. Through careful optimization of parameters such as particle size, surface charge, chemistry, and degradability, nanocarriers can be engineered to achieve sustained and safe drug release profiles [32]. Furthermore, the integration of diagnostic and therapeutic functionalities within a single nanoplatform supports personalized medicine and real-time treatment monitoring. Collectively, these advanced systems offer significant promise for improving healthcare outcomes through safer, more effective, and patient-specific therapies.

4.1 Particle Size and Surface Area

Particle size plays a decisive role in governing the biological performance of nanomaterials. Generally falling within the range of 1–200 nm, these materials exhibit a high surface area-to-volume ratio, which significantly improves drug loading efficiency and promotes interactions with biological membranes [33]. Nanoparticles with smaller dimensions show superior tissue penetration and are capable of traversing biological barriers, including the blood–brain barrier (BBB) [34]. Moreover, their nanoscale dimensions favor cellular uptake through endocytosis, thereby facilitating efficient intracellular delivery of therapeutic agents.

4.2 Surface Charge (Zeta Potential)

The surface charge of nanomaterials plays a crucial role in determining their stability, biodistribution, and cellular internalization. Nanoparticles bearing a positive surface charge exhibit stronger electrostatic interactions with negatively charged cell membranes, thereby enhancing cellular uptake [35]. Nevertheless, an excessively high positive charge can induce cytotoxic effects and promote nonspecific protein adsorption. In contrast, neutral or mildly negatively charged nanoparticles tend to display prolonged blood circulation by reducing rapid recognition and clearance by the reticuloendothelial system (RES) [36]. Consequently, precise modulation of surface charge through appropriate functionalization is essential to achieve balanced biological performance and optimal therapeutic efficacy.

4.3 Surface Functionalization and Modifiability

One of the key advantages of nanomaterials lies in their amenability to surface functionalization with diverse chemical moieties, ligands, polymers, or biomolecules [37]. Such surface modification improves biocompatibility, minimizes immunogenic responses, and enables site-specific drug delivery. The attachment of targeting ligands including antibodies, peptides, and aptamers, allows nanoparticles to actively recognize and bind receptors that are overexpressed on diseased cells, particularly in tumor tissues [38]. Furthermore, surface coating with hydrophilic polymers such as polyethylene glycol (PEG) endows nanoparticles with "stealth" properties, thereby extending systemic circulation by reducing opsonization and phagocytic clearance [39].

4.4 Biocompatibility and Biodegradability

For successful clinical translation, nanomaterials must exhibit both biocompatibility and biodegradability to ensure patient safety. Commonly used platforms such as lipids, polymers (including PLGA and PEG), and select metals like gold and iron oxide are widely regarded as biocompatible [40]. Biodegradable carriers are particularly advantageous, as they break down into non-toxic byproducts following drug release, thereby reducing the risk of long-term tissue accumulation. Careful assessment and optimization of these properties are therefore essential to minimize potential immune or inflammatory reactions and to support safe therapeutic application.

4.5 Drug Loading Capacity and Release Kinetics

Because of their large surface area and adjustable structural features, nanomaterials exhibit high drug encapsulation efficiency. The kinetics of drug release from these systems can be accurately tailored by fine-tuning parameters such as the composition of the carrier, surface functionalization, and responsiveness to the local environment. Notably, stimuli-responsive nanoparticles can be activated by triggers including pH, temperature, enzymes, light, or magnetic fields, enabling site-specific and controlled drug release. This controlled delivery approach helps achieve optimal therapeutic efficacy while reducing undesirable side effects [41].

4.6 Stability and Shelf Life

Nanomaterials are required to exhibit robust physical and chemical stability both under physiological environments and throughout storage. Factors such as temperature, pH, and ionic strength can strongly influence nanoparticle behavior, leading to aggregation, premature drug release, or material degradation. To address these challenges, the use of surface coatings and encapsulating agents is commonly employed to enhance stability and extend shelf life, thereby maintaining reliable and reproducible therapeutic performance [42].

4.7 Targeting and Bio-distribution

The targeting capability of nanomaterials is a hallmark feature that differentiates them from traditional delivery systems. Passive targeting primarily relies on the enhanced permeability and retention (EPR) effect, whereby nanoparticles preferentially accumulate in tumor tissues as a result of their leaky vasculature [43]. In contrast, active targeting is accomplished by decorating nanoparticle surfaces with specific ligands that selectively bind to receptors expressed on target cells. This receptor-mediated approach enhances local drug concentration at the diseased site while simultaneously reducing off-target distribution and systemic toxicity.

4.8 Optical and Magnetic Properties

Some nanomaterials including gold nanoparticles, quantum dots, and iron oxide nanoparticles, exhibit distinctive optical or magnetic characteristics that enable their use in theranostic platforms, where diagnostic and therapeutic functions are integrated into a single system. For instance, magnetic nanoparticles can be directed to specific target sites through the application of external magnetic fields and can concurrently serve as agents for magnetic resonance imaging (MRI) while delivering therapeutic payloads [44].

5 Surface Functionalization of the Nanomaterials

5.1 Various Functionalization Techniques

Surface functionalization represents a key approach for improving the biocompatibility, stability, dispersibility, and targeting capability of nanomaterials in drug delivery and biomedical applications. This strategy involves tailoring the nanoparticle surface with specific chemical or biological entities to enhance their interaction with physiological environments. Functionalization may be achieved through covalent or non-covalent modification routes.

Covalent approaches, including silanization, amidation, and PEGylation, enable the formation of robust and stable linkages for immobilizing polymers, targeting ligands, or biomolecules, thereby supporting prolonged circulation and controlled drug release [45]. In contrast, non-covalent strategies utilize weaker interactions such as hydrogen bonding, π–π stacking, and electrostatic forces, which help preserve the inherent physicochemical properties of materials like graphene and carbon nanotubes [46, 47].

Additional functionalization techniques include polymer coatings to improve biocompatibility, ligand conjugation for receptor-specific active targeting, silanization for modifying silica and metal oxide surfaces, and biomolecule attachment to enable selective bio-recognition [48]. Surfactant-based modification further enhances dispersibility and minimizes nanoparticle aggregation. Collectively, these surface engineering strategies allow precise control over nanomaterial behavior in biological systems, leading to enhanced targeting, regulated drug release, and reduced toxicity. Through judicious selection of functionalization methods, advanced multifunctional and safe nanocarriers can be designed to significantly propel therapeutic and diagnostic progress in nanomedicine.

5.2 PEGylation of Nanomaterials

PEGylation refers to the surface modification of nanomaterials with polyethylene glycol (PEG) chains, a widely adopted approach to enhance physicochemical stability, biocompatibility, and pharmacokinetic behavior. As a core strategy in nanomaterial surface engineering, PEGylation prolongs systemic circulation, lowers immunogenicity, and facilitates controlled drug release (Fig. 9) [12]. Precise control over PEG chain length, grafting density, and conjugation chemistry allows the design of sophisticated stealth nanocarriers tailored for applications in drug delivery, diagnostic imaging, and therapy [39]. Such optimized platforms demonstrate enhanced therapeutic performance while minimizing immune recognition, thereby advancing the development of personalized nanomedicine.

Polyethylene glycol (PEG) is a hydrophilic, non-toxic, and biocompatible polymer that forms a protective hydration layer around nanoparticles when attached to their surface either covalently or through physical adsorption [49]. This hydrated shell

Fig. 9 PEGylation of nanocarriers for efficient drug delivery

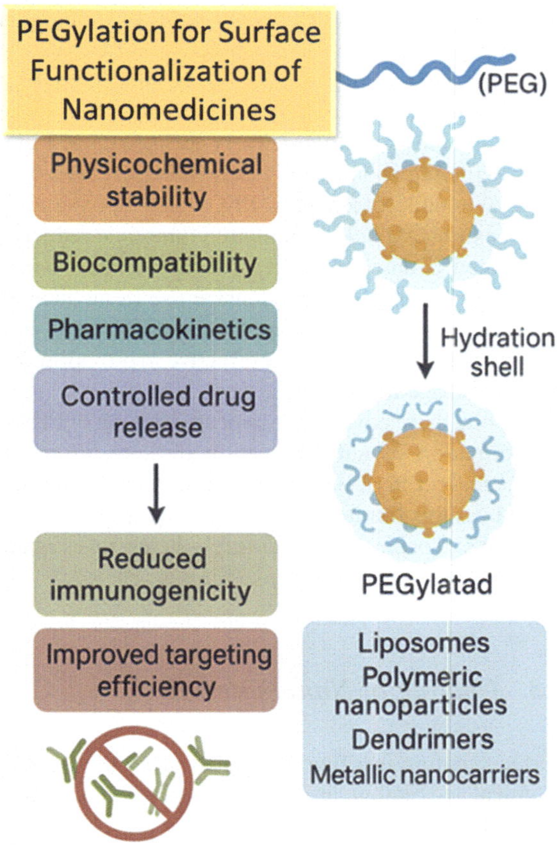

functions as a steric barrier that inhibits protein adsorption and immune opsonization. Consequently, PEGylated nanoparticles can evade recognition and clearance by the reticuloendothelial system (RES), leading to prolonged circulation in the bloodstream [36]. PEGylation is commonly achieved by covalently linking PEG chains to surface functional groups, such as amine, carboxyl, or thiol moieties using EDC/NHS coupling chemistry or thiol–gold interactions [50]. By adjusting the molecular weight and surface density of PEG, nanoparticle stability, surface coverage, and biological interactions can be precisely controlled.

PEGylation method is advantageous for nanomaterial surface fabrication in terms of enhanced circulation time, reduced immunogenicity and toxicity, improved colloidal stability, controlled drug release, better targeting efficiency etc. PEGylated nanoparticles exhibit prolonged residence in systemic circulation, which enhances drug bioavailability and overall therapeutic performance. The presence of a PEG coating also masks nanoparticles from immune surveillance, thereby reducing inflammatory reactions and off-target effects. In addition, PEG suppresses nanoparticle aggregation by diminishing van der Waals and electrostatic interactions, leading to stable and uniform dispersion in biological media. PEGylation further enables regulation of drug release kinetics by modulating surface permeability and hydrophilicity. Owing to these properties, PEG also serves as an effective anchoring platform for targeting ligands, such as antibodies, peptides, or folic acid, facilitating active targeting of diseased cells or tissues [51].

In contemporary drug delivery technologies, PEGylation is widely employed in the design of liposomes, polymeric nanoparticles, dendrimers, and metallic nanocarriers. A notable example is PEGylated liposomal formulations, such as Doxil®, which exhibit extended circulation times and reduced cardiotoxicity relative to non-PEGylated counterparts [13]. Likewise, PEG-functionalized gold and iron oxide nanoparticles have found extensive use in medical imaging, photothermal therapy, and targeted drug delivery application [52]. In polymer-based systems, PEG–PLGA nanoparticles are particularly effective in providing sustained release of hydrophobic therapeutic agents [14].

Despite these advantages, repeated dosing of PEGylated nanocarriers may trigger the formation of anti-PEG antibodies, resulting in rapid systemic clearance, a phenomenon referred to as the accelerated blood clearance (ABC) effect [53, 54]. To address this limitation, alternative hydrophilic polymers, such as polyzwitterions and polysarcosine, are increasingly being investigated as promising substitutes for PEG [55].

5.3 Coating of Nanomaterials

Surface coating is a fundamental strategy in nanomedicine for tuning the physicochemical and biological characteristics of nanomaterials to enable safe, effective, and targeted drug delivery. This approach involves applying a protective or functional layer onto the nanoparticle surface to improve stability, biocompatibility, solubility, and targeting capability, while simultaneously reducing aggregation and immune recognition [56]. The choice of coating material and method is guided by the intended therapeutic purpose, the nature of the nanocarrier, and the physiological environment in which it will operate.

Bare nanomaterials often suffer from limitations such as poor colloidal stability, rapid systemic clearance, and potential toxicity arising from high surface reactivity. Surface coating effectively overcomes these drawbacks by enhancing biocompatibility and limiting nonspecific protein adsorption, improving dispersibility in biological fluids, and supplying functional handles for ligand conjugation and active targeting. In addition, coatings can modulate drug release profiles through controlled diffusion or material degradation, while also reducing immunogenicity and toxicity, thereby extending circulation time and improving safety. As a result, surface coating not only shields nanoparticles but also converts them into multifunctional drug delivery platforms capable of precise, controlled therapeutic action. Appropriately engineered coated nanomaterials therefore hold substantial promise for precision medicine, offering improved efficacy with minimized side effects in applications ranging from cancer and neurological disorders to infectious diseases.

There are various coating techniques on nanomaterials like polymeric coatings, lipid coating, inorganic coatings, and protein and polysaccharide coatings etc. (Fig. 10). Natural and synthetic polymers, such as polyethylene glycol (PEG), [57] chitosan-coated poly(lactic-co-glycolic acid) (PLGA), [58] and polyvinyl alcohol (PVA), [59] are widely used for coating. PEGylation provides a stealth effect, which is preventing immune recognition; [39] while Chitosan enhances muco-adhesion and facilitates cellular uptake and PLGA coatings allow controlled and sustained drug release. This study investigates the solution-state dynamics of 5-fluorouracil encapsulated within PLGA microspheres using one-dimensional selective NMR techniques to probe its molecular mobility and microenvironment, thereby enhancing the understanding and characterization of polymer-based drug delivery systems [15].

Now, lipid bilayers, similar to biological membranes, offer excellent biocompatibility and are used in lipid-coated nanoparticles and liposomes. These coatings protect encapsulated drugs and improve delivery across cell membranes or barriers such as the blood–brain barrier. On the other hand, inorganic compounds like silica (SiO_2) or calcium phosphate ($Ca_3(PO4)_2$) coatings improve the chemical stability and biocompatibility of metallic nanoparticles e.g., gold (Au) or silver (Ag) [60]. Silica shells also allow further surface modification for ligand attachment and drug loading. Furthermore, proteins (e.g., albumin) and polysaccharides (e.g., dextran, hyaluronic acid) enhance biological compatibility and can be tailored for receptor-mediated targeting. Hyaluronic acid-coated nanoparticles, for instance, specifically bind to CD44 receptors overexpressed in cancer cells [61]. Coated surfaces can be further functionalized with ligands, antibodies, or peptides for active targeting. This dual functionalization via coating and ligand attachment enables site-specific accumulation of drugs at diseased tissues while minimizing systemic toxicity.

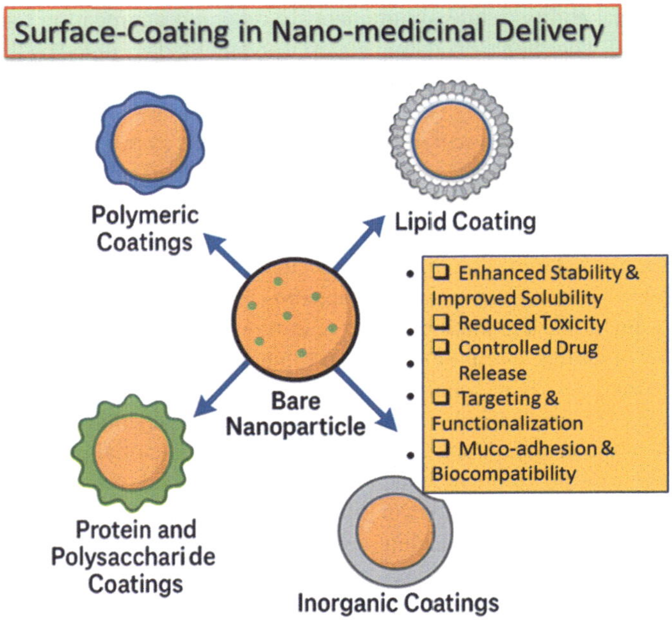

Fig. 10 Diverse surface coating techniques in nanomedicines

5.4 Hybrid and Multifunctional Nanomaterials

Hybrid and multifunctional nanomaterials represent a major advancement in nanomedicine, as they are designed to incorporate multiple complementary functions within a single delivery platform to achieve precise, controlled, and efficient therapeutic action. By integrating components such as polymers, metals, lipids, ceramics, or carbon-based materials, these systems exploit the synergistic advantages of each constituent to improve biocompatibility, structural stability, stimuli responsiveness, and overall therapeutic efficacy. As a new generation of drug delivery systems, they provide exceptional design flexibility, functional adaptability, and biological compatibility. Coupled with advanced surface functionalization strategies and rational hybrid architectures, these nanomaterials enable safe and targeted treatments, particularly for complex diseases including cancer and neurological disorders [62].

Hybrid nanomaterials are created by integrating organic and inorganic components, producing nanocarriers that often outperform single-component systems. For instance, polymer–metal or lipid–silica hybrids merge the biodegradability and structural adaptability of organic materials with the mechanical stability and functional advantages of inorganic components, resulting in multifunctional platforms with enhanced performance [63]. We mention here some benefits of advanced hybrid nanomaterials such as, (a) enhanced drug loading capacity due to diverse surface chemistries, (b) controlled and stimuli-responsive release triggered by pH, temperature, magnetic fields, or enzymes, (c) improved targeting efficiency via surface modification with ligands or antibodies, (d) integration of diagnostic and therapeutic functions i.e., *theranostics* for simultaneous imaging and treatment, (e) reduced toxicity and improved circulation stability through smart surface coatings and functionalization.

A wide range of functionalized nanomaterials has been developed for drug delivery applications, among which polymer–inorganic hybrid nanocarriers stand out as particularly effective systems. These hybrids combine biodegradable polymers, such as PEG, PLGA, or chitosan, with inorganic cores including silica, gold, or iron oxide nanoparticles, thereby integrating the advantages of both material classes [64–66]. A representative example is PEG–gold nanoparticle hybrids, which simultaneously enable drug delivery and photothermal cancer therapy, where the polymer shell improves biocompatibility and circulation behavior, while the metallic core provides imaging and heat-generation capabilities [21].

Another important class comprises lipid–polymer hybrid nanoparticles (LPHNPs), which merge the mechanical stability of polymeric cores with the biocompatibility and membrane-mimicking properties of lipid shells [67]. In these systems, the lipid layer enhances cellular uptake, whereas the polymer matrix supports sustained drug release, making them particularly suitable for the delivery of anticancer agents and genetic therapeutics.

Metal–carbon hybrid nanomaterials represent yet another versatile category, formed by combining metallic nanoparticles (such as gold or iron oxide) with carbon-based nanostructures like graphene or carbon nanotubes. These hybrids exhibit exceptional electrical, optical, and mechanical properties and have demonstrated significant potential in targeted drug delivery, biosensing, and photothermal therapy, contributing to advances in medicinal chemistry. Surface functionalization with polymers or targeting ligands further enhances their solubility and biological compatibility.

In addition, mesoporous silica nanoparticles (MSNs) have recently gained prominence in drug delivery research due to their highly ordered pore structures and large surface areas. Functionalization of MSNs with polymers, lipids, or biomolecules yields multifunctional carriers capable of high drug loading, pH-responsive release, and ligand-mediated targeting [30]. For example, silica–lipid hybrid systems have been successfully employed to co-deliver hydrophobic and hydrophilic drugs with high efficiency.

The surfaces of hybrid nanocarriers can be systematically engineered with functional groups, polymers, or biomolecules to improve their targeting ability and overall therapeutic performance. PEGylation is commonly employed to impart stealth properties, thereby extending systemic circulation time. In parallel, ligand attachment, including antibodies, peptides, or folic acid, enables active targeting of specific cells or tissues. In addition, stimuli-responsive surface coatings allow controlled drug release in response to distinct physiological triggers, such as the acidic microenvironment of tumors.

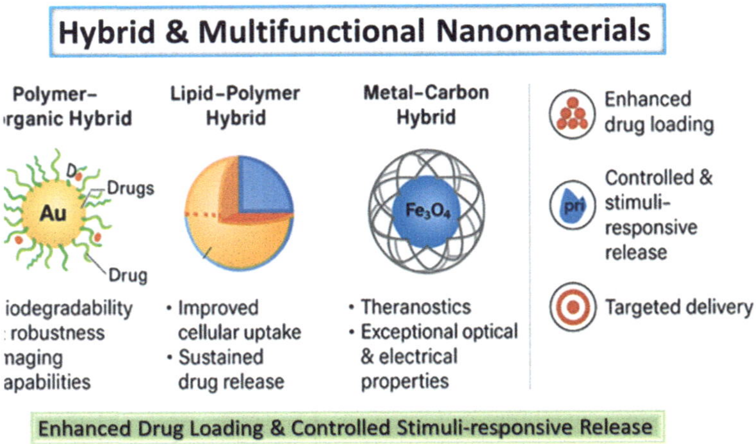

Fig. 11 Diverse hybrid and multi-functional nanomaterials as drug delivery vehicles

Hybrid multifunctional nanomaterials have been widely investigated for applications in cancer treatment, gene delivery, biomedical imaging, and regenerative medicine. For example, gold–polymer hybrid systems facilitate the combination of chemotherapy and photothermal therapy, while iron oxide–silica hybrids are utilized for MRI-guided targeted drug delivery (Fig. 11). Current research is increasingly directed toward the development of biodegradable and smartly responsive hybrid nanocarriers capable of integrating multiple therapeutic functions while minimizing toxicity. These advanced platforms hold significant promise for the realization of personalized, image-guided, and highly efficient drug delivery systems.

5.5 Applications of Functionalized Nanomaterials in Drug Delivery

Functionalized nanomaterials designed with polymers, targeting ligands, biomolecules, and stimuli-responsive elements, have transformed modern drug delivery by enabling accurate targeting, regulated release, and integrated theranostic functions. Their clinical relevance now extends across oncology, vaccine development, gene therapy, infectious disease management, and regenerative medicine. Far from remaining academic concepts, these nanosystems have driven tangible clinical breakthroughs, ranging from mRNA vaccine platforms to deep tumor-penetrating drug carriers. As advances in formulation science, immunology, and scalable manufacturing continue to converge, the coming decade is poised to deliver safer, more intelligent, and highly effective therapeutic systems. Such progress will provide unprecedented control over drug distribution, release timing, and site-specific action, marking a decisive step toward a new era of personalized and precision medicine.

5.5.1 Targeted Cancer Therapy

Targeted nanocarriers utilize a combination of passive and active targeting strategies to enrich therapeutic payloads within tumor tissues. Passive targeting relies on the enhanced permeability and retention (EPR) effect, whereas active targeting is achieved by functionalizing nanoparticle surfaces with antibodies, peptides, folic acid, or small-molecule ligands that selectively recognize tumor-associated receptors [43]. Beyond receptor targeting, surface engineering can also confer tumor-penetration capabilities, such as stimulus-induced size shrinkage or charge conversion in acidic tumor microenvironments, which enhance deep tissue penetration and intracellular uptake [68]. Recent advances emphasize smart, hierarchical nanocarrier designs that integrate prolonged circulation with triggered exposure of targeting ligands, enabling a single platform to effectively navigate successive delivery stages from systemic circulation and tumor accumulation to tissue penetration and cellular internalization (Fig. 12).

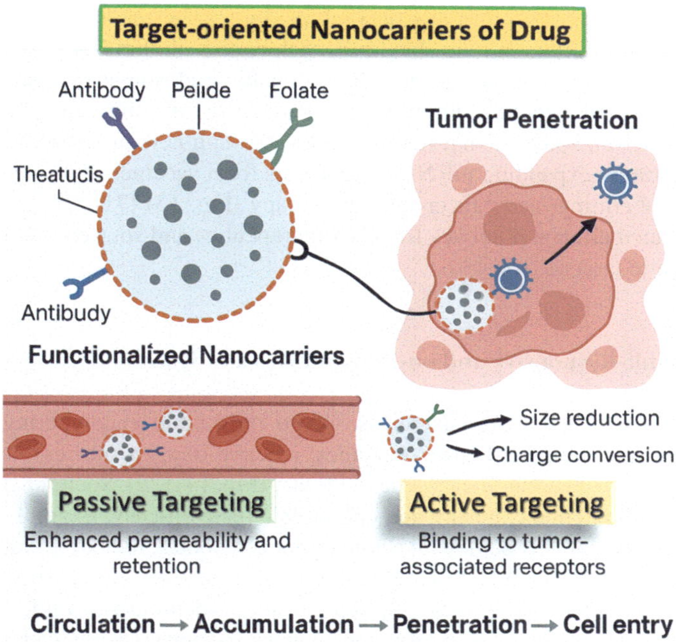

Fig. 12 Targeted cancer cell treatment by advanced nanocarriers

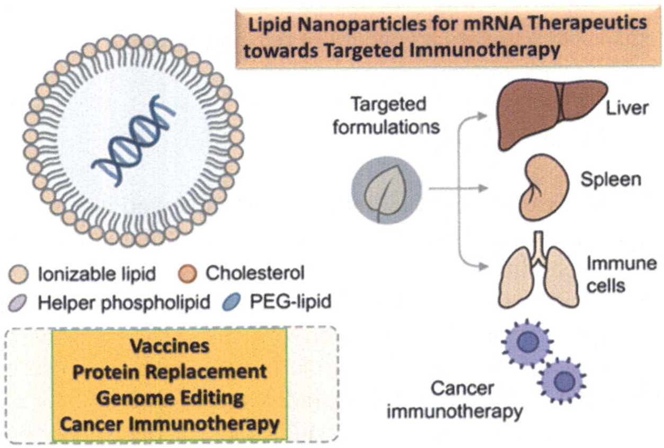

Fig. 13 Targeted cancer cell immunotherapy by lipid nanoparticle-based mRNA vaccines

5.5.2 Nucleic-Acid Delivery (mRNA, siRNA, Gene Editing)

The clinical success of mRNA vaccines during the COVID-19 pandemic cemented lipid nanoparticles (LNPs) as a practical, clinically-viable functionalized nanocarrier [69]. Modern LNPs use ionizable lipids, helper phospholipids, cholesterol, and PEG-lipids to encapsulate and protect RNA, enable endosomal escape, and tune biodistribution [70]. Ongoing innovations focus on organ- and cell-specific LNP formulations (liver, spleen, lung, immune cells) through new lipid chemistries and surface decoration expanding mRNA therapeutics from vaccines to protein replacement, genome editing, and cancer immunotherapy (Fig. 13) [71]. These advances have accelerated clinical pipelines for RNA therapeutics and spurred tailored LNPs for non-hepatic targets.

5.5.3 Stimuli-Responsive and on-Demand Release Systems

Stimuli-responsive nanomaterials release cargo only upon encountering specific internal cues (pH, redox potential, enzymes, reactive oxygen species) or external triggers like light, ultrasound, magnetic fields etc. (Fig. 14) [72]. Such functionalization minimizes systemic exposure and enhances on-site efficacy. Examples include pH-labile linkers that disassemble in acidic tumor lysosomes, redox-cleavable bonds that respond to elevated intracellular glutathione, and enzyme-sensitive coatings that localize release to diseased tissues [73]. Recent studies demonstrate multifunctional carriers that change size, expose targeting ligands, or convert surface charge upon stimulus strategies that reconcile contradictory design needs at different delivery stages [74].

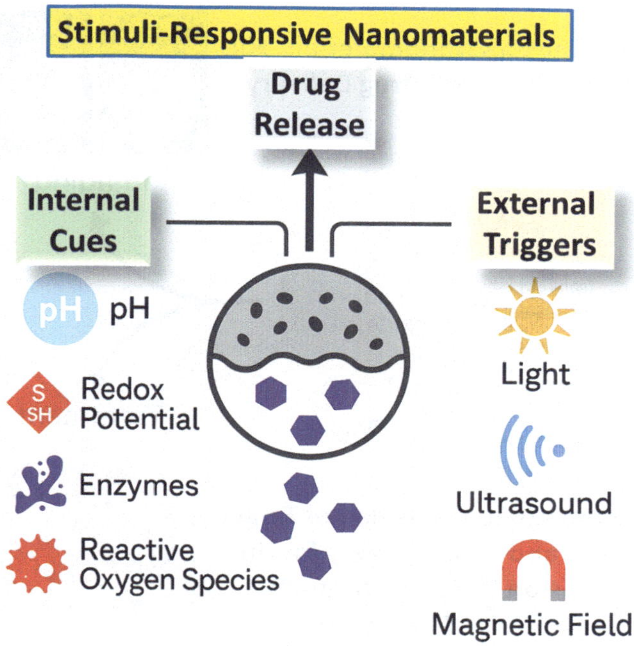

Fig. 14 Stimuli-responsive nanomaterials for on demand targeted drug release

5.5.4 Theranostics: Combining Therapy and Diagnostics

Functionalized nanomaterials enable simultaneous imaging and treatment. Metal like gold, iron oxide or silica cores can be loaded with drugs while their surfaces carry targeting ligands and imaging moieties such as MRI contrast agents, PET isotopes, fluorescent dyes [75]. These theranostic constructs allow noninvasive tracking of biodistribution, real-time monitoring of target engagement, and dose-adjustment based on observed accumulation valuable for precision oncology and image-guided interventions (Fig. 15). Clinical and preclinical studies increasingly show how diagnostic feedback improves patient selection and treatment planning.

5.5.5 Biomimetic Coatings and Immune Evasion

Cell-membrane cloaking (e.g., erythrocyte, platelet, or cancer cell membranes) and polymeric stealth coatings (PEG, polysarcosine, zwitterions) are used to hide nanocarriers from immune surveillance and to exploit natural homing properties [76]. Membrane-coated particles demonstrate prolonged circulation and sometimes intrinsic targeting (e.g., platelet-coated particles localizing to damaged vasculature) [77]. However, immune reactions to coating materials most notably the development

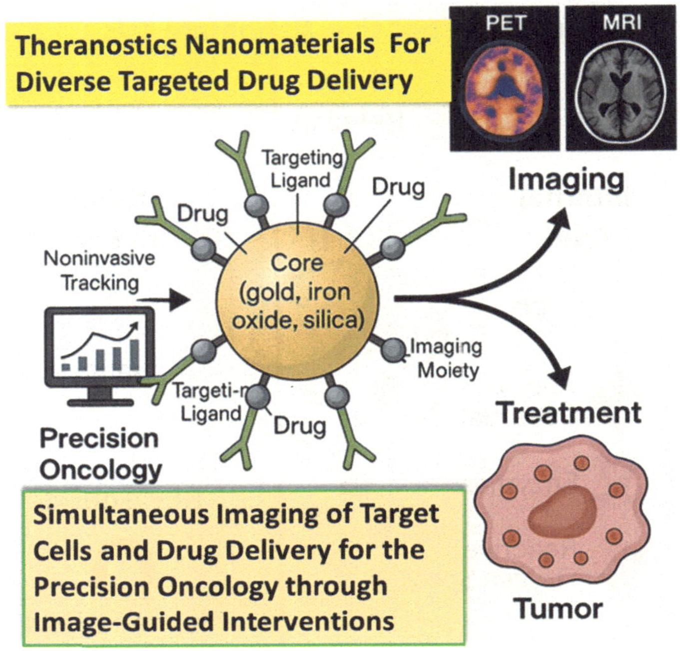

Fig. 15 Theranostic nanoparticles in simultaneous imiging for precision oncology

of anti-PEG antibodies leading to accelerated blood clearance (ABC) have spurred work on alternative stealth polymers and reversible shielding strategies (Fig. 16) [54].

5.5.6 Combination Therapies and Multimodal Platforms

Functionalized nanomaterials make co-delivery of drugs, genes, and immune modulators straightforward, enabling synergistic regimens e.g., chemotherapy plus immunotherapy, photothermal ablation plus checkpoint blockade, or drug plus siRNA combinations [78]. Hybrid carriers such as, lipid–polymer, silica–lipid, metal–polymer integrate complementary properties like drug capacity, imaging contrast, stimuli-responsiveness; improving therapeutic windows and enabling single-dose multimodal interventions (Fig. 17).

Fig. 16 Biomimetic coatings on nanaomaterials for immune evasion

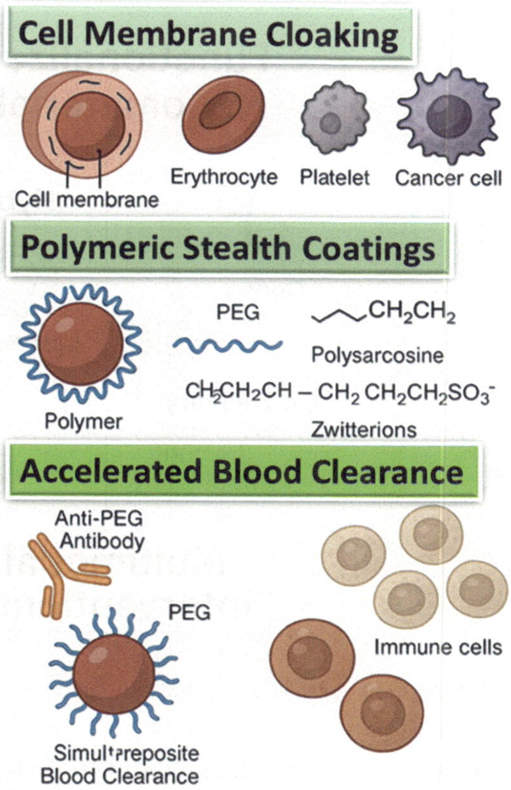

5.5.7 Translational Realities and Regulatory Landscape

Despite striking preclinical successes, clinical translation faces hurdles: manufacturing scale-up with batch-to-batch consistency, long-term safety (off-target accumulation), immunogenicity (anti-PEG), and regulatory complexities for multifunctional constructs [79]. Encouragingly, approved nano-therapeutics (liposomal drugs, Onpattro siRNA LNPs, mRNA vaccines) offer paths forward standardized characterization, quality controls, and clinical endpoints are evolving to match nanomedicine complexity (Fig. 18) [80]. Continued investment in scalable, GMP-compatible formulation technologies and robust safety profiling is essential [81].

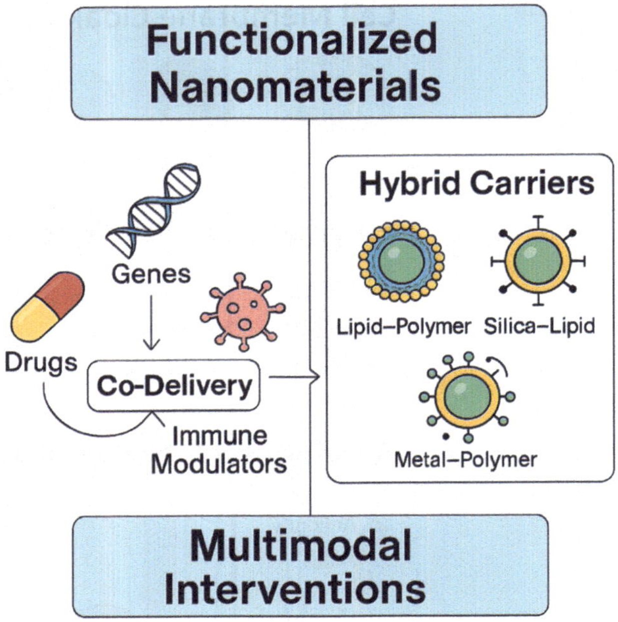

Fig. 17 Hybrid nanomaterials as effective drug vehicles in synergistic regimens

6 Physiological Conditions and their Impact

6.1 Stability and Biocompatibility Consideration

Stability and biocompatibility are central requirements for the successful clinical translation of nanomaterial-based drug delivery systems. Under physiological conditions, nanoparticles encounter blood proteins, immune components, and ionic environments, all of which influence colloidal stability and biodistribution. One of the most critical interfacial processes is protein corona formation, which reshapes nanoparticle identity and strongly impacts immune recognition and cellular uptake. Accordingly, advanced surface engineering strategies aim either to suppress nonspecific protein adsorption or to engineer biomimetic coronas, thereby improving in vivo performance. Maintaining colloidal stability in salt-rich, pH-variable, and protein-dense media is essential to prevent aggregation and reduce off-target interactions. While PEGylation provides effective steric stabilization and extended circulation, repeated administration can induce anti-PEG antibodies and lead to accelerated blood clearance, motivating the development of biodegradable stealth polymers such as polysarcosine and zwitterionic polypeptoids, which exhibit lower immunogenicity and reduced long-term accumulation [54].

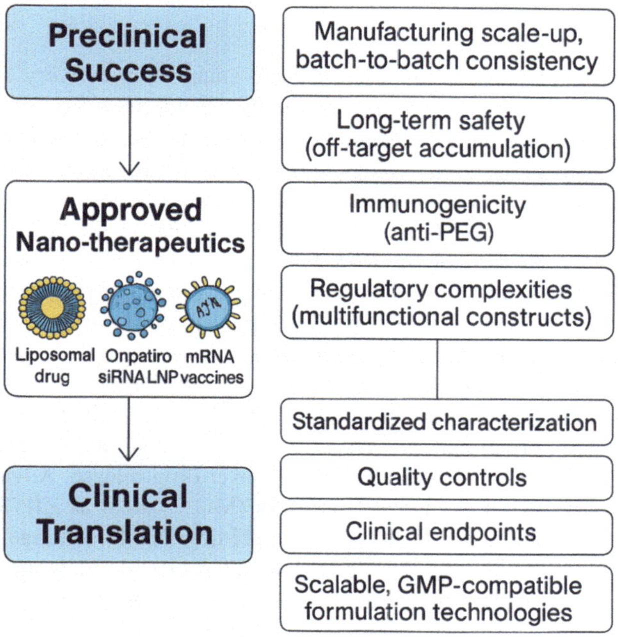

Fig. 18 Nano-therapeutics for advanced drug delivery techniques

Physiological cues including pH gradients, redox conditions, and enzymatic activity are widely exploited to trigger stimuli-responsive drug release, although they also introduce the risk of premature destabilization. Achieving an optimal balance between systemic stability and site-specific responsiveness is therefore particularly important for tumor and mucosal delivery applications [82]. Furthermore, biological heterogeneity, reflected in variable protein corona compositions and the inconsistent nature of the EPR effect, necessitates rational carrier designs that integrate both passive and active targeting strategies. Comprehensive preclinical evaluation, including serum stability assays, corona proteomics, immunogenicity testing, and biodistribution studies, is essential to ensure safety and efficacy. Established regulatory frameworks and validated platforms such as lipid nanoparticles (LNPs) continue to guide scalable and compliant development. Collectively, advances in surface chemistry, stealth polymer design, and responsive nanostructures are steadily enhancing the clinical viability and safety of nanomedicine platforms [83].

6.2 pH Responsiveness

pH-responsive nanomaterials are transforming smart drug delivery strategies by enabling site-selective release within acidic biological environments, including tumor tissues, inflamed regions, and endosomal compartments. These delivery systems are designed to remain stable at physiological pH (~7.4), while undergoing chemical or structural transformations under acidic conditions (pH 4.5–6.9), thereby promoting localized drug release. This pH-triggered behavior enhances therapeutic specificity and significantly reduces systemic toxicity [84]. Key strategies include acid-labile linkages (hydrazone, imine, acetal, ester) that hydrolyze in low pH, and protonation-responsive polymers such as, poly(histidine), poly(β-amino esters), poly(ethylene imine) that swell or disrupt carrier integrity. Hydrazone-linked doxorubicin delivery systems serve as a clear example of tumor-selective drug release, effectively reducing cardiotoxic side effects. More sophisticated dual-responsive nanoplatforms integrate pH sensitivity with additional triggers, such as redox potential or temperature, to achieve enhanced spatial and temporal control over drug release. Advanced carrier systems, including metal–organic frameworks (MOFs) such as ZIF-8 and mesoporous silica nanoparticles (MSNs), employ pH-degradable coatings for targeted cancer therapy, while lipid–polymer hybrid nanoparticles facilitate improved endosomal escape for efficient nucleic acid delivery [85]. It is mentioned earlier that Wei and co-workers developed polymeric micelles responsive to both pH and ultrasound using amphiphilic block copolymers with varying hydrophilic–hydrophobic ratios. They showed that block composition controls micelle formation, stability, drug encapsulation, and enables adjustable, stimulus-induced drug release for enhanced delivery efficiency [29].

In parallel, enteric-coated nanoparticles have enabled oral administration of therapeutics by protecting acid-sensitive drugs in the stomach and releasing them selectively in the small intestine. Despite these advances, challenges related to synthetic reproducibility, large-scale manufacturing, and variability of pH in vivo persist. To address these issues, real-time monitoring and imaging-guided delivery systems are being developed to refine dosing accuracy and therapeutic outcomes. Ongoing progress in polymer chemistry, hybrid nanostructure design, and responsive feedback mechanisms positions pH-responsive nanocarriers as powerful tools for precise, personalized, and safer drug delivery, with a central role anticipated in the future of next-generation nanomedicine.

6.3 Temperature Responsiveness

Temperature-responsive nanomaterials are reshaping smart drug delivery by enabling precise and localized therapeutic release in response to either intrinsic or externally applied temperature changes. These nanocarriers undergo reversible structural transformations, allowing targeted and minimally invasive treatments for conditions such as cancer, infections, and inflammatory diseases. By responding to physiological thermal cues, they improve therapeutic efficacy while limiting undesired side effects. Pathological tissues often display slightly elevated temperatures (~39–40 °C), which can be exploited for selective drug release. In addition, externally induced hyperthermia (40–43 °C) achieved through ultrasound, infrared irradiation, or magnetic field–based techniques offers further control over site-specific delivery [86]. Among the materials employed, thermo-responsive polymers such as PNIPAM, characterized by a lower critical solution temperature (LCST) of approximately 32 °C, are extensively utilized for constructing temperature-sensitive nanocarriers. PNIPAM-based micelles and hydrogels release drugs upon localized heating, showing promise in hyperthermia-assisted cancer therapy [87]. Advanced systems integrate dual triggers temperature and pH using polymers like PNIPAM and polyhistidine for enhanced precision. Magneto-thermal and photo-thermal materials (e.g., gold nanorods, graphene oxide) convert external energy into heat, enabling synergistic therapy [88]. Lipid-based temperature-sensitive liposomes (e.g., ThermoDox®) have entered clinical trials, demonstrating improved drug localization and reduced toxicity [89]. Challenges include in vivo thermal control, overheating risks, and scalable synthesis. Current efforts focus on refining thermal thresholds, enhancing biocompatibility, and integrating imaging-guided monitoring. Emerging biodegradable thermopolymers and hybrid platforms promise safer, scalable, and responsive nanomedicines for personalized therapy.

6.4 Additional Critical Parameters

Physiological conditions critically shape the fate, efficacy, and safety of nanomaterial-based drug delivery systems. Beyond classical design parameters (size, charge, shape), a series of biological barriers and dynamic in-body processes often patient- and disease-specific determine whether a nanocarrier reaches its target, releases cargo appropriately, and is cleared without harm. Understanding and engineering for these additional critical parameters is essential for robust clinical translation.

A dominant early event after systemic administration is formation of the protein corona, a dynamic layer of adsorbed serum proteins that rapidly redefines the nanoparticle's biological identity. Corona composition governs opsonization, cellular recognition routes, circulation half-life, and immunogenic signaling [90]. Contemporary research emphasizes controlling corona formation either by "stealth" coatings that resist nonspecific adsorption or by designing surfaces that recruit a beneficial, controlled corona to improve targeting and reduce clearance. Advanced proteomic characterization and corona-engineering are now routine steps in preclinical development.

Closely linked is the mononuclear phagocyte system (MPS)/reticuloendothelial system (RES) uptake and complement activation [91]. Surface features that promote immunoglobulin and complement binding accelerate hepatic and splenic sequestration, reducing therapeutic index and sometimes triggering adverse reactions. The clinical observation of accelerated blood clearance (ABC) after repeat dosing of PEGylated carriers has catalyzed the search for alternative stealth chemistries (polyzwitterions, polysarcosine) and reversible shielding approaches that minimize anti-polymer antibody responses [76]. Immunogenicity testing, complement activation assays, and dosing-interval design are now integral to formulation strategies.

Tumor and tissue microenvironment heterogeneity presents another major constraint. The enhanced permeability and retention (EPR) effect once considered a universal route to tumor accumulation is highly variable across tumor types, stages, and even intratumoral regions. Low perfusion, dense stroma, and high interstitial pressure can prevent nanoparticle penetration despite apparent accumulation at the tumor periphery. Modern approaches therefore combine EPR-optimizing tactics (vasculature-modulating agents, size-changeable carriers) with active targeting and externally guided accumulation (focused ultrasound, magnetic targeting) to increase delivery reliability.

Mucosal barriers and extracellular matrices mucus in the lung, gut, and reproductive tract, and dense extracellular matrices in solid tumors exclude or trap many nanoparticles. Surface chemistry that minimizes adhesive interactions (neutral, hydrophilic coatings) and particle sizing that favors rapid diffusion have proven effective; the field has produced mucus-penetrating particles and mucoadhesive designs tailored to either cross or persist within mucus depending on the therapeutic goal. High-throughput synthetic mucus models and ex vivo evaluation platforms are increasingly being used to rapidly screen and optimize formulations intended for mucosal drug delivery, significantly accelerating candidate selection [92]. Equally critical are clearance mechanisms and biodegradation behavior, which ultimately determine long-term safety. Processes such as renal filtration, biliary excretion, and enzymatic degradation are strongly influenced by nanoparticle size, morphology, and chemical composition, while non-degradable materials risk long-term accumulation in organs [93]. Consequently, the design of nanocarriers that break down into non-toxic, renally excretable fragments or that are selectively metabolized within target tissues is essential to minimize chronic toxicity. Integrating these material considerations with patient-specific factors, including age, hepatic and renal function, and concomitant medications, further enables personalized dosing strategies and improved safety monitoring.

Overall, effective nanomedicine development demands a systems-level perspective, encompassing protein corona dynamics, immune evasion, tissue-specific transport, barrier penetration, and predictable clearance profiles [94]. Recent advances ranging from corona engineering and non-PEG stealth polymers to synthetic mucus screening platforms and size-adaptive nanocarriers, are collectively improving the robustness of nanocarriers across diverse physiological conditions. The successful translation of these innovations will rely on their integration with rigorous *in vitro–in vivo* correlation studies and patient stratification approaches, ultimately transforming promising nanomedicine formulations into safe, reliable clinical therapies.

7 Limitations and Potential Outlook

7.1 *Synthesis Related Challenges*

The synthesis of nanomaterials for drug delivery stands at the intersection of chemistry, materials science, biology, and engineering. Over the last twenty years, substantial advances have been achieved in the development of nanoparticles, liposomes, dendrimers, polymeric carriers, and hybrid nanostructures designed for precise, controlled, and biocompatible drug delivery. Despite this progress, the translation of these laboratory-scale innovations into clinically effective and industrially scalable technologies remains challenging. These limitations stem from the intrinsic complexity of nanoscale synthesis, difficulties in ensuring reproducibility, biological variability, and stringent regulatory requirements [95]. This essay examines the principal scientific, synthetic, and practical barriers associated with nanomaterial synthesis for drug delivery and highlights recent technological advances aimed at addressing these challenges.

The fabrication of nanomaterials for efficient therapeutic delivery continues to be a complex and rapidly evolving field. Persistent issues such as particle size regulation, scalability, surface stability, reproducibility, and biocompatibility continue to challenge researchers in chemistry and materials science [96]. Nevertheless, ongoing developments in advanced synthetic methodologies, computational modeling, and sustainable green technologies offer strong prospects for overcoming these barriers [97]. Future nanomaterials are expected to arise from integrated, adaptive, and automated synthesis platforms capable of producing safe, reproducible, and multifunctional nanocarriers tailored to the needs of personalized medicine. With sustained innovation and interdisciplinary collaboration, nanotechnology is poised to profoundly transform drug delivery strategies in the decades ahead.

7.1.1 Control over Size, Morphology, and Uniformity

Achieving precise regulation of particle size, morphology, and polydispersity remains one of the most fundamental challenges in nanomaterial synthesis [98]. Nanoparticle dimensions and shape play a decisive role in determining circulation lifetime, biodistribution, cellular internalization, and biological clearance, and even minor deviations can lead to significant changes in therapeutic performance. Maintaining uniform size distributions becomes particularly difficult when processes are scaled from laboratory batches to industrial production. Conventional methods such as emulsion, precipitation, and sol–gel synthesis frequently generate heterogeneous nanoparticle populations due to inconsistencies in nucleation and growth dynamics [99]. To address these limitations, microfluidic-based synthesis has emerged as a promising alternative, enabling continuous nanoparticle generation under highly controlled and uniform conditions. Microfluidic reactors offer rapid and efficient mixing, accurate temperature regulation, and the potential for scalable production of nanoparticles with narrow size distributions [100]. Nevertheless, the widespread industrial adoption of this technology is currently constrained by high capital costs and the lack of standardized large-scale manufacturing protocols.

7.1.2 Surface Functionalization and Chemical Stability

Surface modification or functionalization plays a pivotal role in engineering nanomaterials for improved biocompatibility, stability, and targeted drug delivery. However, achieving precise and reproducible surface chemistry at the nanoscale remains a significant technical challenge. Functionalization strategies such as PEGylation, ligand conjugation, and antibody attachment often involve multi-step reactions, extensive purification, and careful control of surface density to prevent steric crowding or nanoparticle aggregation [101]. In addition, preserving the chemical stability of surface modifications under physiological conditions is particularly demanding. For example, thiol–gold bonds, commonly used in gold nanoparticle functionalization, can be displaced by endogenous thiol-containing biomolecules present in the bloodstream [102]. To overcome these limitations, more robust covalent attachment methods, including click chemistry, silanization, and photoinduced ligation, have been developed to provide stronger and more durable linkages. More recently, the advent of bioorthogonal chemistry, which enables highly selective reactions in biological environments without perturbing native biochemical processes, has transformed surface functionalization by offering clean, efficient, and highly specific conjugation pathways.

7.1.3 Reproducibility and Batch-to-Batch Consistency

Reproducibility continues to be a major challenge in the synthesis of nanomaterials. In contrast to small-molecule drugs with clearly defined chemical structures, nanomaterials are intricate, multicomponent systems that are highly sensitive to subtle variations in synthesis conditions, reagent quality, and surrounding environments. Even minor fluctuations in pH, temperature, stirring speed, or precursor concentration can result in pronounced changes in nanoparticle characteristics, which in turn influence drug loading capacity and release behavior [103]. To overcome these inconsistencies, researchers are turning to automation and artificial intelligence (AI)-assisted synthesis. AI–based algorithms are increasingly used to predict optimal synthesis conditions and dynamically regulate reaction parameters in real time, thereby minimizing human error and enhancing reproducibility [104]. In parallel, machine learning driven materials design is uncovering meaningful relationships between synthetic variables and nanoparticle performance, which is significantly accelerating the discovery, optimization, and development of advanced nanomaterials.

7.1.4 Scalability and Manufacturing Challenges

Transitioning nanomaterial synthesis from the laboratory scale (milligram quantities) to industrial production (kilogram levels) presents substantial technical and economic challenges. Methods that perform efficiently at small scale often lose stability, efficiency, or economic viability when applied to large-volume manufacturing. Key difficulties during scale-up include achieving uniform heat and mass transfer, consistent reactant mixing, and homogeneous particle characteristics [105]. To address these limitations, emerging approaches such as continuous-flow chemistry and supercritical fluid technologies are gaining traction. Continuous-flow reactors offer tightly controlled and reproducible reaction environments, minimize batch-to-batch variability, and enable real-time, in-line monitoring of nanoparticle formation [106]. In parallel, supercritical CO_2 based processes have been successfully employed to fabricate polymeric and lipid nanoparticles with well-defined sizes and high drug encapsulation efficiencies, while avoiding toxic organic solvents and adhering to green chemistry principles [107]. Nevertheless, challenges related to regulatory requirements and high manufacturing costs continue to hinder large-scale implementation and broader commercialization.

7.1.5 Biocompatibility and Toxicity Issues

Ensuring biocompatibility without compromising structural stability or functional efficacy remains a significant challenge in nanomaterial design. Numerous nanoparticles particularly those derived from metals or non-biodegradable polymers have the potential to trigger cytotoxic effects, oxidative stress, or unwanted immune reactions [108]. Moreover, even nanomaterials generally regarded as safe, such as gold or silica, may produce adverse biological responses depending on factors like particle size, surface charge, and surface chemistry. To mitigate these effects, researchers are developing biodegradable nanomaterials such as poly(lactic-co-glycolic acid) (PLGA), polycaprolactone (PCL), and lipid-based carriers that degrade into nontoxic byproducts [109]. Moreover, the emergence of biomimetic nanomaterials, such as cell membrane-coated nanoparticles, has provided a means to camouflage synthetic carriers from immune detection, improving circulation time and reducing toxicity [110]. However, ensuring reproducibility and purity of such biologically derived coatings remains challenging.

7.1.6 Integration of Multifunctionality

Modern nanocarriers are no longer designed for passive drug transport alone; they now integrate multiple functions such as targeting, imaging, sensing, and stimuli-responsive release. Synthesizing such multifunctional nanomaterials requires complex engineering combining polymers, metals, and biological molecules in a single construct without compromising stability or function. For example, constructing a theranostic nanoparticle that combines MRI contrast (using iron oxide), targeted drug delivery (via antibody conjugation), and controlled release (through pH-sensitive polymers) involves multistep synthesis and delicate surface balancing [111]. Maintaining uniformity and avoiding cross-reactions between different functional layers pose significant synthetic challenges. Innovations in modular self-assembly and layer-by-layer nanofabrication are making progress, allowing fine control over composition and architecture, though scalability remains limited [112].

7.1.7 Environmental and Regulatory Constraints

The synthesis of nanomaterials often involves hazardous reagents, organic solvents, or high-energy processes that raise environmental and safety concerns. Developing green and sustainable synthesis routes using water as a solvent, plant extracts as reducing agents, or low-temperature processes is gaining attention. Green nanochemistry not only improves safety but also aligns with regulatory expectations for sustainable manufacturing [113]. Regulatory frameworks for nanomedicines are still evolving, posing an additional challenge. The complex physicochemical characterization required by agencies such as the FDA and EMA covering size, charge, morphology, stability, and impurity profiles, which demands advanced analytical tools and standardized protocols [109]. Without harmonized global standards, translating laboratory-scale synthesis into approved clinical products remains slow and costly.

7.1.8 Interdisciplinary Integration and Predictive Modeling

Another critical challenge is the interdisciplinary gap between synthetic chemists, pharmacologists, and clinicians. Chemists may focus on structural optimization without fully understanding biological barriers, while biologists may not grasp the limitations of synthetic scalability. Bridging this divide requires collaborative frameworks integrating computational modeling, high-throughput screening, and *in vivo* testing [114]. Predictive computational models, powered by molecular dynamics simulations and AI, are now being used to anticipate nanoparticle–biomolecule interactions, stability in serum, and biodistribution patterns [115]. These in silico tools reduce the experimental burden and accelerate the design of nanomaterials tailored for specific biological environments.

7.2 Regulatory Aspects

The rapid expansion of nanotechnology in medicine has outpaced global regulatory frameworks, complicating the approval of nanomaterial-based drug delivery systems. The complex physicochemical properties, dynamic biological behavior, and multi-component nature of nanomedicines challenge conventional safety and efficacy evaluations. Regulatory bodies like the FDA, EMA, and ICH are developing specialized guidelines to address these complexities, emphasizing case-by-case assessments with detailed physicochemical profiling, stability data, surface functionalization, and IVIVC studies [116].

The FDA's 2022 Nanotechnology Guidance mandates comprehensive particle characterization, including aggregation and surface reactivity under physiological conditions [117]. Manufacturing reproducibility is another key focus, with GMP compliance and batch-to-batch consistency essential [81]. Recent innovations, including microfluidic and continuous-flow synthesis, have significantly enhanced process precision, reproducibility, and scalability in nanomaterial production. In parallel, Quality by Design (QbD) strategies are increasingly adopted to identify and control critical quality attributes at early stages of development, thereby improving overall product robustness [118].

Modern toxicological assessment of nanomaterials now extends beyond conventional endpoints to encompass immunogenicity, genotoxicity, biodistribution, and long-term tissue accumulation, with protein corona characterization gaining recognition as an important indicator of biocompatibility. Regulatory frameworks in the European Union, such as REACH and CLP, mandate the submission of nano-specific toxicological data to ensure safety compliance. Meanwhile, AI-assisted predictive toxicology and computational modeling are emerging as efficient and ethically sound alternatives to extensive animal testing. Efforts toward global regulatory harmonization, including initiatives by the ICH nanomedicine working group, seek to standardize terminology and evaluation criteria across regions [119]. Collectively, the integration of advanced manufacturing technologies, AI-driven quality assessment, and unified regulatory guidelines positions nanomedicine for safe, scalable, and compliant clinical translation.

8 Future Prospects

Recent advances in nanomaterial synthesis are reshaping drug delivery by overcoming long-standing challenges related to particle size control, reproducibility, scalability, and biocompatibility [120]. Emerging strategies such as AI-enabled automated synthesis, microfluidic continuous-flow fabrication, green chemistry approaches, and biomimetic surface coatings are improving both manufacturing efficiency and safety profiles. At the same time, the incorporation of omics-based tools, including proteomics and metabolomics, into nanotoxicology is providing deeper insight into cellular and molecular responses, thereby informing the rational design of safer nanomaterials. Looking forward, the evolution of nanomedicine is being driven by three intersecting trends: precision formulation, scalable and sustainable manufacturing, and data-driven material design. Selective organ targeting (SORT) lipid nanoparticles represent a prime example of precision engineering, enabling effective mRNA delivery to extrahepatic organs [121]. In parallel, next-generation stealth polymers, such as polysarcosine and polyzwitterions, are gaining attention for their low immunogenicity and extended circulation times. Microfluidic reactor technologies are revolutionizing nanoparticle production by offering precise control over nucleation and mixing, resulting in uniform particle populations with high drug encapsulation efficiency under GMP-compatible conditions [81]. Furthermore, artificial intelligence and machine learning are accelerating material discovery, optimizing synthesis conditions, and predicting biocompatibility, thereby expediting preclinical development. Adoption of sustainable manufacturing practices, including solvent-free processes, supercritical CO_2 techniques, and biodegradable carrier systems, is also reducing environmental burden and easing regulatory constraints. Future nanocarriers are expected to integrate organ-specific targeting, stimuli-responsive drug release, and theranostic functionality [122]. Achieving successful clinical translation of these innovations will require strong interdisciplinary collaboration and the establishment of harmonized regulatory standards to enable personalized and precision-based therapeutic solutions.

9 Conclusion

In summary, nanomaterial-based drug delivery systems have become a powerful platform for improving therapeutic efficacy, bioavailability, and targeted drug transport. Their successful translation, however, relies on the precise control of physicochemical parameters such as particle size, surface charge, morphology, and surface functionalization, all of which govern biodistribution, cellular internalization, and controlled drug release under physiological conditions. Advanced surface engineering strategies, including polymer coatings, ligand conjugation, and the incorporation of stimuli-responsive components, further enhance stability and site specificity while reducing toxicity and immune reactions. In addition, factors associated with the synthetic environment, such as solvent choice, pH, and processing conditions, are critical for preserving the functional integrity of nanostructures. Achieving an optimal balance between biocompatibility, degradation behavior, and drug loading efficiency is therefore essential for clinical translation. A comprehensive understanding of the interactions between nanomaterial properties and biological systems is key to designing safe and effective nano-carriers. Continued interdisciplinary efforts spanning materials science, pharmacology, and biomedical engineering will be crucial for advancing these platforms toward personalized and precision medicine, ultimately enabling nanomaterials to transition from experimental concepts to dependable therapeutic tools in modern healthcare.

References

1. Ezike, T.C., Okpala, U.S., Onoja, U.L., Nwike, C.P., Ezeako, E.C., Okpara, O.J., Okoroafor, C.C., Eze, S.C., Kalu, O.L., Odoh, E.C., Nwadike, U.G., Ogbodo, J.O., Umeh, B.U., Ossai, E.C., Nwanguma, B.C.: Advances in drug delivery systems, challenges and future directions. Heliyon. **9**(6), e17488 (2023). https://doi.org/10.1016/j.heliyon.2023.e17488
2. Bayda, S., Adeel, M., Tuccinardi, T., Cordani, M., Rizzolio, F.: The history of nanoscience and nanotechnology: from chemical–physical applications to nanomedicine. Molecules. **25**(1), 112 (2020). https://doi.org/10.3390/molecules25010112
3. Durgam, L.K., Oroszi, T.L.: Revolutionizing healthcare: the transformative potential of nanotechnology in medicine. Front. Drug Deliv. **5**, 1–15 (2025). https://doi.org/10.3389/fddev.2025.1556426
4. Adepu, S., Ramakrishna, S.: Controlled drug delivery systems: current status and future directions. Molecules. **26**(19), 5905 (2021). https://doi.org/10.3390/molecules26195905
5. El-Tanani, M., Satyam, S.M., Rabbani, S.A., El-Tanani, Y., Aljabali, A.A., Al Faouri, I., Rehman, A.: Revolutionizing drug delivery: the impact of advanced materials science and technology on precision medicine. Pharmaceutics. **17**, 375 (2025). https://doi.org/10.3390/pharmaceutics17030375
6. Wu, Y., Sun, C., Jin, K., Dong, C.: Theranostic role of advanced Nanotechnological tools in early brain metastases in lung cancer: an updated review. Int. J. Nanomedicine. **20**, 7215–7232 (2025). https://doi.org/10.2147/IJN.S517928
7. Díez-Pascual, A.M.: Surface engineering of nanomaterials with polymers, biomolecules, and small ligands for nanomedicine. Materials. **15**(9), 3251 (2022). https://doi.org/10.3390/ma15093251
8. He, C., Hu, Y., Yin, L., Tang, C., Yin, C.: Effects of particle size and surface charge on cellular uptake and biodistribution of polymeric nanoparticles. Biomaterials. **31**(13), 3657–3666 (2010). https://doi.org/10.1016/j.biomaterials.2010.01.065
9. Islam, S., Mir, M., Ahmed, S., Islam, M.A., Hossain, N., Chowdhury, M.A.: Advances in nanoparticles in targeted drug delivery–A review. Results Surf. Interf. **19**, 100529 (2025). https://doi.org/10.1016/j.rsurfi.2025.100529
10. Zhu, Y.-F., He, M.-Q., Lin, C.-S., Ma, W.-H., Ai, Y., Wang, J., Liang, Q.: Multifunctional Nanocarrier drug delivery systems: from diverse design to precise biomedical applications. Adv. Healthc. Mater., e02178 (2025). https://doi.org/10.1002/adhm.202502178
11. Chelliah, R., Rubab, M., Vijayalakshmi, S., Karuvelan, M., Barathikannan, K., Oh, D.-H.: Liposomes for drug delivery: classification, therapeutic applications, and limitations. Next Nanotechnol. **8**, 100209 (2025). https://doi.org/10.1016/j.nxnano.2025.100209
12. Suk, J.S., Xu, Q., Kim, N., Hanes, J., Ensign, L.M.: PEGylation as a strategy for improving nanoparticle-based drug and gene delivery. Adv. Drug Deliv. Rev. **99**(part A), 28–51 (2016). https://doi.org/10.1016/j.addr.2015.09.012
13. Haftcheshmeh, S.M., Jaafari, M.R., Mashreghi, M., Mehrabian, A., Alavizadeh, S.H., Zamani, P., Zarqi, J., Darvishi, M.H., Gheybi, F.: Liposomal doxorubicin targeting mitochondria: A novel formulation to enhance anti-tumor effects of Doxil® in vitro and in vivo. J. Drug Deliv. Sci. Technol. **62**, 102351 (2021). https://doi.org/10.1016/j.jddst.2021.102351
14. Kumar, D., Pal, S.: NMR relaxation and diffusion studies to probe the motional dynamics of risperidone within PLGA microsphere. Magn. Reson. Lett. **3**, 197–205 (2023). https://doi.org/10.1016/j.mrl.2023.03.005
15. Kumar, D., Pal, S.: Solution dynamics of 5-fluorouracil entrapped in poly lactic-co-glycolic acid (PLGA) microsphere-A study with 1D selective NMR methods. Magn. Reson. Chem. **57**, 118–128 (2019). https://doi.org/10.1002/mrc.4799
16. Singh, P., Gupta, U., Asthana, A., Jain, N.K.: Folate and folate−PEG−PAMAM dendrimers: synthesis, characterization, and targeted anticancer drug delivery potential in tumor bearing mice. Bioconjug. Chem. **19**(11), 2239–2252 (2008). https://doi.org/10.1021/bc800125u

17. Croitoru, G.-A., Pîrvulescu, D.-C., Niculescu, A.G., Grumezescu, A.M., Antohi, A.M., Nicolae, C.-L.: Metallic nanomaterials—targeted drug delivery approaches for improved bioavailability, reduced side toxicity, and enhanced patient outcomes. Romanian J. Morphol. Embryol. **65**(2), 145–158 (2024). https://doi.org/10.47162/RJME.65.2.01
18. Burlec, A.F., Corciova, A., Boev, M., Batir-Marin, D., Mircea, C., Cioanca, O., Danila, G., Danila, M., Bucur, A.F., Hancianu, M.: Current overview of metal nanoparticles' synthesis, characterization, and biomedical applications, with a focus on silver and gold nanoparticles. Pharmaceuticals. **16**(10), 1410 (2023). https://doi.org/10.3390/ph16101410
19. Chaudhary, J., Tailor, G., Kumar, D.: Synthesis and characterization of chromium nanoparticles by thermal decomposition method. Mater. Today Proc. **29**, 321–326 (2020). https://doi.org/10.1016/j.matpr.2020.07.283
20. Kumar, D., Priyadarshini, C.H., Sudha, V., Sherine, J., Harinipriya, S., Pal, S.: Investigation of adsorption behavior of anticancer drug on zinc oxide nanoparticles: A solid state NMR and cyclic voltammetry (CV) analysis. J. Pharm. Sci. **110**, 3726–3734 (2021). https://doi.org/10.1016/j.xphs.2021.08.003
21. Amaral, M.N., Kumar, P., Faísca, P., Ferreira, H.A., Coelho, J.M.P., Gaspar, M.M., Reis, C.P.: Gold nanoparticle-mediated photothermal therapy: expanding the frontiers of cancer treatment and theragnostics. Biomed. Pharmacother. **190**, 118399 (2025). https://doi.org/10.1016/j.biopha.2025.118399
22. Urnukhsaikhan, E., Bold, B.-E., Gunbileg, A., Sukhbaatar, N., Mishig-Ochir, T.: Antibacterial activity and characteristics of silver nanoparticles biosynthesized from Carduus crispus. Sci. Rep. **11**, 21047. https://doi.org/10.1038/s41598-021-00520-2
23. Israel, L.L., Galstyan, A., Holler, E., Ljubimova, J.Y.: Magnetic iron oxide nanoparticles for imaging, targeting and treatment of primary and metastatic tumors of the brain. J. Control. Release. **320**, 45–62 (2020). https://doi.org/10.1016/j.jconrel.2020.01.009
24. Gaur, M., Misra, C., Yadav, A.B., Swaroop, S., Maolmhuaidh, F.Ó., Bechelany, M., Barhoum, A.: Biomedical applications of carbon nanomaterials: fullerenes, quantum dots, nanotubes, nanofibers, and graphene. Materials. **14**(20), 5978 (2021). https://doi.org/10.3390/ma14205978
25. Jin, C., Zheng, H., Chen, J.: Advances in the application of graphene and its derivatives in drug delivery systems. Pharmaceuticals (Basel). **18**(9), 1245 (2025). https://doi.org/10.3390/ph18091245
26. Ghezzi, M., Pescina, S., Padula, C., Santi, P., Del Favero, E., Cantù, L., Nicoli, S.: Polymeric micelles in drug delivery: an insight of the techniques for their characterization and assessment in biorelevant conditions. J. Control. Release. **332**, 312–336 (2021). https://doi.org/10.1016/j.jconrel.2021.02.031
27. Hwang, D., Ramsey, J.D., Kabanov, A.V.: Polymeric micelles for the delivery of poorly soluble drugs: from nanoformulation to clinical approval. Adv. Drug Deliv. Rev. **156**, 80–118 (2020). https://doi.org/10.1016/j.addr.2020.09.009
28. Gao, S., Sui, Z., Jiang, Q., Jiang, Y.: Functional evaluation of Niosomes utilizing surfactants in nanomedicine applications. Int. J. Nanomedicine, 10283–10305 (2024). https://doi.org/10.2147/IJN.S480639
29. Wei, H.-X., Liu, M.-H., Wang, T.-Y., Shih, M.-H., Yu, J., Yeh, Y.-C.: Fabrication of pH- and ultrasound-responsive polymeric micelles: the effect of amphiphilic block copolymers with different hydrophilic/hydrophobic block ratios for self-assembly and controlled drug release. Biomacromolecules. **26**(4), 2116–2130. https://doi.org/10.1021/acs.biomac.4c01202
30. Manzano, M., Vallet-Regí, M.: Mesoporous silica nanoparticles in biomedicine: advances and prospects. Adv. Mater., e12433 (2025). https://doi.org/10.1002/adma.202512433
31. Kazi, R.N.A., Hasani, I.W., Khafaga, D.S.R., Kabba, S., Farhan, M., Aatif, M., Muteeb, G., Fahim, Y.A.: Nanomedicine: the effective role of nanomaterials in healthcare from diagnosis to therapy. Pharmaceutics. **17**(8), 987 (2025)
32. Ly, P.-D., Ly, K.-N., Phan, H.-L., Nguyen, H.H.T., Duong, V.-A., Nguyen, H.V.: Recent advances in surface decoration of nanoparticles in drug delivery. Front. Nanotechnol. **6**, 1–20 (2024). https://doi.org/10.3389/fnano.2024.1456939

33. Dolai, J., Mandal, K., Jana, N.R.: Nanoparticle size effects in biomedical applications. ACS Appl. Nano Mater. **4**(7), 6471–6496 (2021). https://doi.org/10.1021/acsanm.1c00987
34. Bhaskar, S., Tian, F., Stoeger, T., Kreyling, W., de la Fuente, J.M., Grazú, V., Borm, P., Estrada, G., Ntziachristos, V., Razansky, D.: Multifunctional Nanocarriers for diagnostics, drug delivery and targeted treatment across blood-brain barrier: perspectives on tracking and neuroimaging. Part. Fibre Toxicol. **7**, 3 (2010). https://doi.org/10.1186/1743-8977-7-3
35. Barbalinardo, M., Chiarini, F., Teti, G., Paganelli, F., Mercadelli, E., Bartoletti, A., Migliori, A., Piazzi, M., Bertacchini, J., Sena, P., Sanson, A., Falconi, M., Palumbo, C., Cavallini, M., Gentili, D.: Surface charge overrides protein Corona formation in determining the cytotoxicity, cellular uptake, and biodistribution of silver nanoparticles. ACS Appl. Bio Mater. **8**(6), 5032–5043 (2025). https://doi.org/10.1021/acsabm.5c00392
36. Tang, Y., Wang, X., Li, J., Nie, Y., Liao, G., Yu, Y., Li, C.: Overcoming the reticuloendothelial system barrier to drug delivery with a "Don't-eat-us" strategy. ACS Nano. **13**(11), 13015–13026 (2019). https://doi.org/10.1021/acsnano.9b05679
37. Wieszczycka, K., Staszak, K., Woźniak-Budych, M.J., Litowczenko, J., Maciejewska, B.M., Jurga, S.: Surface functionalization—the way for advanced applications of smart materials. Coord. Chem. Rev. **436**, 213846 (2021). https://doi.org/10.1016/j.ccr.2021.213846
38. Singh, R., Srinivas, S.P., Kumawat, M., Daima, H.K.: Ligand-based surface engineering of nanomaterials: trends, challenges, and biomedical perspectives. Open Nano. **15**, 100194 (2024). https://doi.org/10.1016/j.onano.2023.100194
39. Fam, S.Y., Chee, C.F., Yong, C.Y., Ho, K.L., Mariatulqabtiah, A.R., Tan, W.S.: Stealth coating of nanoparticles in drug-delivery systems. Nanomaterials. **10**(4), 787 (2020). https://doi.org/10.3390/nano10040787
40. Kyriakides, T.R., Raj, A., Tseng, T.H., Xiao, H., Nguyen, R., Mohammed, F.S., Halder, S., Xu, M., Wu, M.J., Bao, S.: Biocompatibility of nanomaterials and their immunological properties. Biomed. Mater. **16**, 042005 (2021). https://doi.org/10.1088/1748-605X/abe5fa
41. Fatima, M., Almalki, W.H., Khan, T., Sahebkar, A., Kesharwani, P.: Harnessing the power of stimuli-responsive nanoparticles as an effective therapeutic drug delivery system. Adv. Mater. **36**(24), 2312939 (2024). https://doi.org/10.1002/adma.202312939
42. Albuquerque, L.J.C., Carvalho, J.T.T., Fidelis, C.L.B., Stefanello, T.F., Ceolín, M., Cardoso, M.B., Picco, A.S.: Role of surface coatings in preventing nanoparticle aggregation induced by freeze-drying. ACS Appl. Nano Mater. **8**(33), 16388–16397 (2025). https://doi.org/10.1021/acsanm.5c02655
43. Vagena, I.-A., Malapani, C., Gatou, M.-A., Lagopati, N., Pavlatou, E.A.: Enhancement of EPR effect for passive tumor targeting: current status and future perspectives. Appl. Sci. **15**(6), 3189 (2025). https://doi.org/10.3390/app15063189
44. Sun, C., Lee, J.S.H., Zhang, M.: Magnetic nanoparticles in MR imaging and drug delivery. Adv. Drug Deliv. Rev. **60**(11), 1252–1265 (2008). https://doi.org/10.1016/j.addr.2008.03.018
45. Heuer-Jungemann, A., Feliu, N., Bakaimi, I., Hamaly, M., Alkilany, A., Chakraborty, I., Masood, A., Casula, M.F., Kostopoulou, A., Oh, E., Susumu, K., Stewart, M.H., Medintz, I.L., Stratakis, E., Parak, W.J., Kanaras, A.G.: The role of ligands in the chemical synthesis and applications of inorganic nanoparticles. Chem. Rev. **119**(8), 4819–4880 (2019). https://doi.org/10.1021/acs.chemrev.8b00733
46. Georgakilas, V., Tiwari, J.N., Kemp, K.C., Perman, J.A., Bourlinos, A.B., Kim, K.S., Zboril, R.: Noncovalent functionalization of graphene and graphene oxide for energy materials, biosensing, catalytic, and biomedical applications. Chem. Rev. **116**(9), 5464–5519 (2016). https://doi.org/10.1021/acs.chemrev.5b00620.0
47. Di Crescenzo, A., Ettorre, V., Fontana, A.: Non-covalent and reversible functionalization of carbon nanotubes. Beilstein J. Nanotechnol. **5**, 1675–1690 (2014). https://doi.org/10.3762/bjnano.5.178
48. Shi, D., Narayanan, S., Woeppe, K., Cui, X.T.: Improving the biocompatibility and functionality of neural Interface devices with silica nanoparticles. Acc. Chem. Res. **57**(12), 1684–1695 (2024). https://doi.org/10.1021/acs.accounts.4c00160

49. Shi, L., Zhang, J., Zhao, M., Tang, S., Cheng, X., Zhang, W., Li, W., Liu, X., Peng, H., Wang, Q.: Effects of polyethylene glycol on the surface of nanoparticles for targeted drug delivery. Nanoscale. **13**, 10748–10764 (2021). https://doi.org/10.1039/D1NR02065J
50. Makharadze, D., del Valle, L.J., Katsarava, R., Puiggalí, J.: The art of PEGylation: from simple polymer to sophisticated drug delivery system. Int. J. Mol. Sci. **26**(7), 3102 (2025). https://doi.org/10.3390/ijms26073102
51. Becicka, W.M., Bielecki, P.A., Lorkowski, M.E., Moon, T.J., Zhang, Y., Atukorale, P.U., Covarrubias, G., Karathanasis, E.: The effect of PEGylation on the efficacy uptake of an immunostimulatory nanoparticle in the tumor immune microenvironment. Nanoscale Adv. **3**(17), 4961–4972 (2021). https://doi.org/10.1039/d1na00308a
52. Wolińska, A., Drozd, M., Buchalska, B., Dybko, A., Grabowska-Jadach, I.: Au@Fe_3O_4@PEG nanocubes as photoactive agents in photothermal therapy: an in vitro study. Biomed. Pharmacother. **187**, 118051 (2025). https://doi.org/10.1016/j.biopha.2025.118051
53. Omata, D., Kawahara, E., Munakata, L., Tanaka, H., Akita, H., Yoshioka, Y., Suzuki, R.: Effect of anti-PEG antibody on immune response of mRNA-loaded lipid nanoparticles. Mol. Pharm. **21**(11), 5672–5680 (2024). https://doi.org/10.1021/acs.molpharmaceut.4c00628
54. Miao, G., He, Y., Lai, K., Zhao, Y., He, P., Tan, G., Wang, X.: Accelerated blood clearance of PEGylated nanoparticles induced by PEG-based pharmaceutical excipients. J. Control. Release. **363**, 12–26 (2023). https://doi.org/10.1016/j.jconrel.2023.09.003
55. Maiti, D., Yokoyama, M., Shiraishi, K.: Impact of the hydrophilicity of poly(sarcosine) on poly(ethylene glycol) (PEG) for the suppression of anti-PEG antibody binding. ACS Omega. **9**(32), 34577–34588 (2024). https://doi.org/10.1021/acsomega.4c02655
56. Rezaei, B., Harun, A., Wu, X., Iyer, P.R., Mostufa, S., Ciannella, S., Karampelas, I.H., Chalmers, J., Srivastava, I., Gómez-Pastora, J., Wu, K.: Effect of polymer and cell membrane coatings on Theranostic applications of nanoparticles: A review. Adv. Healthc. Mater. **13**(26), 2401213 (2024). https://doi.org/10.1002/adhm.202401213
57. Kesharwani, P., Kumar, V., Goh, K.W., Gupta, G., Alsayari, A., Wahab, S., Sahebkarm, A.: PEGylated PLGA nanoparticles: unlocking advanced strategies for cancer therapy. Mol. Cancer. **24**, 205 (2025). https://doi.org/10.1186/s12943-025-02410-x
58. Payomhom, P., Panyain, N., Sakonsinsiri, C., Wongtrakoongate, P., Lertsuwan, K., Pissuwan, D., Katewongsa, K.P.: Chitosan-coated poly(lactic-co-glycolic acid) nanoparticles loaded with Ursolic acid for breast cancer therapy. ACS Appl. Nano Mater. **7**(5), 5383–5395 (2024). https://doi.org/10.1021/acsanm.3c06161
59. Rivera-Hernández, G., Antunes-Ricardo, M., Martínez-Morales, P., Sánchez, M.L.: Polyvinyl alcohol based-drug delivery systems for cancer treatment. Int. J. Pharm. **600**, 120478 (2021). https://doi.org/10.1016/j.ijpharm.2021.120478
60. Mofarrah, M., Jafari-Gharabaghlou, D., Farhoudi-Sefidan-Jadid, M., Zarghami, N.: Potential application of inorganic nano-materials in modulation of macrophage function: possible application in bone tissue engineering. Heliyon. **9**(6), e16309 (2023). https://doi.org/10.1016/j.heliyon.2023.e16309
61. Mugundhan, S.L., Mohan, M.: Hyaluronic acid-coated capecitabine nanostructures for CD44 receptor-mediated targeting in breast cancer therapy. RSC Adv. **15**, 12653–12670 (2025). https://doi.org/10.1039/D5RA01275A
62. Srinivasan, M., Rajabi, M., Mousa, S.A.: Multifunctional nanomaterials and their applications in drug delivery and cancer therapy. Nano. **5**(4), 1690–1703 (2015). https://doi.org/10.3390/nano5041690
63. Abdulsalam, L., Abubakar, S., Permatasari, I., Lawal, A.A., Uddin, S., Ullah, S., Ahmad, I.: Advanced biocompatible and biodegradable polymers: A review of functionalization, smart systems, and sustainable applications. Polymers. **17**(21), 2901 (2025). https://doi.org/10.3390/polym17212901
64. Kesharwani, P., Puri, V., Alqahtani, T., Al Shmrany, H., Gupta, G., Goh, K.W., Sahebkar, A.: PEGylated dendrimers for precision cancer therapy: advances in tumor targeting, drug delivery, and clinical translation. Biomater. Adv. **179**, 214493 (2026). https://doi.org/10.1016/j.bioadv.2025.21449

65. Ul-Islam, M., Alabbosh, K.F., Manan, S., Khan, S., Ahmad, F., Ullah, M.W.: Chitosan-based nanostructured biomaterials: synthesis, properties, and biomedical applications. Adv. Ind. Eng. Polym. Res. **7**(1), 79–99 (2024). https://doi.org/10.1016/j.aiepr.2023.07.002
66. Laurent, E., Maric, M.: Organic–inorganic hybrid materials from vegetable oils. Macro Mol. Rapid Commun. **45**(23), 2400408 (2024). https://doi.org/10.1002/marc.202400408
67. Gajbhiye, K.R., Salve, R., Narwade, M., Sheikh, A., Kesharwani, P., Gajbhiye, V.: Lipid polymer hybrid nanoparticles: a custom-tailored next-generation approach for cancer therapeutics. Mol. Cancer. **22**(1), 160 (2023). https://doi.org/10.1186/s12943-023-01849-0
68. Yu, W., Liu, R., Zhou, Y., Gao, H.: Size-tunable strategies for a tumor targeted drug delivery system. ACS Cent. Sci. **6**(2), 100–116 (2020). https://doi.org/10.1021/acscentsci.9b01139
69. Wilson, B., Geetha, K.M.: Lipid nanoparticles in the development of mRNA vaccines for COVID-19. J. Drug Deliv. Sci. Technol. **74**, 103553 (2022). https://doi.org/10.1016/j.jddst.2022.103553
70. Sarmah, S., Baidya, S., De, M.: Recent advances in lipid nanoparticles: nucleic acid therapeutics and targeting strategies. Small. **21**(43), e06812 (2025). https://doi.org/10.1002/smll.202506812
71. Su, K., Shi, L., Sheng, T., Yan, X., Lin, L., Meng, C., Wu, S., Chen, Y., Zhang, Y., Wang, C., Wang, Z., Qiu, J., Zhao, J., Xu, T., Ping, Y., Gu, Z., Liu, S.: Reformulating lipid nanoparticles for organ-targeted mRNA accumulation and translation. Nat. Commun. **15**, 5659 (2024). https://doi.org/10.1038/s41467-024-50093-7
72. Yang, J., Rieux, A.D., Malfanti, A.: Stimuli-responsive nanomedicines for the treatment of non-cancer related inflammatory diseases. ACS Nano. **19**(16), 15189–15219 (2025). https://doi.org/10.1021/acsnano.5c00700
73. Chu, S., Shi, X., Tian, Y., Gao, F.: pH-responsive polymer nanomaterials for tumor therapy. Front. Oncol. **12**, 855019 (2022). https://doi.org/10.3389/fonc.2022.855019
74. Sun, Y., Davis, E.: Nanoplatforms for targeted stimuli-responsive drug delivery: A review of platform materials and stimuli-responsive release and targeting mechanisms. Nano. **11**(3), 746 (2021). https://doi.org/10.3390/nano11030746
75. Singh, P., Pandit, S., Balusamy, S.R., Madhusudanan, M., Singh, H., Haseef, H.M.A., Mijakovic, I.: Advanced nanomaterials for cancer therapy: gold, silver, and iron oxide nanoparticles in oncological applications. Adv. Healthc. Mater. **14**(4), 2403059 (2024). https://doi.org/10.1002/adhm.202403059
76. Lee, N.-H., You, S., Taghizadeh, A., Taghizadeh, M., Kim, H.S.: Cell membrane-cloaked Nanotherapeutics for targeted drug delivery. Int. J. Mol. Sci. **23**(4), 2223 (2022). https://doi.org/10.3390/ijms23042223
77. Yan, H., Shao, D., Lao, Y.-H., Li, M., Hu, H., Leong, K.W.: Engineering cell membrane-based Nanotherapeutics to target inflammation. Adv. Sci. **6**, 1900605 (2019). https://doi.org/10.1002/advs.201900605
78. Amirinezhadfard, E., Tabar, A.N., Xia, L.-J., Yang, W.-C.: Plant-derived nanomaterials in cancer therapy: advances from green synthesis to application of PANoptosis-mediated tumor suppression. Ind. Crop. Prod. **236**, 121850 (2025). https://doi.org/10.1016/j.indcrop.2025.121850
79. Li, X., Li, J., Wei, J., Du, W., Su, C., Shen, X., Zhao, A., Xu, M.: Design strategies for novel lipid nanoparticle for mRNA vaccine and therapeutics: current understandings and future perspectives. MedComm. **6**(10), e70414 (2025). https://doi.org/10.1002/mco2.70414
80. Li, J., Zhang, Y., Yang, Y.-G., Sun, T.: Advancing mRNA therapeutics: the role and future of nanoparticle delivery systems. Mol. Pharm. **21**(8), 3743–3763 (2024). https://doi.org/10.1021/acs.molpharmaceut.4c00276
81. Luostarinen, A., Vuorela, A., Kerkelä, E., Patrikoski, M., Kotovuori, A., Koski, J., Ahoniemi, J., Lähteenmäki, K., Lehtisalo, J., Oja, T., Paavilainen, H., Autio, A., Nyman, M., Nikoskelainen, V., Kergourlay, V., Elbasani, E., van Veen, B., Thotakura, A., Monzo, H., Ojala, P.M., Korhonen, M., Hongisto, H., Laitinen, A.: Establishing a GMP-compliant manufacturing process and phase-appropriate analytics for early development of a FiCAR T-cell product with a novel CAR spacer. Sci Rep. **15**, 8093 (2025). https://doi.org/10.1038/s41598-025-92736-9

82. Nunziata, G., Pollonio, D., Lacroce, E., Rossi, F.: Smart pH-responsive polymers in biomedical applications: nanoparticles, hydrogels, and emerging hybrid platforms. Mater. Today Chem. **49**, 103063 (2025). https://doi.org/10.1016/j.mtchem.2025.103063
83. Amici, A., Pozzi, D., Marchini, C., Caracciolo, G.: The transformative potential of lipid nanoparticle–protein corona for next-generation vaccines and therapeutics. Mol. Pharm. **20**(11), 5247–5253 (2023). https://doi.org/10.1021/acs.molpharmaceut.3c00479
84. Wang, H., Gao, L., Fan, T., Zhang, C., Zhang, B., Al-Hartomy, O.A., Al-Ghamdi, A., Wageh, S., Qiu, M., Zhang, H.: Strategic Design of Intelligent-Responsive Nanogel Carriers for cancer therapy. ACS Appl. Mater. Interfaces. **13**(46), 54621–54647 (2021). https://doi.org/10.1021/acsami.1c13634
85. Wu, W., Luo, L., Wang, Y., Wu, Q., Dai, H.-B., Li, J.-S., Durkan, C., Wang, N., Wang, G.-X.: Endogenous pH-responsive nanoparticles with programmable size changes for targeted tumor therapy and imaging applications. Theranostics. **8**(11), 3038–3058 (2018). https://doi.org/10.7150/thno.23459
86. Yaramiri, A., Asalh, R.A., Asalh, M.A., Al Sawaftah, N., Abuwatfa, W.H., Husseini, G.A.: A comprehensive review of smart thermosensitive Nanocarriers for precision cancer therapy. Int. J. Mol. Sci. **26**(15), 7322 (2025). https://doi.org/10.3390/ijms26157322
87. Throat, S., Bhattacharya, S.: Macromolecular Poly(N-Isopropylacrylamide) (PNIPAM) in cancer treatment and beyond (2024). https://doi.org/10.1155/2024/1444990
88. Cui, X., Ruan, Q., Zhuo, X., Xia, X., Hu, J., Fu, R., Li, Y., Wang, J., Xu, H.: Photothermal nanomaterials: A powerful light-to-heat converter. Chem. Rev. **123**(11), 6891–6952 (2023). https://doi.org/10.1021/acs.chemrev.3c00159
89. Alhussan, A., Ho, L., Zhang, Y., Fan, H., Momeni, A., Brimacombe, C., Cullis, P.R.: Lipid-based nanoparticles external triggered release strategies in cancer nanomedicine. J. Nanobiotechnol. **23**(1), 641 (2025). https://doi.org/10.1186/s12951-025-03704-4
90. Owens III, D.E., Peppas, N.A.: Opsonization, biodistribution, and pharmacokinetics of polymeric nanoparticles. Int. J. Pharm. **307**(1), 93–102 (2006). https://doi.org/10.1016/j.ijpharm.2005.10.010
91. Yona, S., Gordon, S.: From the reticuloendothelial to mononuclear phagocyte system—the unaccounted years. Front. Immunol. **6**, 328 (2015). https://doi.org/10.3389/fimmu.2015.00328
92. Wright, L., Barnes, T.J., Joyce, P., Prestidge, C.A.: Optimisation of a high-throughput model for mucus permeation and nanoparticle discrimination using biosimilar mucus. Pharmaceutics. **14**(12), 2659 (2022). https://doi.org/10.3390/pharmaceutics14122659
93. Choi, H.S., Liu, W., Misra, P., Tanaka, E., Zimmer, J.P., Ipe, B.I., Bawendi, M.G., Frangioni, J.V.: Renal clearance of nanoparticles. Nat. Biotechnol. **25**(10), 1165–1170 (2007). https://doi.org/10.1038/nbt1340
94. Sun, Y., Zhou, Y., Wang, Y.F., Guo, S.: Protein Corona of nanoparticles: isolation and analysis. Chem. Bio. Eng. **1**(9), 757–772 (2023). https://doi.org/10.1021/cbe.4c00105
95. Đorđević, S., Gonzalez, M.M., Conejos-Sánchez, I., Carreira, B., Pozzi, S., Acúrcio, R.C., Satchi-Fainaro, R., Florindo, H.F., Vicent, M.J.: Current hurdles to the translation of nanomedicines from bench to the clinic. Drug Deliv. Transl. Res. **12**(3), 500–525 (2021). https://doi.org/10.1007/s13346-021-01024-2
96. Dalbanjan, N.P., Korgaonkar, K., Eelager, M.P., Gonal, B.N., Kadapure, A.J., Arakera, S.B., Kumar, S.K., P.: In-silico strategies in nano-drug design: bridging nanomaterials and pharmacological applications. Nano TransMed. **4**, 100091 (2025). https://doi.org/10.1016/j.ntm.2025.100091
97. Tailor, G., Mukhtar, M.T., Labaran, A.N., et al.: Green synthesis of iron nanoparticles using Mangifera indica leaves extract: phytochemical analysis and characterizations, for environmental remediation of pollution. Nanotechnol. Environ. Eng. **9**, 77–84 (2024). https://doi.org/10.1007/s41204-023-00353-6
98. Kuddushi, M., Kanike, C., Xu, B.B., Zhang, X.: Recent advances in nanoprecipitation: from mechanistic insights to applications in nanomaterial synthesis. Soft Matter. **21**, 2759–2781 (2025). https://doi.org/10.1039/D5SM00006H

99. Edo, G.I., Mafe, A.N., Ali, A.B.M., Akpoghelie, P.O., Yousif, E., Isoje, E.F., Igbuku, U.A., Zainulabdeen, K., Owheruo, J.O., Essaghah, A.E.A., Umar, H., Ahmed, D.S., Alamiery, A.A.: Eco-friendly nanoparticle phytosynthesis via plant extracts: mechanistic insights, recent advances, and multifaceted uses. Nano TransMed. **4**, 100080
100. Gimondi, S., Ferreira, H., Reis, R.L., Neves, N.M.: Microfluidic devices: A tool for nanoparticle synthesis and performance evaluation. ACS Nano. **17**(15), 14205–14228 (2023). https://doi.org/10.1021/acsnano.3c01117
101. Mao, L., Russell, A.J., Carmali, S.: Moving protein PEGylation from an art to a data science. Bioconjug. Chem. **33**(9), 1643–1653 (2022). https://doi.org/10.1021/acs.bioconjchem.2c00262
102. Jacobs, M., Selvam, A.P., Craven, J.E., Prasad, S.: Antibody-conjugated gold nanoparticle-based Immunosensor for ultra-sensitive detection of troponin-T. SLAS Technol. **19**(6), 546–554 (2014). https://doi.org/10.1177/2211068214538971
103. Patra, J.K., Das, G., Fraceto, L.F., Campos, E.V.R., Rodriguez-Torres, M.D.P., Acosta-Torres, L.S., Diaz-Torres, L.A., Grillo, R., Swamy, M.K., Sharma, S., Habtemariam, S., Shin, H.-S.: Nano based drug delivery systems: recent developments and future prospects. J. Nanobiotechnol. **16**, 71 (2018). https://doi.org/10.1186/s12951-018-0392-8
104. Wang, W., Liu, Y., Wang, Z., Hao, G., Song, B.: The way to AI-controlled synthesis: how far do we need to go? Chem. Sci. **13**, 12604–12615 (2022). https://doi.org/10.1039/D2SC04419F
105. Taylor, C.J., Pomberger, A., Felton, K.C., Grainger, R., Barecka, M., Chamberlain, T.W., Bourne, R.A., Johnson, C.N., Lapkin, A.A.: A brief introduction to chemical reaction optimization. Chem. Rev. **123**(6), 3089–3126 (2023). https://doi.org/10.1021/acs.chemrev.2c00798
106. Santis, P.D., Meyer, L.-E., Kara, S.: The rise of continuous flow biocatalysis—fundamentals, very recent developments and future perspectives. React. Chem. Eng. **5**, 2155–2184 (2020). https://doi.org/10.1039/D0RE00335B
107. Murakami, Y., Inoue, K., Akiyama, R., Orita, Y., Shimoyama, Y.: LipTube: liposome formation in the tube process using supercritical CO_2. Ind. Eng. Chem. Res. **61**(39), 14598–14608 (2022). https://doi.org/10.1021/acs.iecr.2c02095
108. Vignesh, A., Menaka, C., Amal, T.C., Selvakumar, S., Vasanth, K.: Exploring the dual impact of nanoparticles on human Well-being: A comprehensive review of risks and benefits. Next Nanotechnol. **8**, 100223 (2025). https://doi.org/10.1016/j.nxnano.2025.100223
109. Kim, Y., Kwak, J., Lim, M., Lim, S.Y., Chae, S., Ha, S.-J., Won, Y.-W., Kim, H.-O., Lim, K.S.: Advances in PCL, PLA, and PLGA-based Technologies for Anticancer Drug Delivery. Pharmaceutics. **17**(10), 1354 (2025). https://doi.org/10.3390/pharmaceutics17101354
110. Gan, J., Huang, D., Che, J., Zhao, Y., Sun, L.: Biomimetic nanoparticles with cell-membrane camouflage for rheumatoid arthritis. Matter. **7**(3), 794–825 (2024). https://doi.org/10.1016/j.matt.2023.12.022
111. Comanescu, C.: Recent advances in surface functionalization of magnetic nanoparticles. Coatings. **13**(10), 1772 (2023). https://doi.org/10.3390/coatings13101772
112. Rowland, S., Aghakhani, A., Whalley, R.D., Ferreira, A.M., Kotov, N., Gentile, P.: Layer-by-layer nanoparticle assembly for biomedicine: mechanisms, technologies, and advancement via Acoustofluidics. ACS Appl. Nano Mater. **7**(14), 15874–15902 (2024). https://doi.org/10.1021/acsanm.4c02463
113. Ayub, A., Wani, A.K., Malik, S.M., Ayub, M., Singh, R., Chopra, C., Malik, T.: Green nanoscience for healthcare: advancing biomedical innovation through eco-synthesized nanoparticle. Biotechnol. Rep. **47**, e00913 (2025). https://doi.org/10.1016/j.btre.2025.e00913
114. Yang, S., Hu, X., Zhu, J., Zheng, B., Bi, W., Wang, X., Wu, J., Mi, Z., Wu, Y.: Aspects and implementation of pharmaceutical quality by design from conceptual frameworks to industrial applications. Pharmaceutics. **17**(5), 623 (2025). https://doi.org/10.3390/pharmaceutics17050623
115. Yasgar, A., Jain, S., Davies, M., Danchik, C., Niehoff, T., Ran, J., Rai, G., Yang, S.-M., Simeonov, A., Zakharov, A.V., Martinez, N.J.: Integrated approach of machine learning and

high-throughput screening to identify chemical probe candidates targeting aldehyde dehydrogenases. ACS Pharmacol. Transl. Sci. **8**(10), 3568–3584 (2025). https://doi.org/10.1021/acsptsci.5c00395
116. Song, X., Liu, P., Zhang, X., Zhou, Y., Qian, H., Yang, K., Zhang, X., Gao, X., Zheng, J.-J., Shi, X., Zhao, Y.: In silico methods, simulations, and Design of Drug-Delivery Nanocarriers. Adv. Funct. Mater. **e23068** (2025). https://doi.org/10.1002/adfm.202523068
117. DuPont, M.E.: Stability and bioequivalence challenges in generic drug formulation: A regulatory perspective. GSC Biol. Pharm. Sci. **31**(02), 166–178 (2025). https://doi.org/10.30574/gscbps.2025.31.2.0189
118. Rodríguez-Gómez, F.D., Monferrer, D., Penon, O., Rivera-Gil, P.: Regulatory pathways and guidelines for nanotechnology-enabled health products: a comparative review of EU and US frameworks. Front Med (Lausanne). **12**, 1544393. https://doi.org/10.3389/fmed.2025.1544393
119. Yu, L.X., Amidon, G., Khan, M.A., Hoag, S.W., Polli, J., Raju, G.K., Woodcock, J.: Understanding pharmaceutical quality by design. AAPS J. **16**(4), 771–783 (2014). https://doi.org/10.1208/s12248-014-9598-3
120. Caputo, F., Favre, G., Borchard, G., Calzolai, L., Fisicaro, P., Frejafon, E., Günday-Türeli, N., Koltsov, D., Minelli, C., Nelson, B.C., Parot, J., Prina-Mello, A., Zou, S., Ouf, F.-X.: Toward an international standardisation roadmap for nanomedicine. Drug Deliv. Transl. Res. **14**(9), 2578–2588 (2024). https://doi.org/10.1007/s13346-024-01646-2
121. Kurul, F., Turkmen, H., Cetin, A.E., Topkaya, S.N.: Nanomedicine: how nanomaterials are transforming drug delivery, bio-imaging, and diagnosis. Next Nanotechnol. **7**, 100129 (2025). https://doi.org/10.1016/j.nxnano.2024.100129
122. Wang, X., Liu, S., Sun, Y., Yu, X., Lee, S.M., Cheng, Q., Wei, T., Gong, J., Robinson, J., Zhang, D., Lian, X., Basak, P., Siegwart, D.J.: Preparation of selective organ-targeting (SORT) lipid nanoparticles (LNPs) using multiple technical methods for tissue-specific mRNA delivery. Nat. Protoc. **18**(1), 265–291 (2023). https://doi.org/10.1038/s41596-022-00755-x

Biological Significance of Metallic Nanoparticles Synthesized via Plant Extracts

Prashant Kumar, Sunil Luhar, and Jiten Yadav

1 Introduction

Nanotechnology has revolutionized modern science by enabling precise manipulation of materials at the atomic and molecular scale (1–100 nm), unlocking unique physicochemical properties not found in bulk materials [1, 2]. This multidisciplinary field has transformative applications across medicine, energy, environmental remediation, and biotechnology. At the core of this revolution are metal-based nanoparticles (MNPs), which exhibit exceptional optical, catalytic, and biological properties due to their high surface area and quantum effects [3]. Traditional methods of nanoparticle synthesis often rely on the use of toxic agents, energy-consuming operations, and the formation of harmful by-products, which is why the use of eco-friendly methods is highly required. In this regard, the concept of plant-mediated nanoparticle synthesis has been viewed as a scalable and eco-friendly approach, which takes advantage of reducing and stabilizing potentials inherent in phytochemicals present in plants. Plants also provide excellent bioactive components such as polyphenols, flavonoids, terpenoids, and alkaloids which play a significant role in reducing metal ion to stable nanoscale forms. Plant-based methods are more reproducible, have faster reaction rates, and are more feasible at large scale than microbial synthesis methods based on bacteria, fungi, or algae [4]. The synthesis process usually includes the isolation of the

P. Kumar (✉)
Next-Generation Magnet Development Collaboration Unit, RIKEN- Center for Integrative Medical Sciences, Yokohama 230-0045, Japan
e-mail: prashantkhichi92@gmail.com

S. Luhar
Nalecz Institute of Biocybernetics and Biomedical Engineering, Polish Academy of Sciences, Warsaw, Poland

J. Yadav
Department of Chemistry, University Centre for Research and Development, Chandigarh University, Mohali, Punjab, India

G. Tailor et al. (eds.), *Nanomaterials*, Engineering Materials,
https://doi.org/10.1007/978-981-92-0678-0_8

phytochemicals using different plant tissues (e.g., leaves, roots, seeds, or fruits) and the subsequent reaction with the precursors of metal salts to create the nanoparticles through the controlled redox reaction. This is because the biosynthesized nanoparticles are better biocompatible and colloidal due to the capping agents that are naturally occurring and prevent the agglomeration of the particles. In addition to reducing environmental impact, this green strategy fits the description of the circular economy since it can harness renewable and sustainable biomass sources [5, 6]. Green-synthesized nanoparticles have a large biomedical potential in particular. Noble metal nanoparticles including silver and gold exhibit strong antimicrobial properties, which make them very appropriate to be used as wound dressings, antibacterial surface coatings, and infection control measures [7]. Moreover, metal oxide nanoparticles, such as iron oxide and cerium oxide, have been found to have significant potential in targeted drug delivery, cancer therapeutics, as well as in the treatment of neurodegenerative diseases, which can be partly explained by their antioxidant activity and enzyme-like properties [8, 9]. Selenium nanoparticles, known for their low toxicity and high bioavailability, are being explored in anticancer and immunomodulatory therapies [10]. Beyond medicine, these nanoparticles play a crucial role in environmental sustainability, degrading organic pollutants, detecting heavy metals, and enhancing solar cell efficiency. For instance, copper oxide nanoparticles effectively break down textile dyes in wastewater, while zinc oxide nanoparticles are used in UV-blocking coatings and photocatalytic water purification [11, 12]. Despite these advancements, challenges remain in standardizing synthesis protocols, controlling nanoparticle size and morphology, and understanding long-term biocompatibility. Future research should focus on optimizing extraction techniques, exploring underutilized plant species, and integrating artificial intelligence for precise nanoparticle design [13–15]. The convergence of green synthesis with emerging technologies like nanorobotics and smart drug delivery systems could further redefine healthcare and environmental management. By harnessing the full potential of plant-mediated nanotechnology, we can pave the way for a sustainable future where advanced materials are synthesized in harmony with nature, addressing global challenges in health, energy, and environmental preservation [16, 17].

This chapter explores the principles, methodologies, and cutting-edge applications of plant-based green synthesis of metallic nanoparticles, emphasizing its pivotal role as a cornerstone of sustainable nanotechnology. Through continued innovation and interdisciplinary collaboration, green-synthesized nanoparticles hold immense promise in shaping a cleaner, healthier, and more technologically advanced world.

2 Nanotechnology and Nanoparticles: Fundamentals and Synthesis

Nanotechnology as a concept was coined by professor Norio Taniguchi in 1974, an extension of the visionary ideas of physicist Richard Feynman in his 1959 landmark lecture, There's Plenty of Room at the Bottom." [18, 19] Nanotechnology is the manipulation, design and control of materials on the nanoscale of materials, which usually fall in the range of 1–100 nm. At this scale, materials have a strongly high surface to volume ratio leading to a high surface reactivity and different physicochemical properties compared to the bulk material [20, 21]. These exclusive nanoscale properties form the basis of the revolutionary effect of nanotechnology in a broad range of uses. Nanotechnology is widely known as one of the pillars of technological innovation in the twenty-first century with its contribution to physics, materials science, chemistry, biology, engineering, and computer science [22, 23]. Precise engineering of matter at the nanoscale has allowed breakthroughs in a wide variety of applications, in fields such as microelectronics to more complex biomedical applications. Nanoparticles (NPs), which are the basic components in nanotechnology, can be prepared through a wide range of methodologies and these include chemical, physical, and biological methods. The choice of each synthesis route is planned to fabricate nanoparticle properties in terms of size, morphology, dispersibility, and optical behavior to suit a particular application [24]. In general, nanoparticle production methods can be divided into two major methods top-down and bottom-up. The top-down method is whereby the bulk materials are systematically reduced to nanoscale by such processes as mechanical milling or lithographic processing [25]. On the other hand, nanoparticles are built using bottom-up approach using atomic or molecular precursors through chemical reduction, nucleation, and self-assembly. The choice of a suitable synthesis method is determined by the required nanoparticle qualities and capabilities of the application of interest and every method has its own benefits as well as intrinsic drawbacks. Among all the Nanomaterials, metal nanoparticles (MNPs) have received much interest on the basis of outstanding physicochemical characteristics and broadening uses, especially in the biomedical and health sciences [26, 27]. For example, MNPs have shown potential in innovative wound treatment strategies, leveraging their unique physicochemical properties. To align with sustainable practices, environmentally friendly synthesis methods, often referred to as green synthesis, have gained prominence. These methods utilize natural resources, such as plant extracts, to produce nanoparticles with minimal environmental impact. The following sections explore the primary methodologies for nanoparticle synthesis, with a focus on eco-friendly approaches that enhance sustainability [28, 29].

3 Principles of Sustainable and Green Chemistry

Green chemistry, a cornerstone of sustainable development, has emerged as a transformative framework in the chemical sciences over the past two decades. Sustainable development entails fulfilling current needs without jeopardizing the capacity of future generations to meet theirs. This principle is particularly significant for chemical industries, which have historically contributed to environmental degradation and resource depletion [30, 31]. Public perception often associates chemicals with toxicity and risk, driven by the potential hazards of certain substances and processes. While conventional safety measures, such as protective gear, can mitigate risks, their failure may lead to severe consequences, including injury, environmental harm, or fatalities. Green chemistry addresses these issues by promoting the design of inherently safer chemicals and processes that reduce environmental impact and enhance efficiency. This approach focuses on minimizing intrinsic hazards through innovative molecular design and sustainable practices, thereby reducing the likelihood of accidents and exposure. By emphasizing non-toxic, renewable feedstocks and energy-efficient methodologies, green chemistry supports the goals of sustainable development, offering environmentally friendly and economically viable solutions for chemical synthesis and industrial applications. This section examines the core principles of green chemistry and their pivotal role in advancing sustainable nanotechnology, with a particular focus on plant-mediated nanoparticle synthesis [32, 33].

4 Green Synthesis of Nanoparticles

Green chemistry has emerged as a pivotal approach in recent decades, aiming to minimize or eliminate environmentally hazardous waste through sustainable processes and pollution prevention at the molecular and atomic levels. The application of non-toxic reagents, solvents that are environmentally harmless, and sourcing renewable raw materials are the principles of sustainable chemical design highlighted by this paradigm. In the context of the synthesis of metallic nanoparticles (NPs), green chemistry paradigms advocate the utilization of water as a non-toxic solvent, non-toxic reducing agents, biodegradable stabilizing molecules, and the utilization of common lab equipment to create nanoparticles in an an environmentally friendly and energy-efficient manner [34, 35]. Out of the various green synthesis methods, biological methods using plant extracts and microorganisms have become one of the most promising alternatives to the traditional chemical and physical approaches. The main feature of green synthesis is most often a simple one-step process that requires no use of toxic reagents and a small amount of by-products to be formed, which guarantees sustainability of the environment and cost-efficiency. Due to its simplicity, scalability, and high yield of products, this method has received much interest to be used as a viable method of synthesizing nanoparticles at scale. In recent years,

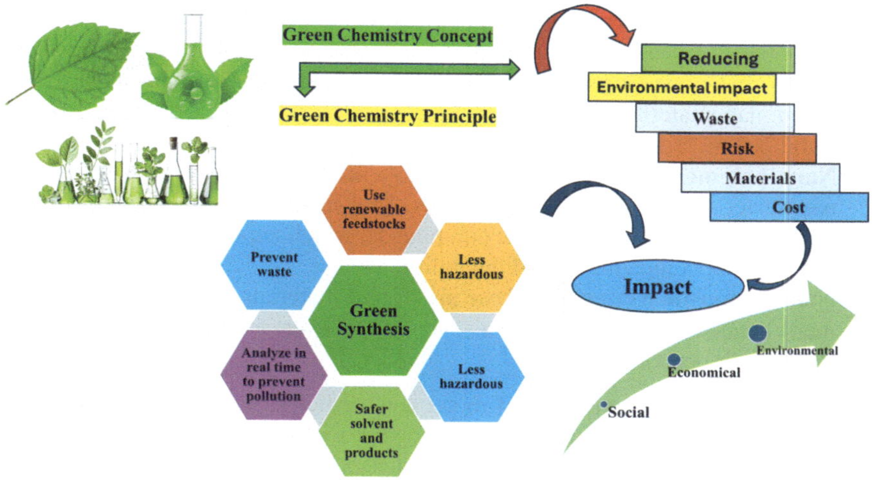

Fig. 1 Diagram illustrating the integration of green chemistry principles in the synthesis of metal nanomaterials

plant-based biosynthesis has been extensively explored, particularly for applications in the health sciences, where nanoparticles are tailored for their unique physicochemical properties and biocompatibility [36]. The following sections delve into the principles and applications of green synthesis, highlighting its role in advancing sustainable nanotechnology as shown in Fig. 1.

5 Classification and Synthesis of Nanoparticles

Nanomaterials, defined as particles or structures with nanoscale dimensions, are categorized based on their origin, composition, and structural arrangement. They may occur naturally (e.g., mineral nanoparticles), arise incidentally (e.g., particles in combustion byproducts like smoke or exhaust), or be intentionally engineered to exhibit specific properties and dimensions. Nanomaterials are broadly classified into two primary categories: inorganic and organic. Additionally, they can be categorized by the dimensionality of their atomic arrangements: one-dimensional (e.g., nanofilms), two-dimensional (e.g., nanowires and nanofibers), and three-dimensional (e.g., carbon nanotubes, CNTs) [37].

Based on structural composition and bonding, nanomaterials are further divided into four distinct types:

i. **Nanometallic Materials**: These consist of nanoparticles or alloys composed of metals, such as silver nanoparticles (AgNPs), gold nanoparticles (AuNPs), and titanium nanoparticles (TiNPs), characterized by primary metallic bonds. Their small size and high surface area confer unique physicochemical properties.

ii. **Nanoceramic Materials**: These encompass oxides, borides, nitrides, or carbides of metals, formed through ionic or covalent bonding. Examples include silica nanoparticles, zinc oxide nanoparticles, copper oxide nanoparticles, and carbon-based nanomaterials like carbon black, fullerenes, and CNTs, which are stabilized by primary covalent bonds.
iii. **Nano-polymer Materials**: These are organic, biochemical, or protein-based nanomaterials formed by covalent bonds. They include nanofibers, supramolecular structures, and macrocyclic compounds, which are critical in applications requiring biocompatibility and flexibility.
iv. **Nanocomposite Materials**: These are hybrid materials combining nanopolymers, nanoceramics, or nanometals, tailored to leverage the synergistic properties of their constituents.

The synthesis and application of metal nanoparticles, particularly AgNPs and AuNPs, are of significant interest in biosensing due to their exceptional optical, electronic, and catalytic properties. These attributes enhance the sensitivity and specificity of biosensors, enabling precise detection of low-concentration biomolecules. For instance, the surface plasmon resonance (SPR) properties of AgNPs and AuNPs facilitate real-time, label-free detection of biological analytes, critical for applications in medical diagnostics, environmental monitoring, and food safety. Moreover, the versatile surface chemistry of metal nanoparticles allows for functionalization with biomolecules, creating robust and selective biosensing platforms that ensure high reliability and accuracy [38, 39].

6 Metal Nanoparticles (NPs) and Their Properties

Metal nanoparticles (NPs) have uncommon physicochemical properties that form the basis of their wide usage in various scientific and technological fields (Fig. 2). Silver nanoparticles (AgNPs) are known to have a broad range of antimicrobial action against pathogenic microorganisms (including bacteria, fungi, and viruses), and therefore, they are very useful in ensuring that medical equipment and surfaces are resistant to infections. The optical properties of gold nanoparticles (AuNPs) include surface plasmon resonance (SPR) and facilitate a wide range of diagnostic imaging, biosensing and targeted drug delivery system uses and, consequently, improve particle precision in therapeutic systems. Platinum nanoparticles (PtNPs) are characterized by the high catalytic activity and are actively used in the fuel cell technologies to enhance the efficiency of energy conversion, and in high-sensitivity sensors to detect the analytes [40, 41]. Due to their superior electrical and thermal conductivity, copper nanoparticles (CuNPs) find use in electronic components, printed electronics conductive inks, antimicrobial healthcare products and as catalysts in a range of chemical reactions. Zinc oxide nanoparticles (ZnONPs) are multifunctional in nature in that they possess ultraviolet (UV)-blocking properties, antibacterial effects and photocatalytic actions, thus suitable in the application of

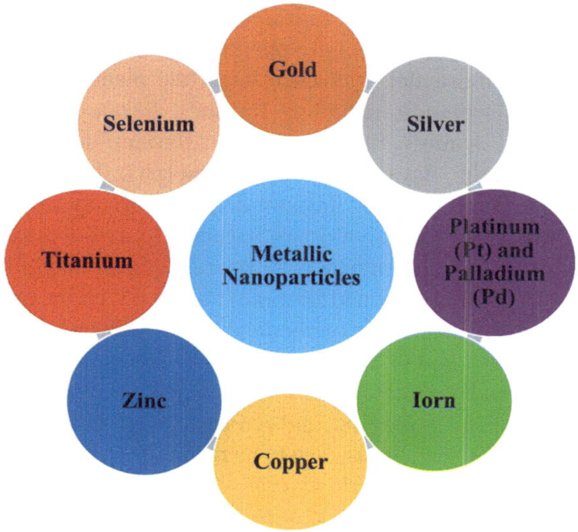

Fig. 2 Different types of metallic nanoparticles

sunscreens, cosmetic formulations and water treatment system to break down pollutants. The metal nanoparticles of iron oxide (FeONPs) have high magnetic properties, which allow their usage in magnetic resonance imaging (MRI), magnetic guided drug delivery, and in the clean-up of the environment to remove contaminants. The palladium nanoparticles (PdNPs) have been reported to have an excellent catalytic rate and are important in increasing the rate of chemical reactions with a wide range of application in not only the heterogeneous catalysis but also in hydrogen storage and gas sensing technology. Taken together, the nanoparticles of metal are still proving to create tremendous progress in biomedical, environmental, and energy-specific areas [42].

7 Green Synthesis of Nanoparticles Using Plant Extracts

Plant extracts have been found to be rich resources of various bioactive compounds in vitamins, alkaloids, carotenoids, phenolic compounds, lipids, carbohydrates, proteins, and enzymes, which are present in different plant tissues, leaves, roots, flowers, fruits, and rhizomes. All these phytochemicals are involved with the biological functionality of plant extracts and the major role of green synthesis of nanoparticles (NPs) synthesis as reducing and stabilizing agent. These biomolecules control important processes in the nanoparticle formation, such as redox reactions, nucleation, particle growth, and surface stabilization, and thus allow the creation of stable and functional nanomaterials [43]. The plants may be subjected to excesses of metal ions which may cause excessive generation of reactive oxygen species (ROS) leading to oxidative stress, metabolic imbalance, and structural defects. In order to counteract the toxicity caused by metals, plants produce chelating and antioxidant compounds and especially phenolic components that have been observed to effectively remove the ROS even when the plant is subjected to high levels of metal ions. Flavonoids, including flavones, flavanones, flavonols, flavanol derivatives, flavanol isomers, flavanol analogs, flavanol-derived polymorphs, flavanol-derived polymorphs, flavanol-derived polymorphs, flavanol-derived polymorphs, flavanol-derived polymorphs, flavanol-derived polymorphs, flavanol-derived polymorphs, flavonol-derived polymorphs, flavanol- As a result, many studies indicated successful production of metal nanoparticles (e.g., silver and gold) and metal oxide nanoparticles (e.g., iron, zinc and zirconium oxides) with the use of plant extracts, and most of them were used in the biomedical field. It is worth noting that biogenic nanoparticles are often found to have higher functional capability than those synthesized through traditional physical or chemical methods such as excellent antibacterial, antifungal, antioxidant, antibiofilm, and cytotoxic effect [44]. The surface modification of bioactive phytochemicals is seen to be the key to synthesizing green nanoparticles that are particularly susceptible to colloidal stability and functional efficacy. Plant extract concentration is a key parameter that determines the nature of nanoparticles like optical absorption intensity, particle size, morphology, dispersion, and agglomeration level. An increase in extract concentrations usually favors the development of smaller and isotropic and mostly spherical nanoparticles with enhanced dispersion and less aggregation. Conversely, smaller amounts of extracts have been shown to produce larger nanoparticles with stronger agglomeration because of a lack of phytochemicals to effectively stabilize and encapsulate particles on the surface. As an illustration, Lin et al. showed that gold nanoparticles (AuNPs) which were synthesized at a higher concentration of Morus lucida leaf extracts were spherical and had a diameter less than 15 nm though those synthesized at a lower concentration in the extract were over 100 nm in diameter and had hexagonal shapes. Such difference in size and shape was explained by the low phytochemical content of low concentrations which resulted in poor stabilization and further agglomeration of nanoparticles [45, 46].

8 Approaches to Nanoparticle Synthesis

Nanoparticle synthesis can be broadly classified into two primary strategies: the bottom-up and top-down approaches, each employing distinct mechanisms to achieve nanoscale dimensions as depicted in Fig. 3.

I. Bottom-Up Approach

The bottom-up approach involves the construction of nanoparticles from smaller building blocks, such as atoms or molecules, through processes like self-assembly or chemical and biological reactions. In this method, atoms aggregate to form nuclei, which subsequently grow into nanoparticles with dimensions typically in the 1–100 nm range. This approach leverages chemical reduction, biological mediation, or

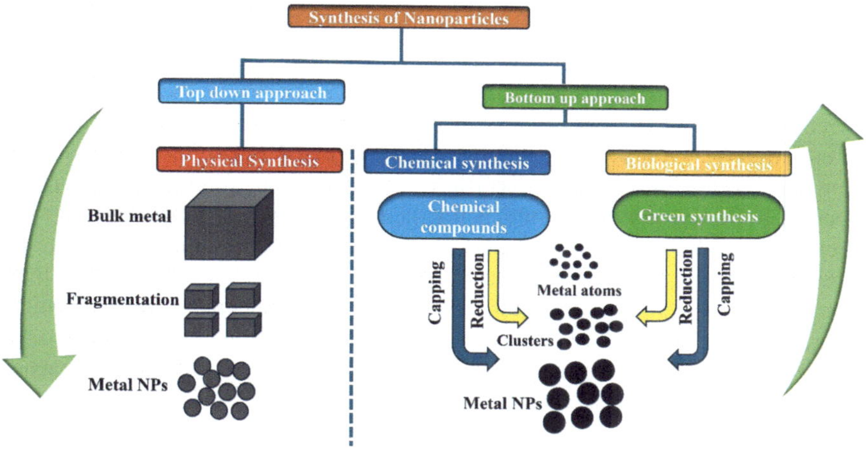

Fig. 3 Synthesis of metallic nanoparticles by top-down and bottom-up approaches

controlled nucleation to produce nanoparticles with tailored properties. For instance, plant-mediated green synthesis, a prominent bottom-up method, utilizes phytochemicals to reduce metal ions and stabilize the resulting nanoparticles, offering an environmentally sustainable pathway [47].

II. Top-Down Approach

In contrast, the top-down approach entails the reduction of bulk materials into nanoscale particles through physical or mechanical techniques, such as milling, crushing, or lithographic methods. These processes systematically break down larger structures into nanoparticles, allowing for precise control over particle size and morphology. The top-down approach is particularly suited for applications requiring uniform nanoparticles derived from macroscopic precursors [48].

III. Hybrid Fabrication Approaches

Hybrid fabrication strategies integrate top-down and bottom-up techniques to create hierarchically organized nanostructured platforms, which are often challenging or infeasible to produce using either method alone. Top-down approaches, such as etching, involve the precise removal of material to shape nanostructures, while bottom-up methods, such as layer growth via ion deposition, rely on the self-assembly or controlled deposition of atoms or molecules. A prominent example of this hybrid methodology is photolithography, where top-down etching is combined with bottom-up layer formation through ion-assisted growth. This synergistic approach enables the fabrication of complex nanostructures with tailored properties, offering significant advantages for applications in nanotechnology, including advanced theragnostic systems [41].

9 Factors Influencing Nanoparticle Properties

The size, shape, and stability of nanoparticles synthesized via either approach can be precisely controlled by optimizing key parameters, including temperature, pH, concentration of precursors (e.g., plant extracts or metal salt solutions), and incubation time. These factors play a critical role in determining the physicochemical properties of the nanoparticles, influencing their suitability for specific applications. For example, Siddiqi and Husen (2016) reviewed the synthesis of palladium and platinum nanoparticles, detailing the bottom-up processes and highlighting their applications in diagnostics, biosensing, catalysis, pharmaceuticals, and medicine. By carefully modulating synthesis conditions, nanoparticles can be engineered for enhanced performance in these fields, underscoring the versatility of both approaches in advancing nanotechnology [49].

10 Types of Metallic Nanomaterials in Green Synthesis

Green synthesis of metallic nanomaterials offers a sustainable alternative to conventional physical and chemical methods, significantly reducing the risk of toxicity associated with hazardous reagents. Recent studies have highlighted a diverse array of metallic nanomaterials, including zinc nanoparticles (Zn NPs), silver nanoparticles (Ag NPs), copper nanoparticles (Cu NPs), chitosan-coated copper nanoparticles (CS-Cu NPs), Pd/Pt nanoparticles, Fe nanoparticles, and polymer-coated nanoparticles, and others. These nanomaterials exhibit enhanced biomedical properties compared to their bulk counterparts, attributed to their high surface-to-volume ratio and unique physicochemical characteristics. The classification of these nanomaterials is based on several criteria, including chemical composition, surface functionalization, morphology, magnetic properties, and other structural attributes, enabling their tailored application in fields such as biomedicine, catalysis, and environmental remediation [50, 51].

10.1 Methods for the Synthesis of Metal Nanoparticles

Metal nanoparticles (NPs) synthesis can be broadly categorized into two main strategies, including top-down and bottom-up approaches. The top-down method is the transformation of bulk materials to nanoscale by physical, chemical or biological means and the bottom-up method is the alternation of nanostructures using atoms or molecules as starting materials via controlled nucleation and growth. These complementary approaches as schematically shown in Fig. 1.2 allow the production of nanoparticles with customized physicochemical characteristics to suit particular application needs. The key physical, chemical, and biological techniques used in the synthesis of metal nanoparticles are described below [52].

10.2 Physical Synthesis Methods

Physical synthesis techniques include methods such as evaporation–condensation, laser ablation, vapor deposition, gamma-ray irradiation, ball milling, and autoclaving. These methods physically break down bulk materials or deposit atoms to form nanoparticles. While effective for producing monodisperse NPs with accurate control over shape and size, physical methods often require expensive, high-energy apparatus, limiting their suitability for large-scale production. Additionally, these approaches may raise environmental concerns due to their energy-intensive nature, making them less favorable for sustainable applications [12].

10.3 Chemical Synthesis Methods

Solvothermal, microwave-assisted, sol–gel, microemulsion, and electrochemical reduction are chemical synthesis pathways that use chemical reducing agents (hydrazine hydrate, sodium citrate, and sodium borohydride) to reduce metal ions into metallic nanoparticles. These techniques provide a high degree of versatility in that nanoparticles can be engineered to be highly controlled in size, morphology, and even crystallinity to be used in the fields of catalysis, electronics and biomedicine. Nevertheless, most of the chemical synthesis methods use chemicals that are toxic and solvents that are dangerous to the environment and human health workers. Also, the high cost and environmental impact of these systems have aroused increasing interest in the creation of alternative systems that are less disastrous and more environmental friendly [53–55].

10.4 Biological Synthesis Methods

Green synthesis also known as biological synthesis involves the use of natural samples like plant extracts, microorganisms or enzymes to synthesize metal nanoparticles in a greener way. Plant extracts, fortified with bioactive molecules, such as polyphenols, flavonoids, and alkaloids, may play multiple roles as a reducing agent, stabilizing agent, reducing metal ion, and preventing nanoparticle agglomeration. An example of such plants includes the extraction of gold and silver nanoparticles in which extracts of plants like Rosa hybrida and Piper betle have been used successfully in the efficient production of nanoparticles [56]. Biological synthesis with bacteria, fungi, or algae is also an alternative; although it usually takes longer reaction times (usually 24–124 h) than plant-based methods. All in all, biological synthesis has the benefits of being less toxic, cheaper and more environmentally friendly making it a promising technology in scalable production of nanoparticles, especially in biomedical practice.

11 Plant Extract-Mediated Synthesis of Nanomaterials

Plants are natural bioreactors that generate a diversity of cost-effective, stable, and biologically active substances that promote the production of nanomaterials (NMs) in an environmental friendly way. These phytochemicals help plants to purify the environment and remove the pollutants in the environment such as the heavy metals that are very toxic even when in low levels. Relative to the microbial synthesis pathways, which necessitate growing and manipulating complex and possibly dangerous cultures, plant-mediated synthesis of nanomaterials has significant benefits of being simple, safe, and easily operable. In addition, the kinetics of synthesis of nanomaterials based on plants are greatly faster than the kinetics of synthesis of other biological processes and, in certain instances, are as efficient as traditional chemistry-based synthesis methods (Fig. 4). Plant extract usage, therefore, has been deemed a valid approach to production of nanomaterials due to its sustainability and eco-friendly nature [57]. Phytochemical compounds, specifically flavonoid, alkaloid, tannin, saponin, phenolic, terpenoid, and vitamins, which are plant in origin are vital in nanoparticle formation, nanoparticle stabilization, and nanoparticle surface functionalization. Plant-produced secondary metabolites used as therapeutic and protection agents have varied biological activities and are also important in potentiating nanoparticle synthesis as reducing and capping agents. The given green synthesis has a number of benefits, such as non-toxicity, cost-effectiveness, and environmental safety, which makes the given approach an appealing alternative to traditional chemical methods. The processes that regulate the synthesis of nanoparticles through plants are regulated by several parameters, such as pH, concentration of extract, and chemical nature of the phytochemicals. As an example, lignin, which is a polyphenolic compound, is both a reducing and stabilizing agent in the synthesis of silver nanoparticles, with its concentration having a direct effect on the size and dispersion of the particles. Likewise, zinc oxide nanoparticles prepared in the Ficus microcarpa leaf extract have shown a high level of larvicidal activity on mosquitoes and a high level of antibacterial activity which indicates their biomedical and environmental applicability. Garcinia genus plants that are especially rich in phenolic metabolites are also very strong in antioxidant and free radical-scavenging properties which makes them very effective in the synthesis of nanoparticles. Garcinia hanburyi extracts have also been effectively utilized in mixes with metal salt precursors like silver nitrate, gold chloride and copper nitrate to form stable nanoparticles at ambient temperature. The functional group interactions and some electrostatic force between phytochemicals and nanoparticle surface are the main mediators of nanoparticle stabilization in these systems. The versatility of plant extracts in nanoparticle synthesis is further

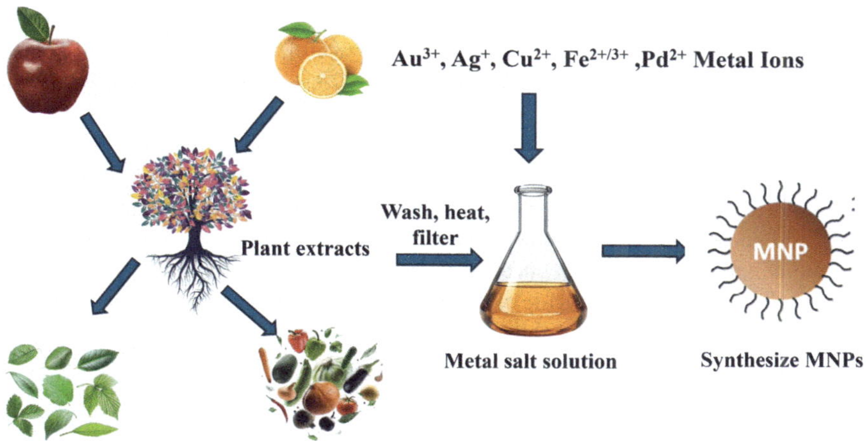

Fig. 4 Schematic illustration of the green synthesis of metal nanoparticles using plant extracts

evidenced by the use of various plant parts, including leaves, stems, roots, and fruits, each contributing distinct phytochemical profiles that influence the properties of the resulting nanomaterials. The integration of plant-mediated synthesis into nanomaterial production not only enhances the sustainability of the process but also opens avenues for developing nanomaterials with tailored properties for applications in medicine, agriculture, and environmental remediation [58, 59].

11.1 Plant Extract-Mediated Synthesis of Gold Nanoparticles

Production of gold nanoparticles (AuNPs) through plant extracts comprises a novel, green, and sustainable, alternative to the traditional chemical synthesis methods, and it takes advantage of the abundance of bioactive compounds found in plants. Phytochemicals, flavonoids, phenolic compounds, terpenoids among other oxygen based biomolecules are also important in reducing trivalent gold ions (Au 3+) to metallic gold (Au 0) and simultaneously stabilizing the resulting nanoparticles by capping them with oxygen. The green synthesis strategy can greatly minimize the environmental load, and it is cost-effective and scaleable, which makes it appealing to a large variety of applications. The effectiveness of plant extracts in the synthesis of AuNP has been proven by numerous studies. As an illustration, in Abdoli et al. leaf extracts of Centaurea cyanus (cornflower) were applied to synthesize AuNPs and phytochemicals of plant origin acted as reducing and stabilizing agents [60]. Equally, Balalakshmi et al. described a cost-effective green synthesis method with Sphaeranthus indicus leaf extracts, where bioactive compounds assisted with gold ion reduction and could prevent aggregation of nanoparticles successfully [61]. Hamelian et al. synthesized AuNPs using *Thymus serpyllum* extracts, emphasizing the eco-friendly nature of the process [62]. Additionally, Desai et al. developed a rapid and sustainable method to produce photoluminescent AuNPs with catalytic and antioxidant properties using *Garcinia indica* (kokum) fruit extracts. The versatility of plant-mediated AuNP synthesis extends to diverse plant species and parts [63]. Hosny and Fawzy employed aqueous extracts of *Persicaria salicifolia* leaves for a one-step, economical biosynthesis of AuNPs, highlighting the immediacy and environmental benefits of the process [64]. Zhao et al. utilized *Tribulus terrestris* extracts, where oxygen-containing phytochemicals promoted the reduction of Au^{3+} ions and stabilized the nanoparticles through encapsulation, preventing aggregation [65]. Gangapuram et al. synthesized spherical AuNPs using *Annona squamosa* fruit waste, which served as both a reducing and stabilizing agent, demonstrating potential applications in treating hazardous industrial effluents [66]. Naraginti and Li produced multifunctional AuNPs and silver nanoparticles using *Actinidia lindl* extracts, evaluating their antibacterial efficacy against *Pseudomonas aeruginosa* [56]. The resulting AuNPs exhibit a wide range of applications, including sensing probes, bioimaging, drug delivery, cancer therapy, catalysis, and environmental remediation. Unlike traditional synthesis methods, which often involve toxic chemicals and high costs, plant-based approaches utilize abundant, non-toxic, and renewable resources, aligning with the principles of green chemistry. The ability of phytochemicals to reduce metal ions and stabilize nanoparticles through functional group interactions or electrostatic forces ensures the production of well-dispersed, stable AuNPs with tailored properties for specific applications [34].

11.2 Green Synthesis of Silver Nanoparticles Using Plant Extracts

Environmentally friendly production of silver nanoparticles (AgNPs) with the aid of plant extracts has been of great interest due to its strong antibacterial activity and capability to align with the principles of sustainable production. This green synthesis method takes advantage of bioactive compounds of plant extracts, namely flavonoids, phenolic compounds, alkaloids, and terpenoids, which are both synergistic as reducing agents, i.e., transforming silver ions (Ag +) into elemental silver (Ag 0), and as stabilizing agents, i.e., preventing agglomeration of nanoparticles. Plant-mediated AgNP synthesis has become a very feasible option by offering a low cost, biocompatible, and environmentally safe alternative mode of physical and chemical synthesis techniques [35]. Many studies have shown that the method of action is effective in various plants. As an example, Giri et al. involved the production of AgNPs with extracts of mature leaves of Eugenia roxburghii and a comparison of their physicochemical properties and antibiofilm activity [66]. In a similar manner, Abdi et al. synthesized AgNPs using aqueous extracts of Rhizophora mucronata stems, roots and leaves, and reported high antibacterial activity [46]. Similarly, Shah et al. utilized *Plantago asiatica* aqueous extracts to generate AgNPs, evaluating their antibacterial and antioxidant activities [67]. Gopu et al. achieved AgNP synthesis with *Momordica charantia* leaf extracts, where the reduction of silver nitrate was confirmed by a visible color change, signaling nanoparticle formation. Additional research underscores the adaptability of this method [68]. Dogiparthi et al. employed *Micrargeria wightii* leaf extracts in a bioreduction process to synthesize AgNPs, noting the method's simplicity and efficiency [65]. Madivoli et al. used *Lantana camara* aqueous extracts to produce AgNPs with pronounced antibacterial activity [16]. Vakili et al. synthesized 100-nm spherical AgNPs using *Biarum chaduchrum* leaf extracts, while Kumavat and Mishra generated AgNPs with antimicrobial properties using *Borago officinalis* leaf extracts, emphasizing the eco-friendly role of the reducing agents. The success of plant-mediated AgNP synthesis lies in the diverse phytochemical profiles of plant extracts, which enable the reduction of silver ions and stabilize nanoparticles through functional group interactions or electrostatic forces [40, 69]. This biological approach reduces reliance on toxic chemicals, adhering to green chemistry principles and positioning it as a sustainable strategy for producing AgNPs with applications in antimicrobial treatments, environmental management, and other fields.

11.3 Plant-Mediated Green Synthesis of Zinc Oxide Nanoparticles

Zinc oxide (ZnO) is a multifunctional inorganic semiconductor that has attracted a lot of attention in the form of nanoparticle, especially owing to the pronounced antibacterial effects of zinc oxide nanoparticles (ZnO NPs) prepared by green methods. Plant-mediated synthesis is based on the use of extracts, which are highly concentration of bioactive molecules, including polyphenols, flavonoids, and terpenoids, which can be used as the reducing and stabilizing agent during nanoparticle synthesis. The simplicity of operation, cost-effectiveness, and a smaller environmental footprint define this green substitute of traditional chemical synthesis, which is an effective method of nanomaterials synthesis with high biomedical importance. Efficiency and viability of plant-based ZnO NP synthesis has been demonstrated by many studies [10]. As an illustration, Khatami et al. produced rod-shaped ZnO nano-particles by employing extracts of Lilium brownii var. viridulum, which had a high anti-Leishmania activity [70]. Similarly, Doan Thi et al. created an effective and environmentally friendly synthesis procedure based on the use of orange peel extracts, which highlights the sustainability of the method and its viability [33]. Thakur et al. employed leaf extracts from *Jatropha curcas* and *Tinospora cordifolia* for green synthesis of ZnO NPs, with those derived from *Jatropha curcas* showing exceptional antibacterial activity against *Staphylococcus aureus* [71]. Additionally, Kombaiah et al. produced rectangular ZnO NPs, ranging from 10 to 90 nm, using *Stevia rebaudiana* extracts, which exhibited strong antimicrobial effects against *Leishmania* spp., *S. aureus*, and *Escherichia coli* [72]. These findings emphasize the capability of green

Table 1 Representative examples of metal-based nanoparticles synthesized using plant extracts

Types of MNPs	Conditions	Properties	Plant	References
Ag	25 °C	35 nm triangular	Pinus thunbergii	[73]
Au	25 °C	~100 nm spherical	Mirabilis jalapa flowers	[74]
Cu	25 °C	20–110 spherical	Euphorbia esula leaves	
Pt &Pb	80 °C	16–20 nm, spherical 6 to 8 nm, irregular	Pinus resinosa bark	[75]
TiO_2	60 °C	100–150 nm spherical	Annona squamosa peel	[76]
In_2O_3	60 °C	5–50 Spherical	Aloe vera	[77]
Pd	95 °C	10–15 nm Spherical	Curcuma longa	[78]
Pt	60 °C	15–19 nm Crystalline spherical and triangular	Diopyros kaki	[79]
CuO	75 °C	4.8 nm spherical	Sterculia urens	[80]
ZnO	60 °C	5–40 nm Spherical	Calotropis procera	[81]
NiO	60 °C	16–52 nm spherical 11 to 59 nm	Zingiber officinale (ginger) Allium sativum (garlic)	[82]
Fe	70 °C	20–80 nm, Spheroidal	Eucalyptus (Leaf)	[83]

synthesis to yield ZnO NPs with customized morphologies and enhanced bioactivity. By harnessing the natural reducing and capping properties of plant-derived phytochemicals, this method reduces reliance on hazardous chemicals, aligning with green chemistry principles and enabling the production of biocompatible nanomaterials for antimicrobial, antiparasitic, and other therapeutic applications (Table 1).

11.4 Green Synthesis of Platinum and Palladium Nanoparticles Using Plant Extracts

Platinum (Pt) and palladium (Pd) nanoparticles green synthesis with plant extracts are of growing interest because of the outstanding functional qualities of these noble metals and the eco-friendliness of plant-based nanoparticle synthesis methods. Platinum, a heavy silver-white alloy, and palladium, which is highly known as catalytic versatile is being explored in nanoparticle form to find application in catalysis, biomedicine and environmental remediation. Plant-based synthesis utilizes bioactive phytochemicals, such as polyols, carbonyl-bearing compounds, and phenolic constituents, which serve as reducing and stabilizing reagents, which is unlikely to be expensive and is environmentally friendly and safe, compared to conventional chemical synthesis strategies. A number of works have proved the effectiveness of this approach. As an example, the Lebasaki et al. created a new process of synthesizing palladium nanoparticles (Pd@Btea) through the usage of aqueous plant extracts, in which polyols and carbonyl compounds served to reduce Pd ion and prevent the aggregation of nanoparticles [72]. Similarly, Nasrollahzadeh and Sajadi (2016a) described the sustainable synthesis of Pd nanoparticles using aqueous *Euphorbia officinalis* leaf extracts, which demonstrated outstanding catalytic activity in Stille and Hiyama cross-coupling reactions without requiring additional reagents, underscoring their efficiency and eco-compatibility [57]. Aygun et al. employed *Nigella sativa* L. extracts to synthesize spherical Pt nanoparticles ranging from 1 to 6 nm, highlighting the precision and adaptability of plant-based synthesis. By leveraging the diverse phytochemical profiles of plant extracts, this green synthesis approach facilitates the reduction of metal ions and stabilizes nanoparticles through functional group interactions, aligning with green chemistry principles. This method minimizes the reliance on toxic chemicals, reduces environmental impact, and enables the production of Pt and Pd nanoparticles with tailored properties for advanced catalytic, biomedical, and environmental applications [84].

11.5 Plant-Mediated Green Synthesis of Copper Nanoparticles

Copper nanoparticles (CuNPs) are a very important nanoparticle with a broad scope of applicability in various fields of catalysis, water purification, data storage, and biomedical technologies. Traditional methods of synthesis of CuNPs frequently use poisonous substances and processes that are hazardous to the environment, which is a serious setback. Green synthesis with plant extracts supplemented with the bioactive molecules (polyphenols, flavonoids, and alkaloids) is a sustainable and eco-friendly alternative in contrast to the other. These phytochemicals can be effectively used in reducing and stabilizing, and this allows the production of CuNPs, which are less toxic and with optimized physicochemical characteristics that can be used in a wide range of applications. There are several studies which confirmed the usefulness of the plant-mediated CuNP synthesis. To illustrate, Nagar and Devra used leaf extracts of Azadirachta indica (neem) to produce CuNPs and examined the impact of the reaction conditions on the morphology and conversion efficiency of nanoparticles [32]. In a comparable study, Rajesh et al. utilized bud and leaf extracts of Syzygium aromaticum (clove) in an easy and ecologically non-hazardous synthesis route, which fully characterised the resultant CuNPs in terms of their structural and morphological, optical and antibacterial characteristics [31]. Nasrollahzadeh et al. produced CuNPs under sustainable conditions using *Plantago depressa* leaf extracts as a natural solvent and biological medium, emphasizing the eco-friendly nature of the process [57]. Similarly, Sharma et al. synthesized CuNPs with *Tinospora cordifolia* leaf extracts, highlighting the role of plant-derived agents in enhancing nanoparticle biocompatibility. These investigations demonstrate the potential of green synthesis to yield CuNPs with tailored properties for healthcare, environmental, and industrial applications. By harnessing the natural reducing and capping capabilities of plant extracts, this approach minimizes environmental harm and toxicity, aligning with green chemistry principles and paving the way for sustainable CuNP production [85].

11.6 Eco-Friendly Synthesis of Titanium Dioxide Nanoparticles

The pursuit of sustainable methods for synthesizing titanium dioxide (TiO_2) nanoparticles has intensified in recent years, driven by the demand for environmentally friendly alternatives to conventional chemical approaches. Green synthesis, employing plant and fungal extracts, provides a biocompatible, reliable, and eco-conscious route to produce TiO_2 nanoparticles with varied sizes and morphologies. By utilizing bioactive compounds such as polyphenols, flavonoids, and proteins as reducing and stabilizing agents, these methods minimize the reliance on toxic chemicals, thereby reducing environmental impact. Numerous studies have validated the effectiveness of biological extracts in TiO_2 nanoparticle synthesis. Hudlikar et al. harnessed latex from *Jatropha curcas* L. to synthesize TiO_2 nanoparticles, demonstrating the potential of plant-based materials [86]. Nabi et al. employed *Cinnamomum verum* (cinnamon) extracts in a top-down approach, starting with bulk TiO_2, to produce TiO_2 nanoparticles [87]. Abisharani et al. synthesized TiO_2 nanoparticles using *Cucurbita pepo* seed extracts combined with titanium trichloride ($TiCl_3$), illustrating the adaptability of plant-derived precursors [88]. Pushpamalini et al. utilized leaf extracts from *Piper betel, Ocimum tenuiflorum, Moringa oleifera*, and *Coriandrum sativum*, paired with titanium tetraisopropoxide, to produce TiO_2 nanoparticles with sizes ranging from 6.7 to 8.3 nm, as confirmed by transmission electron microscopy (TEM) [89]. Additionally, Tarafdar et al.employed the fungus *Aspergillus tubingensis* to synthesize TiO_2 nanoparticles, highlighting the utility of microbial extracts in green synthesis. These findings emphasize the capability of green synthesis to generate TiO_2 nanoparticles with tailored properties for applications in photocatalysis, biomedicine, and environmental remediation [11]. By leveraging the natural reducing and stabilizing properties of plant and fungal extracts, this approach adheres to green chemistry principles, offering a sustainable pathway for nanomaterial production. Titanium dioxide (TiO_2) nanoparticles are highly valued for their distinctive morphologies, adaptable surface chemistry, and broad applications across industries, including textiles, plastics, paper, cosmetics, and food production. Their colloidal forms are particularly effective in environmental applications, such as the photocatalytic degradation of dyes and pollutants in water, providing an eco-friendly approach to water purification. The green synthesis of TiO_2 nanoparticles using plant extracts has gained prominence as a non-toxic, sustainable alternative to conventional chemical methods, minimizing environmental impact and enhancing safety. The TiO_2 nanoparticles synthesis through the plant can be performed in the same way as the synthesis of the TiO_2 nanoparticles using the titanium oxide solution, which is typically reacted with titanium salt solutions under conditions of controlled conditions, to produce the desired TiO_2 nanoparticles. The generation of nanoparticles is usually followed by a distinct color change in the medium of reaction which can be considered as an early warning signal of the formation of nanoparticles. This preliminary observation is then further supported by detailed characterization of its physicochemical properties through ultraviolet–visible (UVU) spectroscopy, Fourier-transform infrared (FTIR) spectroscopy, X-ray diffrahammetry (XRD), scanning electron microscopy (SEM), energy-dispersive spectroscopy (EDS), dynamic light scattering (DLS) and Brunauer–Emmett–Teller (BET) surface analysis. The synthesized TiO_2 nanoparticles tend to have light to dark green color, and their size, morphology, and colloidal stability are controlled by the quality and quantity of phytochemicals the plant extract contains [88, 90].

11.7 Green Synthesis of Iron-Based Nanoparticles- Fe NPs, Nanoscale Zero-Valent Iron (NZVI), and Iron Oxide NPs

Green synthesis of iron-based nanoparticles (Fe NPs) such as nanoscale zero-valent iron (NZVI) and iron oxide nanoparticles have received a lot of attention due to their environmentally friendly synthesis pathways and broad range of application. Plant extracts, which are enriched with bioactive elements including polylphenols, flavonoids, and other natural reducing agents, have been widely studied as being useful in the production of stable and functional Fe NPs. These phytochemicals have a dual activity as reducing and capping agents to promote the formation of nanoparticles as well as improve their colloidal stability and antioxidant activity, and also to greatly decrease the use of toxic chemical reagents. The synthesis of Fe NPs with the use of plant extracts has been proven successful in numerous studies conducted. For instance, extracts from mango (Mangifera indica L.) leaves (Weng et al.,), eucalyptus leaves, grape (Vitis vinifera L.) seeds (Pattanayak and Nayak), pear (Eleocharis tuberosa) tree leaves (Martínez-Cabanas et al., 2016), vine leaves (Machado et al.,), and Terminalia chebula fruit (Mohan Kumar et al.,) have been successfully utilized to produce Fe NPs with strong antioxidant capabilities [6, 91–93]. Notably, Fe NPs synthesized using Urtica dioica leaf extract exhibited exceptional stability, with a zeta potential of approximately −82.6 mV, indicating excellent colloidal stability (Ebrahiminezhad et al.,) [94]. Comparative studies have also found out that phytochemical profile of plant extracts is a determining factor in the physicochemical quality of Fe NPs. An example is where Fe NP synthesis was assessed with extracts of black tea, oolong tea, and green tea, with the enhancement of nanoparticles formation in the presence of green tea extract being more significant because of the high concentration of polyphenol and caffeine in the extracts that act as effective reducing and stabilizing agents [95]. These Fe NPs, which are made using green tea have since been used to provide a sustainable replacement of Fe 2+ ions in Fenton like reactions in the degradation of polluted substances as illustrated by Shahwan et al. [96]. In the same regard, Wang et al. used leaf extracts of Eucalyptus tereticornis, Melaleuca nesophila and Rosmarinus officinalis in the production of iron-polyphenol nanoparticles that could successfully dissolve azo dyes [97]. Moreover, Fe NPs prepared with oolong tea extract demonstrated good performance in dye decomposition process as stated by Huang et al. [98]. Plant-mediated Fe NPs have additionally demonstrated encouraging antimicrobial action in addition to environmental remediation. As an illustration, Naseem et al. were able to produce Fe NPs by heating the leaf extracts of Gardenia jasminoides and Lawsonia inermis under standard heating conditions, whereas Fe NPs produced by Gardenia jasminoides proved to be a powerful inhibitor of Staphylococcus aureus [99]. Similarly, Wei et al. utilized extracts from Eichhornia crassipes, an invasive aquatic weed, to synthesize Fe NPs that effectively removed chromium through reduction, immobilization, and coprecipitation mechanisms [100]. The synthesis of NZVI presents additional challenges compared to other nanoparticles due to its susceptibility to oxidation and agglomeration (Turakhia et al.,) [101]. However, plant extracts rich in polyphenols and proanthocyanidins, such as those derived from grape seeds, have proven effective as stabilizing agents. These compounds mitigate oxidative stress by neutralizing reactive oxygen species, such as peroxyl and hydroxyl radicals, thereby preventing nanoparticle aggregation and enhancing stability (Gao et al.; Manquián-Cerda et al.). The dual role of polyphenols as reducing and capping agents underscores their importance in producing robust NZVI for environmental applications [102]. In conclusion, the green synthesis of nanoscale metals involves a series of well-defined steps that leverage plant-based extracts to produce nanoparticles in an environmentally sustainable manner. The general procedure can be summarized as follows: preparation of plant extract, mixing the extract with a metal salt solution under specific conditions, reduction of metal ions to form nanoparticles, and subsequent purification steps, such as filtration, to isolate the target nanoscale metal. Detailed schematics for obtaining plant extracts and the synthesis process are provided in Figs. 1 and 2, respectively. However, the synthesis conditions must be tailored to the specific metal and plant extract used, as these factors significantly influence the reaction dynamics and nanoparticle characteristics. This customization necessitates further experimental exploration to optimize synthesis protocols for different metal-plant combinations. Additionally, the properties of the resulting nanoparticles, such as their reactivity and functionality, vary depending on the intrinsic chemical behavior of the metal [39].

12 Factors Influencing Green Synthesis and Associated Limitations

The green synthesis of nanoparticles using plant-based extracts is influenced by several factors that can impact the efficiency, reproducibility, and scalability of the process. These factors, along with inherent limitations, must be carefully considered to optimize the synthesis and ensure its practical applicability. The following factors are discussed below.

12.1 Choice and Availability of Raw Materials

The composition and quality of plant-based raw materials significantly affect the green synthesis of nanoparticles. Variability in the phytochemical profiles of plants, influenced by factors such as geographical location, season, and growth conditions, can lead to inconsistencies in nanoparticle size, shape, and functionality. Numerous studies have explored locally available plants to mitigate this challenge. For instance, Aromal et al. utilized fenugreek (*Trigonella foenum-graecum*), widely available in China and the eastern Mediterranean, for the synthesis of gold nanoparticles (Au NPs). Similarly, peppermint (*Mentha piperita*), prevalent in West Asia and Central Europe, has been employed for nanoparticle synthesis. Other species, such as *Cupressus sempervirens* for iron oxide nanoparticles, *Urtica dioica*, and *Malva sylvestris* for iron and cuprous oxide nanoparticles, are abundant across multiple continents and exhibit robust growth independent of specific climatic or soil conditions [42, 49]. These plants offer a reliable source of bioactive compounds for nanoparticle synthesis. However, challenges arise when scaling up production. The availability of raw materials may be limited by regional or seasonal constraints, complicating large-scale synthesis. Additionally, some plant-derived secondary metabolites require further processing, increasing both the complexity and cost of the synthesis process. To address these issues, the selection of raw materials must consider cost-effectiveness, economic feasibility, and practical scalability. Standardizing plant extract preparation and optimizing sourcing strategies are critical to ensuring consistent nanoparticle production [31].

12.2 Influence of pH and Temperature on Green Synthesis of Nanoparticles

The critical parameters that have significant impacts on the reaction kinetics, yield, and quality of nanoparticles synthesized through green routes are the reaction pH and temperature. In spite of mild conditions under which the green synthesis is usually performed to make it environmentally friendly, pH and temperature optimization are important to ensure that nanoparticle sizes, morphology, and operational performance are continually reproducible. The kinetics of growth and nucleation of nanoparticles are directly dependent on temperature. Indicatively, Sneha et al. investigated the impact of temperature on the production of gold nanoparticles (Au NPs) by Piper betle leaf extract [65]. TEM analysis showed that the best temperatures produced uniformly shaped nanoparticles, which were mostly octahedrally and nano-triangular in shape with a narrow size distribution. Green synthesis regimens typically use temperatures of less than 100 °C that favors energy conservation and environmental friendliness. Likewise, reaction pH is an important factor that influences the chemical environment through the regulation of the reduction potential of phytochemicals and the stability of synthesized nanoparticles [33, 37]. Rajput et al. examined the effect of temperature on silver nanoparticle (Ag NP) synthesis using Atropa belladonna leaf extract which established that higher temperatures enhanced molecular kinetic energy and speed of silver ion reduction. Nevertheless, it was discovered that high temperatures had a negative impact on the stability of nanoparticles, which has led to the need to carefully control the temperature. However, excessively high temperatures reduced the likelihood of particle formation and growth due to rapid ion depletion. Thus, an optimal temperature range is critical to balance reaction kinetics and nanoparticle stability. Careful modulation of pH and temperature is therefore essential to tailor the synthesis process for specific metals and plant extracts, ensuring high-quality nanoparticles with desired properties [67].

12.3 Influence of Reaction Time on Green Synthesis of Nanoparticles

Reaction time is a critical parameter in the green synthesis of nanoparticles, significantly influencing their size, shape, formation rate, and stability. Green synthesis methods typically employ mild conditions to align with their eco-friendly principles, but optimizing reaction time is essential to ensure efficient, reproducible, and high-quality nanoparticle production. Variations in reaction time can be strategically used to control the size and morphology of nanoparticles for specific reaction mixtures, tailoring their properties for targeted applications. For instance, Noruzi et al. demonstrated that gold nanoparticles (Au NPs) synthesized using Rosa hybrid petal extract exhibited rapid formation, with the reaction completing in just 5 min. This rapid reaction rate is a hallmark of plant-mediated biosynthesis, which is generally faster than microbial-based synthesis methods [36, 103]. In contrast, nanoparticle synthesis involving microorganisms, such as fungi and bacteria, often requires significantly longer reaction times, ranging from 24 to 124 h. Prolonged reaction times can lead to changes in nanoparticle characteristics, including size, shape, and stability, potentially due to ongoing reduction or aggregation processes. Therefore, precise control of reaction time is crucial to achieving consistent nanoparticle properties and optimizing the efficiency of green synthesis processes.

12.4 Role of Agitation and Mixing in Green Synthesis of Nanoparticles

Agitation and mixing are critical parameters in the green synthesis of nanoparticles, as they facilitate the uniform distribution of reactants, thereby promoting efficient nucleation and controlled growth of nanoparticles. These processes ensure homogeneity in the reaction medium, enhancing the consistency of nanoparticle size and morphology. However, excessive agitation or mixing can disrupt the delicate balance of particle formation, leading to aggregation or coarsening of nanoparticles, which may compromise their quality and functionality. Optimizing agitation conditions is therefore essential to achieve high-quality nanoparticles with desirable properties in green synthesis protocols [35].

12.5 Challenges in the Reproducibility of Green Synthesis of Nanoparticles

The ability to achieve reproducibility is one of the primary issues with the green synthesis of nanoparticles, just as it is with the majority of biologically mediated synthesis mechanisms. Changes in the chemical composition and quality of natural precursors, especially plant extracts, and changes in the experimental parameters, such as pH, temperature, extract concentration, and others, can lead to considerable differences in the yield of nanoparticles, their size, morphology, and functionality. To overcome these issues, it is necessary to use standardized extraction procedures, maintain a high degree of control over the reaction environment, and establish extensive characterization plans to enable reproducible production of nanoparticles. Variations in phytochemical profiles due to factors like plant species, growth conditions, or harvest seasons, as well as differences in parameters such as pH, temperature, or agitation, contribute to challenges in achieving consistent nanoparticle size, morphology, and functionality. Addressing these issues through standardized protocols and rigorous quality control is essential to enhance the reproducibility of green synthesis methods [37].

12.6 Influence of Water Type, Purity, and Impurities on Green Synthesis of Nanoparticles

The type and purity of water used in the green synthesis of nanoparticles play a crucial role in determining the stability, morphology, and properties of the resulting nanoparticles. Green synthesis methods often utilize natural water sources, such as tap water or river water, due to their accessibility and alignment with eco-friendly principles. However, these sources can vary significantly in composition and purity, introducing inconsistencies that affect the reproducibility and quality of nanoparticle synthesis. The presence of ions, organic matter, or other impurities in natural water can interfere with the reduction and stabilization processes, altering nanoparticle characteristics such as size, shape, and stability. To mitigate these challenges, the use of purified water, such as deionized or distilled water, is recommended to ensure a controlled and consistent reaction environment. Alternatively, careful selection and pretreatment of natural water sources can help minimize variability. Furthermore, the presence of extraneous ions and impurities in the reaction medium can disrupt the nucleation and growth of nanoparticles, potentially leading to aggregation or altered surface properties [34, 39]. Optimizing reaction conditions and implementing rigorous purification strategies for both water and plant extracts are essential to reduce the impact of impurities and enhance the reproducibility and quality of nanoparticles produced via green synthesis.

12.7 Challenges in Scaling up Green Synthesis of Nanoparticles

The green synthesis of nanoparticles is typically developed and optimized at a laboratory scale, where small reaction volumes allow for precise control of conditions. However, scaling up these processes to industrial or large-scale production presents significant challenges. The synthesis protocols often require re-optimization to accommodate larger volumes, as factors such as mixing efficiency, heat transfer, and reaction kinetics may differ significantly from laboratory conditions. Additionally, the cost and consistent availability of plant-based raw materials become critical concerns at larger scales, particularly when relying on specific plant species or extracts with variable phytochemical compositions. Addressing these challenges through process engineering, standardized sourcing strategies, and cost-effective material selection is essential to enable the practical implementation of green synthesis for large-scale nanoparticle production [63, 64].

12.8 Impact of Light Exposure on Green Synthesis of Nanoparticles

Light irradiation particularly sunlight is commonly used as a renewable energy during the production of green nanoparticles to enhance redox reactions of metal ions. Though light-driven processes prove to have beneficial impacts on the environment, inappropriate or uncontrollable absorption of light may impose a detrimental effect on the synthesis process through destruction of bioactive phytochemicals that act as reducing and stabilizing agents. This degradation could also decrease the formation efficiency of nanoparticles and destabilize the nanoparticles, resulting in aggregation or disintegration of the particles. Therefore, light intensity and exposure time must be accurately controlled to make the best out of synthesis efficiency and bring out high quality, stable nanoparticles with green synthesis methods [69].

13 Biological Applications of Metallic Nanoparticles

In the current situation, approximately 60% of commercially available pharmaceuticals are derived, either directly or indirectly, from natural sources, including plants, animals, and minerals. This prevalence underscores the potential for leveraging medicinal plants to synthesize biocompatible nanomaterials with synergistic biological effects. By carefully selecting these natural resources, biogenic metal-based nanoparticles (MNPs) can be engineered to enhance therapeutic outcomes, offering distinct advantages over conventional drugs. The metallic components of MNPs offer high surface area and variable oxidation states, which enhance reactivity and functional versatility. The green synthesis of MNPs, utilizing bioactive compounds from natural products, presents significant advantages over traditional therapeutic agents. These include controlled drug release, extended half-life, and enhanced efficacy, particularly in chemotherapeutic applications. Such attributes contribute to the development of advanced nanomedicines optimized for improved absorption, distribution, metabolism, excretion, and reduced toxicity (ADMET) profiles [41]. For example, while silver nanoparticles synthesized via conventional methods may exhibit significant toxicity upon oral administration, their biogenic counterparts, produced using active constituents from natural sources, demonstrate improved biocompatibility. This is achieved through the incorporation of unique chemical entities during particle formation and surface adsorption, which enhance stability, biological activity, and reduce toxicity. Furthermore, the ability to utilize medicinal plant resources in the production of metallic nanoparticle (MNP) has a number of practical benefits that include: large scale production, low cost, and easy access to raw material. The biogenic MNPs have a wide range of biological activities which include antioxidant, anti-inflammatory, antimicrobial, antiviral and anticancer activities. Through exploiting the inherent bioactivity of natural plant-derived compounds, biogenic MNPs represent a prospective catalyst to the creation of next-generation therapeutic systems that are safer with superior efficacies. In this regard, metallic nanoparticles have become one of the most practical tools that have a wide range of applications in the biomedical field such as medical diagnostics, molecular biology, bioimaging, drug delivery, and tumor targeting. Their unique physicochemical features (large surface-to volume ratio, size-on-demand, and customizable surface functionalization) allow them to be designed with innovative diagnostic and therapeutic approaches to selected biomedical applications [43, 104]. The sections below provide an in-depth discussion of the various biomedical uses of MNPs with particular focus being given to their transformative nature in terms of improving healthcare technologies (Figs. 5 and 6).

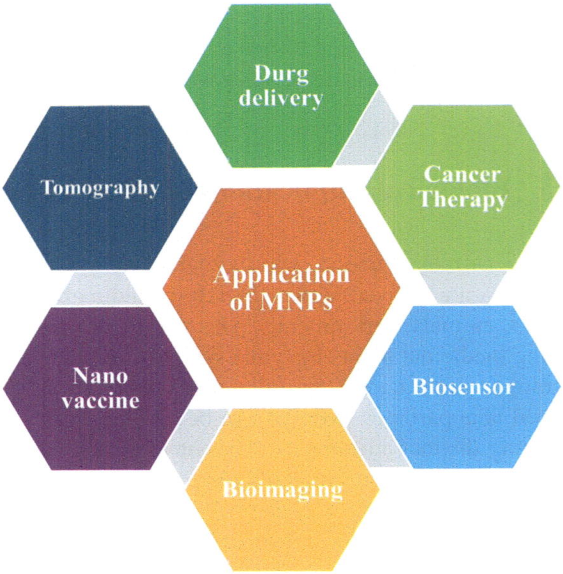

Fig. 5 Biological application of metal nanoparticles

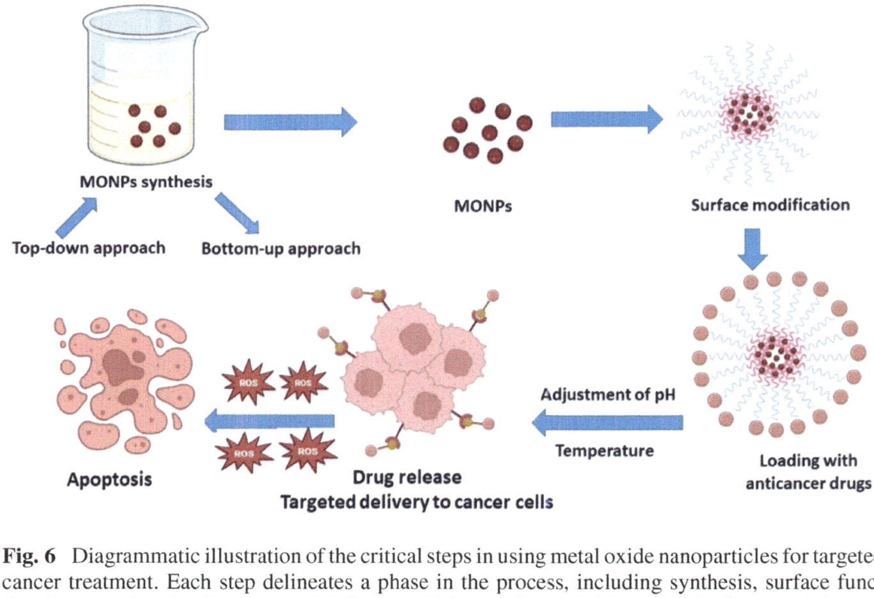

Fig. 6 Diagrammatic illustration of the critical steps in using metal oxide nanoparticles for targeted cancer treatment. Each step delineates a phase in the process, including synthesis, surface functionalization, drug loading, targeted delivery, cellular uptake, and ultimately, the therapeutic impact [106]

13.1 Metallic Nanoparticles in Medical Diagnostics and Molecular Biology

Metallic nanoparticles (MNPs) have exceptional physicochemical properties, which make them very useful in medical diagnostics and molecular biology. Minimization of the material sizes to the nanoscale causes a pronounced surface to volume ratio which gives rise to high chemical reactivity as well as tunable optical, electrical and magnetic properties that are significantly different than those possessed by bulk materials. These attributes enable MNPs to be tailored for advanced biomedical applications, including sensitive diagnostic assays and molecular-level investigations. The two main strategies by which MNPs can be synthesized include the top-down approach and the bottom-up approach. The top-down approach entails the mechanical or chemical disaggregation of bulk materials to nanoparticles, whilst the bottom-up approach synthesizes nanoparticles using precursors in the form of atom or molecules. Thermal decomposition is one of the most remarkable bottom-up approaches with respect to its cost-effectiveness, versatility, and capability to generate monodisperse nanoparticles of a controlled size and morphology. More recently, biological or green synthesis has been of significant interest because of non-toxicity, cost-effectiveness, and sustainability to the environment. Green synthesis reduces the utilization of dangerous chemicals and offers considerable hope of producing MNPs that can be used in biomedical practice by involving using plant extracts or microbial systems [42, 49].

13.2 Bioimaging Applications of Metallic Nanoparticles

Innovations in non-invasive imaging technologies have had a drastic impact on medical research, and the implication of this development is immense to drug development, diagnostics as well as cellular and molecular biology. Imaging techniques, including magnetic resonance imaging (MRI), electro-optic imaging, ultrasonography, positron emission tomography (PET) and molecular bioimaging, have since become invaluable components of visualizing both the structures and processes occurring in biology on a variety of length scales. Here, metallic nanoparticles, especially those that consist of noble metals like gold (Au) and silver (Ag) have come up as significant ingredients used in improving the imaging performance. Their good optical characteristics such as efficient photon emission and scattering in the near infrared spectrum, makes them very effective contrast agents, allowing the process of imaging to be resolved, sensitive, and specific of a signal when used in bioimaging [50].

13.3 Cancer Drug Delivery Using Metallic Nanoparticles

Effective cancer therapeutics are developed with a focus on the high bioavailability and localized drug delivery and less adverse outcomes on normal tissues. The use of Metallic nanoparticles (MNPs) as potent vectors in delivering drugs to cancer cells has been contributed by their size-tunable nature, ability to functionalize surfaces, and their ability to act on the disease in a targeted manner. An example of this is the gold nanoparticles (AuNPs) that have been greatly appreciated due to their low-toxicity, facile surface modulation, and specificity to cancer cells, which can be used to improve therapeutic precision. The anticancer activity of silver nanoparticles (AgNPs) is quite high and is associated with the induction of apoptosis and the appearance of reactive oxygen species (ROS), which destabilizes cancer cells. The nanoparticles of iron oxide (IONPs) have other benefits in terms of magnetic field-targeting, which allows the drugs to be localized at the location of the tumor with minimal harm to other normal tissues. Moreover, zinc oxide (ZnO), copper oxide (CuO), cerium oxide (CeO_2), titanium dioxide (TiO_2), and palladium nanoparticles (Pd NPs) are also other metallic nanoparticles that are in cancer therapy studies. These nanomaterials are involved in therapeutic action in various ways; by inducing apoptosis, creating cytotoxicity by ROS, and silencing tumor growth, and have become effective ingredients in the modern cancer treatment approach [17, 105].

At the cellular level, nanoparticle-mediated drug delivery is governed by endocytic uptake pathways, including clathrin- and caveolae-mediated internalization, followed by intracellular trafficking to endo-lysosomal compartments (Fig. 1.6) [106]. Surface functionalization with targeting ligands enables receptor-specific uptake, while stimuli-responsive coatings facilitate pH-, redox-, or enzyme-triggered drug release within the tumor microenvironment. These mechanisms collectively enhance intracellular drug accumulation, overcome multidrug resistance, and improve therapeutic indices while minimizing off-target toxicity.

13.4 Magnetic Resonance Imaging Enhanced by Metallic Nanoparticles

Magnetic resonance imaging (MRI) is a vital tool in modern medical diagnostics, prized for its high spatial resolution, non-invasive approach, and ability to examine tissues without damage. The incorporation of nanotechnology has further enhanced MRI, particularly through the use of metallic nanoparticles (NPs) as contrast agents. Among these, superparamagnetic iron oxide nanoparticles (SPIONs) stand out for their ability to improve imaging quality at cellular and molecular scales, enabling advanced diagnostic and therapeutic applications. By boosting contrast and sensitivity, SPIONs allow for more precise visualization of biological structures and processes [107, 108]. Mechanistically, SPIONs enhance MRI contrast by inducing local magnetic field inhomogeneities that accelerate transverse (T_2/T_2*) proton relaxation, resulting in signal attenuation in targeted tissues. Surface modification with polymers or biomolecules improves colloidal stability, prolongs circulation time, and enables molecular targeting of pathological sites. These features allow SPION-based probes to bridge anatomical and molecular imaging, supporting early disease detection and image-guided therapeutic interventions.

13.5 Drug Delivery Applications of Metal Nanomaterials

Metal nanomaterials represent a transformative frontier in medical therapeutics and diagnostic technologies. Their unique physicochemical characteristics, including remarkable electrical and thermal stability, an ultrahigh surface-area-to-volume ratio, and intrinsic bioactive properties, position them as ideal platforms for advanced drug delivery systems. Metal nanoparticles (MNPs) significantly improve therapeutic outcomes by enhancing drug permeability and prolonging retention time at target sites. A particularly effective approach involves the nanoencapsulation of bioactive agents within these metallic matrices, which enables spatiotemporally controlled release kinetics and superior bioavailability. These multifunctional capabilities endow metal-based nanocarriers with exceptional versatility, allowing their deployment across diverse therapeutic modalities ranging from oncology to regenerative medicine [109, 110]. The pharmacokinetic behavior of metal-based nanocarriers is strongly influenced by particle size, surface charge, and protein corona formation, which collectively determine biodistribution and clearance pathways. Rational engineering of surface chemistries enables prolonged systemic circulation and reduced reticuloendothelial system uptake. Furthermore, multifunctional metal nanocarriers can integrate therapeutic delivery with imaging or thermal functionalities, enabling synergistic theranostic platforms for precision medicine.

13.6 Metal-Based Nanoparticles as Antimicrobial Agents

The emergence of multi-drug resistant (MDR) pathogenic microorganisms poses a significant global health challenge, driven by the increasing prevalence of microbial infections and the difficulty in maintaining effective tissue antibiotic concentrations while minimizing systemic toxicity. Methicillin-resistant Staphylococcus aureus (MRSA), a prominent MDR bacterium, has evolved from a primarily nosocomial pathogen to a major cause of community-acquired infections, underscoring the urgent need for novel, effective, low-toxicity, and cost-efficient antimicrobial agents. Nanotechnology has introduced "nanobiotics" (NBs), a transformative approach to antimicrobial drug development that leverages nanoparticles to enhance therapeutic outcomes [111, 112]. These nanobiotics enable sustained and synchronized drug delivery, improving efficacy compared to conventional antibiotics. Over the past two decades, NBs have demonstrated advantages such as enhanced drug solubility, reduced toxicity, controlled release, improved pharmacokinetics, and superior biocompatibility. These properties position nanobiotics as a promising strategy to combat antimicrobial resistance. Concerns regarding the potential toxicity and environmental impact of synthetically derived nanoparticle-based antibiotics have prompted exploration into eco-friendly alternatives. Plant-mediated metal-based nanoparticles (MNPs) have emerged as a sustainable solution, offering non-toxic and environmentally benign antimicrobial platforms [113]. For instance, catechin, a naturally occurring antibacterial flavonoid, is widely utilized in the biogenic synthesis of MNPs. Traditionally, MNPs suffer from instability and susceptibility to oxidation, limiting their antimicrobial applications. However, advances in green synthesis have produced stable biogenic MNPs, such as catechin-copper nanoparticles, which effectively target pathogenic bacteria, including Escherichia coli, Staphylococcus aureus, and their MDR strains. Notably, MDR bacteria exhibit slower resistance development to these biogenic MNPs compared to conventional small-molecule antibiotics, enhancing their utility in addressing antimicrobial resistance. The antimicrobial efficacy of biogenic MNPs is attributed to two primary mechanisms: disruption of bacterial membrane potential and integrity, and the generation of reactive oxygen species (ROS), where MNPs act as nanocatalysts. These dual pathways enable potent bactericidal activity, making plant-mediated MNPs a compelling alternative for combating MDR pathogens while aligning with the demand for sustainable and biocompatible therapeutic solutions [114].

13.7 Metallic Nanomaterials as Advanced Platforms for Nano-Vaccine Development

Metallic nanomaterials represent a highly adaptable class of nanostructures with exceptional physicochemical characteristics and immune-modulating capabilities. Extensive research has established that engineered metallic nanoparticles can precisely regulate immunological responses through several key pathways, (a) Facilitation of immune cell recruitment and activation (b) Enhancement of antigen-presenting cell (APC) stimulation (c) Modulation of cytokine and chemokine production profiles. Silver-based nanostructures (AgNPs) have demonstrated particular efficacy in stimulating macrophage activity and other components of innate immunity. In contrast, gold nanoparticles (AuNPs) display multifunctional attributes, combining efficient cellular uptake with the ability to trigger potent cytokine responses. These synergistic properties position AuNPs as ideal candidates for immune system priming through APC activation, a fundamental requirement for effective vaccination strategies [43, 45]. The customizable surface characteristics and dimension-dependent behaviors of metallic nanoparticles render them exceptionally promising for innovative vaccine development. Their superior immunogenic potential, enhanced stability profiles, and capacity for targeted immune stimulation offer distinct advantages over traditional vaccine technologies, paving the way for next-generation immunization approaches. At the mechanistic level, metallic nanoparticles enhance vaccine efficacy by promoting efficient lymphatic transport and sustained antigen presentation within secondary lymphoid organs. Their nanoscale dimensions facilitate endosomal escape and cross-presentation of antigens via MHC class I pathways, thereby strengthening cytotoxic T-cell responses. Additionally, controlled surface functionalization enables co-delivery of antigens and immunostimulatory molecules, allowing precise modulation of immune polarization and long-term immunological memory.

13.8 Plant-Mediated Metal-Based Nanoparticles as Antioxidant Agents

Antioxidants are vital chemical compounds, which relieve oxidative and nitrosative stress, by either trapping reactive free radicals or by preventing the formation of radicals in biological systems. The unregulated formation of these reactive species may lead to oxidative stress on biomolecules and is one of the factors in the development and development of different pathological conditions. Even though, human body has inbuilt antioxidant defensive mechanisms, external supplementation may be needed during instances of high levels of oxidative stress, or antioxidant inadequacy. Artificial antioxidants including butylated hydroxyanisole (BHA) and butylated hydroxytoluene (BHT) have been widely used but due to adverse effects associated with their use, their application is being restricted by the reported adverse effects like hepatotoxicity and possible carcinogenicity. On the contrary, naturally occurring antioxidants (L-ascorbic acid, gallic acid, β-carotene, α-tocopherol, lycopene, resveratrol, and quercetin) are more biocompatible and less toxic and, therefore, can be used in biomedical practices more effectively [115, 116]. Plant-mediated metal-based nanoparticles (MNPs) have emerged as a promising class of antioxidant agents, leveraging the synergistic interplay between natural antioxidants and the metallic nanoparticle framework. These biogenic MNPs are increasingly recognized as innovative tools in biomedicine for managing disease morbidity and mortality. Studies have demonstrated their superior antioxidant activity compared to plant extracts alone or non-biogenic MNPs, attributed to the reducing, stabilizing, and capping properties of plant-derived bioactive compounds within the nanostructure. The antioxidant potential of biogenic MNPs is primarily driven by the reducing capabilities of plant-derived components, such as extracts, fractions, or isolated compounds, which serve as both reducing and stabilizing agents during nanoparticle synthesis. This antioxidant capacity underpins a broad spectrum of biological activities, including anti-inflammatory, antibacterial, antifungal, antiviral, and anticancer effects. By combining the inherent bioactivity of natural products with the unique physicochemical properties of metal nanoparticles, plant-mediated MNPs offer a versatile and effective platform for combating oxidative stress-related disorders, positioning them as a transformative approach in nanomedicine [57].

13.9 Metallic Nanomaterials as Next-Generation Contrast Agents for X-ray Computed Tomography

X-ray computed tomography (CT) stands as a cornerstone of modern diagnostic imaging, though its full diagnostic potential frequently necessitates contrast enhancement for optimal visualization. While iodinated contrast media remain the clinical standard, their inherent limitations have catalyzed the exploration of novel nanoparticle-based alternatives. Engineered metallic nanomaterials have garnered significant attention in this regard, owing to their exceptional X-ray attenuation capabilities and favorable electron density characteristics that markedly improve imaging contrast. Contemporary research has made substantial progress in addressing the biocompatibility challenges associated with these inorganic agents. Innovative surface engineering approaches employing biocompatible coatings have demonstrated particular efficacy in mitigating potential cytotoxic effects while preserving contrast performance. Beyond simply improving safety profiles, these surface modifications enable sophisticated targeting capabilities through the conjugation of molecular recognition elements, opening new possibilities for disease-specific imaging. The advent of metallic nanomaterial-based contrast agents heralds a transformative shift in CT imaging technology. These advanced agents offer numerous advantages over conventional alternatives, including prolonged circulation times, reduced administration volumes, and the exciting potential for theragnostic applications. Ongoing investigations are systematically addressing the critical challenges of clinical implementation, with a dual focus on maximizing diagnostic efficacy while ensuring optimal patient safety [117, 118].

13.10 Plant-Mediated Metal-Based Nanoparticles as Antiviral Agents

Viruses, characterized by their DNA or RNA genomes encapsulated in protein coats, replicate by exploiting host cell machinery in humans, animals, or other organisms. They pose significant public health challenges, causing diseases such as HIV/AIDS, measles, Ebola, influenza, polio, smallpox, and, more recently, infections from severe acute respiratory syndrome coronavirus 2 (SARS-CoV-2). Vaccination remains the cornerstone of viral infection prevention; however, rapid viral mutations often diminish vaccine efficacy, necessitating the development of safe and effective antiviral therapeutics. Developing such drugs is challenging due to the reliance of viruses on host cellular machinery, limitations in in vitro assays, and the scarcity of in vivo models that accurately replicate human viral infections. Natural products such as medicinal plants are a good source of bioactive compounds that have high antiviral capabilities. Inhibitors of the main protease (Mpro) of SARS-CoV-2 and other coronaviruses (indirubin and indigo) have been reported in compounds extracted in leaves and fruits of Couroupita guianensis, highlighting the relevance of plant-derived molecules in antiviral drug discovery [103, 119]. More recent discoveries in nanotechnology have only intensified the pace of antiviral research by allowing the creation of nature-inspired nanoparticles (NPs) to be effectively and specifically used as drug delivery methods. The unique physicochemical characteristics of NPs such as tunable size, morphology and surface chemistry enable strict manipulation of their biological interactions which maximizes the antiviral efficacy. Plant-mediated metal-based nanoparticles (MNPs) have also been introduced in this scenario as an antiviral agent. They are synthesized by combining the natural bioactivity of plant extracts and the multifunctional nature of metal nanoparticles, leading to improved antiviral activity. An example of a MNPs that show excellent virucidal activity against the influenza A virus is the quasi-spherical silver nanoparticles prepared with aqueous extracts of Panax ginseng roots, which indicates the potential of biogenic MNPs as a good antiviral nanotherapeutic agent. Similarly, silver nanoparticles produced with extracts from Lampranthus coccineus and Malephora lutea demonstrated potent antiviral effects against hepatitis A virus (HAV-10), herpes simplex virus type 1 (HSV-1), and coxsackievirus B4 (CoxB4). Additionally, nanotechnology has facilitated the development of biosensors and bioelectronics for rapid and sensitive virus detection, further enhancing its impact on viral research. By leveraging the bioactive compounds in plant extracts for MNP synthesis, this green approach offers a sustainable, biocompatible, and effective strategy for combating viral infections, addressing the urgent need for novel antiviral therapeutics[62, 84, 120].

14 Current Challenges in the Application of Green-Synthesized Metallic Nanoparticles

Synthesis of metallic nanoparticles (MNPs) is a complex process that entails both traditional and green techniques. The high cost of production, complex synthesis procedures and structural complexity are some of the challenges that hinder large scale production. In order to achieve the biomedical potentials of MNPs, existing methods of their synthesis should be reconsidered and optimized in terms of scalability, reproducibility, and cost. Even though green synthesis has evident environmental benefits, it should take care of the stability of the quality of the product, biocompatibility, and removal of toxic by-products. The issue of clinical safety is also a primary one because MNPs can be dose-dependent or long-term toxic. Therefore, thorough toxicological analyses are necessary because metal nanoparticles and their composites become more and more popular in biomedical and biological practices. Ion doping, surface functionalization, and conjugated polymer-metal oxide nanostructure are among the strategies to reduce the toxicity of MNPs, but overall, they should improve the biocompatibility without disrupting the performance of the nanoparticles [121].

i. **Targeted Delivery Challenges**: The therapeutic efficacy of metallic nanoparticles (MNPs) hinges on their ability to precisely deliver active agents to disease sites. While nanoparticles facilitate drug transport across biological barriers, their delivery efficiency is often compromised by physiological obstacles. For orally administered nanoparticles, stability in gastric acid and efficient intestinal absorption are critical determinants of systemic bioavailability. Intravenously delivered nanoparticles must additionally overcome vascular barriers, immune clearance, and tissue-specific penetration challenges. In oncology applications, tumor physiology presents unique hurdles, including dense cellular architecture, extensive stromal tissue, and elevated interstitial fluid pressure, that collectively impede nanoparticle extravasation and distribution. These factors contribute to suboptimal drug accumulation in malignant tissues. Furthermore, the success of active targeting strategies depends critically on the binding affinity and specificity of surface-conjugated ligands, which must be carefully optimized for each therapeutic context[90].
ii. **Analytical and Characterization Challenges**: Comprehensive characterization of green-synthesized MNPs demands sophisticated analytical techniques to evaluate their physicochemical and biological properties. Current analytical methodologies, despite their advancement, frequently lack the resolution to distinguish between therapeutically active nanoparticle formulations and their inactive counterparts. This limitation complicates quality control and standardization processes essential for clinical translation[85].
iii. **Pharmacological and Safety Considerations**: The pharmacokinetic evaluation of MNPs presents unique complexities due to their heterogeneous size distributions and dynamic biological behaviors. Conventional plasma concentration assays often fail to accurately reflect nanoparticle biodistribution or therapeutic activity, complicating bioequivalence assessments. Moreover, substantial knowledge gaps persist regarding the chronic toxicity profiles of metallic nanoparticles, particularly concerning their long-term biodistribution, accumulation patterns, and potential organ-specific effects[31, 44].
iv. **Regulatory Hurdles**: While regulatory agencies have approved select nanoparticle-based therapeutics, the field lacks standardized frameworks for evaluating nanomedicines. Current regulatory pathways rely on case-by-case risk–benefit analyses, creating bottlenecks in the approval process. There is a pressing need for: Development of universal characterization standards, Establishment of safety assessment protocols specific to nanoscale materials, Creation of predictive models for nanoparticle behavior in vivo. The advancement of green-synthesized MNPs toward clinical implementation will require coordinated efforts to address these multifaceted challenges through interdisciplinary collaboration and innovation in characterization technologies, safety assessment methodologies, and regulatory science[50].

15 Conclusion

This chapter has critically analyzed the development of metallic nanoparticles (MNPs), their synthesis methods, physicochemical tunability and the growing uses in biology, especially in nanomedicine and drug delivery. In the past twenty years, chemical, physical, and biologically inspired synthesis methods have facilitated accurate control over the size, morphology, surface chemistry and functionalization of particles. These advancements have highly contributed to stability, biocompatibility and multifunctionality of MNP-based systems, making them platforms of versatility in terms of diagnosis, therapy and theranostics. The capability to design surface characteristics with ligands, polymers, biomolecules and stimuli-responsive coating has played a leading role in increasing the biomedical applicability of MNPs. These changes have made it possible to deliver drugs precisely, regulate drug release, multimodal imaging and single agent theranostic. In cancer therapy, MNPs have become potent agents of enhancement of drug bioavailability, tumour-targeting accumulation, and real-time imaging and monitoring of therapeutic outcomes. The fact that they have been integrated into the process of cancer diagnosis and treatment highlights their increasing significance in precision medicine. In spite of these developments, there are still major obstacles to clinical translation of MNP-based therapeutics. Differences between preclinical success and clinical outcomes are common because of biological barriers, such as rapid clearance by the mononuclear phagocytic system, heterogeneous tumor vasculature, variable enhanced permeability and retention (EPR), and biological variability in specific patients. Furthermore, incomplete knowledge of the nano- bio interactions at the molecular, cellular and systemic levels complicates the prediction of long-term toxicity, biodistribution as well as the pharmacokinetics.

To address these constraints, physicochemical properties of MNPs should be optimized guided by disease-based and patient-based factors and not an overall design principle. Improving translational reliability requires rigorous and standardized preclinical testing, which includes but is not limited to safety, efficacy, biodistribution, immunogenicity, and clearance pathways in clinically relevant animal models. In the end, the key to effective implementation of MNPs in the routine clinical practices will be the connection between basic nanoscience and systems biology, clinical pharmacology, and regulatory science.

16 Future Directions

The future of metallic nanoparticles studies will go beyond traditional one-size-fits-all nanomedicine to more focused engineering and patient-centred nanotherapeutic models. The future developments in materials chemistry, computational chemistry, and high-throughput screening will allow the rational design of MNPs with predictable biological behavior, maximized targeting efficiency, and minimal off-target effects. Specifically, artificial intelligence-supported and data-driven methods can potentially be transformative in aligning the properties of nanoparticles with biological responses, potentially shortening design-translation. New directions involve the creation of stimuli-responsive and adaptive MNPs that can react to pH, redox conditions, enzyme activity, or external stimuli, e.g. magnetic fields and light. These systems are promising in the context of spatiotemporal release of the drug and real-time therapeutic monitoring. Moreover, the combination with MNPs and gene therapy, immunotherapy, and combination treatment could be employed to a considerable extent to improve the efficacy of the therapeutic regimen and decrease the systemic toxicity. In other fields beyond oncology, MNPs will play significant roles in antimicrobial therapy, regenerative medicine, biosensing and implantable medical devices. Their inherent antimicrobial qualities, with the addition of surface engineering, provide a possibility of self-sterilizing surfaces and the future of medicine with medical equipment that is less dependent on traditional antibiotics. Scalable, reproducible and environmentally friendly synthesis pathways, however, are a major demand towards large-scale adoption. It is also crucial that there must be regulatory harmonization and standardized characterization protocols, which will be required to guarantee safety, reproducibility, and clinical acceptance. Nanoparticles should be developed by incorporating long-term toxicity, life-cycle assessment, and post-market surveillance into the pipeline. In general the future interdisciplinary teamwork between chemists, biologists, clinicians, engineers, and regulatory experts will play a crucial role in moving the technologies of metallic nanoparticles to the bedside, which will eventually result in safer, more effective, and personalized treatments.

References

1. Shafiq, T., Uzair, M., Iqbal, M.J., Zafar, M., Hussain, S.J., Shah, S.A.A.: Green synthesis of metallic nanoparticles and their potential in bio-medical applications. Nano Biomed. Eng. **13**, 191–206 (2021). https://doi.org/10.5101/NBE.V13I2.P191-206
2. Marchiol, L.: Synthesis of metal nanoparticles in living plants. Ital. J. Agron. **7**, 274–282 (2012). https://doi.org/10.4081/ija.2012.e37
3. Jeetkar, T.J., Khataokar, S.P., Indurkar, A.R., Pandit, A., Nimbalkar, M.S.: A review on plant-mediated synthesis of metallic nanoparticles and their applications. Adv. Nat. Sci.: Nanosci. Nanotechnol. **13** (2022). https://doi.org/10.1088/2043-6262/ac865d
4. Makarov, V.V., Love, A.J., Sinitsyna, O.V, Makarova, S.S., Yaminsky, I.V., Taliansky, M.E., Kalinina, N.O.: "Green" Nanotechnologies: Synthesis of Metal Nanoparticles Using Plants (2014)
5. Triphati, A., Pirzadah, T.B.: Synthesis methods of nanoparticles and their key applications. In: Synthesis of Bionanomaterials for Biomedical Applications, Elsevier, pp. 57–76 (2023). https://doi.org/10.1016/B978-0-323-91195-5.00001-5
6. Martínez-Barbosa, M.E., Figueroa-Pizano, M.D.: Green synthesis and methodologies of nanomaterials: state of the art. In: Advances in Bionanocomposites: Materials, Applications, and Life Cycle, Elsevier, pp. 17–54 (2023). https://doi.org/10.1016/B978-0-323-91764-3.00012-7

7. Lagashetty, A., Anusha, M., Rachana, Channabasavaraja, M., Shivakumar, Veena, V., Preeti, R., Ganiger, S.K.: Exploring potential biological applications of green derived silver nanoparticles using Codiaeum variegatum leaf extract. Next Res. **2**, 100103 (2025). https://doi.org/10.1016/j.nexres.2024.100103
8. Singh, A., Kumar, P., Pathak, S., Jain, K., Garg, P., Pant, M., Mahapatro, A.K., Singh, R.K., Rajput, P., Kim, S.-K.: Tailored nanoparticles for magnetic hyperthermia: highly stable aqueous dispersion of Mn-substituted magnetite superparamagnetic nanoparticles by double surfactant coating for improved heating efficiency. J. Alloys Compd. **976**, 172999 (2024)
9. Kumar, P., Pathak, S., Singh, A., Verma, R., Khanduri, H., Jain, K., Tawale, J., Wang, L., Pant, R.P.: Augmented magnetic nanoparticle assimilation in rGO sheets for tailored static and dynamic magnetic properties in surface functionalized Co 0.8 Zn 0.2 Fe 2 O 4 nanoferrite–rGO hybrid structures. J. Mater. Chem. C Mater. **12**, 18036–18047 (2024)
10. Habibullah, G., Viktorova, J., Ruml, T.: Current strategies for noble metal nanoparticle synthesis. Nanoscale Res. Lett. **16** (2021). https://doi.org/10.1186/s11671-021-03480-8
11. Lithi, I.J., Ahmed Nakib, K.I., Chowdhury, A.M.S., Sahadat Hossain, M.: A review on the green synthesis of metal (Ag, Cu, and Au) and metal oxide (ZnO, MgO, Co3O4, and TiO2) nanoparticles using plant extracts for developing antimicrobial properties. Nanoscale Adv. **7**, 2446–2473 (2025). https://doi.org/10.1039/d5na00037h
12. Vijayaram, S., Razafindralambo, H., Sun, Y.Z., Vasantharaj, S., Ghafarifarsani, H., Hoseinifar, S.H., Raeeszadeh, M.: Applications of green synthesized metal nanoparticles—a review. Biol. Trace Elem. Res. **202**, 360–386 (2024). https://doi.org/10.1007/s12011-023-03645-9
13. Kumar, P., Khanduri, H., Pathak, S., Singh, A., Basheed, G.A., Pant, R.P.: Temperature selectivity for single phase hydrothermal synthesis of PEG-400 coated magnetite nanoparticles. Dalton Trans. **49**, 8672–8683 (2020). https://doi.org/10.1039/D0DT01318H
14. Kumar, P., Pathak, S., Jain, K., Singh, A., Kuldeep, Basheed, G.A., Pant, R.P.: Low-temperature large-scale hydrothermal synthesis of optically active PEG-200 capped single domain MnFe2O4 nanoparticles. J. Alloys Compd. **904**, 163992 (2022). https://doi.org/10.1016/J.JALLCOM.2022.163992
15. Kumar, P., Pathak, S., Singh, A., Khanduri, H., Wang, X., Basheed, G.A., Pant, R.P.: Optimization of cobalt concentration for improved magnetic characteristics and stability of CoxFe3-xO4 mixed ferrite nanomagnetic fluids. Mater. Chem. Phys. **265**, 124476 (2021)
16. Simon, S., Sibuyi, N.R.S., Fadaka, A.O., Meyer, S., Josephs, J., Onani, M.O., Meyer, M., Madiehe, A.M.: Biomedical applications of plant extract-synthesized silver nanoparticles. Biomedicines **10** (2022). https://doi.org/10.3390/biomedicines10112792
17. Calzoni, E., Cesaretti, A., Polchi, A., Di Michele, A., Tancini, B., Emiliani, C.: Biocompatible polymer nanoparticles for drug delivery applications in cancer and neurodegenerative disorder therapies. J. Funct. Biomater. **10** (2019). https://doi.org/10.3390/jfb10010004
18. Joudeh, N., Linke, D.: Nanoparticle classification, physicochemical properties, characterization, and applications: a comprehensive review for biologists. J. Nanobiotechnology **20** (2022). https://doi.org/10.1186/s12951-022-01477-8
19. Mitchell, M.J., Billingsley, M.M., Haley, R.M., Wechsler, M.E., Peppas, N.A., Langer, R.: Engineering precision nanoparticles for drug delivery. Nat. Rev. Drug Discov. **20**, 101–124 (2021). https://doi.org/10.1038/s41573-020-0090-8
20. Kumar, P., Pathak, S., Singh, A., Khanduri, H., Basheed, G.A., Wang, L., Pant, R.P.: Microwave spin resonance investigation on the effect of the post-processing annealing of CoFe2O4 nanoparticles. Nanoscale Adv. **2**, 1939–1948 (2020). https://doi.org/10.1039/D0NA00156B
21. Pathak, S., Verma, R., Singhal, S., Chaturvedi, R., Kumar, P., Sharma, P., Pant, R.P., Wang, X.: Spin dynamics investigations of multifunctional ambient scalable Fe3O4 surface decorated ZnO magnetic nanocomposite using FMR. Sci. Rep. **11**, 3799 (2021). https://doi.org/10.1038/s41598-021-83394-8
22. Kumar, P., Pathak, S., Singh, A., Jain, K., Khanduri, H., Wang, L., Partridge, J., Pant, R.P.: Effect of post annealing process on structural, magnetic and spin dynamics properties of MnFe2O4 nanoparticles. Mater. Today Proc. (2023)

23. Kumar, P., Pathak, S., Singh, A., Verma, R., Khanduri, H., Jain, K., Tawale, J., Wang, L., Pant, R.P.: Augmented magnetic nanoparticle assimilation in rGO sheets for tailored static and dynamic magnetic properties in surface functionalized Co0.8Zn0.2Fe2O4 nanoferrite–rGO hybrid structures. J. Mater. Chem. C Mater. **12**, 18036–18047 (2024). https://doi.org/10.1039/D4TC03498H
24. Kumar, P., Pathak, S., Singh, A., Jain, K., Khanduri, H., Wang, L., Kim, S.-K., Pant, R.P.: Observation of intrinsic fluorescence in cobalt ferrite magnetic nanoparticles by Mn2+ substitution and tuning the spin dynamics by cation distribution. J. Mater. Chem. C Mater. **10**, 12652–12679 (2022). https://doi.org/10.1039/D2TC02605H
25. Pathak, S., Verma, R., Kumar, P., Singh, A., Singhal, S., Sharma, P., Jain, K., Pant, R.P., Wang, X.: Facile synthesis, static, and dynamic magnetic characteristics of varying size double-surfactant-coated mesoscopic magnetic nanoparticles dispersed stable aqueous magnetic fluids. Nanomaterials **11**, 3009 (2021)
26. Kumar, P., Singh, A., Verma, R., Khanduri, H., Yadav, J., Luhar, S., Tawale, J., Pathak, S., Pant, R.P.: Interfacial spin modulation in MWCNT/Co-Zn ferrite nanocomposites and coupled magnetic and microwave spin resonance phenomena. Surf. Interfaces **77**, 107997 (2025). https://doi.org/10.1016/J.SURFIN.2025.107997
27. Kumar, P., Pathak, S., Singh, A., Khanduri, H., Kuldeep, Jain, K., Tawale, J., Wang, L., Basheed, G.A., Pant, R.P.: Enhanced static and dynamic magnetic properties of PEG-400 coated CoFe2−xErxO4 ($0.7 \leq x \leq 0$) nanoferrites. J. Alloys Compd. **887** (2021). https://doi.org/10.1016/j.jallcom.2021.161418
28. Chopra, H., Bibi, S., Singh, I., Hasan, M.M., Khan, M.S., Yousafi, Q., Baig, A.A., Rahman, M.M., Islam, F., Bin Emran, T., Cavalu, S.: Green metallic nanoparticles: biosynthesis to applications. Front. Bioeng. Biotechnol. **10** (2022). https://doi.org/10.3389/fbioe.2022.874742
29. Sadiq, M.U., Shah, A., Haleem, A., Shah, S.M., Shah, I.: Eucalyptus globulus mediated green synthesis of environmentally benign metal based nanostructures: a review. Nanomaterials **13** (2023). https://doi.org/10.3390/nano13132019
30. Das, D., Bhattacharyya, S., Bhattacharyya, M., Mandal, P.: Green chemistry inspired formation of bioactive stable colloidal nanosilver and its wide-spectrum functionalised properties for sustainable industrial escalation. Results Chem. **4** (2022). https://doi.org/10.1016/j.rechem.2022.100533
31. Jadoun, S., Arif, R., Jangid, N.K., Meena, R.K.: Green synthesis of nanoparticles using plant extracts: a review. Environ. Chem. Lett. **19**, 355–374 (2021). https://doi.org/10.1007/s10311-020-01074-x
32. Nagar, N., Devra, V.: Green synthesis and characterization of copper nanoparticles using Azadirachta indica leaves. Mater. Chem. Phys. **213**, 44–51 (2018). https://doi.org/10.1016/J.MATCHEMPHYS.2018.04.007
33. Al-darwesh, M.Y., Ibrahim, S.S., Mohammed, M.A.: A review on plant extract mediated green synthesis of zinc oxide nanoparticles and their biomedical applications. Results Chem. **7** (2024). https://doi.org/10.1016/j.rechem.2024.101368
34. Ying, S., Guan, Z., Ofoegbu, P.C., Clubb, P., Rico, C., He, F., Hong, J.: Green synthesis of nanoparticles: current developments and limitations. Environ. Technol. Innov. **26**, 102336 (2022). https://doi.org/10.1016/J.ETI.2022.102336
35. Asha, A.B., Narain, R.: Nanomaterials properties. In: Polymer Science and Nanotechnology: Fundamentals and Applications, Elsevier, pp. 343–359 (2020). https://doi.org/10.1016/B978-0-12-816806-6.00015-7
36. Jose Varghese, R., Sakho, E.H.M., Parani, S., Thomas, S., Oluwafemi, O.S., Wu, J.: Introduction to nanomaterials: synthesis and applications. In: Nanomaterials for Solar Cell Applications, Elsevier, pp. 75–95 (2019). https://doi.org/10.1016/B978-0-12-813337-8.00003-5
37. Khan, I., Saeed, K., Khan, I.: Nanoparticles: Properties, applications and toxicities. Arab. J. Chem. **12**, 908–931 (2019). https://doi.org/10.1016/j.arabjc.2017.05.011

38. Green Synthesis, Characterisation and Biomedicinal Applications of Rutaceae Family Plant Mediated Metal Nanoparticles: A Review (n.d.). https://doi.org/10.37896/YMER24.05/30
39. Kumari, S., Raturi, S., Kulshrestha, S., Chauhan, K., Dhingra, S., András, K., Thu, K., Khargotra, R., Singh, T.: A comprehensive review on various techniques used for synthesizing nanoparticles. J. Mater. Res. Technol. **27**, 1739–1763 (2023). https://doi.org/10.1016/j.jmrt.2023.09.291
40. Adeyemi, J.O., Oriola, A.O., Onwudiwe, D.C., Oyedeji, A.O.: Plant extracts mediated metal-based nanoparticles: synthesis and biological applications. Biomolecules **12** (2022). https://doi.org/10.3390/biom12050627
41. Azam, A.: A review of plant-based green nanoparticle synthesis: mechanisms, metals, and applications in nanotechnology. Revista Electronica De Veterinaria 3576–3584, (2024). https://doi.org/10.69980/redvet.v25i1.1657
42. Dikshit, P.K., Kumar, J., Das, A.K., Sadhu, S., Sharma, S., Singh, S., Gupta, P.K., Kim, B.S.: Green synthesis of metallic nanoparticles: applications and limitations. Catalysts **11** (2021). https://doi.org/10.3390/catal11080902
43. Agbaje, L., Elegbede, J.A., Akinola, P.O., Ajayi, V.A.: Biomedical applications of green synthesized-metallic nanoparticles: a review. Pan Afr. J. Life Sci. **3**, 157–182 (2019). https://doi.org/10.36108/pajols/9102/30(0170)
44. Kumar, H., Bhardwaj, K., Kuča, K., Kalia, A., Nepovimova, E., Verma, R., Kumar, D.: Flower-based green synthesis of metallic nanoparticles: applications beyond fragrance. Nanomaterials **10** (2020). https://doi.org/10.3390/nano10040766
45. de Gusmão, A.B., de Albuquerque, P.B.S., de C. Correia, A.C.: Use of metallic nanoparticles synthesized from plant extracts in wound healing—a review. Applied Nano **5**, 205–226 (2024). https://doi.org/10.3390/applnano5040014
46. Lin, Q., Hong, X., Zhang, D., Jin, H.: Biosynthesis of size-controlled gold nanoparticles using M. lucida leaf extract and their penetration studies on human skin for plastic surgery applications. J. Photochem. Photobiol. B **199**, 111591 (2019). https://doi.org/10.1016/J.JPHOTOBIOL.2019.111591
47. Baig, N., Kammakakam, I., Falath, W., Kammakakam, I.: Nanomaterials: a review of synthesis methods, properties, recent progress, and challenges. Mater. Adv. **2**, 1821–1871 (2021). https://doi.org/10.1039/d0ma00807a
48. Bao, Y., He, J., Song, K., Guo, J., Zhou, X., Liu, S.: Plant-extract-mediated synthesis of metal nanoparticles. J. Chem. **2021** (2021). https://doi.org/10.1155/2021/6562687
49. Edo, G.I., Mafe, A.N., Ali, A.B.M., Akpoghelie, P.O., Yousif, E., Isoje, E.F., Igbuku, U.A., Ismael, S.A., Essaghah, A.E.A., Ahmed, D.S., Ozsahin, D.U., Umar, H., Alamiery, A.A.: green biosynthesis of nanoparticles using plant extracts: mechanisms, advances, challenges, and applications. Bionanoscience **15** (2025). https://doi.org/10.1007/s12668-025-01883-w
50. Kulkarni, D., Sherkar, R., Shirsathe, C., Sonwane, R., Varpe, N., Shelke, S., More, M.P., Pardeshi, S.R., Dhaneshwar, G., Junnuthula, V., Dyawanapelly, S.: Biofabrication of nanoparticles: sources, synthesis, and biomedical applications. Front. Bioeng. Biotechnol. **11** (2023). https://doi.org/10.3389/fbioe.2023.1159193
51. El Shafey, A.M.: Green synthesis of metal and metal oxide nanoparticles from plant leaf extracts and their applications: a review. Green Process. Synth. **9**, 304–339 (2020). https://doi.org/10.1515/gps-2020-0031
52. Burlec, A.F., Corciova, A., Boev, M., Batir-Marin, D., Mircea, C., Cioanca, O., Danila, G., Danila, M., Bucur, A.F., Hancianu, M.: Current overview of metal nanoparticles' synthesis, characterization, and biomedical applications, with a focus on silver and gold nanoparticles. Pharmaceuticals **16** (2023). https://doi.org/10.3390/ph16101410
53. Kumar, S., Ohlan, A., Kumar, P., Verma, V.: Improved electromagnetic interference shielding response of polyaniline containing magnetic nano-ferrites. J. Supercond. Nov. Magn. **33**, 1187–1198 (2020)
54. Pathak, S., Jain, K., Kumar, P., Wang, X., Pant, R.P.: Improved Thermal Performance of Annular Fin-Shell Tube System Using Magnetic Fluid (2019)

55. Kumar, P., Khanduri, H., Singh, A., Jain, K., Pathak, S., Gautam, A., Wang, L., Pant, R.P.: Design and development of ferrofluids materials. In: Material Aspects of Ferrofluids. CRC Press, pp. 1–36 (2023)
56. Naraginti, S., Li, Y.: Preliminary investigation of catalytic, antioxidant, anticancer and bactericidal activity of green synthesized silver and gold nanoparticles using Actinidia deliciosa. J. Photochem. Photobiol. B. **170**, 225–234 (2017). https://doi.org/10.1016/J.JPHOTOBIOL.2017.03.023
57. Usman, O., Mohsin Baig, M.M., Ikram, M., Iqbal, T., Islam, S., Syed, W., Al-Rawi, M.B.A., Naseem, M.: Green synthesis of metal nanoparticles and study their anti-pathogenic properties against pathogens effect on plants and animals. Sci. Rep. **14** (2024). https://doi.org/10.1038/s41598-024-61920-8
58. Azad, A., Zafar, H., Raza, F., Sulaiman, M.: Factors influencing the green synthesis of metallic nanoparticles using plant extracts: a comprehensive review. Pharm. Front. **05**, e117–e131 (2023). https://doi.org/10.1055/s-0043-1774289
59. Idris, D.S., Roy, A., Pandit, S., Alghamdi, S., Almehmadi, M., Alsaiari, A.A., Abdulaziz, O., Alsharif, A., Khandaker, M.U., Faruque, M.R.I.: Polymer-based nanocarriers for biomedical and environmental applications. E-Polymers **23** (2023). https://doi.org/10.1515/epoly-2023-0049
60. Abdoli, M., Arkan, E., Shekarbeygi, Z., Khaledian, S.: Green synthesis of gold nanoparticles using Centaurea behen leaf aqueous extract and investigating their antioxidant and cytotoxic effects on acute leukemia cancer cell line (THP-1). Inorg. Chem. Commun. **129**, 108649 (2021). https://doi.org/10.1016/J.INOCHE.2021.108649
61. Balalakshmi, C., Gopinath, K., Govindarajan, M., Lokesh, R., Arumugam, A., Alharbi, N.S., Kadaikunnan, S., Khaled, J.M., Benelli, G.: Green synthesis of gold nanoparticles using a cheap Sphaeranthus indicus extract: Impact on plant cells and the aquatic crustacean Artemia nauplii. J. Photochem. Photobiol. B. **173**, 598–605 (2017). https://doi.org/10.1016/J.JPHOTOBIOL.2017.06.040
62. Hamelian, M., Varmira, K., Veisi, H.: Green synthesis and characterizations of gold nanoparticles using Thyme and survey cytotoxic effect, antibacterial and antioxidant potential. J. Photochem. Photobiol. B. **184**, 71–79 (2018). https://doi.org/10.1016/J.JPHOTOBIOL.2018.05.016
63. Desai, M.P., Sangaokar, G.M., Pawar, K.D.: Kokum fruit mediated biogenic gold nanoparticles with photoluminescent, photocatalytic and antioxidant activities. Process Biochem. **70**, 188–197 (2018). https://doi.org/10.1016/J.PROCBIO.2018.03.027
64. Hosny, M., Fawzy, M.: Instantaneous phytosynthesis of gold nanoparticles via Persicaria salicifolia leaf extract, and their medical applications. Adv. Powder Technol. **32**, 2891–2904 (2021). https://doi.org/10.1016/J.APT.2021.06.004
65. Zhao, P., El-kott, A., Ahmed, A.E., Khames, A., Zein, M.A.: Green synthesis of gold nanoparticles (Au NPs) using Tribulus terrestris extract: Investigation of its catalytic activity in the oxidation of sulfides to sulfoxides and study of its anti-acute leukemia activity. Inorg. Chem. Commun. **131**, 108781 (2021). https://doi.org/10.1016/J.INOCHE.2021.108781
66. Gangapuram, B.R., Bandi, R., Alle, M., Dadigala, R., Kotu, G.M., Guttena, V.: Microwave assisted rapid green synthesis of gold nanoparticles using Annona squamosa L peel extract for the efficient catalytic reduction of organic pollutants. J. Mol. Struct. **1167**, 305–315 (2018). https://doi.org/10.1016/J.MOLSTRUC.2018.05.004
67. Thakare, Y., Kore, S., Sharma, I., Shah, M.: A comprehensive review on sustainable greener nanoparticles for efficient dye degradation. Environ. Sci. Pollut. Res. **29**, 55415–55436 (2022). https://doi.org/10.1007/s11356-022-20127-y
68. Harish, V., Ansari, M.M., Tewari, D., Gaur, M., Yadav, A.B., García-Betancourt, M.L., Abdel-Haleem, F.M., Bechelany, M., Barhoum, A.: Nanoparticle and nanostructure synthesis and controlled growth methods. Nanomaterials **12** (2022). https://doi.org/10.3390/nano12183226
69. Ullah, A., Lim, S.I.: Plant extract-based synthesis of metallic nanomaterials, their applications, and safety concerns. Biotechnol. Bioeng. **119**, 2273–2304 (2022). https://doi.org/10.1002/bit.28148

70. Sabouri, Z., Rangrazi, A., Amiri, M.S., Khatami, M., Darroudi, M.: Green synthesis of nickel oxide nanoparticles using Salvia hispanica L. (chia) seeds extract and studies of their photocatalytic activity and cytotoxicity effects. Bioprocess Biosyst. Eng. **44**, 2407–2415 (2021). https://doi.org/10.1007/s00449-021-02613-8
71. Thakur, S., Shandilya, M., Guleria, G.: Appraisement of antimicrobial zinc oxide nanoparticles through Cannabis Jatropha curcasa Alovera and Tinosporacordifolia leaves by green synthesis process. J. Environ. Chem. Eng. **9**, 104882 (2021). https://doi.org/10.1016/J.JECE.2020.104882
72. Lebaschi, S., Hekmati, M., Veisi, H.: Green synthesis of palladium nanoparticles mediated by black tea leaves (Camellia sinensis) extract: catalytic activity in the reduction of 4-nitrophenol and Suzuki-Miyaura coupling reaction under ligand-free conditions. J. Colloid Interface Sci. **485**, 223–231 (2017). https://doi.org/10.1016/J.JCIS.2016.09.027
73. Velmurugan, P., Lee, S.-M., Iydroose, M., Lee, K.-J., Oh, B.-T.: Pine cone-mediated green synthesis of silver nanoparticles and their antibacterial activity against agricultural pathogens. Appl. Microbiol. Biotechnol. **97**, 361–368 (2013)
74. Vankar, P.S., Bajpai, D.: Preparation of gold nanoparticles from Mirabilis jalapa flowers. Indian J. Biochem. Biophys. **47**, 157 (2010)
75. Coccia, F., Tonucci, L., Bosco, D., Bressan, M., d'Alessandro, N.: One-pot synthesis of lignin-stabilised platinum and palladium nanoparticles and their catalytic behaviour in oxidation and reduction reactions. Green Chem. **14**, 1073–1078 (2012)
76. Aboyewa, J.A., Sibuyi, N.R.S., Meyer, M., Oguntibeju, O.O.: Green synthesis of metallic nanoparticles using some selected medicinal plants from southern africa and their biological applications. Plants **10** (2021). https://doi.org/10.3390/plants10091929
77. Maensiri, S., Laokul, P., Klinkaewnarong, J., Phokha, S., Promarak, V., Seraphin, S.: Indium oxide (In2O3) nanoparticles using Aloe vera plant extract: Synthesis and optical properties. J. Optoelectron. Adv. Mater. **10**, 161–165 (2008)
78. Roopan, S.M., Bharathi, A., Kumar, R., Khanna, V.G., Prabhakarn, A.: Acaricidal, insecticidal, and larvicidal efficacy of aqueous extract of Annona squamosa L peel as biomaterial for the reduction of palladium salts into nanoparticles. Colloids Surf. B Biointerfaces **92**, 209–212 (2012)
79. Song, J.Y., Kwon, E.Y., Kim, B.S.: Biological synthesis of platinum nanoparticles using Diopyros kaki leaf extract. Bioprocess Biosyst. Eng. **33**, 159–164 (2010). https://doi.org/10.1007/s00449-009-0373-2
80. Padil, V.V.T., Černík, M.: Green synthesis of copper oxide nanoparticles using gum karaya as a biotemplate and their antibacterial application. Int. J. Nanomedicine (2013) 889–898
81. Singh, R.P., Shukla, V.K., Yadav, R.S., Sharma, P.K., Singh, P.K., Pandey, A.C.: Biological approach of zinc oxide nanoparticles formation and its characterization. Adv. Mater. Lett. **2**, 313–317 (2011)
82. Haider, A., Ijaz, M., Ali, S., Haider, J., Imran, M., Majeed, H., Shahzadi, I., Ali, M.M., Khan, J.A., Ikram, M.: Green synthesized phytochemically (Zingiber officinale and Allium sativum) reduced nickel oxide nanoparticles confirmed bactericidal and catalytic potential. Nanoscale Res. Lett. **15**, 50 (2020)
83. Wang, T., Jin, X., Chen, Z., Megharaj, M., Naidu, R.: Green synthesis of Fe nanoparticles using eucalyptus leaf extracts for treatment of eutrophic wastewater. Sci. Total Environ. **466**, 210–213 (2014)
84. Aygun, A., Gülbagca, F., Ozer, L.Y., Ustaoglu, B., Altunoglu, Y.C., Baloglu, M.C., Atalar, M.N., Alma, M.H., Sen, F.: Biogenic platinum nanoparticles using black cumin seed and their potential usage as antimicrobial and anticancer agent. J. Pharm. Biomed. Anal. **179**, 112961 (2020). https://doi.org/10.1016/J.JPBA.2019.112961
85. Sharma, P., Pant, S., Dave, V., Tak, K., Sadhu, V., Reddy, K.R.: Green synthesis and characterization of copper nanoparticles by Tinospora cardifolia to produce nature-friendly copper nano-coated fabric and their antimicrobial evaluation. J. Microbiol. Methods **160**, 107–116 (2019). https://doi.org/10.1016/J.MIMET.2019.03.007

86. Hudlikar, M., Joglekar, S., Dhaygude, M., Kodam, K.: Green synthesis of TiO2 nanoparticles by using aqueous extract of Jatropha curcas L. latex. Mater. Lett. **75**, 196–199 (2012). https://doi.org/10.1016/J.MATLET.2012.02.018
87. Al-Zaqri, N., Umamakeshvari, K., Mohana, V., Muthuvel, A., Boshaala, A.: Green synthesis of nickel oxide nanoparticles and its photocatalytic degradation and antibacterial activity. J. Mater. Sci.: Mater. Electron. **33**, 11864–11880 (2022). https://doi.org/10.1007/s10854-022-08149-1
88. Abisharani, J.M., Devikala, S., Dinesh Kumar, R., Arthanareeswari, M., Kamaraj, P.: Green synthesis of TiO2 Nanoparticles using Cucurbita pepo seeds extract. Mater. Today Proc. **14**, 302–307 (2019). https://doi.org/10.1016/J.MATPR.2019.04.151
89. Pushpamalini, T., Keerthana, M., Sangavi, R., Nagaraj, A., Kamaraj, P.: Comparative analysis of green synthesis of TiO2 nanoparticles using four different leaf extract. Mater. Today Proc. **40**, S180–S184 (2021). https://doi.org/10.1016/J.MATPR.2020.08.438
90. Aslam, M., Abdullah, A.Z., Rafatullah, M.: Recent development in the green synthesis of titanium dioxide nanoparticles using plant-based biomolecules for environmental and antimicrobial applications. J. Ind. Eng. Chem. **98**, 1–16 (2021). https://doi.org/10.1016/J.JIEC.2021.04.010
91. Weng, X., Jin, X., Lin, J., Naidu, R., Chen, Z.: Removal of mixed contaminants Cr (VI) and Cu (II) by green synthesized iron based nanoparticles. Ecol. Eng. **97**, 32–39 (2016)
92. Jana, S., Maji, N., Nayak, A.K., Sen, K.K., Basu, S.K.: Development of chitosan-based nanoparticles through inter-polymeric complexation for oral drug delivery. Carbohydr. Polym. **98**, 870–876 (2013)
93. Machado, S., Pacheco, J.G., Nouws, H.P.A., Albergaria, J.T., Delerue-Matos, C.: Characterization of green zero-valent iron nanoparticles produced with tree leaf extracts. Sci. Total Environ. **533**, 76–81 (2015). https://doi.org/10.1016/J.SCITOTENV.2015.06.091
94. Vakili, S., Shabaninejad, Z., Ameri, M., Ebrahiminezhad, A., Shojazadeh, A., Safarpour, H., Khalaf, N., Noorozi, S.: Green synthesis, characterization and antibacterial activity of silver nanoparticles by Biarum chaduchrum leaf extract. Appl. Phys. A **128**, 241 (2022). https://doi.org/10.1007/s00339-022-05384-5
95. Kuang, Y., Wang, Q., Chen, Z., Megharaj, M., Naidu, R.: Heterogeneous fenton-like oxidation of monochlorobenzene using green synthesis of iron nanoparticles. J. Colloid Interface Sci. **410**, 67–73 (2013)
96. Shahwan, T., Sirriah, S.A., Nairat, M., Boyacı, E., Eroğlu, A.E., Scott, T.B., Hallam, K.R.: Green synthesis of iron nanoparticles and their application as a Fenton-like catalyst for the degradation of aqueous cationic and anionic dyes. Chem. Eng. J. **172**, 258–266 (2011)
97. Machado, S., Pinto, S.L., Grosso, J.P., Nouws, H.P.A., Albergaria, J.T., Delerue-Matos, C.: Green production of zero-valent iron nanoparticles using tree leaf extracts. Sci. Total Environ. **445**, 1–8 (2013)
98. Huang, K., Li, Z., Lin, J., Han, G., Huang, P.: Two-dimensional transition metal carbides and nitrides (MXenes) for biomedical applications. Chem. Soc. Rev. **47**, 5109 (2018)
99. Naseem, T., Farrukh, M.A.: Antibacterial activity of green synthesis of iron nanoparticles using Lawsonia inermis and Gardenia jasminoides leaves extract. J. Chem. **2015**, 912342 (2015)
100. Wei, Y., Fang, Z., Zheng, L., Tsang, E.P.: Biosynthesized iron nanoparticles in aqueous extracts of Eichhornia crassipes and its mechanism in the hexavalent chromium removal. Appl. Surf. Sci. **399**, 322–329 (2017)
101. Turakhia, B., Turakhia, P., Shah, S.: Green synthesis of zero valent iron nanoparticles from Spinacia oleracea (spinach) and its application in waste water treatment. J. Adv. Res. Appl. Sci **5**, 46–51 (2018)
102. Jia, B., Gao, L., Sun, J.: Self-assembly of magnetite beads along multiwalled carbon nanotubes via a simple hydrothermal process. Carbon N. Y. **45**, 1476–1481 (2007). https://doi.org/10.1016/j.carbon.2007.03.025
103. Shakoor, A., Ferdous, U.T., Khan, S.A., Gulzar, M.M.: Green synthesis of gold nanoparticles using clerodendrum trichotomum thunberg for antibacterial and anticancer applications. Int. J. Nanomedicine **20**, 2645–2658 (2025). https://doi.org/10.2147/IJN.S503254

104. Mekuye, B., Abera, B.: Nanomaterials: an overview of synthesis, classification, characterization, and applications. Nano Select **4**, 486–501 (2023). https://doi.org/10.1002/nano.202300038
105. Ovais, M., Raza, A., Naz, S., Islam, N.U., Khalil, A.T., Ali, S., Khan, M.A., Shinwari, Z.K.: Current state and prospects of the phytosynthesized colloidal gold nanoparticles and their applications in cancer theranostics. Appl. Microbiol. Biotechnol. **101**, 3551–3565 (2017). https://doi.org/10.1007/s00253-017-8250-4
106. Yassin, M.T., Al-Otibi, F.O., Al-Sahli, S.A., El-Wetidy, M.S., Mohamed, S.: Metal oxide nanoparticles as efficient nanocarriers for targeted cancer therapy: addressing chemotherapy-induced disabilities. Cancers (Basel). **16**, 4234 (2024)
107. Singh, A., Kumar, P., Jain, K., Singh, V.N., Maurya, K.K., Pant, R.P.: Insight on efficient biocompatible magnetic nanoparticles for cancer treatment using magnetic hyperthermia technique. In: Material Aspects of Ferrofluids, CRC Press, pp. 168–183 (2023)
108. Singh, A., Kumar, P., Rajput, P., Jain, K., Pathak, S., Puri, S., Maurya, K.K., Pant, R.P.: Investigating the impact of zinc ion substitution on the rheological properties and hyperthermia potential of cobalt ferrite-based ferrofluids. Mater. Today Proc. (2023)
109. Hsu, C.Y., Rheima, A.M., Kadhim, M.M., Ahmed, N.N., Mohammed, S.H., Abbas, F.H., Abed, Z.T., Mahdi, Z.M., Abbas, Z.S., Hachim, S.K., Ali, F.K., Mahmoud, Z.H., Kianfar, E.: An overview of nanoparticles in drug delivery: properties and applications. S Afr. J Chem. Eng. **46**, 233–270 (2023). https://doi.org/10.1016/j.sajce.2023.08.009
110. Shah, A., Aftab, S., Nisar, J., Ashiq, M.N., Iftikhar, F.J.: Nanocarriers for targeted drug delivery. J. Drug Deliv. Sci. Technol. **62** (2021). https://doi.org/10.1016/j.jddst.2021.102426
111. Kaningini, A.G., Motlhalamme, T., More, G.K., Mohale, K.C., Maaza, M.: Antimicrobial, antioxidant, and cytotoxic properties of biosynthesized copper oxide nanoparticles (CuO-NPs) using Athrixia phylicoides DC. Heliyon **9**, e15265 (2023). https://doi.org/10.1016/J.HELIYON.2023.E15265
112. Kumar, S., Basumatary, I.B., Sudhani, H.P.K., Bajpai, V.K., Chen, L., Shukla, S., Mukherjee, A.: Plant extract mediated silver nanoparticles and their applications as antimicrobials and in sustainable food packaging: a state-of-the-art review. Trends Food Sci. Technol. **112**, 651–666 (2021). https://doi.org/10.1016/J.TIFS.2021.04.031
113. Ermini, M.L., Voliani, V.: Antimicrobial nano-agents: the copper age. ACS Nano **15**, 6008–6029 (2021). https://doi.org/10.1021/acsnano.0c10756
114. Sánchez-López, E., Gomes, D., Esteruelas, G., Bonilla, L., Lopez-Machado, A.L., Galindo, R., Cano, A., Espina, M., Ettcheto, M., Camins, A.: Metal-based nanoparticles as antimicrobial agents: an overview. Nanomaterials **10**, 292 (2020)
115. Badmus, J.A., Oyemomi, S.A., Adedosu, O.T., Yekeen, T.A., Azeez, M.A., Adebayo, E.A., Lateef, A., Badeggi, U.M., Botha, S., Hussein, A.A., Marnewick, J.L.: Photo-assisted bio-fabrication of silver nanoparticles using Annona muricata leaf extract: exploring the antioxidant, anti-diabetic, antimicrobial, and cytotoxic activities. Heliyon **6**, e05413 (2020). https://doi.org/10.1016/J.HELIYON.2020.E05413
116. Khan, S.A., Shahid, S., Ayaz, A., Alkahtani, J., Elshikh, M.S., Riaz, T.: Phytomolecules-coated NiO nanoparticles synthesis using abutilon indicum leaf extract: antioxidant, antibacterial, and anticancer activities. Int. J. Nanomedicine **16**, 1757–1773 (2021). https://doi.org/10.2147/IJN.S294012
117. Kumavat, S.R., Mishra, S.: Green synthesis of silver nanoparticles using Borago officinalis leaves extract and screening its antimicrobial and antifungal activity. Int. Nano Lett. **11**, 355–370 (2021). https://doi.org/10.1007/s40089-021-00345-x
118. Kumavat, S.R., Mishra, S.: Green synthesis of silver nanoparticles using Borago officinalis leaves extract and screening its antimicrobial and antifungal activity. Int. Nano Lett. **11**, 355–370 (2021)
119. Narayanan, D., Geena, M.G., Lakshmi, H., Koyakutty, M., Nair, S., Menon, D.: Poly-(ethylene glycol) modified gelatin nanoparticles for sustained delivery of the anti-inflammatory drug Ibuprofen-Sodium: an in vitro and in vivo analysis. Nanomedicine **9**, 818–828 (2013)

120. Kurutas, E.B.: The importance of antioxidants which play the role in cellular response against oxidative/nitrosative stress: current state. Nutr. J. **15**, 71 (2016). https://doi.org/10.1186/s12937-016-0186-5
121. Abdi, V., Sourinejad, I., Yousefzadi, M., Ghasemi, Z.: Biosynthesis of silver nanoparticles from the mangrove Rhizophora Mucronata: its characterization and antibacterial potential. Iran. J. Sci. Technol., Trans. A: Sci. **43**, 2163–2171 (2019). https://doi.org/10.1007/s40995-019-00739-9

Latest Advances in the Synthesis of Metal and Metal Oxide Nanoparticles

Chesta Mehta, Rakesh Sharma, Hiten Raval, and Garima Shekhawat

1 Introduction

The last decade witnessed a surge in nanotechnology-based research, with a particular focus on practical applications of nanomaterials. Nanomaterials are defined by their structures, wherein their size ranges between 1–100 nm. These materials show unique properties, such as improved catalytic activity, better thermal conductivity, and a large ratio of surface area to volume [1–4]. The sustainable synthesis of various metal and metal oxide nanoparticles has drawn the attention of the global scientific community to nanoscience with a focus on diverse applications, including medicine [5, 6], cosmetics [7, 8], environmental remediation [9, 10], molecular sensing [11] and catalysis [12–14].

Top-down and bottom-up approaches are two widely used methods of nanoparticle formation. Atoms and molecules are utilized as the building blocks in the bottom-up method to create intricate nanostructures [15–18]. In order to produce desired nanostructures with various properties, the top-down methods depend on the shrinking of bulk particles. The three layers that contribute to a nanoparticle are the core, shell, and surface layers. Typically, a range of molecules, including metal ions, surfactants, and polymers, make up the surface layer [19, 20].

C. Mehta (✉)
Department of Chemistry, Mohanlal Sukhadia University, Udaipur, Rajasthan, India
e-mail: chestakmehta@gmail.com

R. Sharma
Department of Applied Chemistry, The Maharaja Sayajirao University of Baroda, Vadodara, Gujarat, India

H. Raval
Department of Chemistry, Government Arts and Science College, Panchmahal, Gujarat, India

G. Shekhawat
Department of Chemistry, Government Meera Girls College, Mohanlal Sukhadia University, Udaipur, Rajasthan, India

G. Tailor et al. (eds.), *Nanomaterials*, Engineering Materials,
https://doi.org/10.1007/978-981-92-0678-0_9

A variety of metal nanoparticles, including Ag, Au, Fe, Pd, Pt, Cu, and metal oxide nanoparticles, including ZnO, TiO_2, CuO, NiO, FeO, have been apparently synthesized utilizing different physical, chemical and biological methods. Due to low production of nanoparticles in the physical methods and use of flammable organic solvents in the chemical approach, the researchers are now concentrating on more environmentally friendly approaches, such as plant-based biosynthesis [21], bacteria [22], fungus [23], marine algae [24], viruses [25], and yeast [26]. Each of these can function as a natural biofactory up to a certain extent to produce specific nanoparticles. Nanoparticles can be spherical [27], triangular [28], hexagonal [29], cubic [30, 31], ellipsoidal [32], and more. They are categorized according to their general shape as either zero, one, two, or three dimensions. Active molecules and chemicals found in all biological entities can function as stabilizing and reducing agents to create nanoparticles with a variety of sizes, shapes, compositions, and physicochemical characteristics [33–36]. The green synthesis of the NPs is still at the laboratory level, and large-scale production remains challenging. Extensive research has been conducted to optimize reaction conditions to achieve precise control over the size and morphology of MNPs and MONPs. Hence, this chapter can be used as a guide for researchers to rationally design NPs for various applications using the latest sustainable alternatives. It can also initiate the research community to focus on the unresolved reaction mechanisms of biosynthesis.

2 Routes for Nanoparticle Synthesis

Nanoparticles can be produced using three distinct methods, namely, physical, chemical, and biological. In physical synthesis, bulk particles or atom beams are broken by mechanical forces to prepare nanoparticles [37]. To synthesize nanoparticles, chemical synthesis entails creating ions via a sequence of chemical processes. Common chemical techniques for creating nanoparticles include chemical vapor deposition (CVD), liquid phase, sol–gel, colloidal, and electrodeposition [38] (as shown in Fig. 1).

Due to low NP yield, higher energy requirements, and higher input costs, physical methods are not as good for producing nanomaterials on a wide scale. Chemical methods of NPs synthesis have become the preferred approach in recent years as they utilize less energy to reduce metals and produce homogenous NPs with great accuracy. Consequently, these traditional methods for synthesizing NPs are risky, time-consuming, and involve the use of hazardous compounds that are powerful and harmful environmental contaminants.

The process of synthesizing various metal nanoparticles using bioactive agents, such as plant materials, microbes, and various bio-waste, including vegetable waste, fruit peel waste, eggshell, agricultural waste, and algae, is known as "green" or "biological" nanoparticle synthesis 38. Since algae are autotrophic in nature, they can easily flourish on minimal media additives. Algal cells contain a range of secondary metabolites and physiologically active substances that act as capping agents during

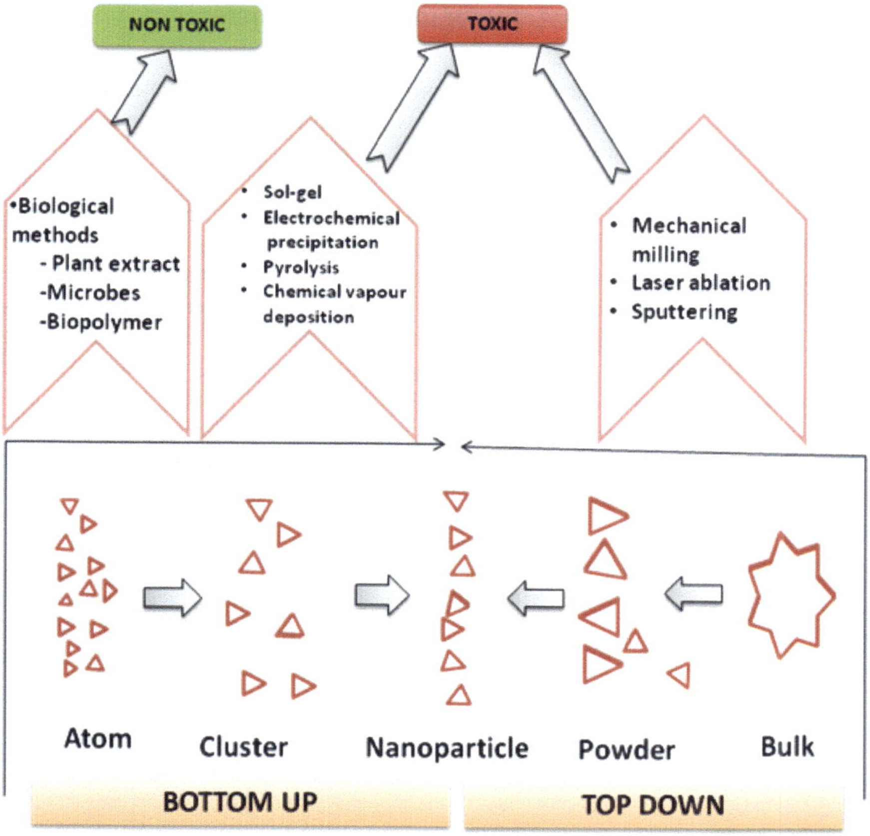

Fig. 1 Several routes for nanoparticle formation

NP formation, making them a unique "nanofactory" for the synthesis of distinct NPs [39, 40]. The bio-fabrication of NPs using fungi is another biological method of producing NPs. Due to their increased ability to tolerate metal concentrations and their high cell wall binding potential with metal ions, fungi can produce more NPs than bacterial cells. It is more economical and efficient to use fungi to produce NPs as they are more likely to accumulate metals than bacteria. Additionally, handling biomass and processing NPs downstream is made simpler by the fungus-based biosynthesis of NPs [41–43]. It is necessary to develop dependable, sustainable green synthesis processes to prevent the production of undesirable or dangerous byproducts. The green synthesis of nanoparticles also has several advantages, including low cost, high stability, quick turnaround time, non-toxic byproducts, and ease of scaling up for large-scale synthesis [44].

3 Biogenic Metal and Metal Oxide Nanoparticles

Metal nanoparticles are created solely from metal precursors. These nanoparticles exhibit unique opto-electrical characteristics, a consequence of their established localized surface plasmon resonance (LSPR) properties [45]. Notably, nanoparticles composed of alkali and noble metals, including Cu, Ag, and Au, possess a broad absorption band within the visible spectrum of solar electromagnetic radiation. The controlled synthesis of metal nanoparticles, with precise control over their surface characteristics, dimensions, and morphology, is fundamental to contemporary advanced materials. The initial stage in the solution-phase formation of nanoparticles is generally considered to be the reduction of metal ions (Fig. 2) from their mono- or divalent oxidation states to their zero-valent forms. This is followed by the reduction of metal ions and subsequent nucleation. Following the initial nucleation event, a kinetically controlled process leads to the growth of larger, thermodynamically stable nanoparticles, achieved through the attachment of smaller, surrounding particles to the low-energy facets of the evolving crystal [46–48].

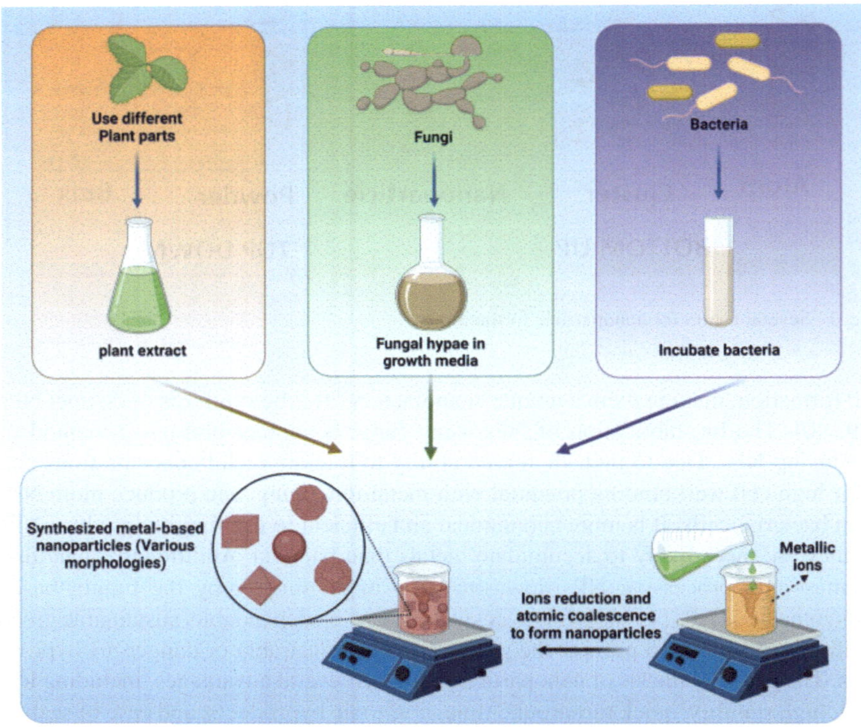

Fig. 2 Biogenic synthesis of metal nanoparticles [49]

The quinones and plastohydroquinone molecules found in plant leaf extract may aid in the metal ion reduction process. This suggests that biomolecules and heterocyclic compounds in plants are capable of carrying out the extracellular MNP synthesis. The phytochemicals of the plants produce nanoparticles, even though the full picture of MONP synthesis using plants is still unclear. Initially, the phytochemicals mediate the reduction of metal ions. The reduced metal ions are bound by oxygen generated by the environment or by decomposing phytochemicals [50]. Metal oxide ions will be drawn to one another by electrostatic attraction, which generates nanoparticles. Phytochemicals stabilize them and keep them from aggregating together (Fig. 3).

Capping agents, such as Polyethylene glycol (PEG) [51], Polyvinylpyrrolidone (PVP) [52], Polyvinyl alcohol (PVA) [53], Bovine serum albumin (BSA) [54], Ethylene diamine tetra acetic acid (EDTA) [55], and Chitosan [56] are sometimes used to stabilize the nanoparticle surface and enhance stability and water solubility, which are crucial to preventing aggregation.

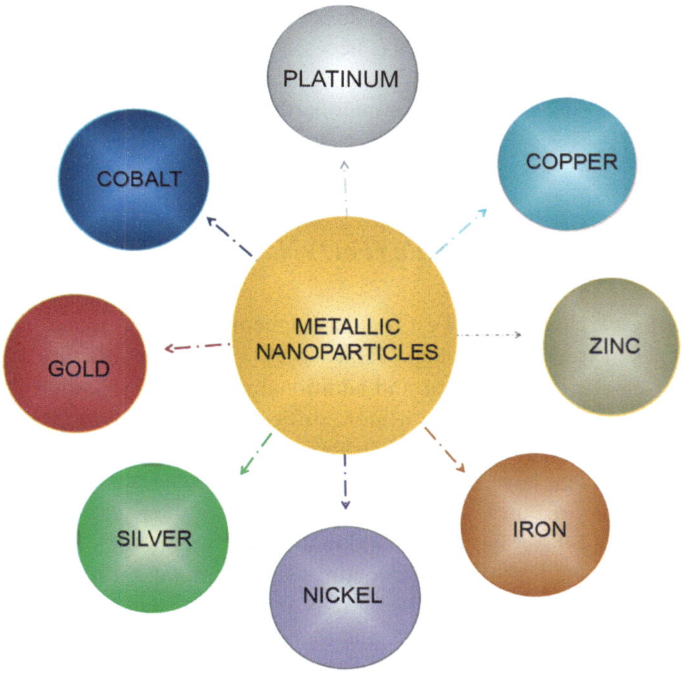

Fig. 3 Different metallic nanoparticles

3.1 Silver (Ag) and Silver Oxide (Ag_2O/AgO) Nanoparticles

Silver nanoparticles have garnered a lot of scientific interest due to significant uses in surface-enhanced Raman scattering, catalysis, and antibacterial properties. Silver has been utilized as an antibacterial for generations. Ionic silver, which is released when silver nanoparticles (Table 1) break down quickly, interacts with key thiol groups to inactivate crucial bacterial enzymes. Silver ions have the ability to harm bacterial cytoplasmic membranes, prevent bacterial DNA replication, reduce intracellular adenosine triphosphate (ATP) levels, and ultimately result in cell death [57].

3.2 Copper (Cu) and Copper Oxide (Cu_2O/CuO) Nanoparticles

Using innovative green techniques, copper nanoparticles with various structural characteristics and potent biological impacts can be synthesized. It has been suggested that nanotoxicity for copper nanoparticles is necessary to continuously study the effects of different nanoparticle characteristics, such as aggregation, size, shape, structure, surface properties, coated agents, pH, water hardness, and aqueous solubility of NPs [83] (Table 2).

3.3 Iron (Fe) and Iron Oxide (FeO) Nanoparticles

Hematite (α-Fe_2O_3), magnetite (Fe_2O_4), and maghemite (γ-Fe_2O_3) are the three most prevalent types of Iron nanoparticles (Table 3) [99]. The microstructure of Iron NPs is dependent on a metal precursor and additional processes, including nucleation, growth, and adsorption, that make it more difficult to control their size and form. Moreover, magnetic iron oxide nanoparticles are attractive particles with exceptional magnetic qualities that have potential applications in biomedicine [100, 101]. Iron oxide nanoparticles are quickly oxidized by air; however, their surface can be altered to offer greater stability than the unmodified form. Nanoparticle surfaces can be functionalized with various inorganic coatings such as metals, silica, sulfides, metal oxides, and other non-metal materials. Many organic substances, such as dextran [102], starch [103], polyethylenimine, and many more, have been used to coat iron nanoparticles.

Table 1 Silver and silver oxide nanoparticles synthesized using plants, bacteria, fungus and algae

Plants

Ag/AgO	Plant species	Size (nm)	Morphology	Application	References
Ag	*Artemisia princeps*	20	Spherical	Anticancer and anti-angiogenic	[58]
Ag	*Amaranthus retroflexus*	10–32	Spherical	Antifungal	[59]
Ag	*Artocarpus hirsutus*	25–40	Spherical	Antibacterial	[60]
Ag	*Artocarpus integer*	5.76–19.1	Spherical	Antibacterial, antitumor	[61]
Ag	*Berberis vulgaris*	20–30	Spherical	Antibacterial	[62]
Ag	*Calendula officinalis*	5–10	Spherical, cubic	Antioxidant	[63]
Ag	*Cnidoscolus aconitifolius*	2–20	Nanosphere	Agriculture	[64]
Ag_2O	*Cyathea nilgiriensis* Holttum	<100	Hexagonal	Antimicrobial, Antifingal, Anticancer	[65]
	Phlomis	25	Spherical	Antibacterial	[66]
Ag	*Malva parviflora*	50.6	Spherical	Antifungal	[67]
Ag_2O	*Centella asiatica and Tridax*	11–12	Spherical	Photocatalytic activity, anti-bacterial and anti-fungal activities	[68]
Ag	*Eclipta prostrata*	35–60	Spherical	Antimalarial	[69]
Ag	*Jasminum nervosum*	4–22	Spherical	Larvicidal activity	[70]

Bacteria

$Ag/Ag_2O/AgO$	Bacterial strain	Size	Morphology	Applications	References
Ag	*E. coli*	5–25	Spherical, monodispersed	Antimicrobial, antifungal	[71]
AgO	*Lactobacillus mindensis*	2–20	Nearly spherical	–	[72]
Ag	*Brevibacterium casei*	10–50	Spherical	–	[73]
Ag	*Bacillus licheniformis*	18–63	Spherical	Antibiofilm	[74]
Ag	*Pseudomonas aeruginosa*	50–85	Spherical	Antibacterial	[75]
Ag	*Enterococcus faecalis* S7	10–16	Spherical	Antioxidant, Antibacterial	[71]

(continued)

Table 1 (continued)

Fungus					
Ag/ Ag_2O/ AgO	Fungal strain	Size	Morphology	Applications	References
Ag	*Aspergillus terreus, Penicillium expansum*	14, 25	Spherical	Antifungal	[76]
Ag	*Aspergillus flavus*	7.13	Spherical	–	[77]
Ag	*Cladosporium sphaerospermum*	15.1 ± 1.0	Spherical	–	[78]
Ag	*Metarhizium anisopliae*	9.47	Spherical	Water treatment	[79]
Ag	*Hormoconis resinae*	20–80	Spherical-triangular	–	[80]
Algae					
Ag/Ag_2O/AgO	Algal strain	Size (nm)	Morphology	Applications	References
Ag	*Acutodesmus dimorphus*	2–20	Spherical	Antioxidant	[81]
Ag	*Caulerpa serrulata*	10 ± 2	Spherical	Catalytic activity, antibacterial activity	[82]

3.4 Zinc (Zn) and Zinc Oxide (ZnO) Nanoparticles

Recently, the biosynthesis of Zinc nanoparticles (Table 4) has gained attention as a substitute for chemically and physically produced ZnO NPs without compromising their characteristics. This process incorporates a variety of bioactive compounds from microorganisms to plants as reductants and stabilizers [49]. A variety of zinc precursors, such as zinc acetate, zinc chloride, zinc nitrate, zinc sulphate, etc., are used for the biogenic ZnO NP synthesis. During the synthesis, these salts are usually supplied as soluble salts that precipitate in the solution or its extract with bioactive components [117].

3.5 Nickel (Ni) and Nickel Oxide (NiO) Nanoparticles

In recent years, green production of nickel nanoparticles (Table 5) has been reported for several applications. Phase stability, high magnetic permeability, high electrical conductivity, low eddy current loss, and a bandgap of 3.6 eV are characteristics of nickel oxide (NiO), a form of transition metal oxide that is inexpensive to produce and ideal for sensing and photocatalysis. Nanocrystalline size and surface effect of

Table 2 Copper and copper oxide nanoparticles synthesized using plants, bacteria, fungus and algae

Plants					
$Cu/Cu_2O/CuO$	Plant species	Size (nm)	Morphology	Applications	References
Cu	*Alstonia scholaris*	–	Irregular	Antibacterial	[84]
CuO	*Portulaca oleracea*	5–30	Spherical	Wastewater treatment	[85]
Cu	*Bauhinia variegate*	8.721	Irregular	Antiobesity Antidiabetic	[86]
Cu	*Morus alba*	50	Spherical	Antimicrobial, food stimulant	[87]
CuO	*Azadirachta indica*	28–35	Spherical	Antibacterial	[88]
Cu	*Olea europaea*	20–50	Spherical	Anticancer agent	[89]
CuO	*Vitex negundo*	100	spherical	Antioxidant activity	[90]
Cu_2O	*Arachis hypogea*	20–40	spherical and hemispherical	Antibacterial	[91]
Bacteria					
$Cu/Cu_2O/CuO$	Bacterial strain	Size (nm)	Morphology	Applications	References
CuO	*Stenotrophomonas* sp. BS95	21.3	Roughly spherical	Antioxidant	[92]
Cu	*Pseudomonas stutzeri*	8–15	Spherical	–	[93]
Cu_2O	*Lactobacillus* sp.	10–20	Spherical	–	[94]
Fungus					
$Cu/Cu_2O/CuO$	Fungal strain	Size (nm)	Morphology	Applications	References
Cu	*Aspergillus flavus*	5–12	Spherical	Antifungal	[95]
Cu	*Aspergillus niger*	50	Spherical	Anticancer, antidiabetic	[96]
Algae					
$Cu/Cu_2O/CuO$	Algal strain	Size (nm)	Morphology	Applications	References
Cu	*Sargassum cervicorne*	2.6–3	Spherical	Dye degradation	[97]
CuO	*Anabaena* sp.	3.6	Rod shaped	Water disinfection	[98]

Table 3 Iron and iron oxide nanoparticles synthesized using plants, bacteria, fungus and algae

Plants					
$Fe/Fe_2O_3/Fe_3O_4$	Plant species	Size (nm)	Morphology	Applications	References
Fe	*Ficus carica*	<20 nm	Spherical	–	[104]
FeO	*Avicennia marina*	10–40	Honeycomb	–	[105]
Fe	*Dodonaea viscosa*	50–60	Spherical	Antibacterial	[106]
Fe	*Gardenia jasminoides*	32	Shattered rock-like	Antibacterial	[107]
Fe_3O_4	*Tridax procumbens*	80–100	Irregular spheres	Antibacterial	[108]
Bacteria					
$Fe/Fe_2O_3/FeO$	Bacterial strain	Size (nm)	Morphology	Applications	References
Fe_3O_4	*Bacillus* sp.	12–32	Cubic	–	[109]
Fungus					
$Fe/Fe_2O_3/FeO$	Fungal strain	Size	Morphology	Applications	References
Fe	*Aspergillus oryzae*	10 and 24.6	Spherical	Agriculture	[110]
Fe_2O_3	*Aspergillus japonicus*	60–70	Cubic	–	[111]
Algae					
$Fe/Fe_2O_3/$ Fe3O4	Algal strain	Size	Morphology	Applications	References
Fe_3O_4	*Hypnea valentiae*	30–50	Spherical	Antioxidant, antitumour	[112]
Fe	*Ulva lactuca*	30–40	Nanosphere	Anticancer, antibacterial	[113]
Fe	Galaxaura rugosa	14.4 to 17.2 nm	Spherical	–	[114]
Yeast					
Fe	*Candida bombicola*	4.5–8.5	–	–	[115]
FeO	*Cryptococcus humicola*	8–9	Spherical	–	[116]

magnetic NiO are extremely fascinating, and its technical uses and chemical stability have increased interest in it. Using *Moringa oleifera* plant extract, NiO NPs can have the best antibacterial qualities against bacterial pathogens and effective cytotoxic action against HT-29 cancer cells [132].

Table 4 Zinc and zinc oxide nanoparticles synthesized using plants, bacteria, fungus and algae

Plants

Zn/ZnO	Plant species	Size	Morphology	Applications	References
ZnO	*Punica granatum*	25 nm	–	Anticancer	[118]
ZnO	*Psidium guajava*	–	Spherical	Antibacterial	[119]
ZnO	*Morus nigra*	–	Spherical	Anticancer	[120]
ZnO	*Limoniastrum monopetalum*	25.3	Nano-rods	Biomedical	[121]
ZnO	*Malus domestica*	12 nm	Hexagonal floral pattern	Antifungal agents in food industry	[117]

Bacteria

Zn/ZnO	Bacterial strain	Size	Morphology	Applications	References
ZnO	*Streptomyces* sp.	20–50	Spherical	Anticancer	[122]
ZnO	S. *nematodiphlia*	15–30	Roughly spherical	Photocatalytic	[123]
ZnO	*Paraclostridium benzoelyticum*	50 nm	Hexagonal	Anti-inflammatory, Antidiabetic Assay	[124]

Fungus

Zn/ZnO	Fungal strain	Size	Morphology	Applications	References
ZnO	*Apergillus niger*	53–69	Spherical	Photocatalytic	[125]
ZnO	*Fusarium keratoplasticum*	10–42	Hexagonal	Antibacterial and anticancer	[126]

Algae

Zn/ZnO	Algal strain	Size	Morphology	Applications	References
ZnO	*Ulva lactuca*	10–50	Asymmetrical	Photocatalytic, antibiofilm and insecticidal activity	[127]
ZnO	*Gracilaria edulis*	65–95	Rod	Anticancer	[128]
ZnO	*Sargassum muticum*	3–57	Hexagonal	Biomedical	[129]

Yeast

Zn/ZnO	Strain	Size	Morphology	Applications	References
ZnO	*Marine yeast*	86.2	Rounded	Antioxidant	[130]
ZnO	*Pichia fermentans*	–	Smooth, elongated	Antimicrobial	[131]

3.6 Platinum

Platinum nanoparticles are extensively studied because of their antimicrobial, antioxidant, and anticancer properties (Table 6).

Table 5 Nickel and nickel oxide nanoparticles synthesized through plants and fungus

Plants

Ni/NiO	Plant species	Size (nm)	Morphology	Applications	References
NiO	*Salvia officinal*	35–50	Nanospheres nanorods	Defluoridation	[133]
NiO	*Musa paradisiaca*	15.26	Cubic-shaped	Antibacterial	[134]
NiO	*Aegle marme-los*	8.5	Agglomerated spherical	In vitro cytotoxicity, photocatalyic	[135]
NiO	*Monsonia burkeana*	20 nm	Spherical	–	[136]
NiO	*Punica granatum*	28.6 nm	Nearly spherical	Dye degradation	[137]
NiO	*Sesbania grandiflora*	19.9–60.4 nm	Rod shape	Cytotoxicity	[138]

Fungus

Ni/NiO	Fungal strain	Size	Morphology	Applications	References
NiO	*Pleurotus fossulatus*	<50	Spherical	Water remediation	[139]
Ni	*Aspergillus terreus*	60	–	Anticancer	[140]

3.7 Gold (Au) Nanoparticles

Colloidal gold nanoparticles have been present since ancient civilizations. Gold nanoparticles have traditionally been utilized mostly in the works of artists and crafters due to their vibrant visible colours. But as research proceeds with the interventions and technology, size, shape, surface chemistry, and optical characteristics of gold nanoparticles are all controllable, which has facilitated some incredibly interesting and novel possibilities (Table 7).

3.8 Cobalt Nanoparticles

Promising catalytic, magnetic, electrical, and chemical capabilities have been demonstrated by cobalt nanoparticles (CoNPs). Presently, the most significant and highly effective NPs in catalysis are CoNPs. Cobalt powder, its oxides, salts, and complexes (Table 8) have been used as homogeneous or heterogeneous catalysts in basic chemical reactions and the petrochemical sector. Typically, magnetic sand polymers are applied to Co NPs to improve biocompatibility, offer functionality, and stop agglomeration [164].

Table 6 Platinum nanoparticles synthesized using plants, bacteria, fungus and algae

Plants

Pt	Plant species	Size	Morphology	Applications	References
Pt	*Fumariae herba*	10–30 nm	Hexagonal and pentagonal	Catalysis	[141]
Pt	*Azadirachta indica*	5–50	Small and large spheres	–	[142]
Pt	*Ocimun sanctum*	23	Irregular	Water electrolysis	[143]
Pt	*Camellia sinensis*	30–60	Flower	Antibacterial ceramics	[144]

Bacteria

Pt	Bacterial strain	Size	Morphology	Applications	References
Pt	*Desulfovibrio desulfuricans*	58	Spherical	Removal of pharmaceutical products	[145]
Pt	*Escherichia coli*	2.3	Clusters and spherical	Electroxidation	[146]

Fungus

Pt	Fungal strain	Size	Morphology	Applications	References
Pt	*Fusarium oxysporum*	15–30	Spherical	–	[147]
Pt	*Neurospora crassa*	**20–90**	Quasi-spherical	–	[148]
Pt	*Fusarium oxysporum*	100–140	Triangular and spherical	–	[149]

Algae

Pt	Algal strain	Size	Morphology	Applications	References
Pt	*Halymenia dilatata*	15 ± 1.7	Spherical	Biomedical	[150]
Pt	*Padina gymnospora*	5–50	Truncated octahedral	Antimicrobial, antioxidant, hemolytic	[151]
Pt	*Caulerpa sertularioides*	6–22	Spherical	Antioxidant	[152]
Pt	*Sargassum polyphyllum*	4.7	Pleomorphic spheroidal	Anticancer	[153]

4 Emerging Interventions for Controlled NP Release and Shape-Size Control

Factors such as temperature, pH, and reaction time and concentration serve a key role in controlling the shape, size, and yield of the synthesized nanoparticles (as shown in Fig. 4). One of the most important variables is temperature since it affects size, synthesis, and shape (Oval, flakes, prismatic, triangular, spherical, octahedral, etc.). Higher yields are the result of increased nucleation center formation and reaction rate as the temperature rises. Nanoparticle size increases with an increase in the

Table 7 Gold nanoparticles synthesized through plants, bacteria, fungus and algae

Plants

Au	Plant species	Size (nm)	Morphology	Applications	References
Au	*Maclura tricuspidata*	19.6–29.7	Spherical/ hexagonal	Anti-metastatic, photocatalytic activity	[154]
Au	*Deinococcus radiodurans*	49.8	Spherical	Dye degradation	[155]
Bimetallic (Au–Ag)	*Azadirachta indica*	50–100	Spherical	–	[156]

Bacteria

Au	Bacterial strain	Size	Morphology	Applications	Reference
Au	*Shewanellao neidensis*	2–50	Spherical	–	[157]

Fungus

Au	Fungal strain	Size	Morphology	Applications	References
Au	*Penicillium chrysogenum*	5–100	Spherical, triangle and rod	–	[158]
Au	*Aspergillus terreus*	10–50	Anisotropic morphology	Antibacterial	[159]
Au	*Endophytic fungi*	22.1	Spherical with slight aggregation	Anticancer	[160]
Au	*Alternaria alternata*	2–30	Spherical and triangular	–	[161]

Algae

Au	Algal strain	Size	Morphology	Applications	References
Au	*Streptomyces griseus isolate*	19–28	Hexagonal	Antimicrobial, anticancer	[162]
Au	*Pithophora oedogonia*	32	Spherical	Electrochemical sensor	[163]

temperature because different temperatures encourage different interactions between the reactants, resulting in diverse morphologies [173, 174].

The nucleation centers are influenced by pH, with higher pH values producing more centers. Another important consequence of pH is that some nanoparticles can only be made in acidic or alkaline media. For example, metal oxide nanoparticles are often manufactured at an acidic or neutral pH, while magnetic nanoparticles are made at an alkaline pH.

Time is crucial in determining the nanoparticle size [175]. The physicochemical properties of metallic nanoparticles, including environmental changes like aggregation, dissolution, or agglomeration, are linked to their toxicity to aquatic life. The findings revealed that polymeric substances, including poly(ethylene glycol), poly(vinyl alcohol), poly(vinylpyrrolidone), poly(methacrylic acid), and polymethylmethacrylate, are effective protective agents for stabilizing nanoparticles. As longer reaction

Table 8 Cobalt and cobalt oxide nanoparticles synthesized using plants, bacteria, fungus and algae

Plants

Co/Co_3O_4	Plant species	Size (nm)	Morphology	Applications	References
Co	*Vitis rotundifolia*	36	Spherical	Photocatalytic activity	[165]
Co_3O_4	*Taraxacum officinale*	50	Spherical	Dye degradation	[166]
CoO	*Calotropis gigantean*	50–80	Spherical	Solar cells	[167]
$CoFe_2O_4$	*Okra (A. esculentus)*	45–55	Spherical	Antibacterial	[168]

Bacteria

Co/Co_3O_4	Bacterial strain	Size (nm)	Morphology	Applications	Reference
Co	*Bacillus thuringiensis*	85..3	Cubic	Antimalarial	[169]

Fungus

Co/Co_3O_4	Fungal strain	Size (nm)	Morphology	Applications	References
Co_3O_4	*Aspergillus brasiliensis*	20–26	Spherical	Antimicrobial	[170]
Co_3O_4	*Aspergillus nidulans*	20.29	Spinel	–	[171]

Algae

Co/Co_3O_4	Algal strain	Size (nm)	Morphology	Applications	Reference
Co_3O_4	Red algae	28.2 ± 7.8	Spherical	Antioxidant, antimicrobial, anticancer, and anticoagulant activity	[172]

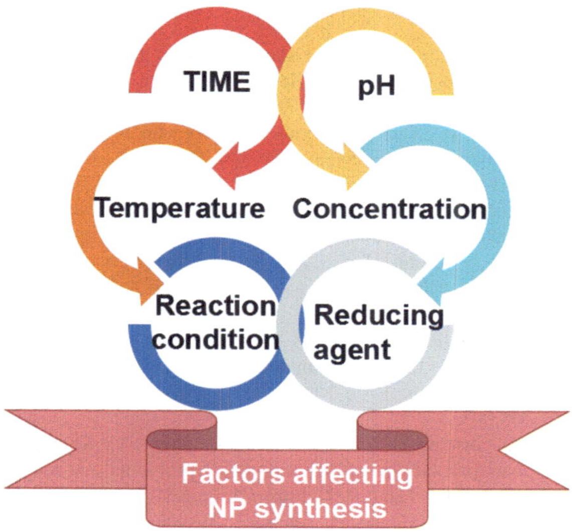

Fig. 4 Different factors affecting nanoparticle formation

times allow reactants to interact for longer periods of time, It has been discovered that longer reaction times favour larger nanoparticles and higher yields [176].

5 Conclusion

In the past few years, a series of studies have been reported in the literature about the synthesis of metal and metal oxide nanoparticles using green synthesis approach based on natural plant extracts. Based on the literature, it is evident that using plant extracts to create environmentally friendly nanoparticles is very successful. This is because the extracts have significant antimicrobial efficacy against a variety of gram-positive and gram-negative bacterial strains in addition to photocatalytic activity and electrochemical sensing. Green synthesis of metallic nanoparticles is limited to the lab scale only; it is essential to improve the synthesis approach for high yield and larger production. Further research has to be focused on the combination of different plant extracts with the quantification of their components for the synthesis of engineered nanoparticles with better characteristic features. Finding reducing, capping, and stabilizing agents and measuring their amounts in specific plant sections are essential to the large-scale production of environmentally beneficial nanoparticles.

References

1. Miyazaki, K., Islam, N.: Nanotechnology systems of innovation—an analysis of industry and academia research activities. Technovation **27**(11), 661–675 (2007). https://doi.org/10.1016/j.technovation.2007.05.009
2. Roco, M.C.: The long view of nanotechnology development: the national nanotechnology initiative at 10 years. J. Nanopart. Res. **13**(2), 427–445 (2011). https://doi.org/10.1007/s11051-010-0192-z
3. Roco, M.C.: National nanotechnology initiative at 20 years: enabling new horizons. J. Nanopart. Res. **25**(10), 197 (2023). https://doi.org/10.1007/s11051-023-05829-9
4. Zhu, S., Meng, H., Gu, Z., Zhao, Y.: Research trend of nanoscience and nanotechnology—a bibliometric analysis of nano today. Nano Today **39**, 101233 (2021). https://doi.org/10.1016/j.nantod.2021.101233
5. Hofmann-Amtenbrink, M., Grainger, D.W., Hofmann, H.: Nanoparticles in medicine: current challenges facing inorganic nanoparticle toxicity assessments and standardizations. Nanomed. Nanotechnol. Biol. Med. **11**(7), 1689–1694 (2015). https://doi.org/10.1016/j.nano.2015.05.005
6. Koksharov, Yu.A., Gubin, S.P., Taranov, I.V., Khomutov, G.B., Gulyaev, Yu.V.: Magnetic nanoparticles in medicine: progress, problems, and advances. J. Commun. Technol. Electron. **67**(2), 101–116 (2022). https://doi.org/10.1134/S1064226922020073
7. Kokura, S., Handa, O., Takagi, T., Ishikawa, T., Naito, Y., Yoshikawa, T.: Silver nanoparticles as a safe preservative for use in cosmetics. Nanomed. Nanotechnol. Biol. Med. **6**(4), 570–574 (2010). https://doi.org/10.1016/j.nano.2009.12.002
8. Lu, P.-J., Huang, S.-C., Chen, Y.-P., Chiueh, L.-C., Shih, D.Y.-C.: Analysis of titanium dioxide and zinc oxide nanoparticles in cosmetics. J. Food Drug Anal. **23**(3), 587–594 (2015). https://doi.org/10.1016/j.jfda.2015.02.009

9. Lin, J.-H., Wu, Z.-H., Tseng, W.-L.: Extraction of environmental pollutants using magnetic nanomaterials. Anal. Methods **2**(12), 1874 (2010). https://doi.org/10.1039/c0ay00575d
10. Song, D., Yang, R., Long, F., Zhu, A.: Applications of magnetic nanoparticles in surface-enhanced Raman scattering (SERS) detection of environmental pollutants. J. Environ. Sci. **80**, 14–34 (2019). https://doi.org/10.1016/j.jes.2018.07.004
11. Wang, J., Munir, A., Zhu, Z., Zhou, H.S.: Magnetic nanoparticle enhanced surface Plasmon resonance sensing and its application for the ultrasensitive detection of magnetic nanoparticle-enriched small molecules. Anal. Chem. **82**(16), 6782–6789 (2010). https://doi.org/10.1021/ac100812c
12. Campelo, J.M., Luna, D., Luque, R., Marinas, J.M., Romero, A.A.: Sustainable preparation of supported metal nanoparticles and their applications in catalysis. Chemsuschem **2**(1), 18–45 (2009). https://doi.org/10.1002/cssc.200800227
13. Dalpozzo, R.: Magnetic nanoparticle supports for asymmetric catalysts. Green Chem. **17**(7), 3671–3686 (2015). https://doi.org/10.1039/C5GC00386E
14. Johnson, B.F.G.: Nanoparticles in catalysis. Top. Catal. **24**(1–4), 147–159 (2003). https://doi.org/10.1023/B:TOCA.0000003086.83434.b6
15. Abid, N., Khan, A.M., Shujait, S., Chaudhary, K., Ikram, M., Imran, M., Haider, J., Khan, M., Khan, Q., Maqbool, M.: Synthesis of nanomaterials using various top-down and bottom-up approaches, influencing factors, advantages, and disadvantages: a review. Adv. Colloid Interface Sci. **300**, 102597 (2022). https://doi.org/10.1016/j.cis.2021.102597
16. Isaacoff, B.P., Brown, K.A.: Progress in top-down control of bottom-up assembly. Nano Lett. **17**(11), 6508–6510 (2017). https://doi.org/10.1021/acs.nanolett.7b04479
17. Kumar, S., Bhushan, P., Bhattacharya, S.: Fabrication of nanostructures with bottom-up approach and their utility in diagnostics, therapeutics, and others. In: Bhattacharya, S., Agarwal, A. K., Chanda, N., Pandey, A., Sen, A. K. (eds.) Environmental, Chemical and Medical Sensors, pp. 167–198. Springer, Singapore (2018). https://doi.org/10.1007/978-981-10-7751-7_8
18. Thiruvengadathan, R., Korampally, V., Ghosh, A., Chanda, N., Gangopadhyay, K., Gangopadhyay, S.: Nanomaterial processing using self-assembly-bottom-up chemical and biological approaches. Rep. Prog. Phys. **76**(6), 066501 (2013). https://doi.org/10.1088/0034-4885/76/6/066501
19. Gawande, M.B., Goswami, A., Asefa, T., Guo, H., Biradar, A.V., Peng, D.-L., Zboril, R., Varma, R.S.: Core-shell nanoparticles: synthesis and applications in catalysis and electrocatalysis. Chem. Soc. Rev. **44**(21), 7540–7590 (2015). https://doi.org/10.1039/C5CS00343A
20. Ghosh Chaudhuri, R., Paria, S.: Core/shell nanoparticles: classes, properties, synthesis mechanisms, characterization, and applications. Chem. Rev. **112**(4), 2373–2433 (2012). https://doi.org/10.1021/cr100449n
21. Logeswari, P., Silambarasan, S., Abraham, J.: Synthesis of silver nanoparticles using plants extract and analysis of their antimicrobial property. J. Saudi Chem. Soc. **19**(3), 311–317 (2015). https://doi.org/10.1016/j.jscs.2012.04.007
22. Moon, J.-W., Rawn, C.J., Rondinone, A.J., Love, L.J., Roh, Y., Everett, S.M., Lauf, R.J., Phelps, T.J.: Large-scale production of magnetic nanoparticles using bacterial fermentation. J. Ind. Microbiol. Biotechnol. **37**(10), 1023–1031 (2010). https://doi.org/10.1007/s10295-010-0749-y
23. Vigneshwaran, N., Ashtaputre, N.M., Varadarajan, P.V., Nachane, R.P., Paralikar, K.M., Balasubramanya, R.H.: Biological synthesis of silver nanoparticles using the fungus *Aspergillus flavus*. Mater. Lett. **61**(6), 1413–1418 (2007). https://doi.org/10.1016/j.matlet.2006.07.042
24. Singaravelu, G., Arockiamary, J.S., Kumar, V.G., Govindaraju, K.: A novel extracellular synthesis of monodisperse gold nanoparticles using marine alga, *Sargassum wightii* Greville. Colloids Surf. B Biointerfaces **57**(1), 97–101 (2007). https://doi.org/10.1016/j.colsurfb.2007.01.010
25. Dujardin, E., Peet, C., Stubbs, G., Culver, J.N., Mann, S.: Organization of metallic nanoparticles using tobacco mosaic virus templates. Nano Lett. **3**(3), 413–417 (2003). https://doi.org/10.1021/nl034004o

26. Fernández, J.G., Fernández-Baldo, M.A., Berni, E., Camí, G., Durán, N., Raba, J., Sanz, M.I.: Production of silver nanoparticles using yeasts and evaluation of their antifungal activity against Phytopathogenic fungi. Process Biochem. **51**(9), 1306–1313 (2016). https://doi.org/10.1016/j.procbio.2016.05.021
27. Butet, J., Dutta-Gupta, S., Martin, O.J.F.: Surface second-harmonic generation from coupled spherical Plasmonic nanoparticles: Eigenmode analysis and symmetry properties. Phys. Rev. B **89**(24), 245449 (2014). https://doi.org/10.1103/PhysRevB.89.245449
28. Kelly, J.M., Keegan, G., Brennan-Fournet, M.E.: Triangular silver nanoparticles: their preparation, functionalisation and properties. Acta Phys. Pol. A **122**(2), 337–345 (2012). https://doi.org/10.12693/APhysPolA.122.337
29. Umar, A., Akhtar, M.S., Dar, G.N., Baskoutas, S.: Low-temperature synthesis of α-Fe_2O_3 hexagonal nanoparticles for environmental remediation and smart sensor applications. Talanta **116**, 1060–1066 (2013). https://doi.org/10.1016/j.talanta.2013.08.026
30. Salazar-Alvarez, G., Qin, J., Šepelák, V., Bergmann, I., Vasilakaki, M., Trohidou, K.N., Ardisson, J.D., Macedo, W.A.A., Mikhaylova, M., Muhammed, M., Baró, M.D., Nogués, J.: Cubic versus spherical magnetic nanoparticles: the role of surface anisotropy. J. Am. Chem. Soc. **130**(40), 13234–13239 (2008). https://doi.org/10.1021/ja0768744
31. Nafday, D., Sarkar, S., Ayyub, P., Saha-Dasgupta, T.: A reduction in particle size generally causes body-centered-cubic metals to expand but face-centered-cubic metals to contract. ACS Nano **12**(7), 7246–7252 (2018). https://doi.org/10.1021/acsnano.8b03360
32. Mendoza-Reséndez, R., Bomati-Miguel, O., Morales, M.P., Bonville, P., Serna, C.J.: Microstructural characterization of ellipsoidal iron metal nanoparticles. Nanotechnology **15**(4), S254 (2004). https://doi.org/10.1088/0957-4484/15/4/026
33. Iravani, S.: Green synthesis of metal nanoparticles using plants. Green Chem. **13**(10), 2638–2650 (2011). https://doi.org/10.1039/C1GC15386B
34. Javed, R., Zia, M., Naz, S., Aisida, S.O., ul Ain, N., Ao, Q.: Role of capping agents in the application of nanoparticles in biomedicine and environmental remediation: recent trends and future prospects. J. Nanobiotechnol. **18**(1), 172 (2020). https://doi.org/10.1186/s12951-020-00704-4
35. Muñoz-Mármol, M., Crespo, J., Fritts, M.J., Maojo, V.: Towards the taxonomic categorization and recognition of nanoparticle shapes. Nanomed. Nanotechnol. Biol. Med. **11**(2), 457–465 (2015). https://doi.org/10.1016/j.nano.2014.07.006
36. Pirsaheb, M., Gholami, T., Seifi, H., Dawi, E.A., Said, E.A., Hamoody, A.-H.M., Altimari, U.S., Salavati-Niasari, M.: Green synthesis of nanomaterials by using plant extracts as reducing and capping agents. Environ. Sci. Pollut. Res. **31**(17), 24768–24787 (2024). https://doi.org/10.1007/s11356-024-32983-x
37. Mafuné, F., Kohno, J., Takeda, Y., Kondow, T., Sawabe, H.: Structure and stability of silver nanoparticles in aqueous solution produced by laser ablation. J. Phys. Chem. B **104**(35), 8333–8337 (2000). https://doi.org/10.1021/jp001803b
38. Aswathi, V.P., Meera, S., Maria, C.G.A., Nidhin, M.: Green synthesis of nanoparticles from biodegradable waste extracts and their applications: a critical review. Nanotechnol. Environ. Eng. **8**(2), 377–397 (2023). https://doi.org/10.1007/s41204-022-00276-8
39. Hashem, A.H., Saied, E., Badr, B.M., Dora, M.S., Diab, M.A., Abdelaziz, A.M., Elkady, F.M., Ali, M.A., Issa, N.I., Hamdy, Z.A., Nafea, M.E., Khalifa, A.N., Adel, A., Hasib, A., Hawela, A.M., El-Gazzar, M.M., Nouh, M.A., Nahool, A.A., Attia, M.S.: Biosynthesis of trimetallic nanoparticles and their biological applications: a recent review. Arch. Microbiol. **207**(3), 50 (2025). https://doi.org/10.1007/s00203-025-04237-y
40. Koul, B., Poonia, A.K., Yadav, D., Jin, J.-O., Koul, B., Poonia, A.K., Yadav, D., Jin, J.-O.: Microbe-mediated biosynthesis of nanoparticles: applications and future prospects. Biomolecules **11**(6) (2021). https://doi.org/10.3390/biom11060886
41. Dhillon, G.S., Brar, S.K., Kaur, S., Verma, M.: Green approach for nanoparticle biosynthesis by fungi: current trends and applications. Crit. Rev. Biotechnol. **32**(1), 49–73 (2012). https://doi.org/10.3109/07388551.2010.550568

42. Hamida, R.S., Ali, M.A., Abdelmeguid, N.E., Al-Zaban, M.I., Baz, L., Bin-Meferij, M.M., Hamida, R.S., Ali, M.A., Abdelmeguid, N.E., Al-Zaban, M.I., Baz, L., Bin-Meferij, M.M.: Lichens—a potential source for nanoparticles fabrication: a review on nanoparticles biosynthesis and their prospective applications. J. Fungi **7**(4) (2021). https://doi.org/10.3390/jof7040291
43. Neethu, S., Radhakrishnan, E.K., Jyothis, M.: Biofabrication of nanoparticles using fungi. In: Panpatte, D.G., Jhala, Y.K. (eds.) Nanotechnology for Agriculture: Advances for Sustainable Agriculture, pp. 53–73. Springer, Singapore (2019). https://doi.org/10.1007/978-981-32-9370-0_4
44. Malhotra, S.P.K., Alghuthaymi, M.A.: Biomolecule-assisted biogenic synthesis of metallic nanoparticles. In: Agri-Waste and Microbes for Production of Sustainable Nanomaterials, pp. 139–163. Elsevier (2022). https://doi.org/10.1016/B978-0-12-823575-1.00011-1
45. Khurana, K., Jaggi, N.: Localized surface Plasmonic properties of Au and Ag nanoparticles for sensors: a review. Plasmonics **16**(4), 981–999 (2021). https://doi.org/10.1007/s11468-021-01381-1
46. An, K., Somorjai, G.A.: Size and shape control of metal nanoparticles for reaction selectivity in catalysis. ChemCatChem **4**(10), 1512–1524 (2012). https://doi.org/10.1002/cctc.201200229
47. Dahoumane, S.A., Jeffryes, C., Mechouet, M., Agathos, S.N., Dahoumane, S.A., Jeffryes, C., Mechouet, M., Agathos, S.N.: Biosynthesis of inorganic nanoparticles: a fresh look at the control of shape, size and composition. Bioengineering **4**(1) (2017). https://doi.org/10.3390/bioengineering4010014
48. Li, Y., Liu, Q., Shen, W.: Morphology-dependent nanocatalysis: metal particles. Dalton Trans. **40**(22), 5811–5826 (2011). https://doi.org/10.1039/C0DT01404D
49. Kazemi, S., Hosseingholian, A., Gohari, S.D., Feirahi, F., Moammeri, F., Mesbahian, G., Moghaddam, Z.S., Ren, Q.: Recent advances in green synthesized nanoparticles: from production to application. Mater. Today Sustain. **24**, 100500 (2023). https://doi.org/10.1016/j.mtsust.2023.100500
50. Kesharwani, J., Yoon, K.Y., Hwang, J., Rai, M.: Phytofabrication of silver nanoparticles by leaf extract of *Datura metel*: hypothetical mechanism involved in synthesis. J. Bionanosci. **3**(1), 39–44 (2009). https://doi.org/10.1166/jbns.2009.1008
51. Pinzaru, I., Coricovac, D., Dehelean, C., Moacă, E.-A., Mioc, M., Baderca, F., Sizemore, I., Brittle, S., Marti, D., Calina, C.D., Tsatsakis, A.M., Şoica, C.: Stable PEG-coated silver nanoparticles—a comprehensive toxicological profile. Food Chem. Toxicol. **111**, 546–556 (2018). https://doi.org/10.1016/j.fct.2017.11.051
52. Ahlberg, S., Antonopulos, A., Diendorf, J., Dringen, R., Epple, M., Flöck, R., Goedecke, W., Graf, C., Haberl, N., Helmlinger, J., Herzog, F., Heuer, F., Hirn, S., Johannes, C., Kittler, S., Köller, M., Korn, K., Kreyling, W.G., Krombach, F., Lademann, J., Loza, K., Luther, E.M., Malissek, M., Meinke, M.C., Nordmeyer, D., Pailliart, A., Raabe, J., Rancan, F., Rothen-Rutishauser, B., Rühl, E., Schleh, C., Seibel, A., Sengstock, C., Treuel, L., Vogt, A., Weber, K., Zellner, R.: PVP-coated, negatively charged silver nanoparticles: a multi-center study of their physicochemical characteristics, cell culture and in vivo experiments. Beilstein J. Nanotechnol. **5**(1), 1944–1965 (2014). https://doi.org/10.3762/bjnano.5.205
53. Hutami Rahayu, L.B., Wulandari, I.O., Santjojo, D.H., Sabarudin, A.: Synthesis and characterization of Fe_3O_4 nanoparticles using polyvinyl alcohol (PVA) as capping agent and glutaraldehyde (GA) as crosslinker. IOP Conf. Ser. Mater. Sci. Eng. **299**(1), 012062 (2018). https://doi.org/10.1088/1757-899X/299/1/012062
54. JyothiKumar, A., Kailasnath, M., Simon, J., Mathew, S.: BSA stabilized gold nanoparticles: synthesis and characterization. Mater. Today Proc. **9**, 111–115 (2019). https://doi.org/10.1016/j.matpr.2019.02.044
55. Reddy, D.A., Murali, G., Vijayalakshmi, R.P., Reddy, B.K.: Room-temperature ferromagnetism in EDTA capped Cr-doped ZnS nanoparticles. Appl. Phys. A **105**(1), 119–124 (2011). https://doi.org/10.1007/s00339-011-6563-1
56. Deng, H.-H., Lin, X.-L., Liu, Y.-H., Li, K.-L., Zhuang, Q.-Q., Peng, H.-P., Liu, A.-L., Xia, X.-H., Chen, W.: Chitosan-stabilized platinum nanoparticles as effective oxidase mimics for

colorimetric detection of acid phosphatase. Nanoscale **9**(29), 10292–10300 (2017). https://doi.org/10.1039/C7NR03399K
57. Khina, A.G., Krutyakov, Yu.A.: Similarities and differences in the mechanism of antibacterial action of silver ions and nanoparticles. Appl. Biochem. Microbiol. **57**(6), 683–693 (2021). https://doi.org/10.1134/S0003683821060053
58. Gurunathan, S., Jeong, J.-K., Han, J.W., Zhang, X.-F., Park, J.H., Kim, J.-H.: Multidimensional effects of biologically synthesized silver nanoparticles in *Helicobacter pylori*, *Helicobacter felis*, and human lung (L132) and lung carcinoma A549 cells. Nanoscale Res. Lett. **10**(1), 35 (2015). https://doi.org/10.1186/s11671-015-0747-0
59. Bahrami-Teimoori, B., Nikparast, Y., Hojatianfar, M., Akhlaghi, M., Ghorbani, R., Pourianfar, H.R.: Characterisation and antifungal activity of silver nanoparticles biologically synthesised by *Amaranthus retroflexus* leaf extract. J. Exp. Nanosci. **12**(1), 129–139 (2017). https://doi.org/10.1080/17458080.2017.1279355
60. Shobana, S., Veena, S., Sameer, S.S.M., Swarnalakshmi, K., Vishal, L.A.: Green synthesis of silver nanoparticles using *Artocarpus hirsutus* seed extract and its antibacterial activity. Curr. Pharm. Biotechnol. **21**(10), 980–989 (2020). https://doi.org/10.2174/13892010216662001 07115849
61. Majeed, S., Bakhtiar, N.F.B., Danish, M., Mohamad Ibrahim, M.N., Hashim, R.: Green approach for the biosynthesis of silver nanoparticles and its antibacterial and antitumor effect against osteoblast MG-63 and breast MCF-7 cancer cell lines. Sustain. Chem. Pharm. **12**, 100138 (2019). https://doi.org/10.1016/j.scp.2019.100138
62. Behravan, M., Hossein Panahi, A., Naghizadeh, A., Ziaee, M., Mahdavi, R., Mirzapour, A.: Facile green synthesis of silver nanoparticles using *Berberis vulgaris* leaf and root aqueous extract and its antibacterial activity. Int. J. Biol. Macromol. **124**, 148–154 (2019). https://doi.org/10.1016/j.ijbiomac.2018.11.101
63. Baghizadeh, A., Ranjbar, S., Gupta, V.K., Asif, M., Pourseyedi, S., Karimi, M.J., Mohammadinejad, R.: Green synthesis of silver nanoparticles using seed extract of *Calendula officinalis* in liquid phase. J. Mol. Liq. **207**, 159–163 (2015). https://doi.org/10.1016/j.molliq.2015.03.029
64. Fabiyi, O.A.: Sustainable management of *Meloidogyne incognita* infecting carrot (*Daucus carota*): green synthesis of silver nanoparticles with *Cnidoscolus aconitifolius*. Vegetos **34**(2), 277–285 (2021). https://doi.org/10.1007/s42535-021-00216-y
65. Pradheesh, G., Suresh, S., Suresh, J., Alexramani, V.: Antimicrobial and anticancer activity studies on green synthesized silver oxide nanoparticles from the medicinal plant *Cyathea nilgiriensis* Holttum. Int. J. Pharm. Investig. **10**(2), 146–150 (2020). https://doi.org/10.5530/ijpi.2020.2.27
66. Allafchian, A.R., Mirahmadi-Zare, S.Z., Jalali, S.A.H., Hashemi, S.S., Vahabi, M.R.: Green synthesis of silver nanoparticles using Phlomis leaf extract and investigation of their antibacterial activity. J. Nanostruct. Chem. **6**(2), 129–135 (2016). https://doi.org/10.1007/s40097-016-0187-0
67. Al-Otibi, F., Perveen, K., Al-Saif, N.A., Alharbi, R.I., Bokhari, N.A., Albasher, G., Al-Otaibi, R.M., Al-Mosa, M.A.: Biosynthesis of silver nanoparticles using *Malva parviflora* and their antifungal activity. Saudi J. Biol. Sci. **28**(4), 2229–2235 (2021). https://doi.org/10.1016/j.sjbs.2021.01.012
68. Rashmi, B.N., Harlapur, S.F., Avinash, B., Ravikumar, C.R., Nagaswarupa, H.P., Anil Kumar, M.R., Gurushantha, K., Santosh, M.S.: Facile green synthesis of silver oxide nanoparticles and their electrochemical, photocatalytic and biological studies. Inorg. Chem. Commun. **111**, 107580 (2020). https://doi.org/10.1016/j.inoche.2019.107580
69. Rajakumar, G., Abdul Rahuman, A.: Larvicidal activity of synthesized silver nanoparticles using *Eclipta prostrata* leaf extract against Filariasis and malaria vectors. Acta Trop. **118**(3), 196–203 (2011). https://doi.org/10.1016/j.actatropica.2011.03.003
70. Lallawmawma, H., Sathishkumar, G., Sarathbabu, S., Ghatak, S., Sivaramakrishnan, S., Gurusubramanian, G., Kumar, N.S.: Synthesis of silver and gold nanoparticles using *Jasminum*

nervosum leaf extract and its larvicidal activity against filarial and arboviral vector *Culex quinquefasciatus* Say (Diptera: Culicidae). Environ. Sci. Pollut. Res. **22**(22), 17753–17768 (2015). https://doi.org/10.1007/s11356-015-5001-x

71. El-Shanshoury, A.E.-R.R., ElSilk, S.E., Ebeid, M.E.: Extracellular biosynthesis of silver nanoparticles using *Escherichia coli* ATCC 8739, *Bacillus subtilis* ATCC 6633, and *Streptococcus thermophilus* ESh1 and their antimicrobial activities. ISRN Nanotechnol. **2011**, 1–7 (2011). https://doi.org/10.5402/2011/385480
72. Dhoondia, Z.H., Chakraborty, H.: Lactobacillus mediated synthesis of silver oxide nanoparticles. Nanomater. Nanotechnol. **2**, 15 (2012). https://doi.org/10.5772/55741
73. Kalishwaralal, K., Deepak, V., Ram Kumar Pandian, S., Kottaisamy, M., BarathManiKanth, S., Kartikeyan, B., Gurunathan, S.: Biosynthesis of silver and gold nanoparticles using *Brevibacterium casei*. Colloids Surf. B Biointerfaces **77**(2), 257–262 (2010). https://doi.org/10.1016/j.colsurfb.2010.02.007
74. Shanthi, S., David Jayaseelan, B., Velusamy, P., Vijayakumar, S., Chih, C.T., Vaseeharan, B.: Biosynthesis of silver nanoparticles using a probiotic *Bacillus licheniformis* Dahb1 and their antibiofilm activity and toxicity effects in *Ceriodaphnia cornuta*. Microb. Pathog. **93**, 70–77 (2016). https://doi.org/10.1016/j.micpath.2016.01.014
75. Paul, D., Sinha, S.N.: Extracellular synthesis of silver nanoparticles using *Pseudomonas aeruginosa* KUPSB12 and its antibacterial activity. Jordan J. Biol. Sci. **7**(4), 245–250 (2014). https://doi.org/10.12816/0008246
76. Ammar, H.A.M., El-Desouky, T.A.: Green synthesis of nanosilver particles by *Aspergillus terreus* HA1N and *Penicillium expansum* HA2N and its antifungal activity against mycotoxigenic fungi. J. Appl. Microbiol. **121**(1), 89–100 (2016). https://doi.org/10.1111/jam.13140
77. Bhangale, H., Sarode, K.M., Patil, A.M., Patil, D.R.: Microbial synthesis of silver nanoparticles using *Aspergillus flavus* and their characterization. In: Pawar, P.M., Ronge, B.P., Balasubramaniam, R., Seshabhattar, S. (eds.) Techno-Societal 2016, pp. 463–470. Springer International Publishing, Cham (2018). https://doi.org/10.1007/978-3-319-53556-2_45
78. Abdel-Hafez, S.I., Nafady, N.A., Abdel-Rahim, I.R., Shaltout, A.M., Mohamed, M.A.: Biogenesis and optimisation of silver nanoparticles by the endophytic fungus *Cladosporium sphaerospermum*. Int. J. Nanomater. Chem. **2**(1), 11–19 (2016). https://doi.org/10.18576/ijnc/020103
79. Alnuaimi, M.Th., Aljanabi, Z.Z., Mahmood, S.I., Ramzy, U.F.: Production of biosynthesized silver nanoparticles using *Metarhizium anisopliae* fungus for the treatment of petroleum pollutants in water. Baghdad Sci. J. **21**(6) (2024). https://doi.org/10.21123/bsj.2023.8577
80. Varshney, R., Mishra, A.N., Bhadauria, S., Gaur, M.S.: A novel microbial route to synthesize silver nanoparticles using fungus Hormoconis resinae
81. Chokshi, K., Pancha, I., Ghosh, T., Paliwal, C., Maurya, R., Ghosh, A., Mishra, S.: Green synthesis, characterization and antioxidant potential of silver nanoparticles biosynthesized from de-oiled biomass of thermotolerant oleaginous microalgae *Acutodesmus dimorphus*. RSC Adv. **6**(76), 72269–72274 (2016). https://doi.org/10.1039/C6RA15322D
82. Aboelfetoh, E.F., El-Shenody, R.A., Ghobara, M.M.: Eco-friendly synthesis of silver nanoparticles using green algae (*Caulerpa serrulata*): reaction optimization, catalytic and antibacterial activities. Environ. Monit. Assess. **189**(7), 349 (2017). https://doi.org/10.1007/s10661-017-6033-0
83. Kowalska-Góralska, M., Dziewulska, K., Kulasza, M.: Effect of copper nanoparticles and ions on spermatozoa motility of sea trout (Salmo trutta m. Trutta L.). Aquat. Toxicol. **211**, 11–17 (2019). https://doi.org/10.1016/j.aquatox.2019.03.013
84. Labaran, A.N., Zango, Z.U., Tailor, G., Alsadig, A., Usman, F., Mukhtar, M.T., Garba, A.M., Alhathlool, R., Ibnaouf, K.H., Aldaghri, O.A.: Biosynthesis of copper nanoparticles using *Alstonia scholaris* leaves and its antimicrobial studies. Sci. Rep. **14**(1), 5589 (2024). https://doi.org/10.1038/s41598-024-56052-y
85. Eid, A.M., Fouda, A., Hassan, S.E.-D., Hamza, M.F., Alharbi, N.K., Elkelish, A., Alharthi, A., Salem, W.M., Eid, A.M., Fouda, A., Hassan, S.E.-D., Hamza, M.F., Alharbi, N.K., Elkelish,

A., Alharthi, A., Salem, W.M.: Plant-based copper oxide nanoparticles; biosynthesis, characterization, antibacterial activity, tanning wastewater treatment, and heavy metals sorption. Catalysts **13**(2) (2023). https://doi.org/10.3390/catal13020348

86. Kumar, M., Kaushik, D., Kumar, A., Krishnan, H., Oz, F., Proestos, C., Hashem, A., Abd_Allah, E.F.: A sustainable approach to prepare green synthesis of copper nanoparticles of *Bauhinia variegata* & *Saussurea lappa*: unveiling in-vitro anti-obesity applications. Heliyon **10**(8) (2024). https://doi.org/10.1016/j.heliyon.2024.e29433
87. Razavi, R., Molaei, R., Moradi, M., Tajik, H., Ezati, P., Shafipour Yordshahi, A.: Biosynthesis of metallic nanoparticles using mulberry fruit (*Morus alba* L.) extract for the preparation of antimicrobial nanocellulose film. Appl. Nanosci. **10**(2), 465–476 (2020). https://doi.org/10.1007/s13204-019-01137-8
88. Sharma, B.K., Mehta, B.R., Shah, E.V., Chaudhari, V.P., Roy, D.R., Mondal Roy, S.: Green synthesis of triangular ZnO nanoparticles using *Azadirachta indica* leaf extract and its shape dependency for significant antimicrobial activity: joint experimental and theoretical investigation. J. Clust. Sci. **33**(6), 2517–2530 (2022). https://doi.org/10.1007/s10876-021-02145-x
89. Sulaiman, G.M., Tawfeeq, A.T., Jaaffer, M.D.: Biogenic synthesis of copper oxide nanoparticles using *Olea europaea* leaf extract and evaluation of their toxicity activities: an in vivo and in vitro study. Biotechnol. Prog. **34**(1), 218–230 (2018). https://doi.org/10.1002/btpr.2568
90. Mohammad, M.S., Shyam, P.: Structural analysis of biogenic copper oxide nanoparticles, and their bio-activity assessment. BioNanoScience **13**(4), 2265–2275 (2023). https://doi.org/10.1007/s12668-023-01168-0
91. Ramesh, C., HariPrasad, M., Ragunathan, V.: Effect of *Arachis hypogaea* L. leaf extract on Barfoed's solution; green synthesis of Cu_2O nanoparticles and its antibacterial effect. Curr. Nanosci. **7**(6), 995–999 (2011). https://doi.org/10.2174/157341311798220781.
92. Talebian, S., Shahnavaz, B., Nejabat, M., Abolhassani, Y., Rassouli, F.B.: Bacterial-mediated synthesis and characterization of copper oxide nanoparticles with antibacterial, antioxidant, and anticancer potentials. Front. Bioeng. Biotechnol. **11** (2023). https://doi.org/10.3389/fbioe.2023.1140010
93. Varshney, R., Bhadauria, S., Gaur, M.S., Pasricha, R.: Characterization of copper nanoparticles synthesized by a novel microbiological method. JOM **62**(12), 102–104 (2010). https://doi.org/10.1007/s11837-010-0171-y
94. Prasad, K., Jha, A.K., Prasad, K., Kulkarni, A.R.: Can microbes mediate nano-transformation? Indian J. Phys. **84**(10), 1355–1360 (2010). https://doi.org/10.1007/s12648-010-0126-8
95. Shende, S., Bhagat, R., Raut, R., Rai, M., Gade, A.: Myco-fabrication of copper nanoparticles and its effect on crop pathogenic fungi. IEEE Trans. NanoBiosci. **20**(2), 146–153 (2021). https://doi.org/10.1109/TNB.2021.3056100
96. Noor, S., Shah, Z., Javed, A., Ali, A., Hussain, S.B., Zafar, S., Ali, H., Muhammad, S.A.: A fungal based synthesis method for copper nanoparticles with the determination of anticancer, antidiabetic and antibacterial activities. J. Microbiol. Methods **174**, 105966 (2020). https://doi.org/10.1016/j.mimet.2020.105966
97. Anwar, S.J., Yusoff, H.M., Bhat, I.U.H., Ern, L.K.: Remediation of dye-contaminated water using brown algae seaweed supported copper nanoparticles. Arab. J. Sci. Eng. **49**(1), 475–496 (2024). https://doi.org/10.1007/s13369-023-08199-5
98. Bhattacharya, P., Swarnakar, S., Ghosh, S., Majumdar, S., Banerjee, S.: Disinfection of drinking water via algae mediated green synthesized copper oxide nanoparticles and its toxicity evaluation. J. Environ. Chem. Eng. **7**(1), 102867 (2019). https://doi.org/10.1016/j.jece.2018.102867
99. Wu, W., Wu, Z., Yu, T., Jiang, C., Kim, W.-S.: Recent progress on magnetic iron oxide nanoparticles: synthesis, surface functional strategies and biomedical applications. Sci. Technol. Adv. Mater. **16**(2), 023501 (2015). https://doi.org/10.1088/1468-6996/16/2/023501
100. Fatima, H., Lee, D.-W., Yun, H.J., Kim, K.-S.: Shape-controlled synthesis of magnetic Fe_3O_4 nanoparticles with different iron precursors and capping agents. RSC Adv. **8**(41), 22917–22923 (2018). https://doi.org/10.1039/C8RA02909A

101. Sharifi Dehsari, H., Halda Ribeiro, A., Ersöz, B., Tremel, W., Jakob, G., Asadi, K.: Effect of precursor concentration on size evolution of iron oxide nanoparticles. CrystEngComm **19**(44), 6694–6702 (2017). https://doi.org/10.1039/C7CE01406F
102. Tassa, C., Shaw, S.Y., Weissleder, R.: Dextran-coated iron oxide nanoparticles: a versatile platform for targeted molecular imaging, molecular diagnostics, and therapy. Acc. Chem. Res. **44**(10), 842–852 (2011). https://doi.org/10.1021/ar200084x
103. Elkhenany, H., Abd Elkodous, M., Ghoneim, N.I., Ahmed, T.A., Ahmed, S.M., Mohamed, I.K., El-Badri, N.: Comparison of different uncoated and starch-coated superparamagnetic iron oxide nanoparticles: implications for stem cell tracking. Int. J. Biol. Macromol. **143**, 763–774 (2020). https://doi.org/10.1016/j.ijbiomac.2019.10.031
104. Chaudhary, J., Tailor, G., Mehta, C., Yadav, M.: An overview of biosynthesized metal nanoparticles via medicinal plant extracts of the Moraceae family. Biocatal. Agric. Biotechnol. **52**, 102812 (2023). https://doi.org/10.1016/j.bcab.2023.102812
105. Karpagavinayagam, P., Vedhi, C.: Green synthesis of iron oxide nanoparticles using *Avicennia marina* flower extract. Vacuum **160**, 286–292 (2019). https://doi.org/10.1016/j.vacuum.2018.11.043
106. Kiruba Daniel, S.C.G., Vinothini, G., Subramanian, N., Nehru, K., Sivakumar, M.: Biosynthesis of Cu, ZVI, and Ag nanoparticles using *Dodonaea viscosa* extract for antibacterial activity against human pathogens. J. Nanopart. Res. **15**(1), 1319 (2013). https://doi.org/10.1007/s11051-012-1319-1
107. Naseem, T., Farrukh, M.A.: Antibacterial activity of green synthesis of iron nanoparticles using *Lawsonia inermis* and *Gardenia jasminoides* leaves extract. J. Chem. **2015**(1), 912342 (2015). https://doi.org/10.1155/2015/912342
108. Senthil, M., Ramesh, C.: Biogenic synthesis of Fe_3O_4 nanoparticles using *Tridax procumbens* leaf extract and its antibacterial activity on *Pseudomonas aeruginosa*. Dig. J. Nanomater. Biostruct. DJNB **7**(4), 1655 (2012)
109. Jubran, A.S., Al-Zamely, O.M., Al-Ammar, M.H.: A study of iron oxide nanoparticles synthesis by using bacteria. Int. J. Pharm. Qual. Assur. **11**(01), 88–92 (2020). https://doi.org/10.25258/ijpqa.11.1.13
110. Tarafdar, J.C., Raliya, R.: Rapid, low-cost, and ecofriendly approach for iron nanoparticle synthesis using *Aspergillus oryzae* TFR9. J. Nanopart. **2013**, 1–4 (2013). https://doi.org/10.1155/2013/141274
111. Bhargava, A., Jain, N., Barathi L.M., Akhtar, Mohd.S., Yun, Y.-S., Panwar, J.: Synthesis, characterization and mechanistic insights of mycogenic iron oxide nanoparticles. In: Diallo, M.S., Fromer, N.A., Jhon, M.S. (eds.) Nanotechnology for Sustainable Development, pp. 337–348. Springer International Publishing, Cham (2014). https://doi.org/10.1007/978-3-319-05041-6_27
112. Baskar, G., Palaniyandi, T., Ravi, M., Viswanathan, S., Wahab, M.R.A., Surendran, H., Govindaraj, M., Sugumaran, A., Almutairi, M.H., Almutairi, B.O.: Biosynthesis of iron oxide nanoparticles from red seaweed *Hypnea valentiae* and evaluation of their antioxidant and antitumor potential via the AKT/PI3K pathway. Process Biochem. **141**, 155–169 (2024). https://doi.org/10.1016/j.procbio.2024.03.010
113. Bensy, A.D.V., Christobel, G.J., Muthusamy, K., Alfarhan, A., Anantharaman, P.: Green synthesis of iron nanoparticles from *Ulva lactuca* and bactericidal activity against Enteropathogens. J. King Saud Univ. Sci. **34**(3), 101888 (2022). https://doi.org/10.1016/j.jksus.2022.101888
114. Shehawy, A.S.E., Ali, E.M., Elsayed, A.: Algae-mediated synthesis and structural characterization of iron nanoparticles using *Galaxaura rugosa* seaweed. In Review (2023). https://doi.org/10.21203/rs.3.rs-3648370/v1
115. Baccile, N., Noiville, R., Stievano, L., Bogaert, I.V.: Sophorolipids-functionalized iron oxide nanoparticles. Phys. Chem. Chem. Phys. **15**(5), 1606–1620 (2013). https://doi.org/10.1039/C2CP41977G
116. Vainshtein, M., Belova, N., Kulakovskaya, T., Suzina, N., Sorokin, V.: Synthesis of magneto-sensitive iron-containing nanoparticles by yeasts. J. Ind. Microbiol. Biotechnol. **41** (2014). https://doi.org/10.1007/s10295-014-1417-4

117. Ali, A., Phull, A.-R., Zia, M.: Elemental zinc to zinc nanoparticles: is ZnO NPs crucial for life? Synthesis, toxicological, and environmental concerns. Nanotechnol. Rev. **7**(5), 413–441 (2018). https://doi.org/10.1515/ntrev-2018-0067
118. Kahil, H., Gad, M.S., Ebraheem, H.: The efficacy of ZnO-NPs prepared via green route against colon and breast cancer cells. Discov. Appl. Sci. **6**(3), 109 (2024). https://doi.org/10.1007/s42452-024-05659-x
119. Fouda, A., Abdel-Rahman, M.A., Eid, A.M., Selim, S., Ejaz, H., Alruwaili, M., Manni, E., Almuhayawi, M.S., Al Jaouni, S.K., Hassan, S.E.-D.: Investigating the potential of green-fabricated zinc oxide nanoparticles to inhibit the foodborne pathogenic bacteria isolated from spoiled fruits. Catalysts **14**(7), 427 (2024). https://doi.org/10.3390/catal14070427
120. Tang, Q., Xia, H., Liang, W., Huo, X., Wei, X.: Synthesis and characterization of zinc oxide nanoparticles from *Morus nigra* and its anticancer activity of AGS gastric cancer cells. J. Photochem. Photobiol. B. **202**, 111698 (2020). https://doi.org/10.1016/j.jphotobiol.2019.111698
121. Naiel, B., Fawzy, M., Mahmoud, A.E.D., Halmy, M.W.A.: Sustainable fabrication of dimorphic plant derived ZnO nanoparticles and exploration of their biomedical and environmental potentialities. Sci. Rep. **14**(1), 13459 (2024). https://doi.org/10.1038/s41598-024-63459-0
122. Balraj, B., Senthilkumar, N., Siva, C., Krithikadevi, R., Julie, A., Potheher, I.V., Arulmozhi, M.: Synthesis and characterization of zinc oxide nanoparticles using marine *Streptomyces* Sp. with its investigations on anticancer and antibacterial activity. Res. Chem. Intermed. **43**(4), 2367–2376 (2017). https://doi.org/10.1007/s11164-016-2766-6
123. Jain, D., Shivani, Bhojiya, A.A., Singh, H., Daima, H.K., Singh, M., Mohanty, S.R., Stephen, B.J., Singh, A.: Microbial fabrication of zinc oxide nanoparticles and evaluation of their antimicrobial and photocatalytic properties. Front. Chem. **8**, 778 (2020). https://doi.org/10.3389/fchem.2020.00778
124. Faisal, S., Abdullah, Rizwan, M., Ullah, R., Alotaibi, A., Khattak, A., Bibi, N., Idrees, M.: *Paraclostridium benzoelyticum* bacterium-mediated zinc oxide nanoparticles and their in vivo multiple biological applications. Oxid. Med. Cell. Longev. **2022**(1), 5994033 (2022). https://doi.org/10.1155/2022/5994033
125. Kalpana, V.N., Kataru, B.A.S., Sravani, N., Vigneshwari, T., Panneerselvam, A., Devi Rajeswari, V.: Biosynthesis of zinc oxide nanoparticles using culture filtrates of *Aspergillus niger*: antimicrobial textiles and dye degradation studies. OpenNano **3**, 48–55 (2018). https://doi.org/10.1016/j.onano.2018.06.001
126. Mohamed, A.A., Fouda, A., Abdel-Rahman, M.A., Hassan, S.E.-D., El-Gamal, M.S., Salem, S.S., Shaheen, T.I.: Fungal strain impacts the shape, bioactivity and multifunctional properties of green synthesized zinc oxide nanoparticles. Biocatal. Agric. Biotechnol. **19**, 101103 (2019). https://doi.org/10.1016/j.bcab.2019.101103
127. Rabecca, R., Doss, A., Mary Kensa, V., Iswarya, S., Mukeshbabu, N., Praveen Pole, R.P., Iyappan, K.: Facile synthesis of zinc oxide nanoparticle using algal extract and their antibacterial potential. Biomass Convers. Biorefinery **15**(20), 27089–27100 (2025). https://doi.org/10.1007/s13399-022-03275-6
128. Priyadharshini, R.I., Prasannaraj, G., Geetha, N., Venkatachalam, P.: Microwave-mediated extracellular synthesis of metallic silver and zinc oxide nanoparticles using macro-algae (*Gracilaria edulis*) extracts and its anticancer activity against human PC3 cell lines. Appl. Biochem. Biotechnol. **174**(8), 2777–2790 (2014). https://doi.org/10.1007/s12010-014-1225-3
129. Azizi, S., Ahmad, M.B., Namvar, F., Mohamad, R.: Green biosynthesis and characterization of zinc oxide nanoparticles using brown marine macroalga *Sargassum muticum* aqueous extract. Mater. Lett. **116**, 275–277 (2014). https://doi.org/10.1016/j.matlet.2013.11.038
130. Aswathy, R., Gabylis, B., Sarkar, A., Rao, B.: Green synthesis and characterization of marine yeast-mediated silver and zinc oxide nanoparticles and assessment of their antioxidant activity. Asian J. Pharm. Clin. Res. **10**, 235 (2017). https://doi.org/10.22159/ajpcr.2017.v10i10.19979
131. Chauhan, R., Reddy, A., Abraham, J.: Biosynthesis of silver and zinc oxide nanoparticles using *Pichia fermentans* JA2 and their antimicrobial property. Appl. Nanosci. **5**(1), 63–71 (2015). https://doi.org/10.1007/s13204-014-0292-7

132. Angel Ezhilarasi, A., Judith Vijaya, J., Kaviyarasu, K., John Kennedy, L., Ramalingam, R.J., Al-Lohedan, H.A.: Green synthesis of NiO nanoparticles using *Aegle marmelos* leaf extract for the evaluation of in-vitro cytotoxicity, antibacterial and photocatalytic properties. J. Photochem. Photobiol. B. **180**, 39–50 (2018). https://doi.org/10.1016/j.jphotobiol.2018.01.023
133. Alterary, S.S.: Potent and versatile biogenically synthesized alumina/nickel oxide nanocomposite adsorbent for defluoridation of drinking water. ACS Omega **9**(22), 23220–23240 (2024). https://doi.org/10.1021/acsomega.3c09076
134. Faizan, M., Fatima, M., Shams, F., Ibrahim, M., Hussain, S., Sabir, K., Salman, S., Ahmed, I., Muzamil Arshad, M., Khan, I., Subhani, M.: Plant extract mediated biogenic synthesis and characterization of nickel oxide nanoparticles and its environmental and antibacterial applications. J. Chem. Environ. **2**(2), 109–121 (2023). https://doi.org/10.56946/jce.v2i2.280
135. Ezhilarasi, A.A., Vijaya, J.J., Kaviyarasu, K., Maaza, M., Ayeshamariam, A., Kennedy, L.J.: Green synthesis of NiO nanoparticles using *Moringa oleifera* extract and their biomedical applications: cytotoxicity effect of nanoparticles against HT-29 cancer cells. J. Photochem. Photobiol. B. **164**, 352–360 (2016). https://doi.org/10.1016/j.jphotobiol.2016.10.003
136. Kganyago, P., Mahlaule-Glory, L.M., Mathipa, M.M., Ntsendwana, B., Mketo, N., Mbita, Z., Hintsho-Mbita, N.C.: Synthesis of NiO nanoparticles via a green route using *Monsonia burkeana*: the physical and biological properties. J. Photochem. Photobiol. B. **182**, 18–26 (2018). https://doi.org/10.1016/j.jphotobiol.2018.03.016
137. Mandal, B.K., Mandal, R., Sikdar, S., Sarma, S., Srinivasan, A., Chowdhury, S.R., Das, B., Das, R.: Green synthesis of NiO nanoparticle using *Punica granatum* peel extract and its characterization for methyl orange degradation. Mater. Today Commun. **34**, 105302 (2023). https://doi.org/10.1016/j.mtcomm.2022.105302
138. Gobinath, E., Dhatchinamoorthy, M., Saran, P., Vishnu, D., Indumathy, R., Kalaiarasi, G.: Synthesis and characterization of NiO nanoparticles using *Sesbania grandiflora* flower to evaluate cytotoxicity. Results Chem. **6**, 101043 (2023). https://doi.org/10.1016/j.rechem.2023.101043
139. Salvadori, M.R., Ando, R.A., Muraca, D., Knobel, M., Oller Nascimento, C.A., Corrêa, B.: Magnetic nanoparticles of Ni/NiO nanostructured in film form synthesized by dead organic matrix of yeast. RSC Adv. **6**(65), 60683–60692 (2016). https://doi.org/10.1039/C6RA07274G
140. Shaban, A., Zakaria, Z., Abdelhamid, I.A., Elhakim, H.K.A., El-Sayed, E.-S.R.: Novel mycofabrication of copper and nickel nanoparticles and evaluation of their effects against antibiotic resistance genes in different bacterial strains and anticancer potentials. Process Biochem. **149**, 192–203 (2025). https://doi.org/10.1016/j.procbio.2024.12.008
141. Dobrucka, R.: Biofabrication of platinum nanoparticles using *Fumariae herba* extract and their catalytic properties. Saudi J. Biol. Sci. **26**(1), 31–37 (2019). https://doi.org/10.1016/j.sjbs.2016.11.012
142. Thirumurugan, A., Aswitha, P., Kiruthika, C., Nagarajan, S., Christy, A.N.: Green synthesis of platinum nanoparticles using *Azadirachta indica*—an eco-friendly approach. Mater. Lett. **170**, 175–178 (2016). https://doi.org/10.1016/j.matlet.2016.02.026
143. Soundarrajan, C., Sankari, A., Dhandapani, P., Maruthamuthu, S., Ravichandran, S., Sozhan, G., Palaniswamy, N.: Rapid biological synthesis of platinum nanoparticles using *Ocimum sanctum* for water electrolysis applications. Bioprocess Biosyst. Eng. **35**(5), 827–833 (2012). https://doi.org/10.1007/s00449-011-0666-0
144. Göl, F., Aygün, A., Seyrankaya, A., Gür, T., Yenikaya, C., Şen, F.: Green synthesis and characterization of *Camellia sinensis* mediated silver nanoparticles for antibacterial ceramic applications. Mater. Chem. Phys. **250**, 123037 (2020). https://doi.org/10.1016/j.matchemphys.2020.123037
145. Martins, M., Mourato, C., Sanches, S., Noronha, J.P., Crespo, M.T.B., Pereira, I.A.C.: Biogenic platinum and palladium nanoparticles as new catalysts for the removal of pharmaceutical compounds. Water Res. **108**, 160–168 (2017). https://doi.org/10.1016/j.watres.2016.10.071
146. Attard, G., Casadesús, M., Macaskie, L.E., Deplanche, K.: Biosynthesis of platinum nanoparticles by *Escherichia coli* MC4100: can such nanoparticles exhibit intrinsic surface enantioselectivity? Langmuir **28**(11), 5267–5274 (2012). https://doi.org/10.1021/la204495z

147. Syed, A., Ahmad, A.: Extracellular biosynthesis of platinum nanoparticles using the fungus *Fusarium oxysporum*. Colloids Surf. B Biointerfaces **97**, 27–31 (2012). https://doi.org/10.1016/j.colsurfb.2012.03.026
148. Castro-Longoria, E., Moreno-Velasquez, S.D., Vilchis-Nestor, A.R., Arenas-Berumen, E., Avalos-Borja, M.: Production of platinum nanoparticles and nanoaggregates using *Neurospora crassa*. J. Microbiol. Biotechnol. **22**(7), 1000–1004 (2012). https://doi.org/10.4014/jmb.1110.10085
149. Govender, Y., Riddin, T., Gericke, M., Whiteley, C.G.: Bioreduction of platinum salts into nanoparticles: a mechanistic perspective. Biotechnol. Lett. **31**(1), 95–100 (2009). https://doi.org/10.1007/s10529-008-9825-z
150. Sathiyaraj, G., Vinosha, M., Sangeetha, D., Manikandakrishnan, M., Palanisamy, S., Sonaimuthu, M., Manikandan, R., You, S., Prabhu, N.M.: Bio-directed synthesis of pt-nanoparticles from aqueous extract of red algae *Halymenia dilatata* and their biomedical applications. Colloids Surf. Physicochem. Eng. Asp. **618**, 126434 (2021). https://doi.org/10.1016/j.colsurfa.2021.126434
151. Ramkumar, V.S., Pugazhendhi, A., Prakash, S., Ahila, N.K., Vinoj, G., Selvam, S., Kumar, G., Kannapiran, E., Rajendran, R.B.: Synthesis of platinum nanoparticles using seaweed *Padina Gymnospora* and their catalytic activity as PVP/PtNPs nanocomposite towards biological applications. Biomed. Pharmacother. **92**, 479–490 (2017). https://doi.org/10.1016/j.biopha.2017.05.076
152. Palanisamy, S., Anjali, R., Jeneeta, S., Mohandoss, S., Keerthana, D., Shin, I.-S., You, S., Prabhu, N.M.: An effective bio-inspired synthesis of platinum nanoparticles using *Caulerpa sertularioides* and investigating their antibacterial and antioxidant activities. Bioprocess Biosyst. Eng. **46**(1), 105–118 (2023). https://doi.org/10.1007/s00449-022-02816-7
153. Ansari, S.M., Saquib, Q., Alwathnani, H.A., Alharbi, S.A., Shaik, M.R., Kuniyil, M., Adil, S.F., Siddiqui, M.A., Ahmad, J., Wahab, R., Ahmad, S.F., Al-Khedhairy, A.A.: One-pot green synthesis of platinum nanoparticles (Pt-NPs) using *Sargassum Polyphyllum* extract displays nanodrug prospects against human cervical, breast, and colon cancer cells. Algal Res. **91**, 104278 (2025). https://doi.org/10.1016/j.algal.2025.104278
154. Park, S.Y., Kim, B., Cui, Z., Park, G., Choi, Y.-W.: Anti-metastatic effect of gold nanoparticle-conjugated *Maclura tricuspidata* extract on human hepatocellular carcinoma cells. Int. J. Nanomed. **15**, 5317–5331 (2020). https://doi.org/10.2147/IJN.S246724
155. Weng, Y., Li, J., Ding, X., Wang, B., Dai, S., Zhou, Y., Pang, R., Zhao, Y., Xu, H., Tian, B., Hua, Y.: Functionalized gold and silver bimetallic nanoparticles using *Deinococcus radiodurans* protein extract mediate degradation of toxic dye malachite green. Int. J. Nanomed. **15**, 1823–1835 (2020). https://doi.org/10.2147/IJN.S236683
156. Shankar, S.S., Rai, A., Ahmad, A., Sastry, M.: Rapid synthesis of Au, Ag, and bimetallic Au core–Ag shell nanoparticles using neem (*Azadirachta indica*) leaf broth. J. Colloid Interface Sci. **275**(2), 496–502 (2004). https://doi.org/10.1016/j.jcis.2004.03.003
157. Ogi, T., Saitoh, N., Nomura, T., Konishi, Y.: Room-temperature synthesis of gold nanoparticles and nanoplates using *Shewanella algae* cell extract. J. Nanopart. Res. **12**(7), 2531–2539 (2010). https://doi.org/10.1007/s11051-009-9822-8
158. Dasaratrao Sawle, B., Salimath, B., Deshpande, R., Dhondojirao Bedre, M., Krishnamurthy Prabhakar, B., Venkataraman, A.: Biosynthesis and stabilization of Au and Au–Ag Alloy nanoparticles by fungus, *Fusarium semitectum*. Sci. Technol. Adv. Mater. **9**(3), 035012 (2008). https://doi.org/10.1088/1468-6996/9/3/035012
159. Balakumaran, M.D., Ramachandran, R., Balashanmugam, P., Mukeshkumar, D.J., Kalaichelvan, P.T.: Mycosynthesis of silver and gold nanoparticles: optimization, characterization and antimicrobial activity against human pathogens. Microbiol. Res. **182**, 8–20 (2016). https://doi.org/10.1016/j.micres.2015.09.009
160. Hammad, S.E., El-Rouby, M.N., Abdel-Aziz, M.M., El-Sayyad, G.S., Elshikh, H.H.: Endophytic fungi-assisted biomass synthesis of gold, and zinc oxide nanoparticles for increasing antibacterial, and anticancer activities. Biomass Convers. Biorefin. **15**(2), 2285–2302 (2025). https://doi.org/10.1007/s13399-023-04954-8

161. Sarkar, J., Ray, S., Chattopadhyay, D., Laskar, A., Acharya, K.: Mycogenesis of gold nanoparticles using a phytopathogen *Alternaria alternata*. Bioprocess Biosyst. Eng. **35**(4), 637–643 (2012). https://doi.org/10.1007/s00449-011-0646-4
162. Hamed, M.M., Abdelftah, L.S.: Biosynthesis of gold nanoparticles using marine *Streptomyces griseus* isolate (M8) and evaluating its antimicrobial and anticancer activity. Egypt. J. Aquat. Biol. Fish. **23**(1), 173–184 (2019). https://doi.org/10.21608/ejabf.2019.26508
163. Li, L., Zhang, Z.: Biosynthesis of gold nanoparticles using green alga with their electrochemical performance for determining Carbendazim in soil. Int. J. Electrochem. Sci. **11**(6), 4550–4559 (2016). https://doi.org/10.20964/2016.06.13
164. Iravani, S., Varma, R.S.: Sustainable synthesis of cobalt and cobalt oxide nanoparticles and their catalytic and biomedical applications. Green Chem. **22**(9), 2643–2661 (2020). https://doi.org/10.1039/D0GC00885K
165. Samuel, M.S., Selvarajan, E., Mathimani, T., Santhanam, N., Phuong, T.N., Brindhadevi, K., Pugazhendhi, A.: Green synthesis of cobalt-oxide nanoparticle using jumbo Muscadine (*Vitis rotundifolia*): characterization and photo-catalytic activity of acid blue-74. J. Photochem. Photobiol. B. **211**, 112011 (2020). https://doi.org/10.1016/j.jphotobiol.2020.112011
166. Rasheed, T., Nabeel, F., Bilal, M., Iqbal, H.M.N.: Biogenic synthesis and characterization of cobalt oxide nanoparticles for catalytic reduction of direct yellow-142 and methyl orange dyes. Biocatal. Agric. Biotechnol. **19**, 101154 (2019). https://doi.org/10.1016/j.bcab.2019.101154
167. Sharma, J.K., Srivastava, P., Singh, G., Akhtar, M.S., Ameen, S.: Green synthesis of Co_3O_4 nanoparticles and their applications in thermal decomposition of ammonium perchlorate and dye-sensitized solar cells. Mater. Sci. Eng. B **193**, 181–188 (2015). https://doi.org/10.1016/j.mseb.2014.12.012
168. Gingasu, D., Mindru, I., Mocioiu, O.C., Preda, S., Stanica, N., Patron, L., Ianculescu, A., Oprea, O., Nita, S., Paraschiv, I., Popa, M., Saviuc, C., Bleotu, C., Chifiriuc, M.C.: Synthesis of nanocrystalline cobalt ferrite through soft chemistry methods: a green chemistry approach using sesame seed extract. Mater. Chem. Phys. **182**, 219–230 (2016). https://doi.org/10.1016/j.matchemphys.2016.07.026
169. Marimuthu, S., Rahuman, A.A., Kirthi, A.V., Santhoshkumar, T., Jayaseelan, C., Rajakumar, G.: Eco-friendly microbial route to synthesize cobalt nanoparticles using *Bacillus thuringiensis* against malaria and dengue vectors. Parasitol. Res. **112**(12), 4105–4112 (2013). https://doi.org/10.1007/s00436-013-3601-2
170. Omran, B.A., Nassar, H.N., Younis, S.A., El-Salamony, R.A., Fatthallah, N.A., Hamdy, A., El-Shatoury, E.H., El-Gendy, N.S.: Novel mycosynthesis of cobalt oxide nanoparticles using *Aspergillus brasiliensis* ATCC 16404-optimization, characterization and antimicrobial activity. J. Appl. Microbiol. **128**(2), 438–457 (2020). https://doi.org/10.1111/jam.14498
171. Vijayanandan, A.S., Balakrishnan, R.M.: Biosynthesis of cobalt oxide nanoparticles using endophytic fungus *Aspergillus nidulans*. J. Environ. Manage. **218**, 442–450 (2018). https://doi.org/10.1016/j.jenvman.2018.04.032
172. Ajarem, J.S., Maodaa, S.N., Allam, A.A., Taher, M.M., Khalaf, M.: Benign synthesis of cobalt oxide nanoparticles containing red algae extract: antioxidant, antimicrobial, anticancer, and anticoagulant activity. J. Clust. Sci. **33**(2), 717–728 (2022). https://doi.org/10.1007/s10876-021-02004-9
173. Patra, J.K., Baek, K.-H.: Green nanobiotechnology: factors affecting synthesis and characterization techniques. J. Nanomater. **2014**(1), 417305 (2014). https://doi.org/10.1155/2014/417305
174. Tsekhmistrenko, S., Цехмістренко, С.І., Bityutskii, V., Бітюцький, В.С., Tsekhmistrenko, O., Цехмістренко, О.С.: Factors Affecting «Green» Nanoparticle Synthesis. (Фактори, що впливають на «зелений» синтез наночастинок). (2021)
175. Zhang, R., Khalizov, A., Wang, L., Hu, M., Xu, W.: Nucleation and growth of nanoparticles in the atmosphere. Chem. Rev. **112**(3), 1957–2011 (2012). https://doi.org/10.1021/cr2001756
176. Tso, C., Zhung, C., Shih, Y., Tseng, Y.-M., Wu, S., Doong, R.: Stability of metal oxide nanoparticles in aqueous solutions. Water Sci. Technol. **61**(1), 127–133 (2010). https://doi.org/10.2166/wst.2010.787

Green Chemistry and Nanotechnology: An Indispensable Modern Yardstick, Symbiotic Relationship and Future Perspective

Brij Bhushan Pandey, Madhu Matang, Vikash M. Ganvit, Hemil Patel, and Rakesh K. Sharma

1 Introduction

The environment can be referred to as the system in which every biotic component on this Earth interact to sustain life [1]. Prior to the industrial revolution, there existed a mutually balanced relationship between the mankind and the environment. Unfortunately, this balance was upset by the growing magnitude of industrial activity [2]. A sharp increase in population along with urbanization have catalysed the exhaustion of natural resources, raising serious concerns about human–environment interactions. Measures are therefore desperately required to regulate the judicious utilization of natural resources and preserve the environment. To address these issues, concept of "Green Chemistry" was born, with the objective of lowering pollution levels in the environment and creating sustainable alternatives for industrial chemicals and products [1–3]. Green chemistry stresses upon developing ecologically safer chemical products and processes through minimal utilization and emission of toxic components. It solely aims at promoting sustainability and conserving natural assets [4]. The present strategy lays emphasis to curb the discharge of detrimental substances within the ecosystem, develop economical and safer synthesis methods, and advocate substitute protocols that have a reduced effect on human health and ecosystems [5].

While initial efforts in the green chemistry movement centred on reducing waste altogether, there is a significant movement toward understanding how to use waste in a truly circular economy [6]. Traditional synthetic routes may be to blame for the crisis of sustainable materials, but the same chemical ingenuity present in their

B. B. Pandey · M. Matang · V. M. Ganvit · R. K. Sharma (✉)
Department of Applied Chemistry, Faculty of Technology and Engineering, The Maharaja Sayajirao University of Baroda, Vadodara, Gujarat, India
e-mail: rksharma-appchem@msubaroda.ac.in

H. Patel
Department of Chemistry, Shrimad Rajchandra Vidyapeeth, Dharampur, Valsad, Gujarat, India

G. Tailor et al. (eds.), *Nanomaterials*, Engineering Materials,
https://doi.org/10.1007/978-981-92-0678-0_10

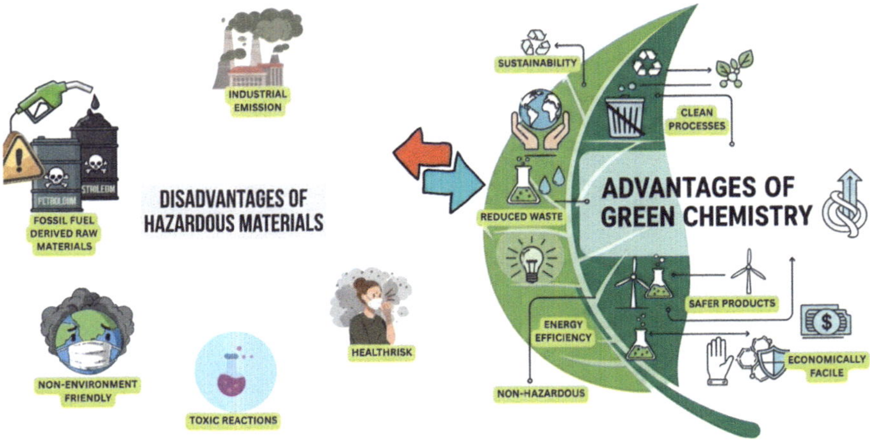

Fig. 1 Comparison of the disadvantages associated with fossil-based materials and the advantages offered by green chemistry approaches

protocols can be used for truly green alternatives [7]. Green chemistry has arisen in recent decades as a way forward in the synthesis of polymers and other materials given the negative outlook provided by reliance on traditional fossil resources and synthetic protocols. The Environmental Protection Act of 1990 can be said to the start of the green chemistry movement, although its roots stretch back to the 1980's [8]. Green chemistry seeks to produce chemicals and materials, "benign by design" where both the feedstock of chemicals is important as well as the harmfulness and toxicity they present to ecosystems [9, 10]. Figure 1 illustrates the hazardous aspects of Fossil based raw materials and advantages of green chemistry.

Since, United Nations Sustainable development goal 3 aims at promoting good health, it is imperative to produce technologies that are environmentally friendly and greener in approach [11, 12]. Green chemistry has found applications in a variety of industries, including personal care, cosmetics, and pharmaceuticals [13]. Improving production efficiency, considering the advantage of the full life cycle of materials, and developing environ mentally friendly materials are key areas of focus [14].

The multiple benefits of green chemistry include less expensive production costs, cheaper disposal costs, and fewer resources needed while compromising the quality of the final product. Researchers and industry may develop innovative, sustainable procedures and products that promote sustainable development targets as the field improves.

The current chapter aims at giving insights upon the how green chemistry is relevant to the current scenario, followed by evolution of nanotechnology. Also, it explains the symbiotic relationship between the two arenas combined together to give rise to a newer field called as green nanotechnology. Furthermore, the chapter delves deeper into various greener substrates utilized to synthesize various class of nanomaterials such as metal-based, carbon-based, ceramic-based, polymeric, lipid-based, etc. Also, insights have been given upon the effective utilization of ecologically friendly nanomaterials for an array of applications, including photocatalytic actions, energy storage units, biomedical applications, waste-water remediation, and coatings applications.

1.1 Green Chemistry and Its 12 Principles

In 1998, Paul Anastas and John Warner presented the 12 principles of green chemistry. The goal of these guidelines is either reduction or complete elimination of the utilization of dangerous solvents in chemical reactions, along with the prevention of toxic waste generated during the process. The 12 principles of green chemistry have been illustrated in Fig. 2. These ideas illustrate a method of approaching chemistry that aims to increase the sustainability of the conventional chemical industry [15].

Fig. 2 12 Principles of green chemistry

The principle of "Waste prevention" asserts that the prevention of waste generation serves as a more efficient approach than remediating the waste after it has been generated. Globally, millions of millions of tons of trash-ranging from household and industrial to agricultural-are generated, every year [16]. To mitigate the possible risks, waste needs to be collected, stored appropriately under proper conditions, followed by subjecting it to recycling and lastly disposing safely. Currently, recycling facilitates the conversion of useless materials into products of high utility. In one such example, researchers used industrially generated waste, including palm oil shells, fuel ash, such example the manufacturing of bricks [17]. Similarly, Tsai and group made use of thermophilic and lipolytic microorganisms to produce biofertilizer out of food waste [18].

The term "Atom economy" occasionally known as "Atom efficiency", illustrates a practice for utilizing atoms within the substrates effectively and its conversion into the final products [4]. It states that an ideal chemical reaction is the one that utilizes all the atoms from the reactants during the process. However, this ideal is not always fully realized in practice. Consequently, it is essential to develop reaction conditions that reduce the formation of by-products, minimize the number of reaction steps, and promote environmental sustainability. For instance, the cycloaddition Diels–Alder reaction, widely regarded as one of the most environmentally friendly reactions in conventional chemistry, involves the reaction of a conjugated diene, with an alkene (behaving like a dienophile). Theoretically, this reaction is 100% efficient since every atom of the reactants is included in the finished product [19].

The concept of reducing hazardous chemicals focuses on developing innovative production methods that encourage the utility and synthesis of ecologically benign substances [20]. One such report effectively used clay and zeolite derived catalysts for selective nitration of aromatic compounds as a newer approach. This approach offers significant advantages, including substantially reduced waste emissions, minimal water usage, and high reaction yields, making it a more sustainable alternative to conventional nitration techniques [21].

The principle of designing safer chemicals is aimed at developing the processes that improve the reactive functional groups within the chemicals that are frequently preferred while minimising their negative consequences. In this regard, Epp and colleagues created Rinskor, a novel herbicide, to manage weeds. Rinskor reduces pesticide residue in food and the environment by being administered at lower rates (5–50 g/ha) [22]. In maritime contexts, plants that inadvertently grow on the surfaces of the ship, provide another illustration of this idea. These microbes are responsible for acid rain and environmental contamination. In order to address this problem, 4,5-dichloro-2-n-octyl-4-isothiazolin-3-one, behaving as a biodegradable substance that complies with green chemistry principles, was created as a less hazardous substitute for extremely poisonous organotin chemicals [21].

The concept of employing safer solvents and auxiliaries emphasizes upon reducing the harmful impacts of traditionally occurring solvents through encouraging the utilization of ecologically benign substitutes. Water based solvents can serve as a viable substitute for organic solvents, thereby helps in reducing the pollution. It is an efficient and eco-friendly solvent comprising of an excellent heat capacity and the capability to improve processes through its salt effect and hydrophobic attributes. It is frequently employed for reverse-phase purification, as a mobile phase in high-performance liquid chromatography (HPLC), and as the aqueous phase during extraction. Although it produces aqueous waste that needs to be properly disposed of and presents a threat of contamination, its impact on the environment is negligible if it is devoid of harmful substances, making it an environmentally sound choice for many applications [23]. As a result, using water-based solvents is a crucial step in improving sustainability and lowering environmental pollution. Narayan and group illustrated the distinctive reactiveness of organic substances in aquatic media and concluded that a range of reactions, including cycloadditions, Ene reactions, and Claisen rearrangements, exhibit enhanced kinetics when performed in water, despite the insolubility of the organic substances [24]. With the increasing demand for more environmentally friendly solvent systems, researchers have sought alternative solutions that offer both sustainability and improved chemical performance. Deep eutectic solvents (DESs) have gained recognition as an effective and ecologically benign substitute for traditionally occurring solvents. These are economically viable, biodegradable, and highly flexible, and are being widely investigated for their utilizations in catalysis, electrochemical approaches. They also facilitate the dissolution of biopolymers. Deep eutectic solvents (DESs) have garnered a lot of interest in the field of green chemistry since their introduction in 2003 [25], mainly because of their excellent solvation capabilities and low toxicity [26, 27]. Modern developments in DESs-mediated protocols reveal their effectiveness as stabilizing agents for metal-derived nanoparticles and in the valorisation of biomass. Owing to their versatile and sustainable attributes, the incorporation of these solvents within the realm of green chemistry practices is anticipated to grow significantly soon.

The idea of energy conservation advocates for the effective utilisation of energy assets while encouraging environmental sustainability. Implementing techniques like employing catalysts, regulating chemical reactions, and substituting traditional solvents with greener alternatives, contributes to energy efficiency. Catalysts, for example, minimise overall energy utilization by enhancing efficiency through lowering reaction temperatures [28]. Improving the energy efficacy within a chemical system serves like a partial response to the broader global energy crisis faced nowadays. The growing global energy requirements has recently intensified the focus on adoption of renewable sources, including bio-fuels, solar energy, wind power, hydroelectric energy, and geothermal resources [29]. Photovoltaic cells, for instance, are now widely employed across different fields to generate power from solar radiation [30]. Ganiyu and coworkers designed a photovoltaic system for treating wastewater electrochemically, aiming to advance environmentally sustainable treatment methods powered by renewable energy sources [31].

The use of renewable raw materials is guided by the principle of promoting sustainability and minimizing environmental impact through the incorporation of renewable resources in chemical processes. Common examples include cellulose, lignin, lactic acid, chitin, etc. [32]. Incorporating such components within a system reduces the ecological damage along with promoting sustainability. One such approach has been utilized by Nature works for producing ecologically safer bottles using corn-derived substrates, thereby providing a sustainable substitute to conventional petroleum-derived plastics.

The principle of reducing derivatives emphasizes minimizing unwanted derivatization steps in chemical processes. Derivatization typically involves introducing new functional moieties within a molecule to alter its reactivity or characteristics; nevertheless, this protocol often calls for further chemicals, making it less desirable when not essential. A representative case of non-covalent derivatization involves modulating hydroquinone diffusion in Polaroid matrices to enhance solubility and enable pH-responsive release without altering molecular integrity [33]. This study achieved reductions in both- release of waste as well as energy utilization. A further illustrative case is the synthesis of 6-aminopenicillanic acid (6-APA), crucial precursor in penicillin manufacturing. Conventional synthetic routes typically require protective group strategies, resulting in elevated material waste and increased operational expenses. In contrast, greener approaches enable the direct production of 6-APA without protective groups, thereby enhancing process efficiency and reducing ecological footprint [34].

A further key principle of green chemistry is the use of catalysts, substances that accelerate the rate of chemical reactions without being consumed or permanently altered in the process. This phenomenon is known as catalysis. In addition to facilitating environmentally friendly chemical transformations, catalysts play a vital role in promoting sustainability and improving economic efficiency. A notable example involves the manufacturing of Ibuprofen, a commonly exploited anti-inflammatory medication. Traditional manufacturing involves a six-step process, generating significant waste and by-products. In contrast, the modern catalytic synthesis reduces the process to three steps, significantly enhancing efficiency and minimizing waste. This improvement exemplifies more ecologically as well as economically facile practices in pharmaceutical manufacturing [35, 36].

Aligned with the concept of degradation design, chemicals should be engineered to reduce their negative effect by decomposing into non-toxic byproducts following the fulfilment of their functional role, thereby preventing environmental persistence. Linear alkylbenzene sulfonate (LAS), could be considered as ecologically safer laundry detergents, discovered by a German chemist; exhibiting surfactant attributes owing to the polar sulfonate moieties and nonpolar alkyl moieties, thereby improving the wastewater mixing and cleansing. Such substances can be properly disposed of in wastewater without causing ecological damage as they are quickly broken down by bacteria through β-oxidation in oxygen-rich conditions. Studies on the environmental degradation of LAS detergents have indicated that approximately 98% of these compounds are decomposed in contaminated water [3]. These results suggest that such components may serve as effective ecologically sustainable substitutes.

In accordance with the principle of monitoring and controlling pollution, real-time analytical methodologies should be implemented to identify and inhibit the formation of toxic compounds during industrial or environmental processes [37]. This principle seeks to reduce the generation of hazardous components by mandating persistent tracking throughout each stage of the production process. Conventional analytical chemistry methods typically involve the use of larger proportion of sample, considerable quantities of solvents, and higher energy utilization. Nonetheless, advancements in innovative methodologies and the availability of precise, portable instrumentation have enabled more sustainable analyses using reduced solvent quantities and minimal sample sizes. Nuclear magnetic resonance (NMR) and infrared (IR) spectroscopy are two methods frequently employed in industry for on-site examination of reaction mixtures. These techniques facilitate the real-time tracking of reactants, intermediates, and final products, thereby improving the mechanistic insight [38].

The concept of inherently safer chemistry for accident prevention highlights the significance of choosing components and designing their applications in such a manner that it reduces the risk of incidents like leakages, explosions, and fires. A significant and sombre incident illustrating the hazards of chemical use happened in January 2009 at the University of California, Los Angeles (UCLA). In this case, the handling of butyllithium, a highly flammable and commonly preferred material resulted in a catastrophic event that claimed the life of a research assistant. Such incidence emphasizes the critical importance of recognizing chemical hazards and prioritizing the substitution of hazardous substances with safer alternatives whenever feasible.

1.2 Nanotechnology and Its Major Applications

Nanotechnology has evolved into a cutting-edge technology that caters to multiple research domains, including engineering, environmental science, food, textiles, and biomedicine. Not only that, it also motivates to developing materials through sustainable approach and then utilizing the for the benefit of mankind. In recent years, the need for nanotechnology has increased significantly [39]. By 2020, the worldwide nanomaterial market generated around $38.5 billion in revenue, which got rose by almost 12% between 2021–2025 [40]. Furthermore, the current technology relies heavily on nanoscale materials to achieve high performance, and their performance is highly influenced by the geometry, structure, and morphology of these materials at the nanoscale [41]. The nanotechnology can be considered as the control of matter at dimensions ranging from 1 to 100 nm. Any material falling within the realm of this dimension can be regarded as nanomaterial and it is widely accepted across various disciplines, including organic chemistry, surface science, physics, molecular engineering, molecular electronics, nanomedicine, and medical diagnostics, among others, and has significantly advanced both science and technology. The initial emergence of this field was unclear, which can be understood by the remarks given by American scientist Ralph Merkle noting, "Nanotechnology is an idea that most people simply didn't believe," it has emerged as a transformative tool across multiple sectors. In today's time, nanotechnology has become a key point to focus upon as its advancement will eventually lead to the upliftment of diverse fields such as medicine, food production, electronics, fuel cells, solar energy, battery technology,

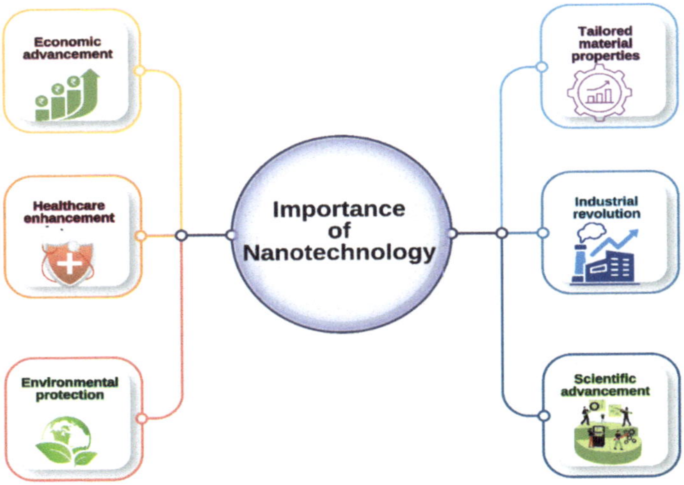

Fig. 3 Importance of nanotechnology

chemical sensing, space exploration, sports equipment, and synthetic fabrics [42]. The importants of nanotechnology is shown in Fig. 3.

1.3 Historical Background and Evolution of Nanotechnology

Despite nanotechnology being explored since ages, its actual advancement has occurred in modern era that surmounts to its utilization extensively [43]. The pioneering work of Michael Faraday in 1857 marks the beginning of nanotechnology. This work showed the world an extensively diverse field. Herein, he explored the interaction between light and matter by studying the synthesis and characteristics of colloidal gold, known as "ruby" gold suspensions and concluded that gold nanoparticles could change the color of a solution depending on specific lighting conditions [44]. Furthermore, several other nanomaterials were developed in the similar line of work. For instance, in 1889, Lea developed a technique for producing finely divided silver, referred to as silver nanoparticles through reduction of substances such as aqueous $AgNO_3$ using agents like citrates. Such materials have the dual functionality whereby they can serve as both, a reducing and stabilizing agent, citrate prevented particle aggregation. This method later became known as the citrate reduction method, which is still widely used today due to its simplicity and effectiveness [45]. Not only has there been advancements in developing nanomaterials but there has also been development in technologies that can effectively be used to confirm the synthesis of such particles. The advancement of electron microscopy marked a pivotal breakthrough in the evolution of modern nanotechnology. In 1935, Ernest Ruska and Max Knoll successfully built the first electron microscope, laying the foundation for nanoscale imaging and analysis [46]. Nanostructured semiconductor field emitters replaced tungsten sources, greatly advancing SEM resolution for material imaging [47]. This made the research flourish which was later accompanied by another significant and transformative advancement in the field of nanotechnology witnessed through invention of the scanning tunnelling microscope by Gerd Binnig and Heinrich Rohrer in 1981, which enabled the acquisition of high-resolution images of nanomaterials. This was followed by the development of atomic force microscope [46–50]. The most remarkable events and discoveries that defined modern nanotechnology is presented in Fig. 4.

In 1952, Radushkevich and Lukyanovichfor the first time described carbon nanotubes through electron microscopy [51]. Richard P. Feynman's lecture, titled "There's Plenty of Room at the Bottom," delivered at the American Physical Society's annual meeting on December 29, 1959, laid the conceptual foundation for the field that would later become known as nanotechnology [52]. Since then, several genuine efforts by researchers have led to the development of the field. One such attempt was witnessed in 1960s, wherein only a few researchers, notably Richard Zsigmondy, anticipated the emerging field of nanoscience and introduced the term "nanometer" to denote the size of gold colloidal particles [53]. Later on, during a conference held in 1974, Norio Taniguchi coined the term "nanotechnology" to refer to the manipulation of materials at the nanoscale [46–50].

Eric Drexler's *Engines of Creation* (1986) introduced nanotechnology as a revolutionary and transformative scientific field [54]. This journey led to development of various other materials. For instance, carbon-based allotropes began to attract significant interest from the scientific community. However, the rapid advancement of nanotechnology and the growing fascination with carbon-based materials were notably accelerated following Sumio Iijima's discovery and structural characterization of carbon nanotubes [55]. This characterization opened a new horizon to the validation of nanomaterials produced. Further, in 2000, The United States National Nanotechnology Initiative (NNI) was officially launched as a federal research and development program aimed at coordinating and advancing nanotechnology across multiple government agencies. This step motivated for developing nanotechnology for better quality of life [56]. To honour such developments in field of nanotechnology, in 1996, F. Curl, Harold W. Kroto, and Richard E. Smalley were awarded the Nobel Prize in Chemistry for their 1985 discovery of fullerenes. Later, in 2010, Andre Geim and Konstantin Novoselov were honoured with the Nobel Prize in Physics for their groundbreaking work on the physical properties of graphene, which they successfully isolated from graphite at room temperature [57, 58]. Carbon dots (CDs) were first discovered serendipitously in 2004 by Xu et al. during the purification of single-walled carbon nanotubes, where they observed small fluorescent carbon nanoparticles as by-products. This embarked the journey of CDs, however, it was for the first time in 2006, when Sun et al. formally introduced the term "carbon dots" and significantly advanced their development by demonstrating that surface passivation with organic molecules enhanced their fluorescence. These two reports in CDs field led to a new discovery within the fluorescent carbon-based nanomaterials as a new

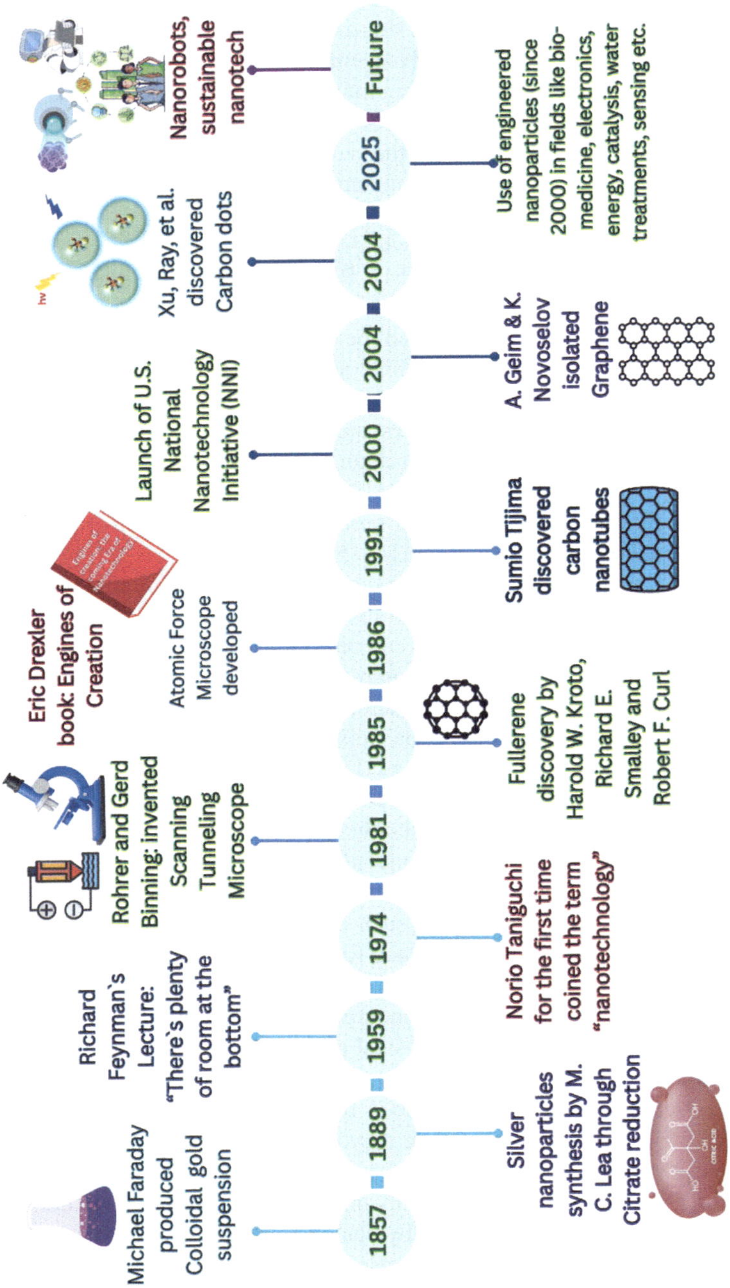

Fig. 4 Progresses in nanotechnology

class of carbon-based nanomaterials with strong, tuneable emission and potential applications in various fields [59, 60].

2 Green Nano-chemistry: A Modern Yardstick

Basically, green nanotechnology explains how green chemistry and nanotechnology have come together as a modern yardstick to promote the newer concept of green nanotechnology. This concept encompasses the sustainable design, synthesis, and application of nanomaterials and nanodevices with minimal environmental footprint. This approach prioritizes the reduction of toxic and hazardous chemical inputs-such as reagents and solvents-while simultaneously enhancing the efficiency of material and energy utilization in chemical processes. By adhering to green chemistry principles and employing safer precursor substances, researchers can develop nanomaterials that are inherently more environmentally benign, thereby mitigating potential ecological and biological risks. Furthermore, green nanotechnology seeks to minimize the adverse environmental consequences traditionally associated with nanotechnological advancements, all while striving to optimize their positive societal impact [61, 62]. Extensive research has examined the use of naturally occurring polyphenolic compounds, citric acid, and various vitamins, along with renewable polymers and silica, for the ecologically safer synthesis approaches along with the functionalization of nanostructures [63]. Such nanostructures offer versatile applications in mitigating the adverse effects of environmental contaminants and are increasingly integrated into technologies aimed at addressing ecological challenges. These include hazardous waste remediation, air filtration, pollutant degradation, suppression of toxic phytopathogens and associated compounds, environmental pollutant detection and monitoring, and purification of contaminated water systems. For example, gold nanoparticles (AuNPs) exhibit strong adsorption capabilities for toxic substances in non-potable water and are employed in the quantitative analysis of pesticide residues in drinking water. Similarly, nanoscale iron has proven effective in assessing and treating polluted soils. Moreover, nanomaterials contribute indirectly to environmental sustainability by enabling lightweight nanocomposites in automotive manufacturing, thereby enhancing fuel efficiency through reduced vehicle mass [64, 65]. This reduction in weight also results in decreased material requirements during manufacturing, thereby conserving natural resources and lowering energy consumption throughout production processes. An additional example of green nanotechnology is the development of organic thin-film solar cells incorporating fullerene compounds, which offer improved cost-effectiveness and energy efficiency compared to conventional silicon-based photovoltaic technologies [62, 66, 67].

Green nanotechnology and its emerging innovations are anticipated to exert substantial influence across a wide range of economic sectors. Although many of these advancements are currently confined to laboratory research or early-stage commercialization, a number of mature green nanotechnological products have already entered the market. It is important to recognize, however, that the designation of a nanotechnology as "green" does not inherently guarantee its sustainability or immunity from potential risks. Despite the considerable environmental, societal, and economic advantages associated with green nanotechnology, each material and application must undergo rigorous evaluation to determine its safety profile and identify any unintended consequences. Consequently, it is essential to systematically compare the risks and costs of green nanotechnological approaches with those of conventional technologies, which may be characterized by high energy demands, toxic chemical usage, or significant ecological harm. Moreover, the potential benefits of green nanotechnology must be critically weighed against the human and environmental costs of failing to address pressing global challenges, such as mitigating carbon emissions and ensuring access to safe drinking water [68]. It is essential to critically evaluate the extent to which green nanotechnology can offer viable solutions to pressing global challenges, while simultaneously mitigating its inherent limitations and potential risks.

3 Fundamentals and Scope of Nanotechnology

Nanomaterials are typically produced through two principal methodologies: the bottom-up and top-down approaches. The former approach involves assembling structures from individual atoms or molecules through a range of chemical techniques, including hydrothermal treatment, sol–gel processing, and redox reactions. Additionally, this method encompasses biologically mediated processes that utilize plant-derived biomass, microbial systems, and biopolymers to facilitate environmentally sustainable fabrication [69]. In contrast, the latter approach involves the reduction of bulk materials to the nanoscale through physical techniques such as ball milling, lithographic patterning, and laser ablation. Each synthesis strategy presents unique benefits and limitations in terms of scalability, economic feasibility, and environmental impact. The selection of an appropriate synthesis method is determined by the targeted physicochemical properties of the nanomaterials and their intended applications across fields such as biomedicine, energy systems, and environmental remediation technologies [70]. The synthesis of nanomaterials facilitated by plant extracts represents a cost-efficient and environmentally benign strategy, capitalizing on the intrinsic reducing, stabilizing, and capping functionalities of phytochemicals such as alkaloids, flavonoids, tannins, and phenolic compounds [71]. This involves the controlled combination of precursor compounds under specific conditions of temperature, pH, and reaction duration. Such parameters critically influence the physicochemical characteristics of the resulting nanoparticles, including their size, morphology, and colloidal stability. For instance, appropriate adjustment of pH and extract concentration can yield nanoparticles with distinct geometries such as spherical, triangular, or rod-like structures. The stability of these nanoparticles is enhanced by naturally occurring capping agents present in the plant extract, which inhibit aggregation. This approach utilizes renewable biological resources to produce biocompatible nanomaterials, aligning with principles of sustainability and environmental stewardship [72].

4 Green Nanotechnology: Sustainable Approaches Through Eco-Friendly Synthesis

4.1 Metal and Metal-Oxide Nanoparticles

Metal nanoparticles (MNPs) and metal oxide nanoparticles (MONPs) have emerged as next-generation, economical disinfecting agents with significant potential in clinical applications, contaminant remediation, and diverse industrial fields. Their unique properties, including antibacterial and antifungal activity, UV filtering ability, high catalytic efficiency, and strong photochemical activity, make them highly valuable [73].

The green synthesis of MNPs and MONPs generally involves the use of plant extracts, microorganisms (bacteria, fungi), and algae as reducing and stabilizing agents, producing nanoparticles from naturally biodegradable matter. Among these, plant extract-mediated synthesis is particularly advantageous because it minimizes the risk of secondary contamination, reduces reaction time, and eliminates the need to maintain microbial cell cultures. Biological systems such as microbes and algae are also widely explored for nanoparticle synthesis due to their eco-friendly nature, low cost, and ability to generate nanostructures with unique functional properties [74]. Below we have given a detailed discussion of nanomaterials either derived from biomass resources or synthesised using greener techniques. The materials have been classified as under.

4.1.1 Plant-Derived Synthesis

Plant-mediated synthesis of nanoparticles offers several benefits over conventional methods: it is simple, cost-effective, relatively reproducible, and often results in stable materials. While microorganisms can also produce nanoparticles, their synthesis is generally slower and more limited in terms of particle size and shape compared to plant-based methods. Green synthesis does not require high pressure, high temperature, or toxic chemicals. As a result, researchers are increasingly moving away from synthetic approaches [75–77]. Plant-based nanoparticles are typically more stable, easier to scale up, and less prone to contamination. This makes plants and their parts (leaves, stems, roots, flowers, fruits, etc.) a promising alternative for nanoparticle synthesis as shown in Fig. 5 [78]. A summary of commonly used plants, their specific parts, and the nanoparticles obtained from them is presented in Table 1. Due to these advantages, plant-mediated synthesis is gaining importance across many fields that emphasize environmental sustainability. Extensive research has been carried out

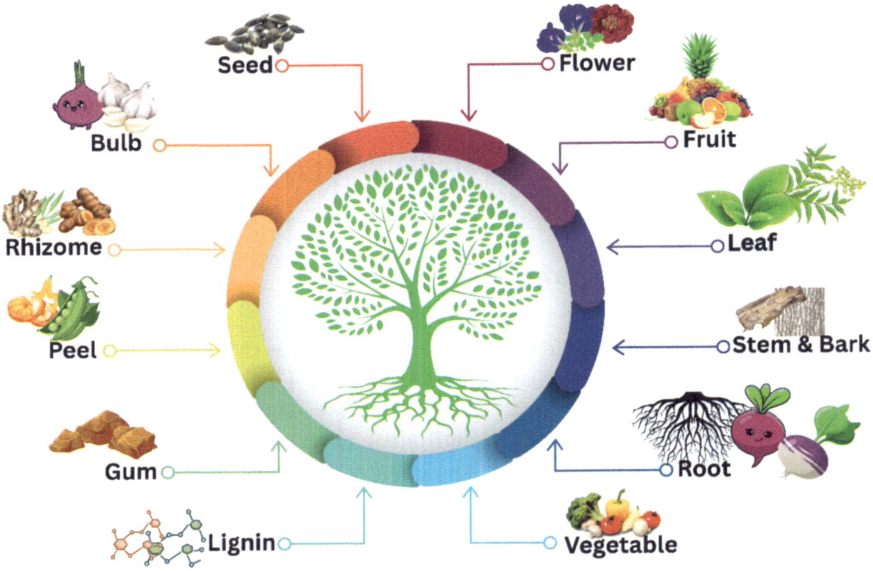

Fig. 5 Plants and their parts as promising alternative for nanoparticle synthesis

using plant extracts because of their scalability, cost-effectiveness, and eco-friendly nature [79, 80].

4.1.2 Microbe-Based Synthesis

Microorganisms synthesize nanoparticles as part of their defence mechanism against toxic concentrations of metal ions in the environment. When studied to such conditions, bacterial cells reduce metal ions to elemental nanoparticles, which help them survive [153, 154]. Fungi are also widely explored because they are easy to handle, cost-effective, and secrete extracellular enzymes that facilitate large-scale nanoparticle production [155, 156]. A summary of commonly studied microorganisms and the types of nanoparticles synthesized is presented in Table 2. However, microbial methods face limitations, including slow synthesis rates, contamination risks during prolonged reactions, and potential issues related to genetic variation. Excess accumulation of nanoparticles inside microbial cells can also cause cell destruction [157].

4.1.3 Algae-Based Synthesis

Algae are eukaryotic aquatic photosites and they break down metallic salts into nanoparticles thanks to the pigments, proteins, carbohydrates, fat, nucleic acid and

Table 1 Nanoparticles synthesized using plant extracts

Nanoparticles	Plants	Applications	Refs.
Ag	**Flower extract**—*Canna indica* L., *Senna auriculata*. **Leaf extract**—*Glauca graham*. **Fruit extract**—*Lagerstroemia speciosa*, *Citrus unshiu*. **Root extract**—*Martynia annua* L. **Rhizome extract**—*Trillium govanianum*, *Hellenia speciosa* **Stem extract**—*Telfaria occidentalis*. **Bark extract**—*Bouea macrophylla* G. **Seed extract**—*Rhus coriaria* L., *Acacia sinuate*. **Bulb extract**—*Urginea indica* **Peel extract**—*Ananas comosus* and *Musa paradisiacal*, *Citrus sinensis*. **Gum extract**—*Ferulagummosa* Boiss., *Senegalia senegal*. **Vegetable extract**—*Lagenaria siceraria* and *Pisum sativum*, *Petroselinum crispum*, *Beta vulgaris* and *Lepidium aucheri*, Mixed vegetable waste	Antibacterial, antioxidant, dye degradation, cytotoxicity, anti-inflammatory, anti-diabetic and anti-glycation and anticancer	[81–102]
Au	**Flower extract**—*Gnidia glauca*, *Madhuca indica*. **Leaf extract**—*Ruta graveolens*. **Root extract**—*Artemisia annua* hairy. **Rhizome extract**—*Zingiber officinale* Rosceo, *Curcuma caesia* **Peel extract**—*Citrus sinensis* **Vegetable extract**—Mixed vegetable waste	Anticancer, chemo-catalysis, catalytic reduction, photocatalytic, anticancer, antimicrobial and antioxidant	[103–109]
ZnO	**Flower extract**—*Jacaranda mimosifolia*, *Butea monosperma*, *Azadirachta indica*. **Leaf extract**—*Justicia schemperiana*, *Syzygium cumini*. **Fruit extract**—*Morinda citrifolia*, *Citrullus lanatus*, *Nauclea latifolia*. **Root extract**—*Andrographis paniculata*. **Stem extract**—*Tinospora cordifolia*. **Bark extract**—*Annona squamosa* L., *Syzygium hemisphericum*. **Seed extract**—*Washingtonia filifera*. **Bulb extract**—*Zingiber officinale* and *Allium sativum*. **Peel extract**—*Citrus aurantium*. **Gum extract**—*Senegalia senegal*	Antimicrobial, bio-reductor, photocatalytic, microwave absorption, dye absorption, anti-inflammatory, anticancer, antihemolytic, anti-inflammatory, antidiabetic, and antioxidant	[110–124]

(continued)

Table 1 (continued)

Nanoparticles	Plants	Applications	Refs.
Cu, $CuFe_2O_4$, Cu_4O_3 and CuO	**Flower extract**—*Hymenocallis littoralis*. **Leaf extract**—*Rosmarinus officinalis* **Fruit extract**—*Momordica cymbalaria, Pedalium murex* L. **Root extract**—*Aloe vera* and *Flacourtia indica*. **Stem extract**—*Balanites aegyptiaca*. **Bark extract**—*Terminalia arjuna* L. **Seed extract**—*Cucurbita pepo*. **Bulb extract**—*Allium sativum* L., *Allium cepa* L. **Vegetable extract**—*Brassica oleracea, Solanum tuberosum*, and *Pisum sativum*	Optimization, antibacterial, antioxidant and anti-inflammatory, photocatalytic, cytotoxicity, Enhancing chromium resistance and Hg degradation	[111–135]
Fe and Fe_2O_4	**Flower extract**—*Hymenocallis littoralis, Madhuca indica*. **Fruit extract**—*Hylocereus undatus*. **Peel extract**—*Punica granatum*. **Vegetable extract**—*Daucus carota* subsp. Sativus	Optimization, antimicrobial, antioxidant, anticancer, and therapeutic	[136–138]
TiO_2	**Flower extract**—*Jasminum officinale*. **Leaf extract**—*Inula viscose, Globularia alypum*. **Root extract**—*Sophora flavescens*	Dye-sensitized solar cell application, antimicrobial, photocatalytic, inflammatory, and anticancer	[139–142]
MgO	**Leaf extract**—*Hyphaene thebaica*	Photocatalytic	[143]
Mn_2O_3	**Fruit extract**—*Hyphaene thebaica*	Photocatalyst for MB dye degradation	[144]
Cr_2O_3	**Root extract**—*Rauvolfia tetraphylla*	Anticancer	[145]
SnO_2	**Seed extract**—*Brassica juncea* L.	Anticancer, antioxidant, and anti-inflammatory	[146]
NiO	**Bulb extract**—*Zephyranthes citrina*	Antioxidant, antimitotic, and anti-diabetic	[147]
Pd	**Gum extract**—*Salmalia malabarica, Limonia acidissima* Groff	Antioxidant, catalytic, and antimicrobial	[148, 149]
Bi_2O_3	**Gum extract**—*Eucalyptus camaldulensis*	Antimicrobial	[150]

(continued)

Table 1 (continued)

Nanoparticles	Plants	Applications	Refs.
Cd	**Leaf extract**—*Nyctanthes arbor-tristis*	Antimicrobial	[151]
CeO_2	**Seed extract**—*Tectona grandis*	Antioxidant, and antibacterial	[152]

Table 2 Nanoparticles synthesized using Microbes

Microbes	Specie	Nanoparticle	Applications	Refs.
Bacteria	*Bacillus safensis, Cytobacillus firmus*	Ag	Antibacterial, larvicidal activity	[158, 159]
	Bacillus subtilis, Serratia nematodiphila	ZnO	Antimicrobial and Photocatalytic	[160, 161]
	Paenibacillus motobuensis	Se	Wound healing activity	[162]
	Geobacillus sp. *strain ID17*	Au	Optimisation study	[163]
Fungi	*Alternaria chlamydospora*	Au	Antibacterial and anticancer	[164]
	Cladosporium halotolerans	Ag	Anti-fungal and cytotoxic effects	[165]
	Humicola sp.	Au_2S	Antipromastigote	[166]
	Pleurotus floridanus	ZnO	Optimisation study	[167]
	Chaetomium cupreum	Fe_2O_3	Anticancer	[168]

secondary metabolites they contain. The algae extract that exists in an aqueous medium at a certain temperature is supplemented with metal solutions of the corresponding pH and concentration, and hence the synthesis of nanoparticles is achieved which may have antimicrobial properties without producing any toxic by-products during the synthesis. Size of nanoparticles is determined by certain parameters such as the incubation time of the solution, the ambient temperature, the pH of the mixture used and the metal ion concentration. Algae also provide an advantage to this synthesis method by virtue of their easy availability and usefulness. In addition, effective biomolecules in the reaction medium in the use of bacteria and plant extracts are less extinguished by the nanoparticles formed [169, 170]. A summary of commonly studied algae and the types of nanoparticles synthesized is presented in Table 3.

Table 3 Nanoparticles synthesized using algae

Specie of algae	Nanoparticle	Applications	Refs.
Gracilaria crassa, Chlorella vulgaris	SiO_2	Antioxidant and antimicrobial	[171, 172]
Ulva fasciata	ZnO	Biopreservation	[173]
Portieriahornemannii	Ag	Antimicrobial	[174, 175]
Sargassum vulgare, Ulva fasciata and *Jania rubens*	Fe_3O_4	Antibiofilm	[176]
Cladophora glomerata (L.), *Pterocladiella capillacea, Anadara gibbosa,* and *Chlorella vulgaris, Turbinaria*	Cu and CuO	Antimicrobial	[177–179]
Asterarcys sp.	Pd	Catalytic	[180]
Nannochloropsis oceanica	Se	Antioxidant	[181]

4.2 Carbon-Based Nanoparticles

Biomass-derived carbon-based nanomaterials (BDCMs) have garnered significant attention as multifunctional and sustainable materials due to their adjustable porosity, expansive surface area, and superior electrochemical performance. In response to the escalating demand for renewable energy technologies, BDCMs have emerged as viable candidates for applications in supercapacitors, lithium-ion batteries, fuel cells, and heterogeneous catalysis. These carbon nanostructures are synthesized from plentiful and renewable biomass resources-including agricultural by-products, forestry residues, and municipal solid waste-offering an economically feasible and environmentally benign alternative to conventional carbon materials derived from fossil fuels. Advanced synthesis techniques such as pyrolysis, hydrothermal carbonization, and chemical activation facilitate the engineering of carbon architectures with precisely controlled structural and surface chemical characteristics, thereby enhancing their functional versatility [182]. Various subclasses of environmentally benign carbon-based nanomaterials have been shown in Fig. 6. We have classified these nanomaterials into various sections as below.

4.2.1 Carbon Nanotubes (CNTs)

The lack of stable and low-cost feedstocks for CNT growth has been a challenge since its discovery [183]. Over a decade, biomass materials such as natural plants (e.g., grass, Typha orientalis, catkin, chlorella, gumwood, reed, jatropha, moss, corn, corn stalks, sawdust, biochar), bio-extracts (e.g., chitosan, wood and cotton fiber), bio-products (e.g., wheat flours, camphor, palm oil, turpentine oil, eucalyptus oil, castor oil, sesame oil), and bio-wastes (e.g., egg yolks, softwood sawdust, potato peels, grain waste, sugarcane bagasse) have been used to produce cost-effective CNTs

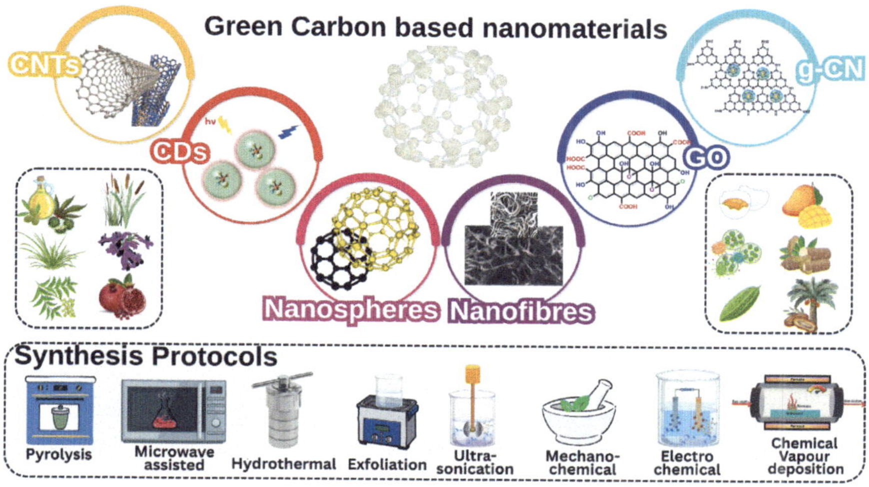

Fig. 6 Carbon based nanomaterials and their ways of synthesis

by catalytic reactions through various synthesis protocols such as pyrolysis, chemical vapour deposition (CVD), mechanochemical, microwave-assisted processes, as summarized in Table 4. Biomass facilitates CNT synthesis due to its abundant carbon storage. In addition, the oxygen within biomass materials helps absorb amorphous carbon, improving CNT nucleation [184–186]. The H_2O molecules decomposed from biomass are found to prevent CNTs from oxidation or damage by removing amorphous carbon [187–189].

4.2.2 Carbon Dots

Carbon dots (CDs) have attracted considerable attention as a new type of fluorescent carbon nanomaterial because of their excellent optical properties, biocompatibility, and high electrical conductivity. Research on CDs has been conducted for nearly two decades and has focused on numerous precursors, various synthesis conditions and properties and applications of CDs. Biomass is critical in the green development of CDs because of its low cost, environmental friendliness, and sustainable properties [213]. Compared with semiconductor quantum dots and organic dyes, carbon dots exhibit remarkable properties such as tuneable photoluminescence, chemical inertness, ease of production, low toxicity, and good biocompatibility. These advantageous characteristics give carbon dots broad potential for applications in several fields, including chemical sensing, bioimaging, photocatalysis, and wastewater treatment. The selection of carbon-rich raw materials is crucial for the efficient production of carbon dots [214]. Several reports have utilized biomass such as pyrolytic lignin [214], amaranth [215], *Syzygium malaccense* [216], *Aegle marmelos* [217], Chinese herbal residue [218], L-arginine [219], mango pulp [220], blue jacaranda petals [221],

Table 4 Green synthesis of carbon nanotubes

Source	Synthesis protocol; conditions	CNT length (μm)	Diameter (nm)	Refs.
Grass	Cyclic pyrolysis; 600 °C, 20 min	1	30–50	[190]
Turpentine oil	Spray pyrolysis; 700 °C, 1 h	300	50–100	[184]
Camphor	CVD; 850 °C, 15 min	–	–	[191]
Eucalyptus oil	Spray pyrolysis; 850 °C, 25 min	–	0.79–1.71	[185]
Wood fiber	Cyclic pyrolysis; 400 °C, 10 min	–	10–20	[192]
Palm oil	CVD; 450–750 °C, 30 min	110	0.6–1.2	[186]
Castor oil	Spray pyrolysis assisted CVD; 850 °C, 15 min	5–10	20–60	[193]
Jatropha	Spray pyrolysis assisted CVD; 800–900 °C, 20 min	20	20–50	[194]
Reed	CVD; 1200 °C, 2 h	10–5	–	[195]
Brown peat moss	Mechanochemical approach; activation 4–36 h	10–70	–	[196]
Gumwood	Microwave assisted pyrolysis; 500 °C, 30 min	0.6–1.6	50	[197]
Typha orientalis	Pyrolysis; 850 °C, 1 h; 900 °C, 1 h	–	35–45	[198]
Catkin	Pyrolysis; 800 °C, 2 h	–	80	[199]
Softwood sawdust	Pyrolysis; 800 °C, 1 h	–	20–50	[200]
Multiple biomass	Mechanochemical approach; activation 4–36 h	–	10–70	[201]
Wood sawdust	Pyrolysis; 750 °C, 3 h	25	–	[202]
Chlorella	Pyrolysis; 900 °C, 1 h	0.1–2	40	[203]
Egg yolks	Pyrolysis; 900–1100 °C, 2 h	–	50–80	[204]
Chitosan	Pyrolysis; 800 °C, 2 h	–	20	[205]
Wheat flour	Pyrolysis; 750 °C, 2 h	3–20	100–200	[206]
Potato peel	Pyrolysis; 600 °C, 3 h; 900 °C, 1 h	–	10–25	[207]
Grain waste	Pyrolysis; 600 °C, 3 h; 900 °C, 1 h	–	15–35	[208]

(continued)

Table 4 (continued)

Source	Synthesis protocol; conditions	CNT length (μm)	Diameter (nm)	Refs.
Cotton fiber	Pyrolysis; 480 °C, 2 h; 900 °C, 2 h	0.5	50	[209]
Pinewood biochar	Microwave assisted pyrolysis; 100–300 W, 80–160 °C, 1–10 min	135–490	12–46	[210]
Pine sawdust	CVD; 500–750 °C, 3 h	–	64.84; 33.61	[211]
Corn stalk waste	CVD; 800 °C, 2 h	–	30.57	[212]

pomegranate juice [222], radish leaves [223], *Azadirachta indica* leaves [224], bitter gourd [225], *Polyalthia bullata* root [226], cassava peel [227], Green alga *Dunaliella salina* [228] through carbonization approaches yielding blue, green, red emissive carbon dots. Furthermore, the heteroatoms doping (nitrogen, boron, sulphur, etc.) or functionalization could alter the intrinsic properties of the CDs by tuning the electronic structures of CDs [216]. The enhanced optical attributes and fluorescent nature has made CDs a prospective candidate in various fields like sensing, photocatalyst, bioimaging, phototherapy, anti-counterfeiting, etc. [213, 216]. The various aspects of the biomass-derived CDs have been reported in Table 5.

4.2.3 Carbon Nanospheres (CNSs)

Generally, CSs are classified according to their dimensions in three categories: (i) fullerene family and well-graphitized onion-like structures (Ø2 – 20 nm); (ii) less graphitized carbon nanospheres (Ø50 nm–1 μm); and (iii) carbon microspheres or carbon beads (>Ø1 μm). They can be further classified according to their morphologies as solid, porous, hollow, core–shell structures, and colloidal spheres [232]. Mono-CSs were synthesized through the conventional hydrothermal approach, employing AA as the precursor. Afterward, a composite consisting of CSs and graphene oxide sheets was utilized in lithium-ion batteries [232]. Wang et al. developed Fe-decorated N/S-codoped porous carbon nanospheres successfully using cuttlefish ink through $Fe_2(SO_4)_3$ assisted hydrothermal carbonization coupled with heat treatment for [233]. Li et al. reported an optimized synthesis protocol for the commercial-scale production of water-soluble nitrogen-doped, polymer-like carbon nanospheres (CNSs) exhibiting a high quantum yield of 38%. This environmentally friendly and straightforward hydrothermal method utilized cocoon silk (*Bombyx mori*) as a natural precursor in aqueous medium. The resulting nanospheres demonstrated strong photoluminescent properties, along with low cytotoxicity and favourable biocompatibility, making them suitable candidates for cellular imaging applications [234]. Blue-emissive carbon nanospheres were synthesized using the pith of tapioca (*Manihot esculenta*) stem, via a straightforward hydrothermal

Table 5 Biomass-derived carbon dots

CDs	Dopant	Biomass source	Synthesis protocol; conditions	Size (nm)	% Yield	Refs.
CDPLs	–	Pyrolytic lignin	Hydrothermal approach; 200 °C, 12 h	34	7.9 ± 0.7	[214]
R-CDs	Nitrogen	Amaranth	Solvothermal approach; 180 °C, 6 h	4.6	42.0	[215]
CDs	–	*Syzygium malaccense* (*S. malaccense*)	Hydrothermal approach; 180 °C, 12 h	4.68	0.32	[216]
N-CDs	Nitrogen			4.50	0.41	
B-CDs	Boron			3.71	0.35	
S-CDs	Sulphur			3.10	0.36	
CQDs	–	Aegle marmelos, lemon juice	Hydrothermal approach; 110–200 °C, 45 min-12 h	3.9 ± 1.1	4.06	[217]
B@HRCDs	Boron	Herbal residue	Solvothermal approach; ethanol, 200 °C, 8 h	2.8	4.0	[218]
rCDs	–	L-arginine and neutral red	Hydrothermal approach; 160 °C, 12 h	2.34	22.9	[219]
BCDs	–	Mango pulp	Hydrothermal approach; 160 °C, 12 h	10	–	[220]
PDs and CPDs	– Nitrogen	Blue jacaranda petals	Room temperature approach; 10%H_2O_2, stirring Followed by functionalization with Schiff base	18.1 (PDs); 3.35 (CPDs)	10.47; 12.15	[221]
CDs	–	Pomegranate juice	Hydrothermal approach; 180 °C, 6 h	5	–	[222]
N-CQDs	Nitrogen	Radish leaves	Microwave assisted pyrolysis; 50 s	4.5	19.76	[223]
CDs	–	*Azadirachta indica* leaves	Pyrolysis 300 °C, 4 h	7.12 ± 1.97	–	[224]
C-dots	–	Bitter gourd	Hydrothermal approach; 180-300 °C, 5 h	5–9	–	[225]

(continued)

Table 5 (continued)

CDs	Dopant	Biomass source	Synthesis protocol; conditions	Size (nm)	% Yield	Refs.
CDs	–	Polyalthia bullata root	Hydrothermal approach; 200 °C, 15 h	3.34 ± 0.76	7.55	[226]
CDs	–	Cassava (*Manihot esculenta*) peel	Hydrothermal approach; 200 °C, 12 h	3.4	24.7 ± 0.5	[227]
N/S-CDs	Nitrogen, Sulphur	Green alga *Dunaliella salina*	Hydrothermal approach; 200 °C, 5 h	3.2 ± 0.5	5.93	[228]
C-dots	–	Sodium citrate and urea	Electrochemical hydrothermal carbonization; 5 V, 1 h	2.4	11.9	[229]
CDs	–	*Escherichia coli* BW25113 (WT)	Hydrothermal approach; 200 °C, 12 h	2.9	15.8	[230]
CDs	–	*Bacillus cereus*	Hydrothermal approach; 200 °C, 12 h	3.3	18.3	[231]

carbonization process. The resulting functionalized fluorescent nanospheres exhibited pronounced luminescence quenching behaviour with high selectivity toward Fe^{3+} ions [235]. It revealed that demonstrated that carbon nanospheres possess strong capabilities for in vivo cellular imaging, while exhibiting minimal cytotoxicity. A report by He and group combined five different biomass sources like rice husk, reed rod, platanus fruit fibers, flax fiber, and walnut shell with polytetrafluoroethylene that was directly pyrolyzed in tubular furnace at 900 °C for 4 h under N_2 flow, resulting in porous hollow carbon nanospheres [236]. These HCNSs were further utilized for supercapacitor applications. Pathania's group offered utilized Garlic peel based mesoporous carbon nanospheres (GCNP) for an effective removal of malachite green dye from aqueous solutions [237]. Another report utilized *Arachis hypogaea* (groundnut) biowaste-based novel nano-adsorbents offering the efficient dye removal cationic dyes such as MB and MG [238]. A single step and simple pyrolysis technique is used to prepare carbon nanospheres (CNSs) from natural biowaste sago hampas in a nitrogen atmosphere without any catalyst [239]. Hydrothermal carbon nanospheres were prepared by hydrothermal graphitization using glucose as the precursor, to improve the properties of water-based drilling fluid [240]. Green synthesized multilayered graphitic carbon nano-onions (CNOs) have been reported using an oil-wick flame pyrolysis approach, employing biowaste (chicken fat) oil as a cost-effective precursor for enhanced energy storage capacity [241]. The prepared

CNOs can provide enhanced ion movement and less resistance for electron transport by interconnecting CNO particles with one another. Furthermore, heteroatom (S,N)-doped CNOs (h-CNOs) were synthesized to optimize the hydrophilic and conductive properties of carbon materials, which eventually exalted the capacitive charge transfer kinetics. The h-CNOs demonstrated superior, highest specific capacitance of 261 F/g, while the undoped CNOs showed a capacitance of 180.6 F/g at a current density of 1 A/g. In addition to capacitance, the h-CNOs also demonstrated a rate capability of 69% and a good cycling stability of 97.5% under high current densities [241]. Another report showed effective utilization of sandalwood bark for nanosphere synthesis which was further composited with polyvinyl alcohol to form nanofibers for biomedical application [242]. A summary of the synthesis of CNSs using various green sources is provided in Table 6.

4.2.4 Carbon Nanofibers

Carbon nanofibers (CNF) are sp^2-based linear, noncontinuous filaments that are different from carbon fibers, which are continuous with diameter of several micrometres [243]. CNF is one of the important carbon nanomaterials that have been used widely, which has similar structure and properties to CNTs, but it has easier production, low cost, and improved functions [244]. Carbon nanofiber is a carbon-based fiber with a thickness of less than 1 μm, existing in many shapes including straight, spiral, and helical forms. CNFs reveal smooth, porous, hollow, helical, and stacked-cup structures, and they have large aspect ratio, low density, high specific modulus, good thermal conductivity, high mechanical strength, electric conductivity, and high specific surface area, which make them excellent nanoscale building blocks for the synthesis of CNF-based functional nanomaterials [243].

CNFs possess a comparable one-dimensional nanostructure, analogous characteristics, and multifunctionality to carbon nanotubes, while exhibiting certain advantages such as lower crystallinity and an increased number of defects. As a result, CNF-based nanomaterials have been widely employed in applications such as energy conversion and storage, catalysis, sensors, adsorption and separation, and biomedical engineering. CNF-based nanomaterials are garnering more attention in multiple fields, especially in energy production/storage and environmental studies, due to their high specific surface area, interconnected porous structure, lightweight characteristics, and considerable mechanical strength [245, 246].

The green synthesis of nanofibers involves the production of nanofibers by natural mechanisms, like microorganisms, fungi, or plant extracts, utilizing environmentally friendly methods that prioritize the reduction of harmful chemicals and the minimization of environmental impact. The approach refers to green chemistry principles, aiming to generate nanofibers that are safe for human use, with advantages such as cheap cost, high efficiency, and eco-friendly characteristics. The environmental impact may be minimised while enhancing chemical stability by the integration of natural or semisynthetic biopolymers with synthetic biopolymers [247].

Table 6 Synthesis of CNSs using various green sources

CNS	Source	Synthesis protocol; conditions	Size (nm)	Refs.
Carbon spheres (CSs)	Ascorbic acid	Microwave assisted hydrothermal approach; 170–220 °C, 10–40 min	1700–2700	[232]
Nitrogen, sulphur codoped-CNSs	Cuttlefish ink	Hydrothermal approach; Autoclave, 180 °C, 24 h	100–130	[233]
ws-N-doped polymer-like carbonaceous nanospheres	Cocoon silk (*Bombyx mori*)	Hydrothermal approach; Autoclave, 200 °C, 72 h	70	[234]
CNs	*Manihot esculenta* stem	Hydrothermal approach; Autoclave, 210 °C, 12 h	23.6	[235]
HCNS-RH; HCNS-RR; HCNS-PFF; HCNS-FF; HCNS-WS	Rice husk (RH), reed rod (RR), platanus fruit fibers (PFF), flax fiber (FF), and walnut shell (WS) with polytetrafluoroethylene	Direct pyrolysis; 900 °C, 4 h	60–100	[236]
GCNP	Garlic peel	Pyrolsis; 600–1000 °C, 1 h	2.9291	[237]
GS-CNSs	*Arachis hypogaea*	Pyrolysis; 800 °C	2.90	[238]
CNSs	Sago hampas	Pyrolysis; 400 °C, 600 °C, 2 h	65 ± 5; 50 ± 5	[239]
HCNs	Glucose	Hydrothermal approach; Autoclave, 180^0C, 12 h	300	[240]
CNOs	Chicken fat oil	Oil-wick flame pyrolysis approach	30–50	[241]
SWCNS	sandalwood bark	Pyrolysis; 750 °C, 2 h	30–50	[242]

Table 7 Biomass used as raw material to fabricate carbon nanofibers

Source	Synthesis protocol; conditions	Diameter (nm)	Refs.
Date palm residues	Precipitating date ethanol vapor, 680 °C	–	[249]
Lignin	Electrospinning, 90 °C	200–253	[250]
Crab shell	Pyrolysis	70	[251]
Straw lignin bio-oil	Catalytic depolymerization, 220 °C, 6 h	137–812	[252]
Palm kernel shell	Pyrolysis at 500 and 600 °C	0.34	[253]
Sucrose	Electrospinning, 950 °C	–	[254]
Chitosan	Electrospinning	25	[255]
Pine nut shell	Pyrolysis, 400–700 °C	400	[248]
Chitosan	Electrospinning, 800 °C	40	[256]
Sugarcane bagasse	Electrospinning, 800 °C, 2 h	117.0	[257]
Chitosan	Electrospinning, 800 °C	2	[258]
High amylose starch	Electrospinning, 260 °C,1 h	250	[259]
Natural fungus	Hydrothermal approach, 800 °C, 2 h	620	[260]
Sawdust	Pyrolysis	100	[261]
Seafood chitin	Pyrolysis	28	[262]
Sugarcane bagasse	Electrospinning, 800 °C for 2 h	130–350	[263]

Various methods, including electrospinning, chemical vapor deposition (CVD), templated synthesis, hydrothermal synthesis, and other hybrid approaches with subsequent chemical treatment, have been extensively employed for producing CNFs [248]. CNF can be easily synthesized using high-temperature carbonization. The CVD synthesis based on the one-dimensional growth of carbon on substrates facilitated by certain catalysts. Moreover, natural polymers, including cellulose, chitin, lignin, and chitosan nanofibers, facilitate the production of carbon nanofibers through direct carbonization. Table 7 summarises the various biomasses for nanofibers synthesis.

4.2.5 Graphene Oxide and Its Various Forms

Graphene is a prominent material in the field of nanomaterials. It gained significant interest from researchers due of the exceptional mechanical, thermal, optical, and electrical properties of the two-dimensional structure composed of a single layer of s sp^2-hybridized carbon atoms [264]. Graphene oxide (GO) is a graphene sheet characterized by carboxylic groups at its edges and epoxide groups and phenolic hydroxyls on its basal plane. The various oxygen-containing functional groups interact with positively charged entities, including organic and inorganic ions. Chemical treatment or thermal annealing can remove the functional groups on GO, which produces reduced graphene oxide (rGO) [265].

The Hummers' method was extensively utilized for the production of graphene oxide; yet, it is marked with several flaws including the creation of hazardous gases (NO_2, N_2O_4), residual nitrates, and low yield [266]. The synthesis of graphene can be categorized into two major techniques: bottom-up and top-down approaches. Bottom-up methods utilize basic carbon molecules to synthesize graphene, whereas top-down approaches involve extracting graphene oxide from a carbon source or graphite. The top-down strategy is recognized as the most prevalent approach for producing graphene derivatives [267].

The green synthesis of nanotechnology, using biomolecules, microbes, and plant extracts as both capping and reducing agents, has been extensively studied in the context of metal-synthesized nanoparticles. This approach helps the conversion of graphene oxide to graphene. Research indicates that several phytochemicals derived from various plant parts, such as leaves, stems, and roots, predominantly consist of biomolecules including amino acids, proteins, carbohydrates, vitamins, flavonoids, alkaloids, pectins, enzymes, and alcohols, which possess the capacity to function as capping and reducing agents in the bioreduction and synthesis of functional graphene from graphene oxide.

In recent years, sustainable green chemistry employs sustainable methods for synthesizing next-generation nanomaterials utilizing plant sources as reducing agents. Numerous researchers have disclosed novel results involving high-yield graphene nanocomposites, highlighting sustainable materials and green chemistry methodologies for development of novel applications. Moreover, graphene materials are seen essential in advancing the green revolution; hence, lots of researchers have given considerable efforts to investigating graphene's potential. The field of green chemistry has developed as a viable option for chemical synthesis, encouraging the reduction of costly harmful substances using biologically-based methods that provide substantial environmental and economic advantages. Following the identification of biological reductants like ascorbic acid and green tea as effective and environmentally friendly alternatives to hazardous chemical reducing agents, significant efforts have been dedicated to investigating additional natural substances for the reduction of graphene oxide. Specifically, bacteria, fungi, yeast, or plants eliminate the need for hazardous chemicals, high temperatures, high pressure, or excessive energy input. Moreover, a wide variety of phytochemicals sourced from nature such as caffeine, spinach, grapes, carrot root, aloevera, sugar cane juice, and protein isolate has been examined for their potential in green reduction processes [268]. Graphene and GO are extensively utilized in applications including energy storage, energy conversion, adsorption, capture of CO_2, heavy metal decontamination, and catalysis. It has generated significant interest for nanomedicine and biomedical applications, in which effectively modified graphene can serve as an exceptional medium for drug delivery in anticancer therapies, diagnostics, biosensing, bioimaging, tissue engineering, cell culture, bioremediation of organic dye pollutants, and antibacterial applications. In conclusion, the synthesis of GO and its derivatives from eco-friendly, sustainable, and cost-effective biomass feedstocks (shown in Table 8) has been highlighted.

Table 8 Biomass used as raw material to fabricate GO and their various forms

Source	Nanoparticle types	Synthesis protocol; conditions	Refs.
Rice bran	GO	Microwave Assisted Synthesis, 300 °C	[264]
Oil palm leaves	GO	Catalytic acid spray (CAS) method	[265]
Green tea extract	RGO	GO aqueous suspension + green tea extract	[269]
Shuteria involucrata	RGO	Modified Hummer's method, 400 °C, 2 h	[270]
Green tea	GO	Modified Hummers method, 80 °C for 8 h	[271]
L-cysteine	RGO	Hummer's method	[272]
Graphite flakes	RGO	Hummer's method	[273]
Rich husk ash	GO	Hummer's method	[274]
Leaf extract of *Lantana camara*	RGO	Hummer's method	[275]
Plant extract of *Salvia spinosa*	GO/RGO	Hummer's method	[276]
Rice straw	GO	Microwave assisted synthesis, 300 °C	[277]
Rice husk, sugarcane bagasse and newspaper	GO Nanoflakes	Pyrolysis, 300–400 °C/ 10 min	[278]
Lignin waste	GO/RGO	Modified Hummer's method	[279]
Coconut shell	GO	Hummer's method	[266]
Cotton cellulose	Graphene quantum dots (GQDs)	Hydrothermal methods	[280]
Rice husk	Graphene quantum dots (GQDs)	Hydrothermal reaction, 200–150 °C	[281]

4.2.6 Graphitic Carbon Nitride (g-CN)

The capacity to utilize solar energy as a driving force for diverse chemical transformations presents a promising strategy to tackle some of the most pressing global challenges. Among various photocatalytic materials, graphitic carbon nitride (g-C_3N_4) has garnered significant attention due to its remarkable physicochemical characteristics. Its intrinsic advantages-such as high thermal and chemical stability, facile synthesis, and an appropriate bandgap for visible-light absorption-render it a highly attractive material for sustainable technologies. These include applications in water treatment, renewable energy generation, air purification, and carbon dioxide mitigation [282, 283]. An Aloe vera, storehouse of phytochemicals such as acemannan, aloin, protein, glucose, etc.—was used to obtain green gel which was dried and burnt in a vacuum furnace using camphor as a fuel agent at 250 °C for 10 min. A hybrid-phase g-C_2N emerged after the organics had been burnt out in a self-propagating spontaneous combustion in camphor at 600 °C in a vacuum furnace for 2 h to remove impurities of carbon (soot) obtained while burning with camphor [284]. This was further utilized for dye degradation. Another study explored the modification of g-C_3N_4 (GCN) using extracts of betel (*Piper betle*) leaves and veld-grape (*Cissus quadrangularis*) stems. The obtained results indicate that these green extracts can effectively exfoliate the GCN-layers, modifying both the 2D structure and the 3D morphological features of GCN, leading to simultaneous chemical and physical modifications in the system, followed by effective dye degradation [285]. Fluorescent graphitic carbon nitride quantum dots (g-C_3N_4 QDs) were synthesized hydrothermally by Zhuang et al. using human urine as precursor, with quantum yield of 15.7% for bioimaging application [286]. Another report indicated successful synthesis of green luminescent graphitic carbon nitride quantum dots (GCNQDs) doped with oxygen and sulfur by microwave treatment of citric acid as green carbonic source and thiourea as dopant, with high fluorescence quantum yield (31.67%). The quantum dots with low cytotoxicity were used for vitro bioimaging of HeLa cells [287]. Deng et al. offered green synthesis of rectangular hollow tubular carbon nitride hydrothermally using water as a green solvent, via in-situ self-assembly strategy to enhance the degradation of tetracycline hydrochloride under visible light irradiation [288]. Similarly, another report showed the exfoliation of bulk g-C3N4 was carried out by aqueous bi-thermal method using green exfoliation technique providing the material with improved band gap for efficient photocatalytic water remediation [289].

4.3 Ceramic Nanoparticles

4.3.1 Nanostructured Hydroxyapatite (Hap)

Phytochemicals have been shown to influence the morphology of hydroxyapatite (HAp) nanoparticles, thereby affecting their behavior in various solvent systems. Natural polysaccharide-based gums, such as those derived from tamarind seeds and guar gum, exhibit notable biocompatibility when functionalized with nano-hydroxyapatite. These biopolymer-mediated HAp composites demonstrate high cell viability and significant drug-loading capabilities. Specifically, guar gum-mediated HAp has achieved a drug-loading capacity of up to 236.18 mg/g with an encapsulation efficiency of 74.79%. Additionally, biosynthesized HAp nanoparticles, particularly those synthesized using pectin, have shown dose-dependent cytotoxic effects against prostate cancer cells. Modified pectin, due to its inherent anticancer properties, holds promise as a synergistic agent when combined with nano-HAp for targeted cancer therapy applications [290, 291]. A key advantage of hydroxyapatite (HAp) nanoparticles lies in their prolonged biodegradation period, which is critical for achieving controlled drug release. Slowly degrading or non-degradable ceramic matrices can effectively retain therapeutic agents over extended periods following administration. This localized and sustained drug delivery facilitated by HAp contributes to lowering systemic drug concentrations, thereby reducing the risk of off-target toxicity. Moreover, it enables precise regulation of drug levels, maintaining concentrations within the therapeutic window—avoiding both subtherapeutic dosing and toxic accumulation [292].

4.3.2 Silica Nanoparticles

Mesoporous silica nanoparticles (MSNs) are widely employed as host matrices for the delivery of therapeutic agents and the encapsulation of bioactive molecules. A major advantage of MSNs is their capacity for surface functionalization with targeting ligands, enabling selective cellular recognition and facilitating site-specific drug delivery. Furthermore, their well-defined and tuneable pore structures significantly enhance their versatility and loading efficiency. The morphology and structural characteristics of MSNs can be precisely controlled through various synthesis techniques, often involving structure-directing agents to tailor the desired porosity and architecture [293]. They also serve as effective carriers for gene delivery to specific cell types. By conjugating MSNs with mannosylated polyethyleneimine (PEI), targeted delivery to macrophages can be achieved, owing to the high expression of mannose receptors on their cell surfaces. Functionalization of MSNs with mannose residues facilitates receptor-mediated endocytosis, enabling the selective and efficient delivery of DNA to macrophage cells [294].

4.4 Polymeric Nanoparticles

4.4.1 Natural Polysaccharides

Starch-based nanomaterials are generally considered non-toxic; however, they have the potential to elicit immune responses. Studies using starch extracted from various microfilms have demonstrated differential effects on immune activation in THP-1 monocytes. While certain starch samples did not exhibit cytotoxicity, they did stimulate the production of pro-inflammatory cytokines such as TNF-α, IL-1β, IL-6, and CCL2 to varying degrees. Additionally, these starch microfilms modulated the expression of multiple immune-related surface receptors on the monocyte cell membrane, indicating their role in influencing innate immune signalling pathways [295]. The semi-crystalline nature and molecular organization of starch significantly influence the formation of nanocrystals during acid hydrolysis. A high amylose content in starch can impede the hydrolysis process, thereby resulting in the production of smaller-sized nanocrystals [296]. Polysaccharides are employed as emulsifying agents to lower the interfacial tension between aqueous and lipid phases, leading to a reduction in droplet size. Additionally, they contribute to the long-term stability of nanoemulsions by providing steric hindrance and electrostatic repulsion. Owing to these properties, polysaccharide-based nanoemulsions are considered efficient delivery systems for functional food ingredients and pharmaceutical compounds [297]. Hydroxypropyl methylcellulose phthalate (HPMCP) serves as a protective carrier for insulin, shielding it from degradation and denaturation under the acidic conditions of the stomach. In studies involving diabetic mice, the group receiving oral insulin-loaded nanoparticles demonstrated a prolonged hypoglycemic response lasting over 24 h, with an effectiveness of approximately 30% relative to the group administered insulin via injection [298].

4.4.2 Hyaluronic Acid-Based Nanoformulations

Hyaluronic acid (HA) is a hydrophilic linear polysaccharide naturally found in the extracellular matrix. It is composed of repeating disaccharide units of D-glucuronic acid and N-acetyl-D-glucosamine, connected via glycosidic linkages. Due to its anionic nature, HA can interact with positively charged molecules, such as chitosan, to form hyaluronic acid-based nanoparticles (HANPs) [299]. In vitro, a high surface potential plays a crucial role in maintaining the uniform dispersion of nanoparticles by preventing their aggregation. When nanocarriers are functionalized with hyaluronic acid (HA) in vivo, the resulting negative surface charge induces electrostatic repulsion with negatively charged plasma proteins, thereby enhancing the nanocarriers' circulation time in the bloodstream. Studies have shown that cationic carriers often exhibit nonspecific interactions with various cell types following systemic administration; however, the use of anionic carriers can reduce such undesirable off-target effects [300]. Self-assembled hyaluronic acid nanoparticles (HA-NPs) demonstrate significant potential for targeting activated macrophages localized at disease sites. These nanoparticles specifically interact with hyaluronic acid receptors, such as CD44, which are overexpressed on activated macrophages. This targeting capability can be leveraged for diagnostic imaging of pathological sites and offers therapeutic potential for inflammatory diseases including type 2 diabetes (T2D), atherosclerosis, and arthritis. HA-NPs may be employed either as standalone agents or as delivery vehicles for chemotherapeutic and immunotherapeutic agents [301].

4.4.3 Cellulose-Derived Nanofibers

Cellulose nanocrystals (CNCs) exhibit inherent hydrophilicity, which poses challenges for their incorporation into hydrophobic polymer matrices. Consequently, CNCs demonstrate greater compatibility with water-soluble or water-dispersible polymers, such as polyvinyl alcohol, starch, and natural rubber. To enhance their applicability as natural nanofillers, CNCs can be dispersed in polar organic solvents like dimethyl sulfoxide (DMSO) and dimethylformamide (DMF). This dispersion can be achieved by subjecting a freeze-dried cellulose suspension to sonication in the presence of a small volume of water. Although some degree of aggregation is observed in organic solvents relative to aqueous media, the resulting suspension remains a promising candidate for the fabrication of nanocomposites based on organic solvent systems [302]. A range of cellular transporters and oxidoreductase enzymes, such as NADH-dependent nitrate reductase, NADPH-dependent sulfite reductase flavoprotein subunit α, and cysteine desulfhydrase, play critical roles in both intracellular and extracellular biocatalytic synthesis processes. Additionally, certain bacterial strains possess the capability to biosynthesize distinctive organic nanoparticles. Notably, bacterial nanocellulose exhibits superior purity, crystallinity, and mechanical properties relative to nanocrystalline cellulose and nano fibrillated cellulose [303].

4.4.4 Chitosan Nanoparticles

Chitosan-based nanoparticles possess notable advantages attributable to their high density of lone-pair electrons, which facilitate robust interactions with substrates containing vacant orbitals. These nanoparticles are utilized in various applications including drug and gene delivery systems, biosensing technologies, and imaging techniques involving fractionated images. The efficacy of chitosan-derived nanoparticles is largely dependent on the uniformity and size of the synthesized microspheres. Particle size plays a crucial role in antigen adsorption and distribution, thereby modulating the subsequent immune response. Additionally, the structural characteristics of the microspheres, including surface micropore dimensions and the rate of antigen release, are key determinants of their functional performance [304]. Chitosan-based aerogels demonstrate improved water absorption capacity and enhanced air permeability. Techniques such as nitrogen adsorption–desorption analysis and scanning electron microscopy are commonly utilized to characterize the morphology and structural properties of these nanoparticles. A significant application of chitosan-based aerogels lies in localized drug delivery, supported by their ability to facilitate rapid drug release. Studies have shown that aerogels loaded with vancomycin exhibit biocompatibility and effective antimicrobial activity against infections in wound sites, as confirmed through fibroblast viability assays and their efficacy against *Staphylococcus aureus* [305]. Chitosan uptake has been shown to induce macrophage activation. It has been employed in the development of DNA-based mucosal vaccines, including a chitosan-formulated DNA influenza vaccine. In one study, a plasmid encoding a peanut allergen gene, pCMVArah2, was incorporated into chitosan nanoparticles. These nanoparticles demonstrated efficient antigen expression and conferred protective immune responses when administered orally to mice [306].

4.4.5 Poly (Lactic-co-Glycolic Acid) Nanoparticles

Cationic polyester nanoparticles enhance encapsulation efficiency and reduce particle size by facilitating the binding and condensation of nucleic acids. Moreover, they accelerate polymer degradation via base-catalysed hydrolysis. These nanoparticles are synthesized by functionalizing polyesters with cationic moieties, including cationic polymers, through techniques such as physical blending, chemical modification, and copolymerization [307]. The researchers developed a PLGA-based nanocomposite encapsulating the anticancer agent doxorubicin (DOX), which was subsequently coated with a porous, branched shell referred to as BGNSH. This composite was further functionalized with a conjugate complex of human serum albumin, indocyanine green, and folic acid (HSA-ICG-FA). The nanostructures were synthesized using the nanoprecipitation technique. To facilitate the formation of a gold (Au) shell around the PLGA core, the PLGA nanoparticles were incubated with chitosan, resulting in a surface charge transition from negative to positive [308]. The in vitro degradation of PLGA nanoparticles (PLGA-NPs) proceeds via a bulk erosion mechanism consisting of three distinct phases. In the initial phase, the polymer undergoes a reduction in molecular weight due to the cleavage of ester bonds, without any observable loss of polymer mass. Subsequently, acidification of the microenvironment leads to the formation of oligomers and the corresponding loss of polymer mass. In the final phase, these oligomers further degrade into monomeric units, culminating in the complete dissolution of the polymer matrix [309]. The fabrication of nanoparticles utilizing PLGA polymers, particularly for the incorporation of antimicrobial agents or their use as delivery vehicles for such compounds, has garnered significant attention. PLGA-based nanoparticles exhibiting antibacterial properties, alongside polyethylene diamine nanoparticles (PNPs), have been applied in a variety of biomedical contexts [310].

4.5 Lipid-Based Nanoparticles

4.5.1 Liposomes

Liposomes serve as adaptable delivery systems capable of encapsulating both hydrophilic and hydrophobic agents. Advances in nanotechnology have led to the development of novel nanoparticulate materials, such as superparamagnetic iron oxide nanoparticles, gold nanoparticles, and quantum dots, which possess unique magnetic and optical properties that make them highly suitable for imaging applications [311]. In addition to liposomes, various other nanoparticle platforms have been developed for targeted non-viral gene delivery, including calcium phosphate nanoparticles, cationic polymers such as polyamidoamine dendrimers, chitosan, polyethyleneimine (PEI), and polymers conjugated with nucleic acids. While liposomes have been extensively studied, these alternative systems also present challenges related to transfection efficiency. Targeting strategies vary depending on the mode of ligand conjugation; for instance, cell-penetrating peptides (CPPs) can be attached to polymers, liposomes, and other nanoparticle carriers. Furthermore, innovative targeted gene delivery vectors continue to emerge, exemplified by galactosylated glycerol propoxylatetriacrylate spermine, which has demonstrated enhanced specificity for hepatocytes through targeting of the asialoglycoprotein receptor [312].

4.5.2 Solid Lipid Nanoparticles (SLNs)

Such nanostructures are formulated from lipids that maintain a solid state at both room and physiological temperatures. These nanoparticles consist of a dense lipid core enveloped by a phospholipid monolayer. SLNs are widely utilized for the delivery of natural bioactive compounds, providing benefits including controlled drug release, targeted delivery, and protection against chemical degradation [313]. Adsorption onto preformed polymeric nanoparticulate carriers is a widely employed method for drug loading. However, achieving high encapsulation efficiency and sustained release can be challenging with very small nanoparticles. As an alternative to encapsulation, drugs can be adsorbed onto the surface of the particles. Proteins, due to their amphipathic nature, tend to adsorb and accumulate at solid–liquid interfaces across diverse biological and non-biological systems. The kinetics and magnitude of protein adsorption are influenced by factors including protein characteristics, concentration, properties of the carrier matrix, and solvent conditions [314]. These nanoparticles (SLNs) have been employed to encapsulate Nigella sativa oil, which is well-known for its anti-inflammatory and antimicrobial effects. These nanoparticles were prepared using high-pressure homogenization and consisted of hydrogenated palm oil combined with Nigella sativa essential oil. The particle sizes ranged between 66 and 142 nm. The average size of the SLNs remained stable throughout storage; although an increase in particle size was observed following the freeze-drying of the nanoparticle suspensions [157].

5 Recent Advances and Applications in Green Nanotechnology

There has been effective utilization of the green synthesised nanomaterials in various realms as depicted in Fig. 7. Few of them have been briefly discussed as below.

5.1 Photocatalytic Applications of Green Nanomaterials

The green synthesis of nanomaterials has emerged as a promising route for developing efficient photocatalysts while adhering to sustainability principles and minimizing ecological risks [315]. Metal oxide nanoparticles (MONPs) such as titanium

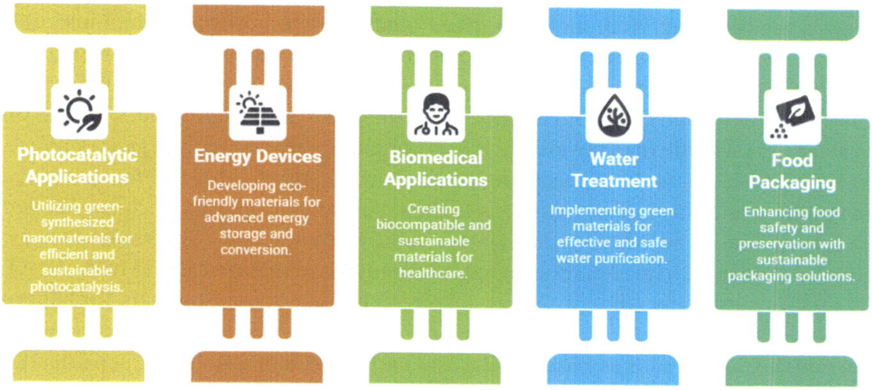

Fig. 7 Applications of green nanomaterials

dioxide (TiO_2) and zinc oxide (ZnO), synthesized via plant extracts and microbial processes, exhibit enhanced photocatalytic activity compared to their chemically produced counterparts due to phytochemical-assisted bandgap modulation, improved charge separation, and surface functionalization [316]. For instance, green-synthesized TiO_2 nanoparticles demonstrate superior dye degradation under visible light, eliminating the need for UV activation, while ZnO nanoparticles obtained from herbal extracts show effective photocatalysis in pollutant degradation and antimicrobial applications [317]. Carbon-based nanostructures including carbon dots (CDs), graphene, and carbon nanotubes (CNTs) produced from renewable biomass such as fruit peels, cellulose, and agricultural residues further contribute to photocatalysis through their excellent electron-accepting properties, high surface area, and ability to enhance charge transfer in hybrid systems [318]. CDs and quantum dots (QDs) derived from natural precursors like lignin and tea leaves also exhibit tuneable bandgaps and strong photoluminescence, making them attractive photocatalysts for environmental remediation and energy conversion applications such as pollutant degradation, solar energy harvesting, and photocatalytic hydrogen generation [319]. Together, these green-synthesized nanomaterials present scalable, cost-effective, and environmentally benign platforms for advanced photocatalytic technologies.

5.2 *Green Material Synthesis for Energy Devices*

The growing global energy demand and environmental challenges have spurred the development of sustainable and eco-friendly materials for energy devices through green synthesis approaches. Unlike conventional methods that rely on toxic chemicals and energy-intensive processes, green synthesis utilizes renewable precursors, mild conditions, and benign solvents such as water, ionic liquids, and supercritical CO_2, aligning with the principles of green chemistry [320]. Various eco-friendly routes have been explored, including plant-mediated synthesis using biomolecules like flavonoids and phenolics for producing metal oxides and carbon nanostructures [321], microbial and enzymatic synthesis employing bacteria, fungi, and algae for controlled nucleation of nanomaterials [322], bio-mass derived polymer of chitin on Zn anode (Zn@chitin) as a novelv [323]. Additionally, solvent-free and green solvent-assisted processes such as mechanochemical and hydrothermal synthesis reduce energy input and waste generation [324]. These sustainable approaches have enabled the fabrication of advanced materials for energy storage and conversion devices: biomass-derived activated carbon and green-synthesized metal oxides (MnO_2, Fe_2O, Co_3O_4) for supercapacitors [322], bio-inspired chitosan-based polymers for battery electrodes [323], Pt-free catalysts and biopolymer proton exchange membranes for fuel cells [324], as well as biomass-derived carbon quantum dots and eco-friendly perovskites for solar cells [321, 323], making green material synthesis a vital pathway for next-generation clean energy technologies.

5.3 Green Materials for Biomedical Applications

The biomedical field increasingly demands materials that are not only biocompatible and functional but also sustainable and environmentally friendly. Conventional material synthesis often involves toxic chemicals and energy-intensive processes that may leave harmful residues or pose biosafety concerns, whereas green materials synthesized via eco-friendly routes using renewable resources, plant extracts, microbial processes, and biomolecules provide sustainable alternatives for biomedical applications [320]. Plant-mediated synthesis, utilizing polyphenols, flavonoids, and alkaloids, has been employed to generate silver, gold, and zinc oxide nanoparticles with antibacterial, anticancer, and imaging potential [325], while microbial systems such as bacteria, fungi, and algae enable mild-condition biosynthesis of nanoparticles and biopolymers, ensuring cost-effectiveness and biocompatibility [326]. Biopolymers like chitosan, cellulose, alginate, gelatin, and silk fibroin, obtained from natural sources, are widely used in drug delivery, wound dressings, and tissue engineering due to their biodegradability and safety [327]. Furthermore, biomass-derived carbon materials, including carbon dots and graphene from agricultural residues, have emerged as promising agents for bioimaging, biosensing, and targeted drug delivery [328].

5.4 Green Materials for Water Treatment

The increasing scarcity of clean water and the persistence of pollutants such as dyes, heavy metals, pesticides, and pharmaceuticals have accelerated the development of sustainable and eco-friendly materials for water treatment. Conventional treatment approaches often rely on chemical coagulants and synthetic adsorbents that may generate secondary pollution, whereas green materials synthesized from renewable and biodegradable resources provide safer and more effective alternatives [329]. Plant-derived biomass including agricultural residues, cellulose, lignin, and fruit peels can be converted into activated carbon, biochar, and nanocomposites with high adsorption capacities for dyes and heavy metals [330]. Biopolymers such as chitosan, alginate, and starch-based materials have been widely employed as natural coagulants, flocculants, and adsorbents, offering advantages of biodegradability and non-toxicity [331]. Metal and metal oxide nanoparticles synthesized through green routes using plant extracts or microbial processes, such as green-synthesized iron oxide, zinc oxide, and silver nanoparticles, have shown excellent antimicrobial activity and catalytic degradation of organic contaminants [326]. Furthermore, biomass-derived carbon dots and graphene-based nanomaterials prepared via eco-friendly processes are emerging as effective photocatalysts and adsorbents for the removal of pharmaceuticals and endocrine-disrupting compounds [332].

5.5 *Food Packaging and Coatings*

The demand for sustainable and eco-friendly solutions in food chemistry has led to the development of green materials that enhance food safety, preservation, packaging, and nutrient delivery while minimizing environmental impact. Traditional food additives, preservatives, and packaging often involve synthetic chemicals or non-biodegradable polymers, which can pose health and ecological risks, whereas green materials derived from natural polymers, plant extracts, and eco-synthesized nanomaterials provide safer and sustainable alternatives [333]. Biopolymers such as chitosan, starch, cellulose, alginate, and carrageenan are widely used in edible films, coatings, and biodegradable packaging due to their biocompatibility, film-forming ability, and antimicrobial properties [334]. Green-synthesized nanoparticles such as silver, zinc oxide, and titanium dioxide produced from plant or microbial sources have shown effectiveness in extending shelf life by inhibiting microbial growth in food products [335]. Natural plant extracts rich in polyphenols, flavonoids, and essential oils serve as antioxidant and antimicrobial agents in active packaging and functional foods [336]. Moreover, green-synthesized carbon dots and nano-emulsions from biomass and edible oils are being explored for nutrient encapsulation, food quality monitoring, and smart packaging applications [337].

6 Challenges and Limitations

To establish the green nanotechnology in the manner of greener synthesis and development of nanomaterials always drive through the many challenges which limits researchers to work with high satisfaction in the field. The unique characteristics of nanomaterials, such as their high surface area, make them ideal for higher process efficiency same time shows some serious drawbacks.

A major limitation in nanoparticle (NP) synthesis using biomass extracts is the tendency for aggregation, driven by strong interparticle interactions, particularly at elevated extract concentrations. This aggregation adversely affects the homogeneity and dispersion of the NPs, thereby diminishing their functional performance. Additionally, these nanomaterials often exhibit reduced thermal and pH stability, which constrains their applicability across diverse environmental and biomedical domains. Moreover, these nanomaterials often exhibit compromised stability under elevated temperature and extreme pH conditions, thereby restricting their practical utility across diverse environmental and biomedical applications [338–340].

Another significant limitation in nanoparticle synthesis is the variability in yield, morphology, and dimensional characteristics. The wide-ranging differences in nanoparticle size and shape can substantially influence their functional properties. Inconsistencies in particle size distribution pose challenges to reproducibility, which is critical for industrial-scale applications. Moreover, fluctuations in the phytochemical profile of biomass-often dependent on batch-specific factors-can lead to unpredictable yields. These inherent variations complicate efforts to achieve standardized production and scalable synthesis protocols [341, 342].

The specificity and functional efficacy of nanomaterials are further compromised by their sensitivity to homologous or competing metal ions, particularly within complex biological and environmental matrices. The widespread deployment of plant-derived nanomaterials is also hindered by various practical and ecological limitations. Additionally, certain carbon-based nanomaterials exhibit diminished surface areas due to their amorphous carbon structure, which can negatively influence their performance under conditions of elevated oxidative stress [343]. In addition, precise evaluation of greenness of the method is also the need of the hour [344]. But now it's impossible to leave the greener technology for future materials development and there many good outcomes also encourage the adoption of such technology.

7 Future Perspectives

Globally harmonized guidelines for phytogenic nanomaterials are essential, particularly in relation to quality control, reproducibility, and risk assessment. Regulatory agencies such as the FDA, EMA, and WHO have underscored the importance of standardized characterization, consistent batch production, and comprehensive safety evaluation. However, disparities in national regulatory frameworks continue to hinder global implementation. Key emerging strategies include establishing validated reference materials and nano-specific risk assessment frameworks, alongside promoting regulatory science education and fostering cross-sector collaboration. These efforts aim to advance safer green nanotechnology and support environmentally sustainable manufacturing practices.

8 Conclusion

The two terms 'green chemistry' and 'nanotechnology' combined as 'green nanotechnology' are not only the sustainable approach towards material development but also the only alternative to fight against the use of fossil raw materials accompanying the high pollution in the environment.

A systematic investigation into the synthesis and characterization of nanomaterials derived from bioactive compounds is essential to elucidate their multifaceted roles across diverse scientific and technological domains. Standardization of extraction protocols is critical to enable consistent and scalable nanoparticle synthesis, in alignment with the increasing emphasis on sustainable and environmentally benign approaches in nanotechnology. Furthermore, rigorous evaluation using advanced characterization techniques-such as UV–Vis spectrophotometry, X-ray diffraction (XRD), Fourier-transform infrared spectroscopy (FTIR), scanning electron microscopy (SEM), transmission electron microscopy (TEM), dynamic light scattering (DLS), and contact angle analysis-is indispensable for deciphering the physicochemical and morphological properties of these nanomaterials. These techniques offer valuable insights into the structural integrity, surface characteristics, and functional potential of nanoparticles synthesized from bioactive sources. Harnessing the intrinsic properties of bioactive compounds through nanotechnological interventions holds promise for pioneering advancements in their effective synthesis, deciphering the mode and mechanism of synthesis, and broadening the scope of applications.

The chapter systematically furnishes the past, present, and future of the development of nanomaterials using green chemistry and greener technology.

References

1. Kurul, F., Doruk, B., Topkaya, S.N.: Principles of green chemistry: building a sustainable future. Discov. Chem. **2**(1), 68 (2025). https://doi.org/10.1007/s44371-025-00152-9
2. Erdoğan, S.: Energy, environment and greenhouse gases. Çankırı Karatekin University J. Econ. Adm. Sci. **10**(1), 277–303 (2020). https://doi.org/10.18074/ckuiibfd.670673
3. Tobiszewski, M., Mechlińska, A., Zygmunt, B., Namieśnik, J.: Green analytical chemistry in sample preparation for determination of trace organic pollutants. Trends Anal. Chem. **28**(8), 943–951 (2009). https://doi.org/10.1016/j.trac.2009.06.001
4. Hurst, G.A.: Systems thinking approaches for international green chemistry education. Curr. Opin. Green Sustain. Chem. **21**, 93–97 (2020). https://doi.org/10.1016/j.cogsc.2020.02.004
5. DeVierno Kreuder, A., House-Knight, T., Whitford, J., Ponnusamy, E., Miller, P., Jesse, N., et al.: A method for assessing greener alternatives between chemical products following the 12 principles of green chemistry. ACS Sustain. Chem. Eng. **5**(4), 2927–2935 (2017). https://doi.org/10.1021/acssuschemeng.6b02399
6. Tuck, C.O., Pérez, E., Horváth, I.T., Sheldon, R.A., Poliakoff, M.: Valorization of biomass: deriving more value from waste. Science **337**(6095), 695–699 (2012). https://doi.org/10.1126/science.1218930
7. Sheldon, R.A.: Green and sustainable manufacture of chemicals from biomass: state of the art. Green Chem. **16**(3), 950–963 (2014). https://doi.org/10.1039/C3GC41935E

8. Sheldon, R.A.: Metrics of green chemistry and sustainability: past, present, and future. ACS Sustain. Chem. Eng. **6**(1), 32–48 (2018). https://pubs.acs.org/doi/10.1021/acssuschemeng.7b03505
9. Anastas, P., Eghbali, N.: Green chemistry: principles and practice. Chem. Soc. Rev. **39**(1), 301–312 (2010). https://doi.org/10.1039/B918763B
10. Sternberg, J., Sequerth, O., Pilla, S.: Green chemistry design in polymers derived from lignin: review and perspective. Prog. Polym. Sci. **113**, 101344 (2021). https://doi.org/10.1016/j.progpolymsci.2020.101344
11. Scanlon, A., Murphy, M., Smolowitz, J., Lewis, V.: United Nations' sustainable development goal 3 target indicators: examples of advanced practice nurses' actions. J. Nurse Pract. **18**(10), 1067–1070 (2022). https://doi.org/10.1016/j.nurpra.2022.03.005
12. Arora, N.K., Mishra, I.: United Nations sustainable development goals 2030 and environmental sustainability: race against time. Environ. Sustain. **2**(4), 339–342 (2019). https://doi.org/10.1007/s42398-019-00092-y
13. Msingh, R., Pramanik, R., Hazra, S., Uni, J.: Role of green chemistry in pharmaceutical industry: a review. J. Univ. Shanghai Sci. Technol. **23**, 291–299 (2021). https://doi.org/10.51201/JUSST/21/121018
14. Mukherjee, S., Das, A., Chowdhury, A., Biswas, A., Roy, S., Majumdar, S., Paul, S.: Harmful effect of personal care products on ecosystem and the possible alternative approach. Biocatal. Agric. Biotechnol. **57**, 103065 (2024). https://doi.org/10.1016/j.bcab.2024.103065
15. Trombino, S., Sole, R., Di Gioia, M.L., Procopio, D., Curcio, F., Cassano, R.: Green chemistry principles for nano-and micro-sized hydrogel synthesis. Molecules **28**(5), 2107 (2023). https://doi.org/10.3390/molecules28052107
16. Mishra, K., Siwal, S.S., Nayaka, S.C., Guan, Z., Thakur, V.K.: Waste-to-chemicals: green solutions for bioeconomy markets. Sci. Total. Environ. **887**, 164006 (2023). https://doi.org/10.1016/j.scitotenv.2023.164006
17. Chin, W.Q., Lee, Y.H., Amran, M., Fediuk, R., Vatin, N., Kueh, A.B.H., Lee, Y.Y.: A sustainable reuse of agro-industrial wastes into green cement bricks. Materials **15**(5), 1713 (2022). https://doi.org/10.3390/ma15051713
18. Tsai, S.H., Liu, C.P., Yang, S.S.: Microbial conversion of food wastes for biofertilizer production with thermophilic lipolytic microbes. Renew. Energy **32**(6), 904–915 (2007). https://doi.org/10.1016/j.renene.2006.04.019
19. Banert, K., Bochmann, S., Ihle, A., Plefka, O., Taubert, F., Walther, T., et al.: Synthesis with perfect atom economy: generation of furan derivatives by 1,3-dipolar cycloaddition of acetylenedicarboxylates at cyclooctynes. Molecules **19**(9), 14022–14035 (2014). https://doi.org/10.3390/molecules190914022
20. Bédard, A.C., Longstreet, A.R., Britton, J., Wang, Y., Moriguchi, H., Hicklin, R.W., et al.: Minimizing E-factor in the continuous-flow synthesis of diazepam and atropine. Bioorg. Med. Chem. **25**(23), 6233–6241 (2017). https://doi.org/10.1016/j.bmc.2017.02.002
21. Choudary, B.M., Sateesh, M., Kantam, M.L., Rao, K.K., Prasad, K.R., Raghavan, K.V., Sarma, J.A.R.P.: Selective nitration of aromatic compounds by solid acid catalysts. Chem. Commun. **1**, 25–26 (2000). https://doi.org/10.1039/a908011b
22. Epp, J.B., Alexander, A.L., Balko, T.W., Buysse, A.M., Brewster, W.K., Bryan, K., et al.: The discovery of Arylex™ active and Rinskor™ active: two novel auxin herbicides. Bioorg. Med. Chem. **24**(3), 362–371 (2016). https://doi.org/10.1016/j.bmc.2015.08.011
23. Zhou, F., Hearne, Z., Li, C.J.: Water—the greenest solvent overall. Curr. Opin. Green Sustain. Chem. **18**, 118–123 (2019). https://doi.org/10.1016/j.cogsc.2019.05.004
24. Narayan, S., Muldoon, J., Finn, M.G., Fokin, V.V., Kolb, H.C., Sharpless, K.B.: "On water": unique reactivity of organic compounds in aqueous suspension. Angew. Chem. **117**(21), 3339–3343 (2005). https://doi.org/10.1002/ange.200462883
25. Zhang, J., Li, S., Yao, L., Yi, Y., Shen, L., Li, Z., Qiu, H.: Responsive switchable deep eutectic solvents: a review. Chin. Chem. Lett. **34**(5), 107750 (2023). https://doi.org/10.1016/j.cclet.2022.107750

26. Janicka, P., Kaykhaii, M., Płotka-Wasylka, J., Gębicki, J.: Supramolecular deep eutectic solvents and their applications. Green Chem. **24**(13), 5035–5045 (2022). https://doi.org/10.1039/D2GC00906D
27. Florindo, C., Branco, L.C., Marrucho, I.M.: Quest for green-solvent design: from hydrophilic to hydrophobic (deep) eutectic solvents. Chemsuschem **12**(8), 1549–1559 (2019). https://doi.org/10.1002/cssc.201900147
28. Kayhan, A.: Birleşmiş milletler çevre programi üzerine bir inceleme. Publ. Priv. Int. Law Bull. **33**(1), 61–90 (2013)
29. Papadis, E., Tsatsaronis, G.: Challenges in the decarbonization of the energy sector. Energy **205**, 118025 (2020). https://doi.org/10.1016/j.energy.2020.118025
30. Brédas, J.L., Norton, J.E., Cornil, J., Coropceanu, V.: Molecular understanding of organic solar cells: the challenges. Acc. Chem. Res. **42**(11), 1691–1699 (2009). https://doi.org/10.1021/ar900099h
31. Ganiyu, S.O., Brito, L.R., de Araújo Costa, E.C., dos Santos, E.V., Martínez-Huitle, C.A.: Solar photovoltaic-battery system as a green energy for driven electrochemical wastewater treatment technologies: application to elimination of Brilliant Blue FCF dye solution. J. Environ. Chem. Eng. **7**(1), 102924 (2019). https://doi.org/10.1016/j.jece.2019.102924
32. Yoruç, H., Binnaz, A., Uğraşkan, V.: Green polymers and applications. AKU J. Sci. Eng. **17**, 318–337 (2017). https://doi.org/10.5578/fmbd.53940
33. Trakhtenberg, S., Warner, J.C.: Green chemistry considerations in entropic control of materials and processes. Chem. Rev. **107**(6), 2174–2182 (2007). https://doi.org/10.1021/cr0509455
34. Anastas, P.T., Warner, J.C.: Green Chemistry: Theory and Practice. Oxford University Press (2000). https://doi.org/10.1093/oso/9780198506980.001.0001
35. Mureşan, A.C.: Ibuprofen: original versus green synthesis. Ann. "Dunarea de Jos" Univ. Galati. Fascicle IX Metall. Mater. Sci. **41**(3), 30–34 (2018). https://doi.org/10.35219/mms.2018.3.05
36. Murphy, M.A.: Early industrial roots of green chemistry and the history of the BHC Ibuprofen process invention and its quality connection. Found. Chem. **20**(2), 121–165 (2018). https://doi.org/10.1007/s10698-017-9300-9
37. Sheldon, R.A.: Fundamentals of green chemistry: efficiency in reaction design. Chem. Soc. Rev. **41**(4), 1437–1451 (2012). https://doi.org/10.1039/C1CS15219J
38. Whiteker, G.T.: Applications of the 12 principles of green chemistry in the crop protection industry. Org. Process Res. Dev. **23**(10), 2109–2121 (2019). https://doi.org/10.1021/acs.oprd.9b00305
39. Singh, T., Shukla, S., Kumar, P., Wahla, V., Bajpai, V.K., Rather, I.A.: Application of nanotechnology in food science: perception and overview. Front. Microbial. **8**, 1501 (2017). https://doi.org/10.3389/fmicb.2017.01501
40. Chausali, N., Saxena, J., Prasad, R.: Recent trends in nanotechnology applications of bio-based packaging. J. Agric. Food Res. **7**, 100257 (2022). https://doi.org/10.1016/j.jafr.2021.100257
41. Nasrollahzadeh, M., Sajjadi, M., Sajadi, S.M., Issaabadi, Z.: Green nanotechnology. In: Interface science and technology, vol. 28, pp. 145–198. Elsevier (2019). https://doi.org/10.1016/B978-0-12-813586-0.00005-5
42. Shrivastava, S., Dash, D.: Applying nanotechnology to human health: revolution in biomedical sciences. J. Nanotechnol. **2009**(1), 184702 (2009). https://doi.org/10.1155/2009/184702
43. Barhoum, A., García-Betancourt, M.L., Jeevanandam, J., Hussien, E.A., Mekkawy, S.A., Mostafa, M., et al.: Review on natural, incidental, bioinspired, and engineered nanomaterials: history, definitions, classifications, synthesis, properties, market, toxicities, risks, and regulations. Nanomaterials **12**(2), 177 (2022). https://doi.org/10.3390/nano12020177
44. Kheirandish, A., Sepehri Javan, N., Mohammadzadeh, H.: Modified Drude model for small gold nanoparticles surface plasmon resonance based on the role of classical confinement. Sci. Rep. **10**(1), 6517 (2020). https://doi.org/10.1038/s41598-020-63066-9
45. Lea, M.C.: On allotropic forms of silver. Am. J. Sci. **3**(223), 47–49 (1889). https://doi.org/10.2475/ajs.s3-38.223.47

46. Niska, K., Zielinska, E., Radomski, M.W., Inkielewicz-Stepniak, I.: Metal nanoparticles in dermatology and cosmetology: interactions with human skin cells. Chem.-Biol. Interact. **295**, 38–51 (2018). https://doi.org/10.1016/j.cbi.2017.06.018
47. Brodusch, N., Demers, H., Gauvin, R.: Field Emission Scanning Electron Microscopy: New Perspectives for Materials Characterization. Springer (2017). https://doi.org/10.1007/978-981-10-4433-5
48. Oinonen, N., Xu, C., Alldritt, B., Canova, F.F., Urtev, F., Cai, S., et al.: Electrostatic discovery atomic force microscopy. ACS Nano **16**(1), 89–97 (2021). https://pubs.acs.org/doi/10.1021/acsnano.1c06840
49. Palmberg, C., Nikulainen, T.: Industrial renewal and growth through nanotechnology? An overview with focus on Finland. Discussion Papers 1020, The Research Institute of the Finnish Economy (2006). https://hdl.handle.net/10419/64005
50. Stephan, P., Black, G.C., Chang, T.: The small size of the small scale market: the early-stage labor market for highly skilled nanotechnology workers. Res. Policy **36**(6), 887–892 (2007). https://doi.org/10.1016/j.respol.2007.02.006
51. Hamid, H.A., Jenidi, Y., Thielemans, W., Somerfield, C., Gomes, R.L.: Predicting the capability of carboxylated cellulose nanowhiskers for the remediation of copper from water using response surface methodology (RSM) and artificial neural network (ANN) models. Ind. Crops Prod. **93**, 108–120 (2016). https://doi.org/10.1016/j.indcrop.2016.05.035
52. Jeevanandam, J., Barhoum, A., Chan, Y.S., Dufresne, A., Danquah, M.K.: Review on nanoparticles and nanostructured materials: history, sources, toxicity and regulations. Beilstein J. Nanotechnol. **9**(1), 1050–1074 (2018). https://doi.org/10.3762/bjnano.9.98
53. Hulla, J.E., Sahu, S.C., Hayes, A.W.: Nanotechnology: history and future. Hum. Exp. Toxicol. **34**(12), 1318–1321 (2015). https://doi.org/10.1177/0960327115603
54. Kaplan, S., Radin, J.: Bounding an emerging technology: para-scientific media and the Drexler-Smalley debate about nanotechnology. Soc. Stud. Sci. **41**(4), 457–485 (2011). https://doi.org/10.1177/0306312711402722
55. Iijima, S.: Helical microtubules of graphitic carbon. Nature **354**(6348), 56–58 (1991). https://doi.org/10.1038/354056a0
56. Morris, J., Willis, J., De Martinis, D., Hansen, B., Laursen, H., Sintes, J.R., et al.: Science policy considerations for responsible nanotechnology decisions. Nat. Nanotechnol. **6**(2), 73–77 (2011). https://doi.org/10.1038/nnano.2010.191
57. Geim, A.K., Novoselov, K.S.: The rise of grapheme. Nat. Mater. **6**, 183–191 (2009). https://doi.org/10.1142/9789814287005_0002
58. Ruess, G., Vogt, F.: Höchstlamellarer Kohlenstoffaus Graphitoxyhydroxyd. Mon. Chem. Verwandte Teile And. Wiss. **78**(3), 222–242 (1948). https://doi.org/10.1007/BF01141527
59. Xu, X., Ray, R., Gu, Y., Ploehn, H.J., Gearheart, L., Raker, K., Scrivens, W.A.: Electrophoretic analysis and purification of fluorescent single-walled carbon nanotube fragments. J. Am. Chem. Soc. **126**(40), 12736–12737. https://pubs.acs.org/doi/10.1021/ja040082h
60. Sun, Y. P., Zhou, B., Lin, Y., Wang, W., Fernando, K.S., Pathak, P., et al.: Quantum-sized carbon dots for bright and colorful photoluminescence. J. Am. Chem. Soc. **128**(24), 7756–7757 (2006). https://pubs.acs.org/doi/10.1021/ja062677d
61. Jafarzadeh, S., Oladzadabbasabadi, N., Dheyab, M.A., Lalabadi, M.A., Sheibani, S., Ghasemlou, M., et al.: Emerging trends in smart and sustainable nano-biosensing: the role of green nanomaterials. Ind. Crops Prod. **223**, 120108 (2025). https://doi.org/10.1016/j.indcrop.2024.120108
62. Jafarzadeh, S., Nooshkam, M., Zargar, M., Garavand, F., Ghosh, S., Hadidi, M., Forough, M.: Green synthesis of nanomaterials for smart biopolymer packaging: challenges and outlooks. J. Nanostruct. Chem. **14**(2), 113–136 (2024). https://doi.org/10.1007/s40097-023-00527-3
63. Naseer, B., Srivastava, G., Qadri, O.S., Faridi, S.A., Islam, R.U., Younis, K.: Importance and health hazards of nanoparticles used in the food industry. Nanotechnol. Rev. **7**(6), 623–641 (2018). https://doi.org/10.1515/ntrev-2018-0076
64. Garavand, F., Khodaei, D., Mahmud, N., Islam, J., Khan, I., Jafarzadeh, S., et al.: Recent progress in using zein nanoparticles-loaded nanocomposites for food packaging applications.

Crit. Rev. Food Sci. Nutr. **64**(12), 3639–3659 (2024). https://doi.org/10.1080/10408398.2022.2133080

65. Ghasemi, S., Assadpour, E., Kharazmi, M.S., Jafarzadeh, S., Zargar, M., Jafari, S.M.: Encapsulation of orange peel oil in biopolymeric nanocomposites to control its release under different conditions. Foods **12**(4), 831 (2023). https://doi.org/10.3390/foods12040831
66. Babaee, M., Garavand, F., Rehman, A., Jafarazadeh, S., Amini, E., Cacciotti, I.: Biodegradability, physical, mechanical and antimicrobial attributes of starch nanocomposites containing chitosan nanoparticles. Int. J. Biol. Macromol. **195**, 49–58 (2022). https://doi.org/10.1016/j.ijbiomac.2021.11.162
67. Jafarzadeh, S., Nafchi, A.M., Salehabadi, A., Oladzad-Abbasabadi, N., Jafari, S.M.: Application of bio-nanocomposite films and edible coatings for extending the shelf life of fresh fruits and vegetables. Adv. Coll. Int. Sci. **291**, 102405 (2021). https://doi.org/10.1016/j.cis.2021.102405
68. Ardila-Fierro, K.J., Hernández, J.G.: Sustainability assessment of mechanochemistry by using the twelve principles of green chemistry. Chemsuschem **14**(10), 2145–2162 (2021). https://doi.org/10.1002/cssc.202100478
69. Ijaz, I., Gilani, E., Nazir, A., Bukhari, A.: Detail review on chemical, physical and green synthesis, classification, characterizations and applications of nanoparticles. Green Chem. Lett. Rev. **13**(3), 223–245 (2020). https://doi.org/10.1080/17518253.2020.1802517
70. Abid, N., Khan, A.M., Shujait, S., Chaudhary, K., Ikram, M., Imran, M., et al.: Synthesis of nanomaterials using various top-down and bottom-up approaches, influencing factors, advantages, and disadvantages: a review. Adv. Coll. Int. Sci. **300**, 102597 (2022). https://doi.org/10.1016/j.cis.2021.102597
71. Devi, L., Kushwaha, P., Ansari, T.M., Kumar, A., Rao, A.: Recent trends in biologically synthesized metal nanoparticles and their biomedical applications: a review. Biol. Trace Elem. Res. **202**(7), 3383–3399 (2024). https://doi.org/10.1007/s12011-023-03920-9
72. Ishak, N.M., Kamarudin, S.K., Timmiati, S.N.: Green synthesis of metal and metal oxide nanoparticles via plant extracts: an overview. Mater. Res. Express **6**(11), 112004 (2019). https://doi.org/10.1088/2053-1591/ab4458
73. Saif, S., Tahir, A., Chen, Y.: Green synthesis of iron nanoparticles and their environmental applications and implications. Nanomaterials **6**(11), 209 (2016). https://doi.org/10.3390/nano6110209
74. Ajitha, B., Reddy, Y.A.K., Reddy, P.S.: Green synthesis and characterization of silver nanoparticles using Lantana camara leaf extract. Mater Sci Eng C **49**, 373–381 (2015). https://doi.org/10.1016/j.msec.2015.01.035
75. Shekhar Samanta, N., Das, P.P., Dhara, S., Purkait, M.K.: An overview of precious metal recovery from steel industry slag: recovery strategy and utilization. Ind. Eng. Chem. Res. **62**(23), 9006–9031 (2023). https://doi.org/10.1021/acs.iecr.3c00604
76. Cardoso, V.F., Francesko, A., Ribeiro, C., Bañobre-López, M., Martins, P., Lanceros-Mendez, S.: Advances in magnetic nanoparticles for biomedical applications. Adv. Healthc. Mater. **7**(5), 1700845 (2018). https://doi.org/10.1002/adhm.201700845
77. Zambonino, M.C., Quizhpe, E.M., Jaramillo, F.E., Rahman, A., Santiago Vispo, N., Jeffryes, C., Dahoumane, S.A.: Green synthesis of selenium and tellurium nanoparticles: current trends, biological properties and biomedical applications. Int. J. Mol. Sci. **22**(3), 989 (2021). https://doi.org/10.3390/ijms22030989
78. Hano, C., Abbasi, B.H.: Plant-based green synthesis of nanoparticles: production, characterization and applications. Biomolecules **12**(1), 31 (2021). https://doi.org/10.3390/biom12010031
79. Khan, F., Shariq, M., Asif, M., Siddiqui, M.A., Malan, P., Ahmad, F.: Green nanotechnology: plant-mediated nanoparticle synthesis and application. Nanomaterials **12**(4), 673 (2022). https://doi.org/10.3390/nano12040673
80. Jeetkar, T.J., Khataokar, S.P., Indurkar, A.R., Pandit, A., Nimbalkar, M.S.: A review on plant-mediated synthesis of metallic nanoparticles and their applications. Adv. Nat. Sci. Nanosci. Nanotechnol. **13**(3), 033004 (2022). https://doi.org/10.1088/2043-6262/ac865d

81. Cheng, H.J., Wang, H., Zhang, J.Z.: Phytofabrication of silver nanoparticles using three flower extracts and their antibacterial activities against pathogen Ralstonia solanacearum strain YY06 of bacterial wilt. Front. Microbiol. **11**, 2110 (2020). https://doi.org/10.3389/fmicb.2020.02110
82. Palanivel, R., Muthumanickam, M.: GC/MS analysis and green-synthesis of silver nanoparticles using Senna auriculata flower extract: antibacterial, antioxidant effects and anticancer effects. Microb. Pathogen. **199**, 107274 (2025). https://doi.org/10.1016/j.micpath.2024.107274
83. Guta, T., Jemal, K.: Antioxidant and antimicrobial potentials of Nicotiana glauca Graham leaves extracts and synthesized silver nanoparticles: a phytochemical approach. Jordan J. Pharm. Sci. **18**(1), 57–76 (2025). https://doi.org/10.35516/jjps.v18i1.2180
84. Varadavenkatesan, T., Nagendran, V., Vinayagam, R., Goveas, L.C., Selvaraj, R.: Green synthesis of silver nanoparticles using *Lagerstroemia speciosa* fruit extract: catalytic efficiency in dye degradation. Mater. Technol. **40**(1), 2463955 (2025). https://doi.org/10.1080/10667857.2025.2463955
85. Mafhala, L., Khumalo, N., Zikalala, N.E., Azizi, S., Cloete, K.J., More, G.K., et al.: Antibacterial and cytotoxicity activity of green synthesized silver nanoparticles using aqueous extract of naartjie (*Citrus unshiu*) fruit peels. Emerg. Contam. **10**(4), 100348 (2024). https://doi.org/10.1016/j.emcon.2024.100348
86. Abbigeri, M.B., Thokchom, B., Singh, S.R., Bhavi, S.M., Harini, B.P., Yarajarla, R.B.: Antioxidant and anti-diabetic potential of the green synthesized silver nanoparticles using *Martynia annua* L. root extract. Nano TransMed **4**, 100070 (2025). https://doi.org/10.1016/j.ntm.2025.100070
87. Bohdanovych, T., Kuzema, P., Anishchenko, V., Duplij, V., Kharchuk, M., Lyzhniuk, V., et al.: Comparison of silver and gold nanoparticles green synthesis by *Artemisia annua* hairy root extracts. Biol. Open **14**(3), bio061739 (2025). https://doi.org/10.1242/bio.061739
88. Manzoor, S.I., Jabeen, F., Patel, R., Rizvi, M.M.A., Imtiyaz, K., Malik, M.A., Dar, T.A.: Green synthesis of biocompatible silver nanoparticles using *Trillium govanianum* rhizome extract: comprehensive biological evaluation and in silico analysis. Mater. Adv. **6**(2), 682–702 (2025). https://doi.org/10.1039/D4MA00959B
89. Vadakkan, K., Hemapriya, J., Ngangbam, A.K., Sathishkumar, K., Mapranathukaran, V.O.: Biofilm inhibition of *Staphylococcus aureus* by silver nanoparticles derived from *Hellenia speciosa* rhizome extract. Microb. Pathog. **196**, 106933 (2024). https://doi.org/10.1016/j.micpath.2024.106933
90. Anjali, Kumar Sonu, P., Sharma, K. P., Kumar, U., Peddireddy, V.: LC–MS profiling of bioflavonoid-catalysed size-controlled synthesis of silver nanoparticles from *Cordia sinensis* leaf and stem extracts for sustainable packaging applications. Appl. Organomet. Chem. **39**(7), e70251 (2025). https://doi.org/10.1002/aoc.70251
91. Obasi, D.E., Nebolisa, N.M., Akinwunmi, A.R., Abimbolu, A.K., Ezeorah, M.C., Areola, O.M., et al.: Eco-friendly and facile production method, natural products chemistry, and pharmacological properties of silver nanoparticles using telfaria occidentalis leaf and stem extracts. Eur. J. Sustain. Dev. Res **9**(2), em0280 (2025). https://doi.org/10.29333/ejosdr/15942
92. Bijang, C.M., Joris, S.N., Hasanela, N., Gaspersz, N., Waremra, A.: Phytosynthesis and antibacterial activity of silver nanoparticles of aqueous extracts of gandaria (*Bouea macrophylla* G.) leaves and stem bark. Trop. J. Nat. Prod. Res. **8**(11), 9315–9321 (2024). https://doi.org/10.26538/tjnpr/v8i11.47
93. Sadeghi-Kiakhani, M., Hashemi, E., Norouzi, M.M., Tehrani-Bagha, A.R.: Bio-synthesis of silver nanoparticles using *Rhus coriaria* L. seed extract on polyamide: characterization, antibacterial efficacy, and colorimetric analysis. Inorg. Chem. Commun. **174**, 113942 (2025). https://doi.org/10.1016/j.inoche.2025.113942
94. Meti, R.S., Neelagund, S.E., Urs, D., Dharmappa, K.K., Kotresh, K.R.: Green synthesis of silver nanoparticles from *Acacia sinuata* seed extract and evaluation of their mosquitocidal and anticancer (Caco-2 and MG-63 cell) activity. Biomass Conver. Biorefin. **15**(1), 175–184 (2025). https://doi.org/10.1007/s13399-023-05161-1

95. Sahu, U., Kunjam, S.: Green synthesis and characterization of silver (Ag) nanoparticles from aqueous bulb extract of *Urginea indica* (Roxb.) Kunth and its antimicrobial activity. Chem. Data Collect. **52**, 101150 (2024). https://doi.org/10.1016/j.cdc.2024.101150
96. Lopes, I.S., de Moraes, B.N.O., de Souza Barreto, S., Le Joncour, L., Couteau, C., Franzolin, M.R., Courrol, L.C.: Green synthesis of antimicrobial silver–copper nanoparticles using Banana and pineapple peel extracts: a sustainable approach for biomaterial sterilization. Mater. Chem. Phys. **333**, 130364 (2025). https://doi.org/10.1016/j.matchemphys.2025.130364
97. Gupta, S., Choudhary, D.K., Sundaram, S.: Green synthesis and characterization of silver nanoparticles using *Citrus sinensis* (Orange peel) extract and their antidiabetic, antioxidant, antimicrobial and anticancer activity. Waste Biomass Valor. **16**(3), 1101–1114 (2025). https://doi.org/10.1007/s12649-024-02782-z
98. Ghasemi, M., Govahi, M., Litkohi, H.R.: (2025). Green synthesis of silver nanoparticles (AgNPs) and chitosan-coated silver nanoparticles (CS-AgNPs) using Ferula gummosa Boiss. gum extract: a green nano drug for potential applications in medicine. Int. J. Biol. Macromol. **291**, 138619. https://doi.org/10.1016/j.ijbiomac.2024.138619
99. Benaissa, S., Benmansour, K., Zenagui, C., Bouchenak Khelladi, M., Belyagoubi, L.: Development of antioxidant and antimicrobial chitosan nanocomposite films incorporating with eco-friendly silver nanoparticles synthesized with Arabic gum. Polym. Bull. 1–27 (2025). https://doi.org/10.1007/s00289-025-05898-x
100. Deepa, Ameen, F., Amirul Islam, M., Dhanker, R.: Green synthesis of silver nanoparticles from vegetable waste of pea *Pisum sativum* and bottle gourd *Lagenaria siceraria*: characterization and antibacterial properties. Front. Environ. Sci. **10**, 941554
101. Mohammad, D.A.E., Taher, E.M.: Antimicrobial activity of silver nanoparticles fabricated from some vegetable plants. J. Phys. Conf. Ser. **1294**(6), 062048 (2019). https://doi.org/10.1088/1742-6596/1294/6/062109
102. Mythili, R., Selvankumar, T., Kamala-Kannan, S., Sudhakar, C., Ameen, F., Al-Sabri, A., et al.: Utilization of market vegetable waste for silver nanoparticle synthesis and its antibacterial activity. Mater. Lett. **225**, 101–104 (2018). https://doi.org/10.1016/j.matlet.2018.04.111
103. Ghosh, S., Patil, S., Ahire, M., Kitture, R., Gurav, D.D., Jabgunde, A.M., et al.: *Gnidia glauca* flower extract mediated synthesis of gold nanoparticles and evaluation of its chemocatalytic potential. J. Nanobiotechnol. **10**(1), 17 (2012). https://doi.org/10.1186/1477-3155-10-17
104. Saha, D., Talukdar, D., Mukherjee, P., Mitra, D., Mukherjee, R., Guha, S., et al.: Green synthesis of gold nano-particles using *Madhuca indica* flower extract and their anticancer activity on head and neck cancer: characterization and mechanistic study. Eur. J. Pharm. Biopharm. **207**, 114625 (2025). https://doi.org/10.1016/j.ejpb.2025.114625
105. Selvarani, J.F.L., Pushpavalli, K.S., Kala, S.M.J., Prakash, K.S.: Eco-friendly synthesis of gold nanoparticles using *Ruta graveolens* Leaf extract and their efficient catalytic reduction of 4-nitroaniline, methylene blue and antibiotic nitrofurantoin. J. Organomet. Chem. 123706 (2025).https://doi.org/10.1016/j.jorganchem.2025.123706
106. Sutarna, T.H., Pratiwi, G.K., Alatas, F., Ratih, H., Hermawati, A.M.: Optimization of green synthesis of nanogold using red ginger rhizome extract and polyvinylpyrrolidone (PVP) K60. J. Phys. Conf. Ser. **2980**(1), 012027 (2025). https://doi.org/10.1088/1742-6596/2980/1/012027
107. Das, A.K., Borah, M., Kalita, J.J., Bora, U.: Cytotoxic potential of Curcuma caesia rhizome extract and derived gold nanoparticles in targeting breast cancer cell lines. Sci. Rep. **14**(1), 17223 (2024). https://doi.org/10.1038/s41598-024-66175-x
108. Ghoreishi, S.M., Mortazavi-Derazkola, S.: Eco-friendly synthesis of gold nanoparticles via tangerine peel extract: unveiling their multifaceted biological and catalytic potentials. Heliyon **11**(1) (2025). https://doi.org/10.1016/j.heliyon.2024.e40104
109. Mythili, R., Selvankumar, T., Srinivasan, P., Sengottaiyan, A., Sabastinraj, J., Ameen, F., et al.: Biogenic synthesis, characterization and antibacterial activity of gold nanoparticles synthesised from vegetable waste. J. Mol. Liq. **262**, 318–321 (2018). https://doi.org/10.1016/j.molliq.2018.04.087

110. Sharma, D., Sabela, M.I., Kanchi, S., Mdluli, P.S., Singh, G., Stenström, T.A., Bisetty, K.: Biosynthesis of ZnO nanoparticles using *Jacaranda mimosifolia* flowers extract: synergistic antibacterial activity and molecular simulated facet specific adsorption studies. J. Photochem. Photobiol. B Biol. **162**, 199–207 (2016). https://doi.org/10.1016/j.jphotobiol.2016.06.043
111. Alam, K., Din, I.U., Tariq, S., Hayat, K., Khan, F.U., Khan, M., Mohamed, H.I.: Green synthesis and characterization of zinc oxide nanoparticles biosynthesized from *Butea monosperma* flowers and *Glycyrrhiza glabra* roots and their antioxidant and antibacterial properties. Appl. Biochem. Biotechnol. **197**(3), 1630–1649 (2025). https://doi.org/10.1007/s12010-024-05102-2
112. Yuvarani, A., Vaideeswaran, R., Rani, I.S., Mary, I.R.: Hydrothermal assisted green synthesis of zinc oxide nanostructures using Azadirachta indica flower extract and their antibacterial activity. J. Environ. Nanotechnol. **14**(2), 105–111 (2025). https://doi.org/10.13074/jent.2025.06.2511402
113. Yizengaw, D.E., Godie, E.M., Manayia, A.H.: Green synthesis and characterization of ZnO nanoparticles using *Justicia schemperiana* leaf extract and its antibacterial and antioxidant activity. Inorg. Chem. Commun. **174**, 114071 (2025). https://doi.org/10.1016/j.inoche.2025.114071
114. Evawati, K., Sutoyo, S.: Synthesis and characterization of ZnO nanoparticles using the duwet (Syzygiumcumini) leaves extract as bioreductor. J. Pijar MIPA **20**(1), 168–172 (2025) https://doi.org/10.29303/jpm.v20i1.7928
115. Nguyen, N.T.T., Phan, A.N.Q., Van Tran, T., Nguyen, T.T.T.: Antibacterial and photocatalytic activities of fruit-extract-mediated synthesized ZnO nanoparticles. Nanoscale Adv. **7**(11), 3514–3527 (2025). https://doi.org/10.1039/D5NA00211G
116. Haji, B.S., Barzinjy, A.A., Abbas, A.O., Kaygili, O., Mousa, M.S.: Green synthesis of ZnO nanoparticles using *Citrullus lanatus* fruit extract and their potential for microwave absorption. Nano-Struct. Nano-Objects **43**, 101502 (2025). https://doi.org/10.1016/j.nanoso.2025.101502
117. Abegunde, S.M., Olasehinde, E.F., Adebayo, M.A.: Green synthesis of ZnO nanoparticles using Nauclea latifolia fruit extract for adsorption of Congo red. Hybrid Adv. **5**, 100164 (2024). https://doi.org/10.1016/j.hybadv.2024.100164
118. Raj, L.A., Pavithra, R., Namasivayam, S.K.R.: Green route synthesis of highly stable zinc oxide nanoparticles using root extract of *Andrographis paniculata* and evaluation of their potential activities. Plant Nano Biol. 100162 (2025). https://doi.org/10.1016/j.plana.2025.100162
119. Sharma, R., Sharma, R., Singh, R.R., Kumari, A.: Evaluation of biogenic zinc oxide nanoparticles from *Tinospora cordifolia* stem extract for photocatalytic, anti-microbial, and antifungal activities. Mater. Chem. Phys. **297**, 127382 (2023). https://doi.org/10.1016/j.matchemphys.2023.127382
120. Saravanan, K., Sekar, V., Alshehri, M.A., Panneerselvam, C., Kathirvel, P.: Phytoassisted synthesis of biogenic ZnO nanoparticles using *Annona squamos*a L. bark extract: characterization and its application studies. Green Chem. Lett. Rev. **17**(1), 2395915 (2024). https://doi.org/10.1080/17518253.2024.2395915
121. Sushmitha, C.H., Krishnakumar, G., Navada, K.M.: Solution combustion method for synthesis of ZnO NPs from *Syzygium hemisphericum* bark extract and a comparative analysis of the same with the crude bark extract for biomedical applications. Chem. Pap. **78**(6), 3443–3462 (2024). https://doi.org/10.1007/s11696-024-03319-3
122. Jalil, P.J., Mhamedsharif, R.M., Shnawa, B.H., Hamad, S.M., Aspoukeh, P., Wsu, K.W., et al.: Biosynthesis of ZnO nanoparticles using washingtonia filifera seed extract and assessment of their anti-inflammatory and antimicrobial efficacy. J. Clust. Sci. **36**(1), 31 (2025). https://doi.org/10.1007/s10876-024-02761-3
123. Kebede Urge, S., Tiruneh Dibaba, S., Belay Gemta, A.: Green synthesis method of ZnO nanoparticles using extracts of *Zingiber officinale* and garlic bulb (*Allium sativum*) and their synergetic effect for antibacterial activities. J. Nanomater. **2023**(1), 7036247 (2023). https://doi.org/10.1155/2023/7036247

124. Erdoğan, Ö., Paşa, S., Demirbolat, G.M., Birtekocak, F., Abbak, M., Çevik, Ö.: Synthesis, characterization, and anticarcinogenic potent of green-synthesized zinc oxide nanoparticles via *Citrus aurantium* aqueous peel extract. Inorg. Nano-Met. Chem. **55**(1), 67–75 (2025). https://doi.org/10.1080/24701556.2023.2240768
125. Mokshith, D.G., Kalpashree, M.M., Krishna, D.: Natural synthesis of nanoparticles using flower extract of *Hymenocallis littoralis* (Jacq) Salisb and evaluation of its antimicrobial activity. Int. J. Res. Appl. Sci. Eng. Technol. **10**(10), 1199–1207 (2022). https://doi.org/10.22214/ijraset.2022.47158
126. Moroda, M.D., Deressa, T.L., Tiwikrama, A.H., Chala, T.F.: Green synthesis of copper oxide nanoparticles using *Rosmarinus officinalis* leaf extract and evaluation of its antimicrobial activity. Next Mater. **7**, 100337 (2025). https://doi.org/10.1016/j.nxmate.2024.100337
127. Parveen, S.T., Riazunnisa, K.: Phytofabrication of copper nanoparticles using *Momordica cymbalaria* fruit extract and their antibacterial, antioxidant and anti-inflammatory properties. Hybrid Adv. **9**, 100421 (2025). https://doi.org/10.1016/j.hybadv.2025.100421
128. Raghavendra, N., Nagaswarupa, H.P., TR, S.S., Mylarappa, M., Basavaraju, N., Surendra, B.S.: Green synthesis of $CuFe_2O_4$ nanoparticles: multifunctional applications in electrochemistry, sensing and photocatalysis. Next Mater. **8**, 100863 (2025). https://doi.org/10.1016/j.hybadv.2025.100421
129. Teklu, B., Kadiri, S.K., Vidavalur, S.: Green synthesis of copper oxide nanoparticles using *Balanites aegyptiaca* stem bark extract and investigation of antibacterial activity. Results Chem. **6**, 101152 (2023). https://doi.org/10.1016/j.rechem.2023.101152
130. Ramalakshmi, C., Suresh, S., Murugan, A.M., Alyami, N.M., Aljawdah, H.M., Alharbi, S.A., Mariselvam, R.: Green synthesis of copper nanoparticles from *Terminalia arjuna* (L.) bark extract: characterization and potential for mercury degradation. Bull. Chem. Soc. Ethiopia **38**(6), 1703–1713 (2024). https://doi.org/10.4314/bcse.v38i6.16
131. Khan, A.U., Khan, A.A., Malik, N., Her, J., Gupta, M., Panda, S., et al.: Synthesis of Cu_4O_3 nanoparticles using pumpkin seed extract: optimization, antimicrobial, and cytotoxicity studies. Nanotechnol. Rev. **14**(1), 20250203 (2025). https://doi.org/10.1515/ntrev-2025-0203
132. Kumar, A., Sofi, A.H., Hamza, B., Rubab, S., Shah, M.A.: Eco-friendly synthesis of antibacterial CuO nanoparticles using garlic bulb extract. Green Mater. **13**(2), 111–120 (2025). https://doi.org/10.1680/jgrma.24.00139
133. Naseem, Z., Naveed, M., Asif, M., Alamri, S., Nawaz, S., Siddiqui, M.H., Mustafa, A.: Enhancing chromium resistance and bulb quality in onion (*Allium cepa* L.) through copper nanoparticles and possible health risk. BMC Plant Biol. **24**(1), 777 (2024). https://doi.org/10.1186/s12870-024-05460-3
134. Ullah, H., Ullah, Z., Fazal, A., Irfan, M.: Use of vegetable waste extracts for controlling microstructure of CuO nanoparticles: green synthesis, characterization, and photocatalytic applications. J. Chem. **2017**(1), 2721798 (2017). https://doi.org/10.1155/2017/2721798
135. Behera, A., Ranjith, N., Balasubramani, S., Ryntathiang, I., Dharmalingam Jothinathan, M.K.: Evaluating the toxicity profile of green-synthesized ferric nanoparticles using madhuca indica in a zebrafish model. Biomed. Mater. Dev. 1–15 (2025). https://doi.org/10.1007/s44174-025-00348-8
136. Oyshi, S.A., Jahan, R.A., Aktar, F., Sultan, M.Z., Chowdhury, A.A., Chowdhury, J.A., et al.: Preparation and evaluation of the biosynthetic procedure of iron oxide and magnesium oxide nanoparticles using *Hylocereus undatus* fruit peel extract and their anticancer properties. RSC Adv. **15**(19), 15366–15374 (2025). https://doi.org/10.1039/D4RA07411D
137. Hasoon, R.N., Abd, A.N., Majeed, A.H.: Eco-friendly synthesis and characterization of iron oxide (α-Fe_2O_3) nanoparticles using macromolecules from pomegranate peel extract. Macromol. Symp. **414**(1), 2400231 (2025). https://doi.org/10.1002/masy.202400231
138. UncuKirtiş, E.B., Yiğit Koçak, D., Eldervis, U., Tuna, S., Bayraç, C.: Green synthesis of iron oxide nanoparticles from extracts of *Daucus carota* subsp. sativus whole vegetable, peel, pomace, and juice and their application as antibacterial agents and Fenton-like catalysts. Int. J. Food Sci. Technol. **60**(1), vvae035 (2025). https://doi.org/10.1093/ijfood/vvae035

139. Sofyan, N., Ridhova, A., Angellinnov, F., M'rad, M., Yuwono, A.H., Dhaneswara, D., et al.: Jasmine flowers extract mediated green synthesis of tio_2 nanoparticles and their photocurrent characteristics for dye-sensitized solar cell application. Mater. Renew. Sustain. Energy **14**(2), 1–18 (2025). https://doi.org/10.1007/s40243-025-00320-y
140. Moghni, N., Khalaf, H., Menseri, O., Boutoumi, H., Boudali, R., Dif, F., Boucheffa, Y.: One–pot green synthesis of TiO_2 nanoparticles using Inula Viscosa leaf extract as an efficient photocatalyst for organic dyes removal. J. Photochem. Photobiol. A Chem. **461**, 116158 (2025). https://doi.org/10.1016/j.jphotochem.2024.116158
141. Sellami, H., Akinyemi, M.O., Gdoura-Ben Amor, M., Onwudiwe, D.C., Mthiyane, D.: Structural and optical characterization of TiO_2 nanoparticles synthesized using *Globularia alypum* leaf extract and the antibacterial properties. Discov. Appl. Sci. **7**(8), 1–19 (2025). https://doi.org/10.1007/s42452-025-07521-0
142. Al-darwesh, M.Y., Babakr, K.A., Qader, I.N., Mohammed, M.A.: Antimicrobial, anti-inflammatory, and anticancer potential of green synthesis TiO_2 nanoparticles using *Sophora flavescens* root extract. Chem. Pap. **79**(2), 1207–1221 (2025). https://doi.org/10.1007/s11696-024-03853-0
143. Muhaymin, A., Mohamed, H.E.A., Hkiri, K., Safdar, A., Azizi, S., Maaza, M.: Green synthesis of magnesium oxide nanoparticles using *Hyphaene thebaica* extract and their photocatalytic activities. Sci. Rep. **14**(1), 20135 (2024). https://doi.org/10.1038/s41598-024-71149-0
144. Munir, G., Mohamed, H.E.A., Hkiri, K., Ghotekar, S., Maaza, M.: Phyto-mediated fabrication of Mn_2O_3 nanoparticles using *Hyphaene thebaica* fruit extract as a promising photocatalyst for MB dye degradation. Inorg. Chem. Commun. **167**, 112695 (2024). https://doi.org/10.1038/s41598-024-71149-0
145. Saha, S., Dutta, G., Mohanto, S., Prabhu, A., Ghosh, B., Pal, S., et al.: Sustainable synthesis of Cr_2O_3 nanoparticles utilizing *Rauvolfia tetraphylla* root extract for lung cancer treatment. J. Drug Deliv. Sci. Technol. **104**, 106531 (2025). https://doi.org/10.1016/j.jddst.2024.106531
146. Jeyaraj, S., Mary Saral, A.: Synthesis of SnO_2 nanoparticles using *Brassica juncea* (L.) czern seed extract: characterization and biological applications. ChemistrySelect **10**(7), e202404490 (2025). https://doi.org/10.1002/slct.202404490
147. Deepti, S., Bhavana, A., Inchana, A., Rashmi, K.V., Lalithamba, H.S., Vidya, S.M., et al.: Eco-friendly fabrication of zirconium-and neodymium-doped NiO nanoparticles using *Zephyranthes citrina* bulb extract: structural, optical, biomedical, and agronomic assessments. J. Aust. Ceram. Soc. 1–18 (2025). https://doi.org/10.1007/s41779-025-01225-9
148. Seku, K., Pejjai, B., Osman, A.I., Hussaini, S.S., Al Abri, M., Kumar, N.S., et al.: Ecofriendly synthesis of *Salmalia malabarica* gum stabilized palladium nanoparticles: antibacterial and catalytic properties. Biomass Conver. Biorefin. **15**(4), 6055–6067 (2025). https://doi.org/10.1007/s13399-024-05443-2
149. Seku, K., Pejjai, B., Osman, A.I., Hussaini, S.S., Al-Abri, M., Swathi, R., et al.: Microwave-assisted synthesis of *Limonia acidissima* Groff gum stabilized palladium nanoparticles for colorimetric glucose sensing. J. Colloid Interface Sci. **659**, 718–727 (2024). https://doi.org/10.1016/j.jcis.2024.01.046
150. Samaila, I.S.I., Yusuf, M.M.Y.M.M.: Green synthesis of bismuth oxide nanoparticles and its in vitro antimicrobial study using red river gum (*Eucalyptus camaldulensis*) leaf. Chem Class J. **9**(1), 287–297 (2025). https://doi.org/10.33003/chemclas_2025_0901/023
151. Chaudhary, J., Tailor, G.: Bio-fabrication, characterization and antimicrobial action of cadmium nanoparticles using Nyctanthes arbor-tristis (Leaves). In: Annual meeting of the Asian Polymer Association, pp. 111–124. Springer, Singapore (2024). https://doi.org/10.1007/978-981-97-7209-4_9
152. Shetty, A.T., Shetty, D.N., Raghavendra, N., Rani, S.: Highly efficient green synthesized cerium oxide nanoparticles from *Tectona grandis* seed extract for Rhodamine B dye degradation and biological applications. Next Mater. **8**, 100796 (2025). https://doi.org/10.1016/j.nxmate.2025.100796
153. Narayanan, K.B., Sakthivel, N.: Biological synthesis of metal nanoparticles by microbes. Adv. Colloid Interface Sci. **156**(1–2), 1–13 (2010). https://doi.org/10.1016/j.cis.2010.02.001

154. Carmona, M., Poblete-Castro, I., Rai, M., Turner, R.J.: Opportunities and obstacles in microbial synthesis of metal nanoparticles. Microb. Biotechnol. **16**(5), 871–876 (2023). https://doi.org/10.1111/1751-7915.14254
155. Elazab, N.T., Younis, S.A., Abdelgalil, S.A.: Biogenic synthesis of nanoparticles mediated by fungi. In: Plant Mycobiome: Diversity, Interactions and Uses, pp. 241–265 (2023). https://doi.org/10.1007/978-3-031-28307-9_10
156. Velamakanni, R.P., Gothalwal, R., Velamakanni, R.S., Ayinampudi, S.R., Vuppugalla, P., Merugu, R.: Fungi-mediated green synthesis of nanoparticles and their renewable energy applications. In: Green Nano Solution for Bioenergy Production Enhancement, pp. 201–224. Springer Nature Singapore, Singapore (2022). https://doi.org/10.1007/978-981-16-9356-4_8
157. Devatha, C.P., Thalla, A.K.: Green synthesis of nanomaterials. In: Synthesis of Inorganic Nanomaterials, pp. 169–184. Woodhead Publishing (2018). https://doi.org/10.1016/B978-0-08-101975-7.00007-5
158. Wilson, J.J., Harimuralikrishnaa, T., Ponmanickam, P., Lakshmi, M.P.: Bacterial silver nanoparticles: method, mechanism of synthesis and application in mosquito control. In: Mosquito Research—Recent Advances in Pathogen Interactaction, Immunity, and Vector Control Strategies. IntechOpen (2023). https://doi.org/10.5772/intechopen.104144
159. Sudarsan, S., Kumar Shankar, M., Kumar Belagal Motatis, A., Shankar, S., Krishnappa, D., Mohan, C.D., et al.: Green synthesis of silver nanoparticles by Cytobacillus firmus isolated from the stem bark of *Terminalia arjuna* and their antimicrobial activity. Biomolecules **11**(2), 259 (2021). https://doi.org/10.3390/biom11020259
160. Hamk, M., Akçay, F.A., Avcı, A.: Green synthesis of zinc oxide nanoparticles using *Bacillus subtilis* ZBP4 and their antibacterial potential against foodborne pathogens. Prep. Biochem. Biotechnol. **53**(3), 255–264 (2023). https://doi.org/10.1080/10826068.2022.2076243
161. Jain, D., Shivani, Bhojiya, A.A., Singh, H., Daima, H.K., Singh, M., et al.: Microbial fabrication of zinc oxide nanoparticles and evaluation of their antimicrobial and photocatalytic properties. Front. Chem. **8**, 778 (2020). https://doi.org/10.3389/fchem.2020.00778
162. Long, Q., Cui, L.K., He, S.B., Sun, J., Chen, Q.Z., Bao, H.D., et al.: Preparation, characteristics and cytotoxicity of green synthesized selenium nanoparticles using *Paenibacillus motobuensis* LY5201 isolated from the local specialty food of longevity area. Sci. Rep. **13**(1), 53 (2023). https://doi.org/10.1038/s41598-022-26396-4
163. Correa-Llantén, D.N., Muñoz-Ibacache, S.A., Castro, M.E., Muñoz, P.A., Blamey, J.M.: Gold nanoparticles synthesized by *Geobacillus* sp. strain ID17 a thermophilic bacterium isolated from Deception Island, Antarctica. Microb. Cell Factor. **12**(1), 75 (2013). https://doi.org/10.1186/1475-2859-12-75
164. Ameen, F., Al-Maary, K.S., Almansob, A., AlNadhari, S.: Antioxidant, antibacterial and anticancer efficacy of *Alternaria chlamydospora*-mediated gold nanoparticles. Appl. Nanosci. **13**(3), 2233–2240 (2023). https://doi.org/10.1007/s13204-021-02047-4
165. Ameen, F., Al-Homaidan, A.A., Al-Sabri, A., Almansob, A., AlNAdhari, S.: Anti-oxidant, anti-fungal and cytotoxic effects of silver nanoparticles synthesized using marine fungus *Cladosporium halotolerans*. Appl. Nanosci. **13**(1), 623–631 (2023). https://doi.org/10.1007/s13204-021-01874-9
166. Syed, A., Al Saedi, M.H., Bahkali, A.H., Elgorgan, A.M., Kharat, M., Pai, K., et al.: αAu2S nanoparticles: fungal-mediated synthesis, structural characterization and bioassay. Green Chem. Lett. Rev. **15**(1), 61–70 (2022). https://doi.org/10.1080/17518253.2021.1999509
167. Rafeeq, C.M., Paul, E., Saagar, E.V., Ali, P.M.: Mycosynthesis of zinc oxide nanoparticles using *Pleurotus floridanus* and optimization of process parameters. Ceram. Int. **47**(9), 12375–12380 (2021). https://doi.org/10.1016/j.ceramint.2021.01.091
168. Wani, N.A., Khanday, W.I., Tirumale, S.: Biosynthesis of iron oxide nanoparticles using ethyl acetate extract of *Chaetomium cupreum* and their anticancer activity. Matrix Sci. Pharma **4**(2), 23–30 (2020). https://doi.org/10.4103/MTSP.MTSP_6_20
169. Siddiqi, K.S., Husen, A.: Fabrication of metal and metal oxide nanoparticles by algae and their toxic effects. Nanoscale Res. Lett. **11**(1), 363 (2016). https://doi.org/10.1186/s11671-016-1580-9

170. Karaduman, I., Güngör, A.A., Nadaroğlu, H., Altundaş, A., Acar, S.: Green synthesis of γ-Fe_2O_3 nanoparticles for methane gas sensing. J. Mater. Sci. Mater. Electron. **28**(21), 16094–16105 (2017). https://doi.org/10.1007/s10854-017-7510-5
171. Palanimuthu, V., Periakaruppan, R., Romanovski, V., Bharathi, A., Vijai Selvaraj, K.S., Anukeerthana, S., et al.: Synthesis and structural characterization of SiO2 nanoparticles using extract of *Gracilaria crassa* via green chemistry approach. ChemistryOpen **14**(2), e202400356 (2025). https://doi.org/10.1002/open.202400356
172. Manikandan, R., Thomas, J.: Sustainable approaches in green synthesis of silica nanoparticles using extracts of Chlorella and its application. Appl. Biochem. Biotechnol. **196**(11), 7928–7939 (2024). https://doi.org/10.1007/s12010-024-04949-9
173. Alsaggaf, M.S., Diab, A.M., ElSaied, B.E., Tayel, A.A., Moussa, S.H.: Application of ZnO nanoparticles phycosynthesized with *Ulva fasciata* extract for preserving peeled shrimp quality. Nanomaterials **11**(2), 385 (2021). https://doi.org/10.3390/nano11020385
174. Fatima, R., Priya, M., Indurthi, L., Radhakrishnan, V., Sudhakaran, R.: Biosynthesis of silver nanoparticles using red algae *Portieria hornemannii* and its antibacterial activity against fish pathogens. Microb. Pathog. **138**, 103780 (2020). https://doi.org/10.1016/j.micpath.2019.103780
175. Patel, D., Patel, B., Yadav, V.K., Sudhakar, M.P., Alharbi, S.A., Salmen, S.H., et al.: Silver nanoparticles synthesized from marine algae *Spatoglossum asperum*: antioxidant properties and seed germination enhancement. J. Hazard. Mater. Adv. **16**, 100478 (2024). https://doi.org/10.1016/j.hazadv.2024.100478
176. Salem, D.M., Ismail, M.M., Tadros, H.R.: Evaluation of the antibiofilm activity of three seaweed species and their biosynthesized iron oxide nanoparticles (Fe_3O_4-NPs). Egypt. J. Aquat. Res. **46**(4), 333–339 (2020). https://doi.org/10.1016/j.ejar.2020.09.001
177. Kumar, V., Singh, Y., Kaushal, S., Kumar, R.: Bioinspired synthesis of copper oxide nanoparticles using aqueous extracts of *Cladophora glomerata* (L.) Kuetz and their potential biomedical applications. Bioprocess Biosyst. Eng. **48**(4), 633–646 (2025). https://doi.org/10.1007/s00449-025-03133-5
178. Baskar, G., Palaniyandi, T., Ravi, M., Viswanathan, S., Wahab, M.R.A., Surendran, H., et al.: Biosynthesis of iron oxide nanoparticles from red seaweed *Hypnea valentiae* and evaluation of their antioxidant and antitumor potential via the AKT/PI3K pathway. Process Biochem. **141**, 155–169 (2024). https://doi.org/10.1016/j.procbio.2024.03.010
179. Choudhary, S., Kumawat, G., Kher, K., Baroliya, P.K., Gupta, A.K., Tripathy, M.K.: Green synthesis of palladium nanoparticles using Asterarcys sp. and their applications. Nano Trans Med **3**, 100046 (2024). https://doi.org/10.1016/j.ntm.2024.100046
180. Yassein, A.S., Elamary, R.B., Alwaleed, E.A.: Biogenesis, characterization, and applications of Spirulina selenium nanoparticles. Microb. Cell Factor. **24**(1), 39 (2025). https://doi.org/10.1186/s12934-025-02656-6
181. Vasanthakumar, S., Manikandan, M., Arumugam, M.: Green synthesis, characterization and functional validation of bio-transformed selenium nanoparticles. Biochem. Biophys. Rep. **39**, 101760 (2024). https://doi.org/10.1016/j.bbrep.2024.101760
182. Khandaker, T., Islam, T., Nandi, A., Anik, M.A.A.M., Hossain, M.S., Hasan, M.K., Hossain, M.S.: Biomass-derived carbon materials for sustainable energy applications: a comprehensive review. Sustain. Energy Fuels **9**(3), 693–723 (2025). https://doi.org/10.1039/D4SE01393J
183. Saxena, R.: Biomass valorization for production of futuristic materials—a review. Int. J. Novel Res. Dev. **8**(6), c291–c309 (2023). https://doi.org/10.1016/j.mtsust.2022.100138
184. Afre, R.A., Soga, T., Jimbo, T., Kumar, M., Ando, Y., Sharon, M.: Growth of vertically aligned carbon nanotubes on silicon and quartz substrate by spray pyrolysis of a natural precursor: turpentine oil. Chem. Phys. Lett. **414**(1–3), 6–10 (2005). https://doi.org/10.1016/j.cplett.2005.08.040
185. Ghosh, P., Afre, R.A., Soga, T., Jimbo, T.: A simple method of producing single-walled carbon nanotubes from a natural precursor: Eucalyptus oil. Mater. Lett. **61**(17), 3768–3770 (2007). https://doi.org/10.1016/j.matlet.2006.12.030

186. Suriani, A.B., Azira, A.A., Nik, S.F., Nor, R.M., Rusop, M.: Synthesis of vertically aligned carbon nanotubes using natural palm oil as carbon precursor. Mater. Lett. **63**(30), 2704–2706 (2009). https://doi.org/10.1016/j.matlet.2009.09.048
187. He, L., Hu, S., Jiang, L., Liao, G., Zhang, L., Han, H., et al.: Co-production of hydrogen and carbon nanotubes from the decomposition/reforming of biomass-derived organics over Ni/ α-Al_2O_3 catalyst: performance of different compounds. Fuel **210**, 307–314 (2017). https://doi.org/10.1016/j.fuel.2017.08.080
188. Hata, K., Futaba, D.N., Mizuno, K., Namai, T., Yumura, M., Iijima, S.: Water-assisted highly efficient synthesis of impurity-free single-walled carbon nanotubes. Science **306**(5700), 1362–1364 (2004). https://doi.org/10.1126/science.1104962
189. Khalilov, U., Vets, C., Neyts, E.C.: Molecular evidence for feedstock-dependent nucleation mechanisms of CNTs. Nanoscale Horiz. **4**(3), 674–682 (2019). https://doi.org/10.1039/C8NH00323H
190. Kang, Z., Wang, E., Mao, B., Su, Z., Chen, L., Xu, L.: Obtaining carbon nanotubes from grass. Nanotechnology **16**(8), 1192 (2005). https://doi.org/10.1088/0957-4484/16/8/036
191. Kumar, M., Okazaki, T., Hiramatsu, M., Ando, Y.: The use of camphor-grown carbon nanotube array as an efficient field emitter. Carbon **45**(9), 1899–1904 (2007). https://doi.org/10.1016/j.carbon.2007.04.023
192. Goodell, B., Xie, X., Qian, Y., Daniel, G., Peterson, M., Jellison, J.: Carbon nanotubes produced from natural cellulosic materials. J. Nanosci. Nanotechnol. **8**(5), 2472–2474 (2008). https://doi.org/10.1166/jnn.2008.235
193. Awasthi, K., Kumar, R., Raghubanshi, H., Awasthi, S., Pandey, R., Singh, D., et al.: Synthesis of nano-carbon (nanotubes, nanofibres, graphene) materials. Bull. Mater. Sci. **34**(4), 607 (2011). https://doi.org/10.1007/s12034-011-0170-9
194. Kumar, R., Yadav, R.M., Awasthi, K., Shripathi, T., Sinha, A.S.K., Tiwari, R.S., Srivastava, O.N.: Synthesis of carbon and carbon–nitrogen nanotubes using green precursor: jatropha-derived biodiesel. J. Exp. Nanosci. **8**(4), 606–620 (2013). https://doi.org/10.1080/17458080.2011.577102
195. Nersisyan, H.H., Lee, T.H., Lee, K.H., Maeng, D.Y., Lee, J.H.: Fabrication of tunable carbon micro-and nanotubes using reed as bio-template. Mater. Lett. **107**, 79–82 (2013). https://doi.org/10.1016/j.matlet.2013.05.102
196. Onishchenko, D.V., Reva, V.P.: Physicochemical properties of carbon nanotubes obtained from peat moss. J. Eng. Phys. Thermophys. **86**(2), 363–369 (2013). https://doi.org/10.1007/s10891-013-0842-y
197. Shi, K., Yan, J., Lester, E., Wu, T.: Catalyst-free synthesis of multiwalled carbon nanotubes via microwave-induced processing of biomass. Ind. Eng. Chem. Res. **53**(39), 15012–15019 (2014). https://pubs.acs.org/doi/10.1021/ie503076n
198. Wang, G., Wang, W.H., Wang, L.K., Yao, W.T., Yao, P.F., Zhu, W.K., et al.: A N-, Fe- and Co-tridoped carbon nanotube/nanoporous carbon nanocomposite with synergistically enhanced activity for oxygen reduction in acidic media. J. Mater. Chem. A **3**(34), 17866–17873 (2015). https://doi.org/10.1039/C5TA03523F
199. Li, M., Xiong, Y., Liu, X., Han, C., Zhang, Y., Bo, X., Guo, L.: Iron and nitrogen co-doped carbon nanotube@ hollow carbon fibers derived from plant biomass as efficient catalysts for the oxygen reduction reaction. J. Mater. Chem. A **3**(18), 9658–9667 (2015). https://doi.org/10.1039/C5TA00958H
200. Thompson, E., Danks, A.E., Bourgeois, L., Schnepp, Z.: Iron-catalyzed graphitization of biomass. Green Chem. **17**(1), 551–556 (2015). https://doi.org/10.1039/C4GC01673D
201. Reva, V.P., Filatenkov, A.E., Yagofarov, V.U., Gulevskii, D.A., Kuryavyi, V.G., Mansurov, Y.N.: Analysis of the formation of multi-layer carbon nanotubes in the process of mechanical activation of the pyrolysis products of vegetable raw materials. IOP Conf. Ser. Mater. Sci. Eng. **127**(1), 012008 (2016). https://doi.org/10.1088/1757-899X/127/1/012008
202. Bernd, M.G.S., Bragança, S.R., Heck, N., da Silva Filho, L.C.: Synthesis of carbon nanostructures by the pyrolysis of wood sawdust in a tubular reactor. J. Mater. Res. Technol. **6**(2), 171–177 (2017). https://doi.org/10.1016/j.jmrt.2016.11.003

203. Wang, G., Deng, Y., Yu, J., Zheng, L., Du, L., Song, H., Liao, S.: From chlorella to nestlike framework constructed with doped carbon nanotubes: a biomass-derived, high-performance, bifunctional oxygen reduction/evolution catalyst. J. Mater. Res. Technol. **9**(37), 32168–32178 (2017). https://pubs.acs.org/doi/10.1021/acsami.7b10668
204. Zhang, J., Zhao, H., Li, J., Jin, H., Yu, X., Lei, Y., Wang, S.: In situ encapsulation of iron complex nanoparticles into biomass-derived heteroatom-enriched carbon nanotubes for high-performance supercapacitors. Adv. Energy Mater. **9**(4), 1803221 (2019). https://doi.org/10.1002/aenm.201803221
205. Zhang, Y., Lu, L., Zhang, S., Lv, Z., Yang, D., Liu, J., et al.: Biomass chitosan derived cobalt/nitrogen doped carbon nanotubes for the electrocatalytic oxygen reduction reaction. J. Mater. Chem. A **6**(14), 5740–5745 (2018). https://doi.org/10.1039/C7TA11258K
206. Gao, Z., Song, N., Zhang, Y., Schwab, Y., He, J., Li, X.: Carbon nanotubes derived from yeast-fermented wheat flour and their energy storage application. ACS Sustain. Chem. Eng. **6**(9), 11386–11396 (2018). https://pubs.acs.org/doi/10.1021/acssuschemeng.8b01292
207. Osman, A.I., Blewitt, J., Abu-Dahrieh, J.K., Farrell, C., Al-Muhtaseb, A.A.H., Harrison, J., Rooney, D.W.: Production and characterisation of activated carbon and carbon nanotubes from potato peel waste and their application in heavy metal removal. Environ. Sci. Pollut. Res. **26**(36), 37228–37241 (2019). https://doi.org/10.1007/s11356-019-06594-w
208. Osman, A.I., O'Connor, E., McSpadden, G., Abu-Dahrieh, J.K., Farrell, C., Al-Muhtaseb, A.A.H., et al.: Upcycling brewer's spent grain waste into activated carbon and carbon nanotubes for energy and other applications via two-stage activation. J. Chem. Technol. Biotechnol. **95**(1), 183–195 (2020). https://doi.org/10.1002/jctb.6220
209. Yang, M., Yuan, Y., Li, Y., Sun, X., Wang, S., Liang, L., et al.: Dramatically enhanced electromagnetic wave absorption of hierarchical CNT/Co/C fiber derived from cotton and metal-organic-framework. Carbon **161**, 517–527 (2020). https://doi.org/10.1016/j.carbon.2020.01.073
210. Nemati, B., Kamali, M., Aminabhavi, T.M., Dewil, R., Ferreira, P., Silva, M.R., et al.: Production of high yield and high-quality carbon nanotubes from biomass wastes under microwave irradiation—mechanism and properties. Sustain. Mater. Technol. e01573 (2025). https://doi.org/10.1016/j.susmat.2025.e01573
211. Su, H., Huang, X., Zhang, Z., Ye, Y., Wang, D.: Synthesis of carbon nanotubes by chemical vapor deposition using nickel supported on nitric acid-treated pine sawdust. Biomass Bioenergy **193**, 107579 (2025). https://doi.org/10.1016/j.biombioe.2024.107579
212. Liu, F., Yuan, M., Feng, P., Yu, S., Hu, Z., Xiang, H., Zhu, M.: Synthesis of nitrogen-doped carbon nanotubes from biomass polysaccharides and lignin in waste corn stalk as host materials for lithium-sulfur batteries. Int. J. Biol. Macromol. **304**, 140813 (2025). https://doi.org/10.1016/j.ijbiomac.2025.140813
213. Fang, M., Wang, B., Qu, X., Li, S., Huang, J., Li, J., et al.: State-of-the-art of biomass-derived carbon dots: preparation, properties, and applications. Chin. Chem. Lett. **35**(1), 108423 (2024). https://doi.org/10.1016/j.cclet.2023.108423
214. Aparicio, R.R., de Lima, G.G., Perissutti, G.E., Bassi, M.D.J., Jacumazo, J., Schiavon, M.A., et al.: Pyrolytic lignin: a promising precursor for the green synthesis of fluorescent carbon nanoparticles. ACS Omega **10**(11), 11054–11062 (2025). https://pubs.acs.org/doi/10.1021/acsomega.4c09764
215. Ma, L., Liao, S., Liu, J., Ma, Y.: Green synthesis of high-yield red fluorescent carbon dots from Amaranth for sensitive water detection in organic solvents. Microchem. J. **209**, 112654 (2025). https://doi.org/10.1016/j.microc.2025.112654
216. Arul, V., Chandrasekaran, P., Sivaraman, G., Sethuraman, M.G.: Biogenic preparation of undoped and heteroatoms doped carbon dots: effect of heteroatoms doping in fluorescence, catalytic ability and multicolour in-vitro bio-imaging applications-a comparative study. Mater. Res. Bull. **162**, 112204 (2023). https://doi.org/10.1016/j.materresbull.2023.112204
217. Kansay, V., Sharma, V.D., Chandan, G., Chakrabarti, S., Bera, M.K.: Facile green synthesis of heteroatoms doped Aegle marmelos-derived carbon quantum dots and its tuneable fluorescence characteristics influenced by solvent polarity. Carbon Lett. **34**(3), 1045–1054 (2024). https://doi.org/10.1007/s42823-023-00659-0

218. Zhang, L., Luo, W., Chen, Y., Zheng, J., Cao, L., Duan, L., et al.: Green synthesis of boron-doped carbon dots from Chinese herbal residues for Fe3+ sensing, anti-counterfeiting, and photodegradation applications. J. Clean. Prod. **422**, 138577 (2023). https://doi.org/10.1016/j.jclepro.2023.138577
219. Chen, J., Li, T., Lin, C., Hou, Y., Cheng, S., Gao, B.: Green synthesis of red-emitting carbon dots for bioimaging, sensing, and antibacterial applications. Spectrochim. Acta A **328**, 125458 (2025). https://doi.org/10.1016/j.saa.2024.125458
220. Xu, Z., Gan, Y., Zeng, J., Chen, J., Fu, A., Zheng, X., Li, W.: Green synthesis of functionalized fluorescent carbon dots from biomass and their corrosion inhibition mechanism for copper in sulfuric acid environment. Chem. Eng. J. **470**, 144425 (2023). https://doi.org/10.1016/j.cej.2023.144425
221. Lan, L., Bao, R., Kong, X., Liu, L., Wang, J., Yi, J.: Room-temperature synthesis of carbon polymer dots from biomass for advanced sensing and fluorescent applications. J. Environ. Chem. Eng. **12**(6), 114514 (2024). https://doi.org/10.1016/j.jece.2024.114514
222. El-Said, W.A., Akhdhar, A., Al-Bogami, A.S., Saleh, T.S.: Design and green synthesis of carbon dots/gold nanoparticles composites and their applications for neurotransmitters sensing based on emission spectroscopy. Spectrochim. Acta Part A **327**, 125402 (2025). https://doi.org/10.1016/j.saa.2024.125402
223. Magdy, G., El-Deen, A.K., Radwan, A.S., Belal, F., Elmansi, H.: Ultrafast microwave-assisted green synthesis of nitrogen-doped carbon dots as turn-off fluorescent nanosensors for determination of the anticancer Nintedanib: monitoring of environmental water samples. Talanta Open **11**, 100423 (2025). https://doi.org/10.1016/j.talo.2025.100423
224. Gagana, M., Krushna, B.R., Sharma, S.C., Sharmila, S., Meenakshi, R., Devikala, A., et al.: Green synthesis of carbon dots encapsulated MoO_3: La3+ for enhanced photocatalytic degradation, dactyloscopy and real-time FP detection using YOLOv8x. J. Taiwan Inst. Chem. Eng. **170**, 106032 (2025). https://doi.org/10.1016/j.jtice.2025.106032
225. Renuga, V., Vinoliya, X., Dinesh, A., Suthakaran, S., Ayyar, M., Mohanavel, V., et al.: Green synthesis of biocompatible fluorescent carbon dots from bitter gourd for effective metal sensing and biological applications. Sens. Bio-Sens. Res. **47**, 100751 (2025). https://doi.org/10.1016/j.sbsr.2025.100751
226. Raypah, M.E., Jamlos, M.F., Giwa, H.S., Ahmad, H.F., Abd Hamid, H.: Green synthesis of carbon dots derived from biomass of *Polyalthia bullata* root extract as colorimetric sensor for selective detection of Fe3+ ions. Inorg. Chem. Commun. **174**, 113944 (2025). https://doi.org/10.1016/j.inoche.2025.113944
227. Wechakorn, K., Khaopueak, P., Seesuea, C., Leung, F.K.C., Tantipanjaporn, A.: Facile green synthesis of multifunctional carbon dots from cassava peel for heavy metal ion sensing, photocatalytic degradation, antioxidant activity, and fluorescence bioimaging. Mater. Sci. Eng. B **320**, 118435 (2025). https://doi.org/10.1016/j.mseb.2025.118435
228. Li, Y., Liu, F., Cai, J., Huang, X., Lin, L., Lin, Y., et al.: Nitrogen and sulfur co-doped carbon dots synthesis via one step hydrothermal carbonization of green alga and their multifunctional applications. Microchem. J. **147**, 1038–1047 (2019). https://doi.org/10.1016/j.microc.2019.04.015
229. Hou, Y., Lu, Q., Deng, J., Li, H., Zhang, Y.: One-pot electrochemical synthesis of functionalized fluorescent carbon dots and their selective sensing for mercury ion. Anal. Chim. Acta **866**, 69–74 (2015). https://doi.org/10.1016/j.aca.2015.01.039
230. Qin, K., Zhang, D., Ding, Y., Zheng, X., Xiang, Y., Hua, J., et al.: Applications of hydrothermal synthesis of *Escherichia coli* derived carbon dots in in vitro and in vivo imaging and p-nitrophenol detection. Analyst **145**(1), 177–183 (2020). https://doi.org/10.1039/C9AN01753D
231. Zhang, S., Zhang, D., Ding, Y., Hua, J., Tang, B., Ji, X., et al.: Bacteria-derived fluorescent carbon dots for highly selective detection of p-nitrophenol and bioimaging. Analyst **144**(18), 5497–5503 (2019). https://doi.org/10.1039/C9AN01103J
232. Abdulwali, N., van der Zalm, J., Thiruppathi, A.R., Khaleel, A., Chen, A.: Microwave-assisted green synthesis of monodispersed carbon micro-spheres and their antibacterial activity. Appl. Surf. Sci. **642**, 158579 (2024). https://doi.org/10.1016/j.apsusc.2023.158579

233. Wang, D., Zhou, Q., Fu, H., Lian, Y., Zhang, H.: A $Fe_2(SO_4)_3$-assisted approach towards green synthesis of cuttlefish ink–derived carbon nanospheres for high-performance supercapacitors. J. Colloid Interface Sci. **638**, 695–708 (2023). https://doi.org/10.1016/j.jcis.2023.02.024
234. Li, W., Zhang, Z., Kong, B., Feng, S., Wang, J., Wang, L., et al.: Simple and green synthesis of nitrogen-doped photoluminescent carbonaceous nanospheres for bioimaging. Angew. Chem. Int. Ed. **52**(31), 8151–8155 (2013). https://doi.org/10.1002/anie.201303927
235. Nima, A.M., Amritha, P., Lalan, V., Subodh, G.: Green synthesis of blue-fluorescent carbon nanospheres from the pith of tapioca (*Manihot esculenta*) stem for Fe(III) detection. J. Mater. Sci. Mater. Electron. **31**(23), 21767–21778 (2020). https://doi.org/10.1007/s10854-020-04689-6
236. He, D., Gao, Y., Wang, Z., Yao, Y., Wu, L., Zhang, J., et al.: One-step green fabrication of hierarchically porous hollow carbon nanospheres (HCNSs) from raw biomass: formation mechanisms and supercapacitor applications. J. Colloid Interface Sci. **581**, 238–250 (2021). https://doi.org/10.1016/j.jcis.2020.07.118
237. Pathania, D., Bhat, V.S., Shivanna, J.M., Sriram, G., Kurkuri, M., Hegde, G.: Garlic peel based mesoporous carbon nanospheres for an effective removal of malachite green dye from aqueous solutions: detailed isotherms and kinetics. Spectrochim. Acta Part A **276**, 121197 (2022). https://doi.org/10.1016/j.saa.2022.121197
238. Sharma, A., Shivanna, J.M., Alodhayb, A.N., Hegde, G.: Efficient cationic dye removal from water through *Arachis hypogaea* skin-derived carbon nanospheres: a rapid and sustainable approach. Nanoscale Adv. **6**(12), 3199–3210 (2024). https://doi.org/10.1039/D4NA00254G
239. Manaf, S.A.A., Roy, P., Sharma, K.V., Ngaini, Z., Malgras, V., Aldalbahi, A., et al.: Catalyst-free synthesis of carbon nanospheres for potential biomedical applications: waste to wealth approach. RSC Adv. **5**(31), 24528–24533 (2015). https://doi.org/10.1039/C4RA14693J
240. Zhong, H.Y., Li, S.S., Li, D.Q., Jin, J.B., Chen, C.Z., Qiu, Z.S., Huang, W.A.: Hydrothermal carbon nanospheres as environmentally friendly, sustainable and versatile additives for water-based drilling fluids. Pet. Sci. **22**(2), 1997–2019 (2025). https://doi.org/10.1016/j.petsci.2025.03.021
241. Nallapureddy, J., Sreekanth, T.V.M., Pallavolu, M.R., Srinivasa Babu, P.S., Nallapureddy, R.R., Jung, J.H., Joo, S.W.: Strategic way of synthesizing heteroatom-doped carbon nano-onions using waste chicken fat oil for energy storage devices. ACS Appl. Mater. Interfaces, **16**(18), 23334–23343 (2024). https://pubs.acs.org/doi/10.1021/acsami.4c02753
242. Roopesh, M., Davis, D., Jyothi, M.S., Vandana, M., Thippeswamy, B.S., Hegde, G., et al.: Wound healing efficacy of curcumin-loaded sandalwood bark-derived carbon nanosphere/PVA nanofiber matrix. RSC Adv. **13**(35), 24320–24330 (2023). https://doi.org/10.1039/D3RA04181F
243. Zhang, B., Kang, F., Tarascon, J.-M., Kim, J.-K.: Recent advances in electrospun carbon nanofibers and their application in electrochemical energy storage. Prog. Mater. Sci. **76**, 319–380 (2016). https://doi.org/10.1016/j.pmatsci.2015.08.002
244. Zhou, X., Liu, B., Chen, Y., Guo, L., Wei, G.: Carbon nanofiber-based three-dimensional nanomaterials for energy and environmental applications. Mater. Adv. **1**(7), 2163–2181 (2020). https://doi.org/10.1039/D0MA00492H
245. Zhou, X., Wang, Y., Gong, C., Liu, B., Wei, G.: Production, structural design, functional control, and broad applications of carbon nanofiber-based nanomaterials: a comprehensive review. Chem. Eng. J. **402**, 126189 (2020). https://doi.org/10.1016/j.cej.2020.126189
246. Wang, Z., Wu, S., Wang, J., Yu, A., Wei, G.: Carbon nanofiber-based functional nanomaterials for sensor applications. Nanomaterials **9**(7), 1045 (2019). https://doi.org/10.3390/nano9071045
247. Verma, S.K., Das, A.K., Gantait, S., Panwar, Y., Kumar, V., Brestic, M.: Green synthesis of carbon-based nanomaterials and their applications in various sectors: a topical review. Carbon Lett. **32**(2), 365–393 (2022). https://doi.org/10.1007/s42823-021-00294-7
248. Zhang, J., Tahmasebi, A., Omoriyekomwan, J.E., Yu, J.: Direct synthesis of hollow carbon nanofibers on bio-char during microwave pyrolysis of pine nut shell. J. Anal. Appl. Pyrolysis **130**, 142–148 (2018). https://doi.org/10.1016/j.jaap.2018.01.016

249. Alminderej, F.M., Albadri, A.E.A.E., El-Ghoul, Y., El-Sayed, W.A., Younis, A.M., Saleh, S.M.: Sustainable and green synthesis of carbon nanofibers from date palm residues and their adsorption efficiency for eosin dye. Sustainability **15**(13), 10451 (2023). https://doi.org/10.3390/su151310451
250. Zhao, Y., Liu, Y., Tong, C., Ru, J., Geng, B., Ma, Z., Liu, H., Wang, L.: Flexible lignin-derived electrospun carbon nanofiber mats as a highly efficient and binder-free counter electrode for dye-sensitized solar cells. J. Mater. Sci. **53**(10), 7637–7647 (2018). https://doi.org/10.1007/s10853-018-2059-0
251. Liu, H.-J., Wang, X.-M., Cui, W.-J., Dou, Y.-Q., Zhao, D.-Y., Xia, Y.-Y.: Highly ordered mesoporous carbon nanofiber arrays from a crab shell biological template and its application in supercapacitors and fuel cells. J. Mater. Chem. **20**(20), 4223–4230 (2010). https://doi.org/10.1039/B925776D
252. Du, B., Chen, C., Sun, Y., Yu, M., Liu, B., Wang, X., Zhou, J.: Lignin bio-oil-based electrospun nanofibers with high substitution ratio property for potential carbon nanofibers applications. Polym. Test. **89**, 106591 (2020). https://doi.org/10.1016/j.polymertesting.2020.106591
253. Omoriyekomwan, J.E., Tahmasebi, A., Zhang, J., Yu, J.: Formation of hollow carbon nanofibers on bio-char during microwave pyrolysis of palm kernel shell. Energy Convers. Manag. **148**, 583–592 (2017). https://doi.org/10.1016/j.enconman.2017.06.022
254. Chakravarty, A., Maiti, S., Mahanty, S., De, G.: Green synthesis of electrospun porous carbon nanofibers from sucrose and doping of Ag nanoparticle with improved electrical and electrochemical properties. ChemistrySelect **2**(7), 2265–2276 (2017). https://doi.org/10.1002/slct.201601920
255. Yang, I.-H., Szabó, L., Sasaki, M., Uto, K., Henzie, J., Lin, F.-H., Samitsu, S., Ebara, M.: Biobased chitosan-derived self-nitrogen-doped porous carbon nanofibers containing nitrogen-doped zeolites for efficient removal of uremic toxins during hemodialysis. Int. J. Biol. Macromol. **253**, 126880 (2023). https://doi.org/10.1016/j.ijbiomac.2023.126880
256. Szabó, L., Xu, X., Ohsawa, T., Uto, K., Henzie, J., Ichinose, I., Ebara, M.: Ultrafine self-N-doped porous carbon nanofibers with hierarchical pore structure utilizing a biobased chitosan precursor. Int. J. Biol. Macromol. **182**, 445–454 (2021). https://doi.org/10.1016/j.ijbiomac.2021.04.023
257. Chen, W., Meng, X.-T., Wang, H.-H., Zhang, X.-Q., Wei, Y., Li, Z.-Y., Li, D., Zhang, A.-P., Liu, C.-F.: A feasible way to produce carbon nanofiber by electrospinning from sugarcane bagasse. Polymers **11**(12), 1968 (2019). https://doi.org/10.3390/polym11121968
258. Szabó, L., Xu, X., Uto, K., Henzie, J., Yamauchi, Y., Ichinose, I., Ebara, M.: Tailoring the structure of chitosan-based porous carbon nanofiber architectures toward efficient capacitive charge storage and capacitive deionization. ACS Appl. Mater. Interfaces **14**(3), 4004–4021 (2022). https://doi.org/10.1021/acsami.1c20199
259. Wang, H., Wang, H., Ruan, F., Feng, Q., Wei, Y., Fang, J.: High-porosity carbon nanofibers prepared from polyacrylonitrile blended with amylose starch for application in supercapacitors. Mater. Chem. Phys. **293**, 126896 (2023). https://doi.org/10.1016/j.matchemphys.2022.126896
260. Deng, L., Zhong, W., Wang, J., Zhang, P., Fang, H., Yao, L., Liu, X., Ren, X., Li, Y.: The enhancement of electrochemical capacitance of biomass-carbon by pyrolysis of extracted nanofibers. Electrochim. Acta **228**, 398–406 (2017). https://doi.org/10.1016/j.electacta.2017.01.099
261. Liu, W.-J., Tian, K., He, Y.-R., Jiang, H., Yu, H.-Q.: High-yield harvest of nanofibers/mesoporous carbon composite by pyrolysis of waste biomass and its application for high durability electrochemical energy storage. Environ. Sci. Technol. **48**(23), 13951–13959 (2014). https://doi.org/10.1021/es504184c
262. Duan, B., Gao, X., Yao, X., Fang, Y., Huang, L., Zhou, J., Zhang, L.: Unique elastic N-doped carbon nanofibrous microspheres with hierarchical porosity derived from renewable chitin for high rate supercapacitors. Nano Energy **27**, 482–491 (2016). https://doi.org/10.1016/j.nanoen.2016.07.034

263. Chen, W., Wang, H., Lan, W., Li, D., Zhang, A., Liu, C.: Construction of sugarcane bagasse-derived porous and flexible carbon nanofibers by electrospinning for supercapacitors. Ind. Crops Prod. **170**, 113700 (2021). https://doi.org/10.1016/j.indcrop.2021.113700
264. Hashmi, A., Singh, A.K., Jain, B., Singh, A.: Muffle atmosphere promoted fabrication of graphene oxide nanoparticle by agricultural waste. Fuller. Nanotub. Carbon Nanostruct. **28**(8), 627–636 (2020). https://doi.org/10.1080/1536383X.2020.1728744
265. Fathy, M., Hosny, R., Keshawy, M., Gaffer, A.: Green synthesis of graphene oxide from oil palm leaves as novel adsorbent for removal of Cu(II) ions from synthetic wastewater. Graphene Technol. **4**(1), 33–40 (2019). https://doi.org/10.1007/s41127-019-00025-w
266. Omar, H., Malek, N.S.A., Nurfazianawatie, M.Z., Rosman, N.F., Bunyamin, I., Abdullah, S., Khusaimi, Z., Rusop, M., Asli, N.A.: A review of synthesis graphene oxide from natural carbon based coconut waste by Hummer's method. Mater. Today Proc. **75**, 188–192 (2023). https://doi.org/10.1016/j.matpr.2022.11.427
267. Cao, K., Tian, Z., Zhang, X., Wang, Y., Zhu, Q.: Green preparation of graphene oxide nanosheets as adsorbent. Sci. Rep. **13**(1), 9314 (2023). https://doi.org/10.1038/s41598-023-36595-2
268. Gkika, D.A., Maroulas, K.N., Kyzas, G.Z.: Various reduced graphene oxide green synthetic routes: comparing the cost procedures. ACS Omega **10**(32), 36221–36237 (2025). https://doi.org/10.1021/acsomega.5c04090
269. Vatandost, E., Ghorbani-HasanSaraei, A., Chekin, F., Naghizadeh Raeisi, S., Shahidi, S.-A.: Green tea extract assisted green synthesis of reduced graphene oxide: application for highly sensitive electrochemical detection of sunset yellow in food products. Food Chem. X **6**, 100085 (2020). https://doi.org/10.1016/j.fochx.2020.100085
270. Ranjith, R., Renganathan, V., Chen, S.-M., Selvan, N.S., Rajam, P.S.: Green synthesis of reduced graphene oxide supported TiO_2/Co_3O_4 nanocomposite for photocatalytic degradation of methylene blue and crystal violet. Ceram. Int. **45**(10), 12926–12933 (2019). https://doi.org/10.1016/j.ceramint.2019.03.219
271. Lin, Z., Weng, X., Ma, L., Sarkar, B., Chen, Z.: Mechanistic insights into Pb(II) removal from aqueous solution by green reduced graphene oxide. J. Colloid Interface Sci. **550**, 1–9 (2019). https://doi.org/10.1016/j.jcis.2019.04.078
272. Nhlane, D., Richards, H., Etale, A.: Facile and green synthesis of reduced graphene oxide for remediation of Hg(II)-contaminated water. Mater. Today Proc. **38**, 737–742 (2021). https://doi.org/10.1016/j.matpr.2020.04.163
273. Joshi, S., Siddiqui, R., Sharma, P., Kumar, R., Verma, G., Saini, A.: Green synthesis of peptide functionalized reduced graphene oxide (rGO) nano bioconjugate with enhanced antibacterial activity. Sci. Rep. **10**(1), 9441 (2020). https://doi.org/10.1038/s41598-020-66230-3
274. Liou, T.-H., Tseng, Y.K., Liu, S.-M., Lin, Y.-T., Wang, S.-Y., Liu, R.-T.: Green synthesis of mesoporous graphene oxide/silica nanocomposites from rich husk ash: characterization and adsorption performance. Environ. Technol. Innov. **22**, 101424 (2021). https://doi.org/10.1016/j.eti.2021.101424
275. Thiyagarajulu, N., Arumugam, S.: Green synthesis of reduced graphene oxide nanosheets using leaf extract of *Lantana camara* and its in-vitro biological activities. J. Clust. Sci. **32**(3), 559–568 (2021). https://doi.org/10.1007/s10876-020-01814-7
276. Yang, J., Xia, X., He, K., Zhang, M., Qin, S., Luo, M., Wu, L.: Green synthesis of reduced graphene oxide (RGO) using the plant extract of *Salvia spinosa* and evaluation of photothermal effect on pancreatic cancer cells. J. Mol. Struct. **1245**, 131064 (2021). https://doi.org/10.1016/j.molstruc.2021.131064
277. Tohamy, H.-A.S., Anis, B., Youssef, M.A., Abdallah, A.E.M., El-Sakhawy, M., Kamel, S.: Preparation of eco-friendly graphene oxide from agricultural wastes for water treatment. Desalin. Water Treat. **191**, 250–262 (2020). https://doi.org/10.5004/dwt.2020.25652
278. Naik, M.J.P., Debbarma, J., Saha, M., Bhargava, A.: Graphene oxide nanoflakes from various agrowastes. Mater. Wiss. Werkst. **51**(3), 368–374 (2020). https://doi.org/10.1002/mawe.201900061

279. Li, J., Yan, Q., Zhang, X., Zhang, J., Cai, Z.: Efficient conversion of lignin waste to high value bio-graphene oxide nanomaterials. Polymers **11**(4), 623 (2019). https://doi.org/10.3390/polym11040623
280. Chen, W., Shen, J., Lv, G., Li, D., Hu, Y., Zhou, C., Liu, X., Dai, Z.: Green synthesis of graphene quantum dots from cotton cellulose. ChemistrySelect **4**(10), 2898–2902 (2019). https://doi.org/10.1002/slct.201803512
281. Wang, Z., Yu, J., Zhang, X., Li, N., Liu, B., Li, Y., Wang, Y., Wang, W., Li, Y., Zhang, L., Dissanayake, S., Suib, S.L., Sun, L.: Large-scale and controllable synthesis of graphene quantum dots from rice husk biomass: a comprehensive utilization strategy. ACS Appl. Mater. Interfaces **8**(2), 1434–1439 (2016). https://doi.org/10.1021/acsami.5b10660
282. Bhanderi, D., Lakhani, P., Modi, C.K.: Graphitic carbon nitride (gC3N4) as an emerging photocatalyst for sustainable environmental applications: a comprehensive review. RSC Sustain. **2**(2), 265–287 (2024). https://doi.org/10.1039/D3SU00382E
283. Tiwari, B., Ram, S.: Biogenic synthesis of graphitic carbon nitride for photocatalytic degradation of organic dyes. ACS Omega **4**(6), 10263–10272 (2019). https://pubs.acs.org/doi/10.1021/acsomega.9b00989
284. Vijayarangan, R., Mohan, S., Selvaraj, M., Assiri, M.A., Ilangovan, R.: Surface-engineered anisotropic g-C3N4 photocatalysts via green-exfoliation for visible light-driven water remediation. Surf. Interfaces **44**, 103782 (2024). https://doi.org/10.1016/j.surfin.2023.103782
285. Zhuang, Q., Guo, P., Zheng, S., Lin, Q., Lin, Y., Wang, Y., Ni, Y.: Green synthesis of luminescent graphitic carbon nitride quantum dots from human urine and its bioimaging application. Talanta **188**, 35–40 (2018). https://doi.org/10.1016/j.talanta.2018.05.060
286. Li, H., Shao, F.Q., Huang, H., Feng, J.J., Wang, A.J.: Eco-friendly and rapid microwave synthesis of green fluorescent graphitic carbon nitride quantum dots for vitro bioimaging. Sens. Actuators B Chem. **226**, 506–511 (2016). https://doi.org/10.1016/j.snb.2015.12.018
287. Deng, Q., Li, R., Chen, A., Zhong, Y., Yin, X., Zhang, Y., Yang, R.: Green synthesis of rectangular hollow tubular carbon nitride via in-situ self-assembly strategy to enhance the degradation of tetracycline hydrochloride under visible light irradiation. Environ. Res. **238**, 117252 (2023). https://doi.org/10.1016/j.envres.2023.117252
288. Pattnaik, S.P., Behera, A., Acharya, R., Parida, K.: Green exfoliation of graphitic carbon nitride towards decolourization of Congo-Red under solar irradiation. J. Environ. Chem. Eng. **7**(6), 103456 (2019). https://doi.org/10.1016/j.jece.2019.103456
289. Palani, N., Mendonce, K.C., Altaf, R.R.S., Mohan, A., Surya, P., Rajadesingu, S.: Advancing one-pot green synthesis of plant-derived nanomaterials for extraction techniques, characterization, multifunctional biomedical applications, challenges and future perspectives. Mater. Today Commun. **46**, 112780 (2025). https://doi.org/10.1016/j.mtcomm.2025.112780
290. Manoj, M., Yuan, A.: A plant-mediated synthesis of nanostructured hydroxyapatite for biomedical applications: a review. RSC Adv. **10**(67), 40923–40939 (2020). https://doi.org/10.1039/D0RA08529D
291. Mondal, S., Dorozhkin, S.V., Pal, U.: Recent progress on fabrication and drug delivery applications of nanostructured hydroxyapatite. Wiley Interdiscip. Rev. Nanomed. Nanobiotechnol. **10**(4), e1504 (2018). https://doi.org/10.1002/wnan.1504
292. Niculescu, V.C.: Mesoporous silica nanoparticles for bio-applications. Front. Mater. **7**, 36 (2020). https://doi.org/10.3389/fmats.2020.00036
293. Farjadian, F., Roointan, A., Mohammadi-Samani, S., Hosseini, M.: Mesoporous silica nanoparticles: synthesis, pharmaceutical applications, biodistribution, and biosafety assessment. Chem. Eng. J. **359**, 684–705 (2019). https://doi.org/10.1016/j.cej.2018.11.156
294. Torres, F.G., Troncoso, O.P., Pisani, A., Gatto, F., Bardi, G.: Natural polysaccharide nanomaterials: an overview of their immunological properties. Int. J. Mol. Sci. **20**(20), 5092 (2019). https://doi.org/10.3390/ijms20205092
295. Bilal, M., Gul, I., Basharat, A., Qamar, S.A.: Polysaccharides-based bio-nanostructures and their potential food applications. Int. J. Biol. Macromol. **176**, 540–557 (2021). https://doi.org/10.1016/j.ijbiomac.2021.02.107

296. Rehman, A., Jafari, S.M., Tong, Q., Riaz, T., Assadpour, E., Aadil, R.M., et al.: Drug nanodelivery systems based on natural polysaccharides against different diseases. Adv. Colloid Interface Sci. **284**, 102251 (2020). https://doi.org/10.1016/j.cis.2020.102251
297. Qiu, A., Wang, Y., Zhang, G., Wang, H.: Natural polysaccharide-based nanodrug delivery systems for treatment of diabetes. Polymers **14**(15), 3217 (2022). https://doi.org/10.3390/polym14153217
298. Plucinski, A., Lyu, Z., Schmidt, B.V.: Polysaccharide nanoparticles: from fabrication to applications. J. Mater. Chem. B **9**(35), 7030–7062 (2021). https://doi.org/10.1039/D1TB00628B
299. Lei, C., Liu, X.R., Chen, Q.B., Li, Y., Zhou, J.L., Zhou, L.Y., Zou, T.: Hyaluronic acid and albumin based nanoparticles for drug delivery. J. Control. Rel. **331**, 416–433 (2021). https://doi.org/10.1016/j.jconrel.2021.01.033
300. Rao, N.V., Rho, J.G., Um, W., Ek, P.K., Nguyen, V.Q., Oh, B.H., et al.: Hyaluronic acid nanoparticles as nanomedicine for treatment of inflammatory diseases. Pharmaceutics **12**(10), 931 (2020). https://doi.org/10.3390/pharmaceutics12100931
301. Pandey, J.K., Nakagaito, A.N., Takagi, H.: Fabrication and applications of cellulose nanoparticle-based polymer composites. Polym. Eng. Sci. **53**(1), 1–8 (2013). https://doi.org/10.1002/pen.23242
302. Bamal, D., Singh, A., Chaudhary, G., Kumar, M., Singh, M., Rani, N., et al.: Silver nanoparticles biosynthesis, characterization, antimicrobial activities, applications, cytotoxicity and safety issues: an updated review. Nanomaterials **11**(8), 2086 (2021). https://doi.org/10.3390/nano11082086
303. Zhao, D., Yu, S., Sun, B., Gao, S., Guo, S., Zhao, K.: Biomedical applications of chitosan and its derivative nanoparticles. Polymers **10**(4), 462 (2018). https://doi.org/10.3390/polym10040462
304. Rashki, S., Asgarpour, K., Tarrahimofrad, H., Hashemipour, M., Ebrahimi, M.S., Fathizadeh, H., et al.: Chitosan-based nanoparticles against bacterial infections. Carbohydr. Polym. **251**, 117108 (2021). https://doi.org/10.1016/j.carbpol.2020.117108
305. Garg, U., Chauhan, S., Nagaich, U., Jain, N.: Current advances in chitosan nanoparticles based drug delivery and targeting. Adv. Pharm. Bull. **9**(2), 195 (2019). https://doi.org/10.15171/apb.2019.023
306. Zhao, J., Weng, G., Li, J., Zhu, J., Zhao, J.: Polyester-based nanoparticles for nucleic acid delivery. Mater. Sci. Eng. C **92**, 983–994 (2018). https://doi.org/10.1016/j.msec.2018.07.027
307. Rocha, C.V., Gonçalves, V., da Silva, M.C., Bañobre-López, M., Gallo, J.: PLGA-based composites for various biomedical applications. Int. J. Mol. Sci. **23**(4), 2034 (2022). https://doi.org/10.3390/ijms23042034
308. Cappellano, G., Comi, C., Chiocchetti, A., Dianzani, U.: Exploiting PLGA-based biocompatible nanoparticles for next-generation tolerogenic vaccines against autoimmune disease. Int. J. Mol. Sci. **20**(1), 204 (2019). https://doi.org/10.3390/ijms20010204
309. Shakya, A.K., Al-Sulaibi, M., Naik, R.R., Nsairat, H., Suboh, S., Abulaila, A.: Review on PLGA polymer based nanoparticles with antimicrobial properties and their application in various medical conditions or infections. Polymers **15**(17), 3597 (2023). https://doi.org/10.3390/polym15173597
310. Al-Jamal, W.T., Kostarelos, K.: Liposomes: from a clinically established drug delivery system to a nanoparticle platform for theranostic nanomedicine. Acc. Chem. Res. **44**(10), 1094–1104 (2011). https://pubs.acs.org/doi/10.1021/ar200105p
311. Zylberberg, C., Gaskill, K., Pasley, S., Matosevic, S.: Engineering liposomal nanoparticles for targeted gene therapy. Gene Ther **24**(8), 441–452 (2017). https://doi.org/10.1038/gt.2017.41
312. Wickramasinghe, A.S.D., Kalansuriya, P., Attanayake, A.P.: Nanoformulation of plant-based natural products for type 2 diabetes mellitus: from formulation design to therapeutic applications. Curr. Ther. Res. **96**, 100672 (2022). https://doi.org/10.1016/j.curtheres.2022.100672

313. Bayón-Cordero, L., Alkorta, I., Arana, L.: Application of solid lipid nanoparticles to improve the efficiency of anticancer drugs. Nanomaterials **9**(3), 474 (2019). https://doi.org/10.3390/nano9030474
314. Pisoschi, A.M., Pop, A., Cimpeanu, C., Turcuş, V., Predoi, G., Iordache, F.: Nanoencapsulation techniques for compounds and products with antioxidant and antimicrobial activity—a critical view. Eur. J. Med. Chem. **157**, 1326–1345 (2018). https://doi.org/10.1016/j.ejmech.2018.08.076
315. Krishna, P.G., Chandra Mishra, P., Naika, M.M., Gadewar, M., Ananthaswamy, P.P., Rao, S., et al.: Photocatalytic activity induced by metal nanoparticles synthesized by sustainable approaches: a comprehensive review. Front. Chem. **10**, 917831 (2022). https://doi.org/10.3389/fchem.2022.917831
316. AlMohamadi, H., Awad, S.A., Sharma, A.K., Fayzullaev, N., Tavara-Aponte, A., Chiguala-Contreras, L., et al.: Photocatalytic activity of metal-and non-metal-anchored ZnO and TiO_2 nanocatalysts for advanced photocatalysis: comparative study. Catalysts **14**(7), 420 (2024). https://doi.org/10.3390/catal14070420
317. Zhai, Y., Zhu, Z., Dong, S.: Carbon-based nanostructures for advanced catalysis. ChemCatChem **7**(18), 2806–2815 (2015). https://doi.org/10.1002/cctc.201500323
318. Syed, N., Huang, J., Feng, Y.: CQDs as emerging trends for future prospect in enhancement of photocatalytic activity. Carbon Lett. **32**(1), 81–97 (2022). https://doi.org/10.1007/s42823-021-00282-x
319. Yang, X., Guo, Y., Liang, S., Hou, S., Chu, T., Ma, J., et al.: Preparation of sulfur-doped carbon quantum dots from lignin as a sensor to detect Sudan I in an acidic environment. J. Mater. Chem. B **8**(47), 10788–10796 (2020). https://doi.org/10.1039/D0TB00125B
320. Sharma, R., Tripathi, A.: Green synthesis of nanoparticles and its key applications in various sectors. Mater. Today Proc. **48**, 1626–1632 (2022). https://doi.org/10.1016/j.matpr.2021.09.512
321. Muhammad, A., Jatoi, A.S., Mazari, S.A., Abro, R., Mubarak, N.M., Ahmed, S., et al.: Recent advances and developments in advanced green porous nanomaterial for sustainable energy storage application. J. Porous Mater. **28**(6), 1945–1960 (2021). https://doi.org/10.1007/s10934-021-01138-5
322. Chen, W., Tan, Y., Guo, C., Zhang, X., He, X., Kuang, W., et al.: Biomass-derived polymer as a flexible "zincophilic–hydrophobic" solid electrolyte interphase layer to enable practical Zn metal anodes. J. Colloid Interface Sci. **669**, 104–116 (2024). https://doi.org/10.1016/j.jcis.2024.04.234
323. Rajpurohit, N.A., Bhakar, K., Nemiwal, M., Kumar, D.: Design and synthesis of hybrid nanostructures for sustainable energy and environmental remediation. Arab. J. Geosci. **15**(2), 137 (2022). https://doi.org/10.1007/s12517-022-09456-x
324. Ahmed, S., Ahmad, M., Swami, B.L., Ikram, S.: Green synthesis of silver nanoparticles using *Azadirachta indica* aqueous leaf extract. J. Radiat. Res. Appl. Sci. **9**(1), 1–7 (2016). https://doi.org/10.1016/j.jrras.2015.06.006
325. Singh, J., Dutta, T., Kim, K.H., Rawat, M., Samddar, P., Kumar, P.: 'Green' synthesis of metals and their oxide nanoparticles: applications for environmental remediation. J. Nanobiotechnol. **16**(1), 84 (2018). https://doi.org/10.1186/s12951-018-0408-4
326. Pattanashetti, N.A., Heggannavar, G.B., Kariduraganavar, M.Y.: Smart biopolymers and their biomedical applications. Procedia Manuf. **12**, 263–279 (2017). https://doi.org/10.1016/j.promfg.2017.08.030
327. Lim, S.Y., Shen, W., Gao, Z.: Carbon quantum dots and their applications. Chem. Soc. Rev. **44**(1), 362–381 (2015). https://doi.org/10.1039/C4CS00269E
328. Anastas, P.T., Zimmerman, J.B.: The United Nations sustainability goals: how can sustainable chemistry contribute? Curr. Opin. Green Sustain. Chem. **13**, 150–153 (2018). https://doi.org/10.1016/j.cogsc.2018.04.017
329. Gupta, V.K.: Application of low-cost adsorbents for dye removal—a review. J. Environ. Manage. **90**(8), 2313–2342 (2009). https://doi.org/10.1016/j.jenvman.2008.11.017

330. Crini, G., Badot, P.M.: Application of chitosan, a natural aminopolysaccharide, for dye removal from aqueous solutions by adsorption processes using batch studies: a review of recent literature. Prog. Polym. Sci. **33**(4), 399–447 (2008). https://doi.org/10.1016/j.progpolymsci.2007.11.001
331. Goyat, R., Saharan, Y., Singh, J., Umar, A., Akbar, S.: Synthesis of graphene-based nanocomposites for environmental remediation applications: a review. Molecules **27**(19), 6433 (2022). https://doi.org/10.3390/molecules27196433
332. Ballesteros-Vivas, D., Socas-Rodríguez, B., Mendiola, J.A., Ibáñez, E., Cifuentes, A.: Green food analysis: current trends and perspectives. Curr. Opin. Green Sustain. Chem. **31**, 100522 (2021). https://doi.org/10.1016/j.cogsc.2021.100522
333. Gupta, V., Biswas, D., Roy, S.: A comprehensive review of biodegradable polymer-based films and coatings and their food packaging applications. Materials **15**(17), 5899 (2022). https://doi.org/10.3390/ma15175899
334. Prabhu, S., Poulose, E.K.: Silver nanoparticles: mechanism of antimicrobial action, synthesis, medical applications, and toxicity effects. Int. Nano Lett. **2**(1), 32 (2012). https://doi.org/10.1186/2228-5326-2-32
335. Abd-Alhadi, R., Abou-Ghorrah, S., Al Oklah, B.: Physical properties, antioxidant and antimicrobial activity of chitosan edible films containing essential oils. J. Nutr. Food Secur. **8**(2), 212–220 (2023). https://dor.isc.ac/dor/20.1001.1.24767417.2023.8.2.9.7
336. Roy, S., Rhim, J.W.: Preparation of carbohydrate-based functional composite films incorporated with curcumin. Food Hydrocoll. **98**, 105302 (2020). https://doi.org/10.1016/j.foodhyd.2019.105302
337. Praharaj, C., Patel, G.K., Nara, S.: Plant biomass-derived nanomaterials for biosensing and bioimaging applications. Ind. Crops Prod. **230**, 121074 (2025). https://doi.org/10.1016/j.indcrop.2025.121074
338. Terea, H., Selloum, D., Rebiai, A., Bouafia, A., Ben Mya, O.: Preparation and characterization of cellulose/ZnO nanoparticles extracted from peanut shells: effects on antibacterial and antifungal activities. Biomass Convers. Biorefin. **14**(16), 19489–19500 (2024). https://doi.org/10.1007/s13399-023-03959-7
339. Kumari, P., Meena, A.: Green synthesis of gold nanoparticles from Lawsoniainermis and its catalytic activities following the Langmuir-Hinshelwood mechanism. Colloids Surf. A Physicochem. Eng. Asp. **606**, 125447 (2020). https://doi.org/10.1016/j.colsurfa.2020.125447
340. Gokulakrishnan, J., Koppole, K., Rugmini, R., Sekhar, K.C.: Green synthesis of carbon dots-functionalized silver nanoparticles for SERS-based detection of thiram fungicide. Appl. Phys. A **129**(11), 778 (2023). https://doi.org/10.1007/s00339-023-07038-6
341. Ameen, F., Altuner, E.E., Tiri, R.N.E., Gulbagca, F., Aygun, A., Sen, F., et al.: Highly active iron (II) oxide-zinc oxide nanocomposite synthesized *Thymus vulgaris* plant as bioreduction catalyst: characterization, hydrogen evolution and photocatalytic degradation. Int. J. Hydrogen Energy **48**(55), 21139–21151 (2023). https://doi.org/10.1016/j.ijhydene.2022.11.229
342. Palisoc, S., Dungo, J.M., Natividad, M.: Low-cost supercapacitor based on multi-walled carbon nanotubes and activated carbon derived from *Moringa oleifera* fruit shells. Heliyon **6**(1), e03202 (2020). https://doi.org/10.1016/j.heliyon.2020.e03202
343. Gedda, G., Sankaranarayanan, S.A., Putta, C.L., Gudimella, K.K., Rengan, A.K., Girma, W.M.: Green synthesis of multi-functional carbon dots from medicinal plant leaves for antimicrobial, antioxidant, and bioimaging applications. Sci. Rep. **13**(1), 6371 (2023). https://doi.org/10.1038/s41598-023-33652-8
344. Mansi, G., Chavda, V., Ranga, S., Raghav, S.: Green chemistry strategies for industrial product formulation. In: Towards Green Chemical Processes: Strategies and Innovations, pp. 165–192. Springer Nature Switzerland, Cham (2025). https://doi.org/10.1007/978-3-031-94090-3_8

Green Nanomaterials in Dentistry

Mobina Heidari, Morteza Banakar, Faezeh Heidari, and Chin Wei Lai

1 Introduction

Modern dentistry is undergoing a paradigmatic shift toward less mechanical and passive therapeutic modalities of minimally invasive, bioactive and regenerative methods of therapy. This change stems out of the growing intensity of the need for restorative procedures, which do not merely repair its functionality but also play a proactive role in the stimulation of innate biological healing and tissue regeneration [1]. The previously held paradigm of removing infected tissue and replacing it by biologically inert materials is gradually being replaced by methods of using the natural ability of the organism to recover with the involvement of the application of complex biomaterials. Such an interdisciplinary nexus of convergence between materials science, cellular biology, and clinical dentistry has the potential to develop the treatment of complicated dental diseases, such as caries, periodontal disease, and bone regeneration [2].

M. Heidari
Student Research Committee, School of Dentistry, Shahid Sadoughi University of Medical Sciences, Yazd, Iran

M. Banakar (✉)
Department of Medical Biotechnology, Shahid Sadoughi University of Medical Sciences, Yazd, Iran
e-mail: Dr.mbanakar@gmail.com

F. Heidari
Department of Pediatric Dentistry, Faculty of Dentistry, Shahid Sadoughi University of Medical Sciences, Yazd, Iran

C. W. Lai
Nanotechnology and Catalysis Research Centre, Universiti Malaya, Kuala Lumpur, Malaysia
e-mail: cwlai@um.edu.my

G. Tailor et al. (eds.), *Nanomaterials*, Engineering Materials,
https://doi.org/10.1007/978-981-92-0678-0_11

1.1 Defining Nanomaterials in Dentistry

Nanotechnology is a revolution in the science of dental materials as it enables an unprecedented command of material characteristics on the molecular level [3]. Dental products of the past, including composites, adhesives, ceramics, and implants, have been made to exhibit high levels of performance by the introduction of nanomaterials, which are usually in the scale of 1–1000 nm. An example of nanocomposites is nanoscale filler particles, which provide better mechanical strength, wear, and aesthetic properties as compared to the macrofilled counterparts. Furthermore, nanohydroxyapatite (nHA) particles, which resemble the mineral structure of natural enamel closely, have a strong activity in enamel remineralization, and reducing dentinal hypersensitivity [4]. These advances are only the first steps of the effects of nanotechnology on dental therapies and restorative substances.

1.2 The Requirement for "Green" Synthesis

Although the advantages granted by nanotechnology in the field of dentistry are very strong, the environmental issues and the very toxicity of the natural synthesized nanomaterials pose the challenge to reassess the synthesis methodologies critically. Extracting physical and chemical fabrication methods are typified by a high energy usage, use of hazardous reagents, and the production of toxic by-products hence undermining sustainability and biocompatibility [5]. By contrast, the principles of green chemistry, which focus on the prevention of waste, the reduction of hazardous syntheses, and utilization of renewable feedstocks help to provide a more sustainable approach to the synthesis of nanomaterials [6]. Using plant extracts, microorganisms, or their derivatives to make things is a greener option that often makes the nanomaterials biocompatible and useful [7].

1.3 Chapter Outline and Scope

The chapter presents an in-depth analysis of the synthesis, use, and future advancement of green nanomaterials in the area of dentistry. The introduction section explains the different green synthesis techniques that use plant extracts, microorganisms, and additional biological sources highlighting their relative strengths over the traditional methods. These sections are followed by a critical evaluation of the applicability of these nanomaterials to major areas in dental specialty, such as restorative dentistry, endodontics, periodontics, implantology, and oral cancer treatment. A critical review is conducted on each of the studied works, according to their therapeutic effectiveness, inherent advantages, and possible limitations. The final discussion concerns existing issues of standardization, scalability and future regulatory paths of research and clinical translation. Special emphasis is also placed on the explanation of the biological basis, clinical value and sustainability of green nanomaterials in modern day practice of dentistry.

2 Green Synthesis Routes for Dental Nanomaterials

The principle-based green chemistry nanomaterials make the cornerstone of the sustainable nanotechnology. This paradigm employs the power of biological systems to engineer or create nanostructures with new physicochemical properties and, therefore, makes them efficient in dental applications. It is a safe and eco-friendly substitute to the traditional chemical and physical fabrication processes. The key green synthesis directions are outlined in the next sections.

2.1 Phytosynthesis (Using Plant Extracts)

The use of plant extracts to synthesize nanomaterials can be termed as phytosynthesis, which is one of the most promising and fully explored green synthesis approaches. The capitalization on the diverse phytochemical arsenal, such as flavonoids, terpenoids, alkaloids, phenols, and terpenes, inherent in the biomass of plants utilized as reducing and stabilizing agents in the formation of nanoparticles is employed in the methodology [8]. The protocol involves the decrease of metal salt precursors like $AgNO_3$ to obtain silver nanoparticles and $HAuCl_4$ to obtain gold nanoparticles. The presence of hydroxyl and carbonyl functional groups in the phytochemical matrix enhances the electron transfer process, which reduces metal ions to the zero-valent state, and, therefore, enables the nucleation and manageable growth to nanoscale particles. At the same time, biomolecules on the nanoparticles' surface act as capping reagents and eliminate aggregation, as well as conferring colloidal stability [9]. Phytosynthesis has many benefits: it is cheap, fast, and can be scaled to large-scale production, can be run at ambient pressure and temperature thus minimizing energy consumption. Furthermore, the resultant nanoparticles are normally coated with biocompatible organic moieties making them especially suitable in bio applications [10]. As an example, silver nanoparticles produced using Azadirachta indica (Neem) leaf extract have pronounced bactericidal effects against cariogenic microorganisms, including *Streptococcus mutans,* and thus have a potential in being incorporated in dental adhesives and composite resins [11]. In the same way, the gold nanoparticles prepared using Cinnamomum zeylanicum (Cinnamon) bark extract have inherent optoelectronic characteristics which may be utilized in biosensing to detect oral cancer [12]. The schematic representation of the green synthesis process of nanoparticles in the case of plant extracts is shown in Fig. 1.

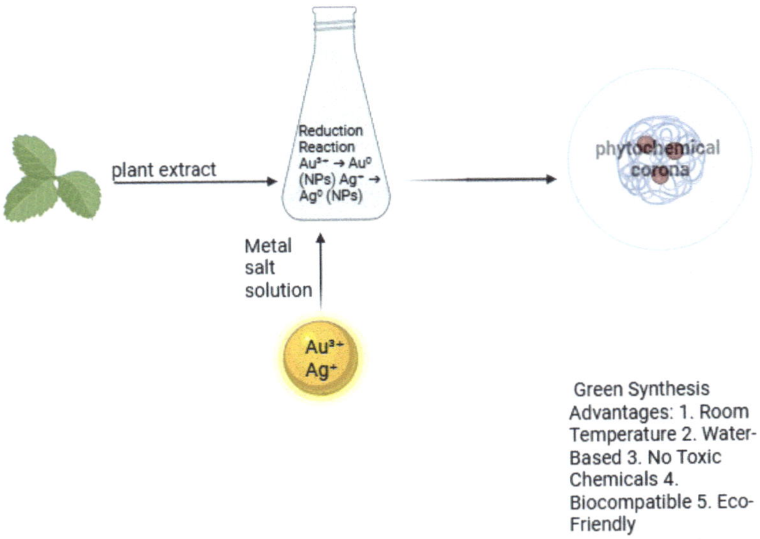

Fig. 1 Schematic representation of the green synthesis of nanoparticles using plant extract

2.2 *Biogenic Synthesis (Microorganisms)*

Biogenic synthesis utilizes bacteria, fungi, yeast, and algae as bioreactors for nanoparticle production. This method uses the natural metabolic abilities of microorganisms to turn metal ions into nanomaterials with a specific shape and high purity.

Bacteria: There are a wide variety of bacteria with enzymatic systems of both intracellular and extracellular nanoparticle production. Bacteria like Shewanella oneidensis are metal-reducing, so that they can incorporate metal ions, and the reductases reduce them down to nanoscale sizes. Secreted biomolecules and enzymes help in extracellular synthesis. The mode is appreciated due to the ability to scale through large fermentation processes and allows genetic modification of strains to increase output and selectivity [13].

Fungi: The fungi-mediated nanoparticle synthesis, or Mycosynthesis is particularly beneficial in generating stable monodispersed nanoparticles. Fungi release large amounts of extracellular reductases and proteins, which makes their synthesis and downstream recovery efficient in comparison with an intracellular bacterial system. The filamentous nature of growth of fungi offers a large surface area for catalytic reaction and their ability to grow on cheap substrates makes the process cost-effective [14]. Fusarium oxysporum and Aspergillus niger strains have been able to develop silver and titanium dioxide nanoparticles with strong antimicrobial properties that are applicable in endodontic irrigation and also in the surface coating of implants.

Yeast and Algae: Yeasts, such as *Saccharomyces cerevisiae* can synthesize nanoparticles both intracellular and extracellularly. Algae, including microalgae and macroalgae, are rich in pigments and polysaccharides which are reducing and capping agents and can be used to easily synthesize various metallic nanoparticles. Marine algae is specifically a renewable, cheap resource in terms of sustainability [15].

2.3 Synthesis by Natural Polymers and Agricultural Waste Products

With the emergence of natural polymers and agricultural waste streams, which are environmentally benign sources of nanomaterials synthesis, the field of green chemistry has made a major step forward. This approach falls under the principles of a circular economy, whereby the waste biomass is used to produce useful nanomaterials to use in the dental industry [16]. Cellulose and chitosan are natural biopolymers which have been shown to have better performance as reducing and stabilizing agents. A biodegradable polysaccharide derived from chitin, chitosan, contains a large number of amino and hydroxyl groups that cause chitosan to reduce metal ions and to stabilize the obtained nanoparticles. Its chelating capability also allows it to be used to nucleate and grow in a controlled manner, resulting in nanoparticles with a tight size distribution, which are perfect in dental drug delivery systems [17]. Likewise, cellulose-derived nanocellulose is also a renewable scaffolding to produce cathodoluminescence nanoparticles. Nanocellulose has large surface area and reactive hydroxyl groups, which facilitate efficient adsorption and reduction of metal ions to generate stable hybrid nanomaterials with improved mechanical properties to be utilized in dental composites [18]. Agro industrial waste products represent a new waste to wealth technology that both deals with environmental issues and sustainable resources. The citric wastes are abundant in ascorbic acid and flavonoids that are very effective in reducing metal ions to nanoparticles. In a few examples, pomegranate peel extract has been shown to enhance the synthesis of silver nanoparticles with better antimicrobial effect against oral pathogens than the chemically synthesized equivalents [19]. Similarly, lignocellulosic wastes like sugarcane bagasse, rice husks, and other agricultural byproducts also have silica and bioactive substances, which may be utilized in the green synthesis of practical nanomaterials, including silica nanoparticles to treat the mouth [20].

This kind of synthesis process not only has less of an effect on the environment, but it also turns trash into money, making it a sustainable way to employ materials. The resulting nanomaterials are generally more biocompatible and active because they contain natural bioactive chemicals that stay attached to the surfaces of the nanoparticles [7].

2.4 Characterization of Green-Synthesized Nanomaterial

Extensive characterization is essential in the clarification of the physicochemical characterizations and quality evaluation of the green-synthesized nanomaterials to be used in the dental field. Various methods of analysis give supplementary information on the properties which control nanomaterial performance and safety [21].

2.4.1 UV–Visible Spectroscopy (UV–Vis)

UV 2-Visible spectroscopy is a standard method of characterization that follows the surface plasmon resonance (SPR) band, which indicates the formation of nanoparticles. The spectral location and SPR peak shape provide a first-hand source of information regarding the size of the particle, its shape, and stability. An example is the spherical silver nanoparticles, which have a SPR peak at 420'nm but when it comes to gold nanoparticles, the absorbance is in the 520–550'nm spectral range [22]. The technique is very useful in terms of optimization of synthesis parameters and tracing of reaction kinetics. 2.4.2. Fourier-Transform Infrared Spectroscopy (FTIR) the FTIR analysis determines the functional groups on the surface of the nanoparticle and thus, determines the biomolecules which cause reduction and stabilization. The highest absorptions observed in the diagnostic range of hydroxyl (3200–3600 cm^{-1}), carbonyl (1600–1800 cm^{-1}), and amine (1500–1650 cm^{-1}) groups helps to identify the phytochemicals that capped the nanoparticles, which is an important factor in characterizing their interactions with the biological environment and stability [23].

2.4.2 X-Ray Diffraction (XRD)

XRD is used to identify the crystal structure and purity of the phase of nanomaterial. The resulting diffraction patterns can be used to estimate the size of the crystallites using the Debye Scherrer equation and the location of the peaks can be used to determine the crystal lattice using the position of the peaks. In the case of metal nanoparticles, the common characteristic peaks are those of a face-centred cubic structure [24].

2.4.3 Electron Microscopy Techniques

TEM and scanning electron microscopy (SEM) are direct methods to observe the size distribution, morphology and state of dispersion of nanoparticles. The TEM in high-resolution can be used to provide visualization and crystalline structure at an atomic scale, and SEM can provide data about the three-dimensional surface topography. These methods are crucial in measuring size and shape of nanoparticles, which is of vital significance in the dental practice [25].

2.4.4 Energy-Dispersive X-Ray Spectroscopy (EDS)

EDS normally accompanies electron microscopy, and provides elements composition analyses. Typical X-ray peaks determine the elements and the percentage content and as such, validate effective reduction of metal precursors and identify possible impurities [26].

Further characterization with dynamic light scattering (DLS) of hydrodynamic size distribution, zeta potential analysis of surface charge analysis, thermal stability using thermogravimetric analysis (TGA) gives a detailed picture of properties of the nanomaterial needed in dental practices.

3 Green Nanomaterials and Its Dentally Significant Properties

Nanotechnology has provided unmatched possibilities in diagnosis, treatment and prevention solutions in the field of dentistry. Nevertheless, the ecofriendliness and toxicity issues of the nanomaterials traditionally prepared through synthesis, which tend to involve the reductant toxic reagents and reactions of high energy, have also driven the transition towards more ecofriendlier synthesis methods. Green synthesis, which is based on reducing and stabilizing properties inherent in biological systems, including plants, bacteria, fungi, and agricultural wastes, is a paradigm shift. This method is based on the concept of green chemistry and normally results in nanoparticles with increased bioactivity and biocompatibility, due to the phytochemical corona which coats its surfaces. In the next section, the particular green nanomaterials which are likely to revolutionize dental practice are discussed along with their mode of synthesis, their properties and their mechanism of action.

3.1 Metallic Nanoparticles

Nanodentistry revolves around the green synthesis of metal nanoparticles, which increases their therapeutic and diagnostic capabilities.

3.1.1 Green Silver Nanoparticles (AgNPs)

Green Silver Nanoparticles (AgNPs) have been developed to overcome the problem of silver as a pollutant (Silver nanoparticles). Antimicrobial activity of silver is highly reported, and its efficacy is significantly enhanced when developed as a green-synthesized AgNPs, especially when used with plant extracts like Azadirachta indica (neem) or Syzygium aromaticum (clove). The high surface-area to volume ratio of

AgNPs will allow the release of Ag + ions, which are the main biocidal activity agents to be controlled [27]. The antibacterial effect is multi-modal: Ag + ions bind thiol groups of respiratory enzymes, prevent the electron transport and ATP production [28]; they permeabilise membranes, increasing their permeability and leading to the ultimate lysis of cells [29]; and create reactive oxygen species (ROS), oxidising proteins or DNA [27]. In the dental applications, this activity is used as a broad spectrum attack on polymicrobial oral biofilms. Incorporation of the green AgNPs in dental composite resins and adhesive systems results in production of so-called bioactive restorations, which inhibit proliferation of cariogenic bacteria like Streptococcus mutans at the restoration-tooth interface, which in turn reduce secondary caries [28]. Moreover, AgNP-covered dental implants and periodontal membranes confer antimicrobial behavior, reduce microbial attachment, reduce the risk of peri-implantitis, and improve the process of osseointegration [29].

3.1.2 The Green Gold Nanoparticles (AuNPs)

Biocompatibility and adjustable optical properties have been lauded in gold nanoparticles made through, e.g., Cinnamomum verum bark extract. Their inertness is reduced to their chemical nature, which makes them appropriate drug carriers of therapeutics in the oral cavity. AuNPs have localized surface plasmon resonance (LSPR), which also gives them vivid colors that are very sensitive to the environmental conditions. Aggregation on target binding Aggregation on target binding (i.e., Periodontal disease biomarker recognition) results in a detectable change of the color of red to blue [30]. This is the basis of the chair-side, colorimetric, rapid biosensors in biosensing of the conditions such as periodontitis or oral squamous cell carcinoma by studying saliva [31].

Green AuNPs have therapeutic uses; they can serve as effective drug-targeting nanocarriers; they can be functionalized to deliver anti-inflammatory or chemotherapeutic drugs to periodontal pockets or oral cancer cells. Their high light-absorbing properties also make them good photosensitizers in photothermal ablation whereby the localized targets are thermally ablated by laser irradiation and specific tumor tissue ablated [32].

3.1.3 Green ZnO Nanoparticles (ZnONPs)

Green Zinc Oxide Nanoparticles, which can be made from things like Aloe vera gel, have a dual-active profile that makes them very useful for dental applications. They kill bacteria, especially those that live in the mouth, by releasing Zn^{2+} ions, which throw off cellular homeostasis and enzyme systems, and by making hydrogen peroxide (H_2O_2) and other ROS on the nanoparticle surface when they come into contact with moisture or light [33].

In addition to their antimicrobial activity, ZnONPs are effective UV radiation blockers. This property enhances the stability of dental polymers. Upon incorporation into dental adhesives and composites, ZnONPs absorb UV light, thereby preventing photo-degradation of the resin matrix, which leads to yellowing and loss of mechanical strength over time [34]. This preserves the aesthetic and functional integrity of tooth-colored restoratives. They have applications in orthodontics too, with ZnONP-coated archwires being able to restrict plaque accumulation, and in preventive dentistry as mouth rinse additives for enhanced antiplaque activity.

3.2 Metal Oxide Nanoparticles

Metal oxide nanoparticles possess unique photocatalytic and biomimetic properties that are now being harnessed for future dental applications, ranging from self-sterilizing surfaces to bioactive regeneration.

3.2.1 Green Titanium Dioxide Nanoparticles (TiO_2NPs)

Titanium dioxide nanoparticles, for example, prepared by green synthesis techniques such as Eucalyptus globulus leaf extract, are effective photocatalysts. Light energy (primarily UV-A) activates TiO_2 nanoparticles, generating electron–hole pairs. Reactive sites on them interact with water and oxygen to create effective ROS, e.g., hydroxyl radicals (•OH) and superoxide anions (O_2^-), which non-selectively oxidize organic pollutants and microbial cell components [35].

This photocatalytic property is utilized in dentistry to create "self-cleaning" surfaces. TiO_2NPs coatings for orthodontic appliances, dental implants, and acrylic denture bases can, upon irradiation with ambient light, repeatedly break down adhered organic detritus and biofilms, thereby inhibiting the development of infections and discoloration [36]. Science is currently exploring doping methods using compounds such as nitrogen or fluorine to shift the photocatalytic excitation of TiO_2 into the visible range, bringing this technology closer to the clinic.

3.2.2 Green Nano-Hydroxyapatite (nHAp)

Nano-Hydroxyapatite (nHAp), the principal inorganic constituent of enamel and dentin, is the ultimate biomimetic material. Green synthesis of nHAp is done by extracting calcium and phosphate ions from natural, sustainable sources such as eggshells, fish scales, or bovine bone. The product is chemically and structurally comparable to the biological apatite of human teeth and has excellent biocompatibility and osteoconductivity [37]. The critical application of green nHAp in dentistry is in enamel remineralization. Its nanoparticles can penetrate and seal minute defects within the enamel, reducing dentin hypersensitivity and the early stages of dental caries by forming a new mineral layer that mimics natural enamel [38]. In dentin regeneration, nHAp is an ideal scaffold material for bone tissue engineering. Its porosity encourages the attachment, growth, and differentiation of osteogenic cells, thereby promoting periodontal bone regeneration in periodontal defects and peri-implant bone surrounding implants. Its derivation from biowaste renders it a component of sustainability and converts waste into a high-value biomedical material.

3.3 *Organic and Polymeric Nanoparticles*

Compared with their inorganic counterparts, natural organic and polymeric nanoparticles offer unique advantages for dental applications, including inherent biocompatibility, biodegradability, and, in most cases, therapeutic activity. Their molecular structure allows easy functionalization, making them highly attractive as drug-delivery vehicles and structural composites in restorative and regenerative dentistry. Two of the brightest hopes among them are cellulose nanocrystals and chitosan nanoparticles.

3.3.1 Chitosan Nanoparticles

Chitosan is a linear polysaccharide obtained primarily by deacetylation of crustacean shells (e.g., shrimp, crab) and represents a case of green nanotechnology, as it adds value to seafood waste. Green synthesis of CsNPs is typically carried out by ionic gelation with natural cross-linkers such as sodium tripolyphosphate, in the absence of harsh chemicals [39]. The cationic nature of chitosan, imparted by the primary amine groups, forms the basis of its useful properties in dental applications. Among the key applications is their inherent antimicrobial and anti-biofilm activity. The positive charge of CsNPs electrostatically binds to the negatively charged cell membranes of most oral pathogens, including Streptococcus mutans and Porphyromonas gingivalis. These interactions destabilize membrane integrity, leading to cell lysis and death [40]. CsNPs can also penetrate biofilms, which renders them more effective against established dental plaque. These qualify them for incorporation into dental adhesives, composite resins, and periodontal dressings to prevent secondary caries and periodontal infections [41]. Their most eminent feature is their intense mucoadhesiveness. CsNPs' positive charge electrostatically binds strongly to the negatively charged sialic acid residues of mucin, which is the primary component of the oral mucosa. This feature drastically extends CsNP residency time at the application site and makes them ideal for drug delivery through localized oral administration [42]. For instance, CsNPs encapsulating antifungal drugs such as nystatin can prolong release for the treatment of oral candidiasis, and those encapsulating anti-inflammatory drugs (e.g., curcumin) can be delivered into periodontal pockets for the treatment of gingivitis and periodontitis.

3.3.2 Cellulose Nanocrystals

Composites fillers are reinforcing in nature and in form of rod-like nanoparticles known as cellulose nanocrystals (CNCs) or nanocrystalline cellulose, which are reinforcing in nature and are produced out of cellulose, the most common biopolymer on the planet. Environmentally benign feedstocks are wood pulp, cotton and agricultural residues. The extraction is usually carried out through the process of acid hydrolysis, i.e., the use of citric acid as an environmentally-friendly substitute of concentrated mineral acids, to break up the amorphous cellulose domains, thus producing strong, crystalline nanocrystals [43].

CNCs are most commonly studied in the context of dental materials science, in the form of reinforcements of polymeric composites based on nanofillers. Their high tensile strength and modulus are a result of their well ordered crystalline structure and their improved mechanical properties. Added to dental adhesives and resin-based composites, CNCs can significantly enhance mechanical performance, including fracture toughness, flexural strength, and wear resistance [44]. This reinforcement reduces the shrinkage during polymerization of the polymer and consequent failure modes that plague dental restorations. Moreover, surface hydroxyl groups and a high surface area of CNCs promote the surface alteration, consequently, improving dispersion in hydrophobic resin systems, which is a key matter in efficient reinforced systems. The use of CNCs in pulp-regeneration hydrogel scaffolds and guided tissue regeneration membranes is also researched on (where their biocompatibility and mechanical resilience are especially beneficial) [45]. CNCs can help to create more durable restorative therapy by increasing the lifespan of dental restoration. Table 1 summarizes the key green nanomaterials, their biological sources, and primary dental properties.

3.4 Preventive Dentistry: Green Nanomaterials New Paradigm

Prevention of dentistry is the foundation of the oral health and its goal is to prevent the development and deterioration of dental caries and periodontal disease. With the development of nanotechnology, preventive dentistry has been transformed because it is possible now to design agents that are more biologically active, permeable, and targeted. However, the traditional production of nanomaterials often incorporates poisonous by-products, energy waste and violent chemicals, which are very dangerous to human health and the environment. Green nanotechnology provides a more sustainable way out through the design of biocompatible nanoparticles using plant residues, microorganisms or other processes that are ecologically clean. These environmentally friendly nanomaterials are progressively being included in ordinary oral health care products, including toothpastes and mouth rinses, and thus they achieve the goal of amplifying preventive care and reduce the ecological effect [53].

Table 1 Summary of key green nanomaterials, their sources, and primary dental properties

Nanomaterial	Example (green source)	Key properties for dentistry	References
Silver NPs (AgNPs)	Neem leaf extract	Broad-spectrum antimicrobial, anti-biofilm, used in composites and implant coatings	[46]
Gold NPs (AuNPs)	Cinnamon bark extract	Diagnostic (LSPR-based sensors), biocompatible, photothermal therapy, drug delivery	[47]
Zinc Oxide NPs (ZnONPs)	Aloe vera gel	Antibacterial, anti-inflammatory, UV-blocking, used in composites and orthodontics	[48]
Titanium Dioxide NPs (TiO_2NPs)	Eucalyptus leaf extract	Photocatalytic antibacterial, self-cleaning surfaces, and biofilm prevention	[49]
Nano-Hydroxyapatite (nHAp)	Eggshells, fish scales, bovine bone	Enamel remineralization, osteoconductive scaffold, bone regeneration	[50]
Chitosan NPs (CsNPs)	Crustacean shells (shrimp, crab)	Antimicrobial, mucoadhesive, drug delivery (antifungal, anti-inflammatory)	[51]
Cellulose Nanocrystals (CNCs)	Wood pulp, cotton, and agricultural residues	Reinforcing filler for composites/adhesives, mechanical strength, hydrogel scaffolds for regeneration	[52]

3.4.1 Nano-Hydroxyapatite (nHAp)

Dental caries is a remineralisation and demineralisation of the tooth structure, which is dynamic. The gold standard in remineralisation has long been thought to be traditional fluoride treatment but the issue with fluoride therapy has been the risk of fluorosis due to overload and the inability of fluoride to mimic the natural mineral composition of enamel. Nano-hydroxyapatite (nHAp), which is a biomimetic agent derived through the green process, is one of the new agents. The chemical formula of nHAp is $Ca_{10}(PO4)_6(OH)_2$, which is identical to the chemical formula of apatite crystals that are found in the dentin and enamel, which means that it can fully integrate into the tooth lattice. Green synthetic pathways, typically using calcium-rich biological feedstock, e.g. eggshells, fish scales or plant extracts, give remarkably pure nanoparticles with excellent biocompatibility [54]. The mode of action of nHAp is complex. Surface deposition and occlusion: nHAp nanoparticles contained in toothpaste may directly deposit on the enamel surfaces, directly occluding dentinal tubules that are exposed in this case, reducing hypersensitivity by forming a protective layer of minerals [55]. Biomimetic remineralisation: These nanoparticles are highly reactive, which is due to the high surface-to-volume ratio. They act as nanocrystallites that grow and catalyze the development of new layers of apatites and thus seal the sub-surface micro-lesions and white-spot lesions before cavitation. This is a better method than surface deposition because it provides a replacement of subsurface mineral depletion [56]. Protein pellicle modification: nHAp is able to infiltrate the developed salivary pellicle to create a mineralised layer providing increased resistance to acid attacks by cariogenic bacteria. Unlike fluoride, which results in a structurally altered form of fluorapatite layer, which is more acid resistant, as well as unlike natural hydroxyapatite but is no longer natural, green-synthesised nHAp recreates the true structure of the enamel in a truly biomimetic way.

3.4.2 Silver Nanoparticles (AgNPs) and Zinc Oxide Nanoparticles (ZnONPs)

The major etiological factor of periodontal disease and caries is plaque biofilm. The proactive measures are required to prevent its pathogenicity. Broad spectrum chemical antimicrobials like chlorhexidine are effective but may produce adverse effects, including the staining of teeth and the distortion of tastes which undermine patient compliance. As an alternative to plaque-controlling, green-synthesised metallic nanoparticles can be used, namely silver (AgNPs) and zinc oxide (ZnONPs) that offers an environmentally benign alternative.

3.4.3 Green-Synthesised Silver Nanoparticle (AgNPs)

AgNPs have a well-known and multifactorial antimicrobial efficacy. Membrane disruption: AgNPs interact with cell membranes and walls of microbes, and disrupt permeability and cause cell lysis. ROS generation: These stimulate ROS generation, which causes oxidative damage to proteins, lipids and DNA. Ion release: Silver ions ($Ag+$) released on the surface of the nanoparticle can block cell respiration and enzyme activity [57]. The use of organisms like Azadirachta indica (neem) or Aloe vera in a process known as plantmediated phytosynthesis reduces the silver ions to nanoparticles, as well as coating the particles with other phytochemicals. This capping has been reported to reduce cytotoxicity to human cells, improve stability and antimicrobial potency, and reduce cytotoxicity of AgNPs, which has an overall improved safety profile against chemically synthesised AgNPs [58].

3.4.4 Zinc Oxide Nanoparticles, Green-Synthesised (ZnONPs)

ZnONPs are a powerful antimicrobial tool that possesses several benefits: (i) they have a broad spectrum of action, (ii) cytotoxicity mediated by ROS, and (iii) pH-dependent solubility. In particular, (i) ZnONPs have the ability to act against a variety of cariogenic bacteria, including Streptococcus mutans, and periodontal pathogens. (ii) To a large extent, their bactericidal effect is executed by oxidative stress pathways that produce hydrogen peroxide and other species of ROS. ZnONPs in acid microenvironments found in carious lesions are more easily dissolved, thereby providing a localized antibacterial effect through the release of Zn^{2+} ions.

These green metal nanoparticles can be used in mouthwash as well as toothpastes to provide the antibacterial action by preventing the establishment of plaque biofilms and reducing the bacterial load without causing the side effects found in using traditional antimicrobials. Improved formulation stability also increases shelf life and stable efficacy.

3.5 Restorative Dentistry: Enhancing Longevity and Biofunctionality

Restorative dentistry aims at restoring damaged tooth structure as well as restoring morphological and functional integrity. The main challenge to the survival of restoration is premature failure of the prostheses often caused by secondary caries and fracture, which is a result of incompetent mechanical characteristics and microleaking of bacteria at the interface of the tooth and the restoration. The addition of green nanomaterials to restorative matrices (resin composites and dental adhesives) presents a new approach of addressing these shortcomings. Using nanoparticles, which have renewable sources, it is now possible to create not only mechanically superior restorations, but also biologically active ones, and hence, increase their clinical service life [60]. The interaction between green nanomaterials and dental composites and adhesives to boost their mechanical strength and bioactivity properties is shown in Fig. 2.

3.5.1 Green Nanocomposites

Reinforcing from Within Dental resin composites are ubiquitous in modern dentistry but suffer from inherent limitations: polymerization shrinkage, susceptibility to wear, and lack of bioactivity. The incorporation of nanofillers has been a significant advancement, and using green nanofillers is the next logical step towards sustainability and improved performance.

3.5.2 Nano-Hydroxyapatite (nHAp) as a Bioactive Filler

The addition of green-synthesized nHAp to composite resins does more than act as a filler. It introduces biomimetic and bioactive properties. Studies have shown that nHAp particles, due to their similarity to tooth mineral, can form a strong interfacial bond with the surrounding resin matrix and even with the natural tooth structure. This improves the fracture toughness and wear resistance of the composite, making it more durable under masticatory loads [60]. More importantly, nHAp-filled composites

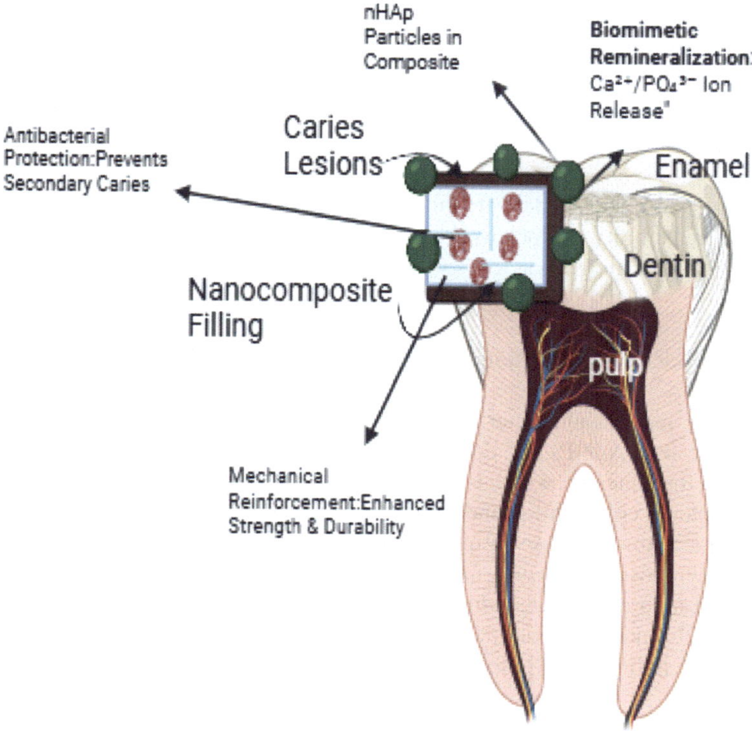

Fig. 2 Schematic illustration of green nanomaterials (e.g., nHAp, CNCs, AgNPs) enhancing the biofunctionality and longevity of dental restorations

can release calcium and phosphate ions in a cariogenic acidic environment. This ion release promotes remineralization of the adjacent tooth structure, effectively creating a "halo" of protection against secondary caries at the critical restoration margin—the most common site of failure [61].

3.5.3 Cellulose Nanocrystals (CNCs) as Renewable Reinforcements

The ultimate in green nanotechnology is cellulose nanocrystals that are synthesized by using the wood pulp, cotton, or another plant-based material. These are long nanoparticles that exhibit high tensile strength and stiffness. In low loadings in a resin matrix, they are strong reinforcing agents, significantly increasing the flexural strength, modulus and fracture resistance of the composite by effectively conducting the stress across a material [62]. Furthermore, every aspect of the hydrophilic properties of CNCs can be adjusted to enhance contact with the polymeric network and, thus, reduce the effect of water sorption, which is the crucial element of dental composite degradation. Some studies also state that CNCs that are surface modified are able to produce antimicrobial effects further providing an additional layer of defense against cariogenic biofilms [63].

3.5.4 Dental Adhesives

An adhesive interface is a delicate thin layer dividing tooth structure and restorative material. The inability to block its breakdown, which is often caused by the effects of bacterial enzymes and saliva, leads to microleakage, staining, and repetitive decay.

3.5.5 Antibacterial Adhesion Silver Nanoparticles (AgNPs)

An interface that is antibacterial and long-lasting is obtained by incorporating green-synthesized AgNPs into dental adhesive systems. Such nanoparticles are well dispersed in the adhesive, which is guaranteed to maintain a long antimicrobial effect to avoid bacterial colonization and acid formation of the hybrid layer. Such an ability will be critical towards preventing long-term corrosion of the resin-dentin bond. The green synthesis pathway is the key one since it often results in AgNPs that are capped with natural moieties which promote the dispersion of the nanoparticles in the hydrophobic adhesive resin, which reduces the possible level of cytotoxicity, thus retaining the vitality of the pulp [64].

3.5.6 Chitosan Nanoparticles (CSNPs) for Multifunctional Protection

Chitosan is a biocompatible, biodegradable and inherently antimicrobial polymer that is extracted out of crustacean exoskeletons. In nanoscale engineering, it binds to negatively charged bacterial membranes of the cell strongly thereby inhibiting the formation of biofilms. Addition of CSNPs to the adhesive formulations has two advantages:

1. **Antibacterial Effect**: CSNPs are effective in eliminating viable populations of both *S.mutans* and *Lactobacilli* at the adhesive interface.
2. **Matrix Metalloproteinase (MMP) Inhibition**: CSNPs inhibit endogenous MMPs in dentin, which are responsible for collagen breakdown in the hybrid layer. Chitosan nanoparticles inhibit these MMPs and therefore preserve the integrity of the bonds, significantly improving durability [65].

The synergistic interaction of these green nanomaterials is driving the development of a new group of these so-called smart restorative materials which do not passively occupy voids, but actually maintain oral health.

3.6 Endodontics: Redesigning Disinfection and Sealing

The ultimate objective of endodontic treatment is the total eradication of microbial infection in the radicles and the establishment of an impermeable barrier that inhibits subsequent infections. The intricate architecture of root canals, characterized by dentinal tubules and accessory canals, frequently serves as a reservoir for persistent biofilms, thereby diminishing treatment efficacy. Green nanomaterials are changing the field of endodontics by making irrigants and obturation materials work better, which greatly improves the outcome of the procedure [66].

3.6.1 Nano-Based Irrigants: Biofilm Penetration and Elimination

His conventional irrigants like sodium hypochlorite (NaOCl) are effective but constrained by tissue toxicity, incomplete eradication of biofilm within the dentinal tubules and unpleasant taste. Green methods of nanoparticles synthesis, offer a target biocompatible nanoparticle pore that can be used as a canal decontaminant. The green silver nanoparticles (AgNPs) have been shown to have better antibacterial and antibiofilm effects against endodontic pathogens such as the Enterococcus faecalis. They can penetrate dentinal tubules and the multifaceted root canal system that is not accessible to traditional irrigants due to their very small size. Their mechanism of action is through the progressive release of Ag + ions that disrupt cell membranes and enzymes of the microbials, and thus they maintain the antibacterial effect after the initial irrigation stage [67]. The positive charge of chitosan nanoparticles (CSNPs) and their affinity to bacterial cell walls cause rapid association with the pathogen, which leads to the lysis of cells. Their adhesive qualities allow them to seal over the dentinal surface to provide remaining antimicrobial effect and discourage the re-colonization of bacteria [68].

3.6.2 Nanomodified Sealers

The attainment of the complete three-dimensional seal of the root canal system is pivotal to the success of endodontic procedures. Nanomodification of sealers aims at enhancing their physicochemical properties besides giving them biofunctionality. Nano-hydroxy apatite (nHAp) or nanobioglass incorporated into sealer compositions improves the biocompatibility and contact with periradicular tissues. These bioactive nanoparticles aid in repair of cementum and periodontal ligaments to provide a biological seal that is supplemented with a mechanical one. Also, they enhance the sealer flowability, which allows them to better adapt to canal walls and enhances the compressive strength [69). Antibacterial nanoparticles are added to sealers in order to overcome microbial leakage- the major failure mode. The leaching of green AgNPs and CSNPs with the set sealer provide an antibacterial halo at the apex of the root and protect the survival of trapped microbes and the invasion of new bacteria. This active sealing process is superior to passive sealers because it provides continuous reinfection prevention, and the integrity of endodontic treatment is ensured in the long run [70].

3.7 *Periodontology and Implantology*

The formation of bacterial biofilms is an extremely difficult problem with regard to the management of periodontal disease and the future reliability of dental implants. Periodontitis, which is a chronic inflammatory pathology, is known to destroy the tooth-supporting structures, but the counterpart of such inflammatory process is known as peri-implantitis, the inflammatory process that destroys implant-surrounding tissues and is a major cause of implant failure. Green nanomaterials have offered new therapeutic opportunities and targeted therapeutic opportunities and improved implantation functionality that are beyond the ability of the traditional modalities.

3.7.1 Irrigants Based on Nanoparticles: Penetration and Elimination of Biofilms

Conventional irrigants, such as sodium hypochlorite (NaOCl), although effective, have limitations, including tissue toxicity, inability to remove biofilm from dentinal tubules in toto, and a sour taste. Nanoparticles synthesized using green methods offer a targeted, biocompatible means for decontaminating root canals. Plant-based green silver nanoparticles (AgNPs) exhibited excellent antibacterial and antibiofilm activity against endodontic pathogens, including Enterococcus faecalis. Because of their tiny particle size, they can penetrate dentinal tubules and intricate root canal anatomy that are not readily accessible to conventional irrigants. Their activity is through the slow release of Ag^{+} ions, which affect microbial cell membranes and enzymatic activities, resulting in long-term antibacterial activity even after the first irrigation period [67]]. Chitosan nanoparticles (CSNPs) are highly antimicrobial because they are positively charged and bind to bacteria's negatively charged cell walls, leading to cell bursting and death. Their mucoadhesive properties allow them to seal the dentinal surface, providing residual antimicrobial protection and blocking bacterial recolonization [68].

3.7.2 Nanomodified Sealers

The success of endodontic therapy depends on an effective three-dimensional seal of the root canal system. Nanomodification of root canal sealers attempts to enhance their physicochemical properties and incorporate biofunctionality.

The addition of nano-hydroxyapatite (nHAp) or nanobioglass to sealer formulations increases their biocompatibility and interaction with periradicular tissues. These bioactive nanoparticles promote the regeneration of cementum and periodontal ligaments, creating a biological seal in combination with a mechanical seal. Moreover, they can improve sealer flowability, which, in turn, can better conform to the root canal walls, as well as their compressive strength [69].

To counter the primary cause of failure—microbial leakage—antibacterial nanoparticles are added to sealers. Green AgNPs and CSNPs leached from the set sealer create an antibacterial zone at the root apex, preventing survival of entrapped microbes and penetration of new bacteria. This "active sealing" mechanism is superior to passive sealers, as it provides consistent protection against reinfection, thereby ensuring the long-term integrity of the endodontic treatment [70].

3.8 *Implantology and Periodontology*

Bacterial biofilm formation represents a significant obstacle to periodontal disease treatment and to the long-term success of dental implants. While periodontitis, as a chronic inflammatory process, leads to destruction of the tooth-supporting structures, peri-implantitis, the same pathologic process of implant-surrounding tissues, represents one of the leading causes of implant failure. Green nanomaterials offer new perspectives for targeted therapy and enhanced implant function beyond the limitations of conventional therapies.

3.8.1 Local Drug Delivery Systems

Systemic antibiotic treatment in periodontal diseases is generally limited by low drug concentrations at the target site, potential toxicity, and the risk of causing antimicrobial resistance.

Green nanomaterial LDDS provides a directed and controlled delivery pathway. Chitosan nanoparticles (CSNPs) are highly promising because they are biocompatible, biodegradable, and naturally mucoadhesive and antimicrobial.

Chitosan, a biopolymer, can be green-synthesized into nanoparticles and blended with antimicrobial agents (tetracycline, metronidazole) or anti-inflammatory agents. When administered subgingivally into periodontal pockets, these nanoparticles can adhere to mucosal tissue, delivering the therapeutic drug in a controlled manner over a prolonged period. This creates a very high local concentration of the drug at the biofilm interface, effectively breaking down pathogenic communities without causing significant systemic effects [71]. The cationic nature of chitosan enables it to bind to negatively charged bacterial cell membranes, enhancing the drug's intrinsic antibacterial activity and synergizing with its therapeutic effect [72]. This focused approach facilitates more effective biofilm removal and better regulation of periodontal inflammation than traditional non-sustained-release fibers or gels.

3.8.2 Dental Implant Coatings

Various materials have been investigated for the success of dental implants, especially nanomaterials (Fig. 3). The success of a dental implant relies on two critical considerations: osseointegration (the direct structural and functional connection between bone and the implant surface) and the inhibition of bacterial colonization that leads to peri-implantitis. Green-synthesized nanocoatings are being designed to fulfill both requirements in one.

Implants coated with green-synthesized silver nanoparticles (AgNPs) exploit their broad-spectrum and strong antibacterial activity. The coatings create a barrier that prevents initial bacterial adherence and biofilm formation on the transmucosal and implant abutment surfaces, thereby significantly reducing the risk of peri-implantitis [73].

To also facilitate bone integration, nanomaterials such as titanium dioxide nanoparticles (TiO_2NPs) are used. When applied as a nanocoated layer on the implant

Fig. 3 Modern strategies for managing biofilm on dental implants [75]

surface, TiO_2NPs enhance nanoscale surface roughness and hydrophilicity, thereby increasing protein adsorption and facilitating the attachment, growth, and differentiation of osteoblasts (bone cells) [74]. Advanced studies focus on creating multifunctional coatings that can integrate these functionalities. For instance, a coating with green AgNPs for bacteriostatic protection, along with nano-hydroxyapatite (nHA) to mimic bone mineral and stimulate osteogenesis, is a future-generation method for preparing genuinely bioactive and infection-resistant dental implants [75].

3.9 Oral Cancer Diagnostics and Therapy

The phrase oral cancer, or oral squamous cell carcinoma (OSCC), remains a critical issue of concern to the general population since it has a significant mortality and morbidity rate, which is often cause by the late identification of the disease. The use of green nanotechnology is bringing transformative change in early detection and specific tumoral treatment, thus, has the potential to significantly increase the prognostic outcome.

3.9.1 Biosensing Early Diagnosis is One of the Foundations of Effective Oncological Care

Traditional forms of diagnosis which are based on visual observation and biopsy are subjective in nature and are only likely to detect lesions at the late stages. The technologies of biosensing based on the principles of green-synthesized gold nanoparticles (AuNPs) also offers a new and non-invasive method able to identify select oncologic biomarkers with high sensitivity in saliva, blood, and other biological samples. AuNPs possess exceptional optical properties, among them, a well-developed surface plasmon resonance (SPR), which gives them a vivid coloration in solution and enables them to switch colors on aggregation or during a specific binding event. To identify OSCC biomarkers (e.g., CD44, IL-6, and telomerase) surface functionalization using antigen-specific antibodies or DNA aptamers can be used to green-synthesize AuNPs. The binding of the target molecules generates a detectable or visualizable color change or alteration in electric current, and this allows a rapid, point of care diagnostic test or an electrochemical biosensor. This technology has the potential to deliver low cost and highly specific early screening, with the option of enabling intervention during the pre-malignant or early malignant stages.

3.9.2 Targeted Drug Delivery

Traditional methods of chemotherapy to cure oral cancer are marred with systemic poisoning, untargeted destruction of normal cells, and extreme side effects. The functionalized green nanoparticles which are used in the targeted drug delivery systems have become the most innovative approach that can be used in order to address these constraints. These systems are designed to passively and actively homing in tumor tissue to accumulate in the tumor tissue. The aberrant and permeable tumor vasculature has provided passive targeting of nanoparticles of a given size through the Enhanced Permeability and Retention (EPR) effect. More importantly, nanoparticles created greenly (e.g., by using plant extracts) can be targeted actively by conjugating high-affinity ligands to their surfaces, targeting overexpressed cancer cell surface-receptors (e.g., folate receptors). This functionalization enables the nanoparticle carrier to be loaded with chemotherapeutic agents (e.g., doxorubicin, paclitaxel) and subsequently internalized by the malignant cells targeting the concentration of the cytotoxic effect in the tumor. This strategy will reduce or minimize exposure of drugs to normal tissues, decrease side effects, and enhance therapeutic index of strong anticancer agents. Table 2 is a summary of the various uses of green nanomaterials in different fields of dentistry.

4 Biocompatibility, Toxicity, and Regulatory Considerations

The path of any novel material from the laboratory to the clinic is predicated upon an exceedingly careful and rigorous evaluation of its biosafety profile. For dentistry's green nanomaterials, such an evaluation is especially advanced. While the green synthetic route is claimed to be non-toxic and free of harmful chemicals, the intrinsic biological activity of nanomaterials must be rigorously investigated to ensure they do not trigger unwanted reactions in patients. This investigation must extend beyond a mere comparison with their chemically synthesized counterparts to encompass a detailed understanding of their interactions with very complex biological systems.

4.1 Assessment of the Biosafety of Green Nanomaterials

The primary hypothesis for green-synthesized nanomaterials is that phytochemicals (alkaloids, flavonoids, terpenoids) in plant extracts used during synthesis serve as capping and stabilizing agents. The capping layer by phytochemicals is hypothesized to not only enhance the colloidal stability of nanoparticles but also reduce their direct reactivity with cellular compounds, potentially lowering cytotoxicity and improving biocompatibility relative to nanoparticles synthesized with synthetic stabilizing agents such as sodium citrate or polyvinylpyrrolidone (PVP).

Table 2 Summary of key applications of green nanomaterials in modern dentistry

Dental specialty	Green nanomaterial	Specific application	Key mechanism of action/benefit	References
Preventive dentistry	Nano-hydroxyapatite (nHAp)	Biomimetic toothpaste for enamel remineralization and dentin hypersensitivity treatment	Deposits onto enamel and occludes dentinal tubules; acts as a nanocrystallite seed for biomimetic remineralization	[54–56]
Preventive dentistry	Silver NPs (AgNPs), Zinc Oxide NPs (ZnONPs)	Additives in mouthwashes and toothpastes for plaque control	Broad-spectrum antimicrobial activity via membrane disruption, ROS generation, and ion release (Ag^+, Zn^{2+})	[57–59]
Restorative dentistry	nHAp, Cellulose Nanocrystals (CNCs)	Reinforcing and bioactive fillers in dental composite resins	nHAp: Releases Ca^{2+}/PO_4^{3-} ions for remineralization; CNCs: Enhance mechanical strength and fracture toughness	[28, 61–63]
Restorative dentistry	AgNPs, Chitosan NPs (CsNPs)	Incorporation into dental adhesives to fortify the tooth-restoration interface	AgNPs: Provide sustained antibacterial effect. CsNPs: Inhibit MMPs to prevent bond degradation and offer antimicrobial action	[64, 65]
Endodontics	AgNPs, CsNPs	Nanoparticle-based irrigants for root canal disinfection	Small size allows penetration into dentinal tubules; provides sustained antimicrobial and anti-biofilm activity	[66–68]

(continued)

Table 2 (continued)

Dental specialty	Green nanomaterial	Specific application	Key mechanism of action/benefit	References
Endodontics	nHAp, AgNPs, CsNPs	Nanomodified root canal sealers	nHAp: Bioactive and osteoconductive. AgNPs/CsNPs: Impart 'active sealing' with antibacterial properties	[69, 70]
Implantology and periodontology	CsNPs	Localized drug delivery systems for periodontal pockets	Mucoadhesion ensures prolonged retention and controlled release of antimicrobial/anti-inflammatory drugs	[71, 72]
Implantology and periodontology	AgNPs, TiO_2NPs, nHAp	Multifunctional coatings for dental implants	AgNPs: Prevent bacterial adhesion. TiO_2NPs: Photocatalytic self-cleaning. nHAp: Enhance osseointegration	[29, 73–75]
Oral cancer diagnosis and therapy	Gold NPs (AuNPs)	Colorimetric biosensors for early detection of salivary biomarkers	Localized Surface Plasmon Resonance (LSPR) causes a visible color shift upon target binding	[30, 76, 77]
Oral cancer diagnosis and therapy	AuNPs, other green NPs	Targeted drug delivery systems for chemotherapeutic agents	Active targeting (e.g., with ligands) and passive targeting (EPR effect) concentrate drugs in tumor tissue, reducing systemic toxicity	[78, 79]

4.1.1 In Vitro Studies

Cell culture assays are the preferred screening method for nanomaterial toxicity. In all studies, green-synthesized nanoparticles exhibit greater tolerance across various cell lines relevant to dentistry, including human gingival fibroblasts, periodontal ligament cells, and osteoblasts. For instance, comparative cytotoxicity of chemically synthesized and AgNPs synthesized from Azadirachta indica (neem) extract was compared in one study and revealed that while both classes had dose-dependent decrease of cell viability, the green AgNPs required a very higher concentration to demonstrate the same extent of cytotoxicity [80]. The proposed mechanism is that the natural capping molecules form a corona that cushions the sudden release of Ag^+ ions, the principal vectors of silver toxicity. This controlled ion release may maintain antimicrobial activity while minimizing mammalian cell damage. The same trends are seen with green-synthesized AuNPs and ZnO NPs, where plant metabolites appear to give a shielding effect.

4.1.2 In Vivo Studies

The more complex physiological environment of animal models is used to assess systemic toxicity, immunogenicity, organ deposition, and inflammatory responses. Findings from in vivo experiments typically, but not always, validate the improved biosafety of green nanomaterials. For example, subgingival application of chitosan nanoparticle-based antibiotic formulations capped with green AgNPs in rat models resulted in effective biofilm destruction without significant inflammation or damage to the surrounding periodontal tissues, a drawback reported for high-concentration application of an older conventional synthetic AgNP preparation [81]. Histopathological examination of animal tissue exposed to green nanomaterials tends to be less suggestive of inflammation, necrosis, or fibrous capsule formation compared to tissue exposed to chemically similar yet synthetically stabilized nanoparticles. But care should be taken not to overgeneralize. The biosafety of a green nanomaterial cannot be ensured; it is very susceptible to a variety of variables:

Plant Source: Phytochemical profile of plant extracts differs significantly among taxa and even among an individual species that has been subjected to divergent environmental factors. A nanoparticle derived using a particular plant extract could be very biocompatible, whereas another formulation can be highly toxic because of the specific bioactive ingredients found.

Conditions of Synthesis: The size and morphology of the particle, including the surface and charge (zeta potential) and the nature of the stabilizing capping layer; the aforementioned features are directly dependent on parameters like pH, temperature, and reaction time; these parameters are all associated with biological interactions and possible toxicity.

Dose and Route of Exposure: The dose makes the poison!- the maxim of Paracelsus is still applicable; even non-toxic substances may become toxic with an increase in dose. In addition, the administration route (e.g. direct pulp intention or subgingival position) varies the level of safety studies that are needed.

Despite the existing in vitro and in vivo evidence indicating the green nanomaterials have a desirable biosafety profile over that of their chemically synthesized counterparts, the area is still in its infancy. The strong heterogeneity of synthesis approaches requires the consideration of cases on a case-by-case basis. Scientists should also standardise reporting and characterization procedures so that comparisons can be made. The wide adoption of such materials in the dental field should be preceded by extensive, chronic toxicity tests and a thorough pharmacokinetic and biodistribution.

4.2 Green Capping Agent Role: Enhanced Biocompatibility and Reduced Cytotoxicity

The first difference between chemically and green-synthesised nanoparticles (NPs) is their surface chemistry. Traditionally, stabilisers, including sodium citrate, cetyltrimethylammonium bromide (CTAB) or polyvinylpyrrolidone (PVP), are used; these can be hard to completely eliminate, and also themselves cause cytotoxicity. Alternatively, green synthesis takes advantage of a glut of phytochemicals in the form of alkaloids, flavonoids, terpenoids, phenolic acids, and proteins, found in plant extracts. These biomolecules are capable of performing two roles: reducing the metal ions to nanoscale and then capping the formed NPs, creating a stabilising corona. It is this phytochemical corona that forms the foundation of the noted improvement in biocompatibility and decrease of cytotoxicity of the green-synthesised nanomaterials.

This increased biocompatibility has a multifactorial basis, which is mechanistic. First of all, the phytochemical capping layer serves as a buffer that restricts the direct interaction of the reactive metal core of the NP with cellular elements. The release of free ions (e.g., Ag^{+}, Zn 2^{+}) and the formation of reactive oxygen species (ROS) through surface catalysis are often the cause of metal NP cytotoxicity. The biomolecular corona regulates ion-release kinetics and offers an antioxidant capability that eliminates surplus ROS prior to their ability to cause oxidative stress, lipid peroxidation and DNA damage. As an illustration, polyphenols and flavonoids, which capped green-synthesised silver nanoparticles (AgNPs), are strong antioxidants and reduce the generation of ROS and establishes a more biocompatible contact surface with biological systems.

Secondly, the natural capping layer has the potential to impart to NPs the qualities of stealth, reducing the areas of nonspecific adsorption of proteins and immune surveillance, which suppresses the inflammatory reaction. It is particularly applicable in the field of dentistry as the inflammatory process of both gingiva and pulp should be reduced to achieve the best result. It has been shown using experimental evidence that aloevera compounded green-synthesised zinc oxide nanoparticles (ZnONPs) do not activate cytotoxicity on human gingival fibroblasts as readily as the chemically synthesised counterparts leading to disruption of membranes and apoptosis. Equally, neem-encased AgNPs exhibit strong antimicrobial effect on oral pathogens but also exhibit low toxicity against human oral keratinocytes- a property that is normally not exhibited by other conventional antimicrobials. This biocompatibility design is a pore to direct use in sensitive oral conditions. Green NPs can be included in dental adhesives, composite resins, and intracanal medicaments, which will decrease the chances of adverse pulpal reactions or hypersensitivity and enhance long-term safety of nano-enhanced dental materials.

4.3 Present-Day Regulatory Environment and Clinical Translation Problems

Although the results of green-synthesised nanomaterials have shown promising in vitro and preclinical results, the challenge of successfully navigating regulatory barriers in transforming the material to clinical applications remains relatively high. The main issue is in the fact that botanical extracts are natural and heterogeneous in nature, which does not align with the high standardisation requirements imposed on them by the regulatory bodies, including the U.S. Food and Drug Administration (FDA) and the European Medicines Agency (EMA) as well.

There is a problem with batch-to-batch reproducibility. The phytochemical composition of plant extract depends on the area of plant growing, the time when it is harvested, the soil, the method through which it is extracted. An example is that AgNPs prepared using an extract of Ocimum sanctum (holy basil) can be of a specified size, shape and capping profile, but a different harvest can produce particles with different characteristics. This variability has a direct effect on biological activity, safety, and pharmacokinetics thus affecting regulatory compliance. Agencies require strict presentation of conformity, purity and stability-quality that is hard to assess in a multicomponent naturally occurring system.

Detailed characterisation also stands in the way of regulation. In order to qualify, a nanomaterial should be well characterised with respect to size, size distribution, morphology, surface charge, composition, and capping layer. Although the metal core can be easily analysed, the various biomolecules that make the green corona could not be easily identified and quantified. Quality control is complicated by the need of advanced techniques, e.g. mass spectrometry-chromatography. Lacking a thorough molecular characterization of the capping layer, it is impossible to build a clear structure–activity relationship or identify the possible negative effects.

Lastly, the regulatory framework itself is under development in order to reflect the details of nanomedicine, not to mention green nanomedicine. The nanomaterial regulation by FDA is at the initial phase and there is no established approval route of green-derived formulations. Regulators are used to the evaluation of well-characterised, single-molecule entities or fully-characterised mixtures, and not botanical complexes and nanoscale particles.

Safety demonstration involves extensive toxicological assessment (cytotoxicity, genotoxicity, immunotoxicity, chronic toxicity) that must take into account the particular nature of the nanomaterial and its biological corona [91]. Apart from that, environmental impacts from the production and disposal of these new nanomaterials must be assessed to properly approve their "green" label, a process known as life-cycle assessment (LCA).

To address such challenges, the scientific community must focus on plant cultivation, extraction processes, and the standardization of synthesis parameters. Highly sophisticated purification and characterization methods should be established to delineate the green corona further. Increased interaction and discussion among nanotechnologists, regulatory scientists, and industry partners are most urgently needed to develop new, fit-for-purpose standards that enable the safe and effective transfer of these potential green nanomaterials from the research lab to the dental clinic.

Among the numerous fields revolutionized by nanotechnology, dentistry is especially well-positioned to benefit significantly from the discovery of novel materials with enhanced mechanical, antibacterial, and bioactive properties. The introduction of green nanotechnology, which uses biotemplates for the environmentally benign synthesis of nanomaterials, represents a paradigm shift toward greener, more biocompatible dentistry. This concluding section recognizes the existing reality of adversity, predicts the lines in future studies, and extrapolates the revolutionary contribution of green nanomaterials towards ushering in a new era in dental medicine.

4.4 Benefit-Risk Analysis

Evaluation of the Enhanced Effectiveness Compared to Long-Term Potential of Harm. The future of introducing green-synthesised nanomaterials to dentistry is outstanding as it allows enhancing the level of therapeutic efficacy and maintaining biocompatibility and sustainability. The benefits are significant and are better antimicrobial and antibiofilm performance, better acceptance of the material by the bone, longer-lasting delivery of drugs, increased mechanical strength of the dental products, and low dependency on toxic chemicals, as opposed to traditional nanomaterials. Together, these advantages have the potential of radically transforming restorative, preventive, as well as surgical dentistry. However, there are also threats that should be taken into consideration. The biological interactions of green nanomaterials in the long-term are also not well comprehended especially during chronic exposure in the oral region, tissue retention, low-grade inflammatory response, and activation of new immune responses. The inconsistency in biomolecular routes will also create ambiguity in regard to reproducibility and safety outcomes, and the financial and ecological consequences of mass production are other considerations. With such considerations, the risk–benefit analysis report shows that though short-term studies and laboratory-based ones have shown massive positive outcomes, the ability of the results to be sustained over long durations has not been properly analyzed. The positive features warrant further studies and preclinical trials in the controlled conditions, but the use of green nanomaterials should be avoided so far until solid proof is provided by the large, longitudinal studies. The green nanomaterials will finally be translated into clinical use, as the solution to the problems listed above, and this is bound to ensure that the potential therapeutic benefits of these nanomaterials do not surpass the harmful effects they have on the health of the patients and ecological sustainability, in the long-term.

5 Challenges, Future Perspectives

5.1 Challenges

While there are extremely promising advantages of green-synthesized nanomaterials, their clinical translation from the lab to everyday clinical dentistry is awaiting solutions to many significant challenges. Scalability of Green Synthesis at Industrial Scale. The first challenge is to scale up the synthesis process from laboratory-scale small lots to sizes that are sufficient for industrial scale. Any microbe- or plant-assisted syntheses are, more often than not, engineered for specific conditions that cannot be replicated at larger scales without compromising yield, morphology, or nanomaterial properties [53]. Seasonal variation in plant extracts, the requirement for sterility conditions in microbial culture, and the removal of nanoparticles from biological matrices are all posing engineering and economic difficulties for mass production.

In close connection to scalability is the problem of reproducibility. Biological extracts, regardless of their plant, algae, and fungal origins, may have a diverse chemical composition depending on the geographical location, harvesting time, climatic conditions, and methods of extracts. This natural heterogeneity can be the source of batch-to-batch discrimination in nanoparticle size, morphology, crystallinity, and surface chemistry [18]. Absence of standardisation presents a major threat in clinical practices, where precision is a major priority. Therefore, strict procedures in the production, formulation and characterisation of biological precursors are essential.

Since green nanomaterials have been proven to be desirable biocompatible and therapeutic in vitro, there is scanty in-vivo data and long term clinical evidence. The oral cavity is an inhospitable location, being prone to constant mechanical forces, pH fluctuations, and microbiome overpopulation. The long-term effect of chronic contact with oral tissues, possible biomagnification, extended-period inflammatory response, and general biodistribution and biodegradation mechanisms are not well defined [92]. Strict toxicological profiling and long-term clinical researches are critically required to test the safety and efficacy. In addition, the up-front expenses of creating, researching and standardisation, mass-scale production of green nanomaterials are higher than that of conventional synthesised dental products. These extra expenses tend to be transferred to the end product and make the green alternatives less economically viable in the short-run. Mass adoption will therefore not only be determined by the ability to improve performance in a demonstrable way but also by the ability to generate cost-effectiveness with regard to technological breakthrough and economies of scale.

5.2 Future Directions

In order to overcome these problems and fully utilize the promise of green nanomaterials, the future work should be carefully aimed at some of the key fields. Research on New Biological Sources. The pool of biological material that can be synthesized is enormous and still unexploited. Future research should focus on the untapped and extremophilic organisms. An example is marine ecosystems, which are an untapped and rich source of algae, sponges and molluscs, which produce novel metabolites with high reducing and capping properties that could produce nanomaterials with a sequence of properties previously not realized [93]. Similarly, agricultural and food-processing by-products (fruit peels, seed shells and crop biomass) could be used as low-value and renewable raw materials to waste disposal and the production of nanomaterials. Preparation of Multi-purpose Smart Nanomaterials. Green nanomaterials are destined to have more than just single uses. There are scientific attempts toward the structure of the so-called smart or responsive systems. As an example, nanoparticles may be designed to delivery of antimicrobial agents (e.g., ions or drugs) when exposed to a hypoxic microclimate in cariogenic bacterial biofilms [94]. Similarly, nanomaterials can be programmed to be remineralised only upon an increase of concentrations of calcium and phosphate in the ambient conditions beyond a specific concentration threshold, which mimics natural biological activities, and increases therapeutic efficacy. Green Nanotechnology Digital Dentistry Fusion. Green nanotechnology and digital dentistry (especially additive manufacturing (3D printing)) have a lot of potential when converged. The research should focus on the formation of new types of green nanocomposite resin to 3D printing. Patient-specific dental implants, crowns, dentures, and bone scaffolds to regenerate can be produced by inclusion of the green-synthesised nanoparticles, like zirconia, cellulose nanocrystals or bioactive glass, in to printable polymers, which, in turn, can be endowed with a significantly better mechanical strength, antibacterial properties, and bioactivity [95]. This technology would make possible very customised and accurate dental treatment.

The gap between clinical practice and laboratory research can only be reduced through large-scale randomised controlled trials. Multidisciplinary cooperation is a must to produce Level 1 evidence on the effectiveness, safety and longevity of dental restorations, implants and therapeutics made of green nanomaterials. The data that will be the result of this will be very useful in regulatory acceptance—e.g. by FDA or EMA—and in creating trust among dental professionals and patients.

6 Conclusion

Green nanomaterials are an antecedent to high-technology, sustainable dentistry. Their biosynthetic pathways offer an excellent alternative to traditional manufacturing, and therefore the technological advances are adapting to the worldwide need to manage the environment. In addition, these materials also show significant potential in augmenting antimicrobial properties of restorative composites, enhancing implantation of implants, precision drug delivery, and enhancing mechanical performance of dental polymers. However, the difficulties encountered are the ability to produce at scale, standardisation, long-term safety, and cost-effectiveness, among others, which restrain widespread clinical use. These barriers have to be overcome using interdisciplinary approach that incorporates materials science, microbiology, clinical dentistry, and process engineering. It will be necessary to accelerate the process of identifying new biogenic sources, designing multifunctional systems, integrating them into digital processes, and validating their results in large-scale randomized trials. The continuous intensive studies along these lines would be the key to the complete fulfillment of the potential of green nanomaterials and eventually the dawn of a new era of dental therapeutics that is not just highly effective and bio-compatible, but also sustainable and eco-friendly over generations.

References

1. Banakar, M., Sijanivandi, S.S.: Green dentistry. In: Encyclopedia of green materials 2022 Oct 23, pp. 1–6 Singapore: Springer Nature Singapore
2. Chen, L., Zhu, C., Fan, D., Liu, X., Jin, Z.: A bioactive nano-calcium phosphate paste for mineralization of dentin surface. Mater. Sci. Eng., C **108**, 110391 (2020)
3. Allaker, R.P.: The use of nanoparticles to control oral biofilm formation. J. Dent. Res. **89**(11), 1175–1186 (2018)
4. Zafar, M.S., Fareed, M.A., Riaz, S., Latif, M., Habib, S.R.: Customized therapeutic surface coatings for dental implants. Coatings **10**(6), 568 (2020)
5. Banakar, M., Shahbazi, Z., Mousavi, S.M., Hashemi, S.A., Lai, C.W.: Herbal green nanomaterials and their applications. In Encyclopedia of green materials 2022 Oct 20, pp. 1–8. Singapore: Springer Nature Singapore
6. Anastas, P.T., Warner, J.C.: Green chemistry: Theory and Practice. Oxford University Press (1998)
7. Sharma, D., Kanchi, S., Bisetty, K.: Biogenic synthesis of nanoparticles: a review. Arab. J. Chem. **12**(8), 3576–3600 (2019)
8. Mittal, A.K., Chisti, Y., Banerjee, U.C.: Synthesis of metallic nanoparticles using plant extracts. Biotechnol. Adv. **31**(2), 346–356 (2013)
9. Banakar, M, Shahbazi, Z, Mousavi, S.M., Hashemi, S.A., Lai, C.W.: Antimicrobial and antiviral properties of herbal green materials. In: Encyclopedia of Green Materials 2024 Dec 27, pp. 87–95. Singapore: Springer Nature Singapore
10. Iravani, S.: Green synthesis of metal nanoparticles using plants. Green Chem. **13**(10), 2638–2650 (2011)
11. Kumar, A., Ghosh, S., Gupta, A.: Green synthesis of silver nanoparticles from Azadirachta indica and its efficacy against oral pathogens. J. Cluster Sci. **33**(1), 215–226 (2022)
12. Patil, M.P., Kim, G.D., Kim, J.O.: Eco-friendly approach for nanoparticles synthesis and mechanism behind antibacterial activity of silver and anticancer activity of gold nanoparticles. Appl. Microbiol. Biotechnol. **105**(3), 905–919 (2021)
13. Srivastava, P., Mukhopadhyay, M.: Green synthesis of silver nanoparticles using latex of Jatropha curcas and its novel antimicrobial activity. J. Nanomater. **2015**, 639816 (2015)
14. Mughal, B., Zaidi, S.Z.J., Zhang, X., Hassan, S.U.: Biogenic nanoparticles: synthesis, characterisation and applications. Appl. Sci. **11**(6), 2598 (2021)

15. Khanna, P., Kaur, A., Goyal, D.: Algae-based metallic nanoparticles: synthesis, characterization and applications. J. Microbiol. Methods **163**, 105656 (2019)
16. Banakar, M., Shahbazi, Z., Mousavi, S.M., Hashemi, S.A., Lai, C.W.: Green materials sterilization solutions. In: Encyclopedia of Green Materials 2022 Dec 21, pp. 1–10. Singapore: Springer Nature Singapore
17. Divya, K., Smitha, V., Jisha, M.S.: Antifungal, antioxidant and cytotoxic activities of chitosan nanoparticles and its use as an edible coating for vegetables. Int. J. Biol. Macromol. **115**, 572–577 (2018)
18. Trache, D., Tarchoun, A.F., Derradji, M., Hamidon, T.S., Masruchin, N., Brosse, N., et al.: Nanocellulose: from fundamentals to advanced applications. Front. Chem. **8**, 2020 (2020)
19. Prasher, P., Sharma, M.: Silver Nanoparticles: Synthesis. Bentham Science Publishers, Functionalization and Applications (2022Mar 13)
20. Ramesh, P., Saravanan, K., Manogar, P.: Green synthesis and characterization of silica nanoparticles using sugarcane bagasse. J. Nanomater. **2021**, 7814586 (2021)
21. Thakkar, K.N., Mhatre, S.S., Parikh, R.Y.: Biological synthesis of metallic nanoparticles. Nanomedicine: Nanotechnol., Biol. Med. **6**(2), 257–62 (2010)
22. Mie, G.: Beiträge zur Optik trüber Medien, speziell kolloidaler Metallösungen. Ann. Phys. **330**(3), 377–445 (1908)
23. Coates, J.: Interpretation of infrared spectra, a practical approach. John Wiley & Sons, Encyclopedia of Analytical Chemistry (2006)
24. Cullity, B.D., Stock, S.R.: Elements of X-ray diffraction. Prentice Hall (2001)
25. Goldstein, J.I., Newbury, D.E., Michael, J.R., Ritchie, N.W.: Scanning electron microscopy and X-ray microanalysis. Springer (2017)
26. Newbury, D.E., Ritchie, N.W.: Is scanning electron microscopy/energy dispersive X-ray spectrometry (SEM/EDS) quantitative? Scanning. **35**(3), 141–168 (2013)
27. Dakal, T.C., Kumar, A., Majumdar, R.S., Yadav, V.: Mechanistic basis of antimicrobial actions of silver nanoparticles. Front. Microbiol. **7**, 1831 (2016)
28. Melo, M.A., Guedes, S.F., Xu, H.H., Rodrigues, L.K.: Nanotechnology-based restorative materials for dental caries management. Trends Biotechnol. **31**(8), 459–467 (2013)
29. Zhao, L., Wang, H., Huo, K., Cui, L., Zhang, W., Ni, H., et al.: Antibacterial nano-structured titania coating incorporated with silver nanoparticles. Biomaterials **32**(24), 5706–5716 (2011)
30. Saha, K., Agasti, S.S., Kim, C., Li, X., Rotello, V.M.: Gold nanoparticles in chemical and biological sensing. Chem. Rev. **112**(5), 2739–2779 (2012)
31. Cheong, Y., Kim, Y.J., Kang, H., Choi, H.W., Lee, J.H., Kim, H.S.: Rapid and sensitive detection of periodontitis pathogens by a surface-enhanced Raman scattering-based immunoassay. Anal. Chem. **90**(12), 7754–7760 (2018)
32. Huang, X., Jain, P.K., El-Sayed, I.H., El-Sayed, M.A.: Plasmonic photothermal therapy (PPTT) using gold nanoparticles. Lasers Med. Sci. **23**(3), 217–228 (2008)
33. Zhang, L., Jiang, Y., Ding, Y., Povey, M., York, D.: Investigation into the antibacterial behaviour of suspensions of ZnO nanoparticles (ZnO nanofluids). J. Nanopart. Res. **9**(3), 479–489 (2007)
34. Li, Q., Chen, S.L., Jiang, W.C.: Durability of nano-ZnO modified ethylene-vinyl acetate copolymer and its application to exterior finishes. J. Appl. Polym. Sci. **134**(20), 44832 (2017)
35. Pelaez, M., Nolan, N.T., Pillai, S.C., Seery, M.K., Falaras, P., Kontos, A.G., et al.: A review on the visible light active titanium dioxide photocatalysts for environmental applications. Appl. Catal. B **125**, 331–349 (2012)
36. Sun, M., Li, W., Zhang, Y.: Enhanced photocatalytic disinfection of microorganisms by TiO_2 nanoparticles. Environ. Sci. Nano **5**(11), 2575–2588 (2018)
37. Goto, T., Sasaki, K.: Synthesis of hydroxyapatite from eggshells and its application in adsorption of lead ions from aqueous solution. J. Asian Ceram. Soc. **4**(2), 137–144 (2016)
38. Roveri, N., Iafisco, M.: Evolving applications of biomimetic nano-hydroxyapatite in dentistry. Nanotechnol. Rev. **1**(1), 1–15 (2010)
39. Ahmad, N., Gopinath, P., Roy, R.: Chitosan nanoparticles: a comprehensive review on properties, preparation methods and biomedical applications. Int. J. Biol. Macromol. **184**, 546–563 (2021)

40. Kong, M., Chen, X.G., Xing, K., Park, H.J.: Antimicrobial properties of chitosan and mode of action: a state of the art review. Int. J. Food Microbiol. **144**(1), 51–63 (2010)
41. Shukla, S.K., Mishra, A.K., Arotiba, O.A., Mamba, B.B.: Chitosan-based nanomaterials: a state-of-the-art review. Int. J. Biol. Macromol. **59**, 46–58 (2013)
42. Sogias, I.A., Williams, A.C., Khutoryanskiy, V.V.: Why is chitosan mucoadhesive? Biomacromol **9**(7), 1837–1842 (2008)
43. Trache, D., Tarchoun, A.F., Derradji, M., Hamidon, T.S., Masruchin, N., Brosse, N., et al.: Nanocellulose: from fundamentals to advanced applications. Front. Chem. **8**, 392 (2020)
44. Banakar, M., Rokhshad, R., Shoorgashti, R., Mousavi, S.M., Lai, C.W.: Nanofillers in dentistry. In: Handbook of Nanofillers 2024 Mar 22, pp. 1–19. Singapore: Springer Nature Singapore
45. Domingues, R.M., Silva, M., Gershovich, P., Betta, S., Babo, P., Caridade, S.G., et al.: Development of injectable hyaluronic acid/cellulose nanocrystals composite hydrogels for tissue engineering applications. Biomacromol **15**(10), 3805–3815 (2014)
46. Kumari, S.A., Patlolla, A.K., Madhusudhanachary, P.: Biosynthesis of silver nanoparticles using azadirachta indica and their antioxidant and anticancer effects in cell lines. Micromachines (Basel). **13**(9) (2022)
47. Vijayakumar, S.G.S.: Gold nanoparticles as a diagnostic and therapeutic tool in dentistry: a systematic review. J. Nanomater. **2021**, 1–12 (2021)
48. Pushpalatha, C., Suresh, J., Gayathri, V.S., Sowmya, S.V., Augustine, D., Alamoudi, A., et al.: Zinc oxide nanoparticles: a review on its applications in dentistry. Front Bioeng Biotechnol. **10**, 917990 (2022)
49. Mansoor, A., Khurshid, Z., Khan, M.T., Mansoor, E., Butt, F.A., Jamal, A., et al.: Medical and dental applications of titania nanoparticles: an overview. Nanomaterials (Basel). **12**(20) (2022)
50. Zhao, R., Meng, X., Pan, Z., Li, Y., Qian, H., Zhu, X., et al.: Advancements in nanohydroxyapatite: synthesis, biomedical applications and composite developments. Regen Biomater. **12**, rbae129 (2025)
51. Mo, X., Zhang, X., Fang, L., Zhang, Y.: Research progress of wood-based panels made of thermoplastics as wood adhesives. Polymers **14**(1), 98 (2022)
52. Moraes, G., Zambom, C., Siqueira, W.L.: Nanoparticles in dentistry: a comprehensive review. Pharmaceuticals (Basel) **14**(8) (2021)
53. Makarov, V.V., Love, A.J., Sinitsyna, O.V., et al.: Green" nanotechnologies: synthesis of metal nanoparticles using plants. Acta Naturae. **6**(1), 35–44 (2014)
54. Sadat-Shojai, M., Khorasani, M.T., Dinpanah-Khoshdargi, E., Jamshidi, A.: Synthesis methods for nanosized hydroxyapatite with diverse structures. Acta Biomater. **9**(8), 7591–7621 (2013)
55. Tschoppe, P., Zandim, D.L., Martus, P., Kielbassa, A.M.: Enamel and dentine remineralization by nano-hydroxyapatite toothpastes. J. Dent. **39**(6), 430–437 (2011)
56. Omid, R., Banakar, M., Mousavi, S.M., Lai, C.W.: Bio-based nanomaterials in dentistry. In: Biobased Nanomaterials: Applications in Biomedicine, Food Industry, Agriculture, and Environmental Sustainability 2024 May 14, pp. 249–313. Singapore: Springer Nature Singapore
57. Rai, M., Yadav, A., Gade, A.: Silver nanoparticles as a new generation of antimicrobials. Biotechnol. Adv. **27**(1), 76–83 (2009)
58. Ahmed, S., Ahmad, M., Swami, B.L., et al.: A review on plants extract mediated synthesis of silver nanoparticles for antimicrobial applications: a green expertise. J. Adv. Res. **7**(1), 17–28 (2016)
59. Agarwal, H., Kumar, S.V., Rajeshkumar, S.: A review on green synthesis of zinc oxide nanoparticles—an eco-friendly approach. Resour.-Effic. Technol. **3**(4), 406–413 (2017)
60. Khvostenko, D., Mitchell, J.C.: Bioactive and therapeutic dental polymeric materials. Dent. Mater. (2019)
61. Esteves, C.V., et al.: Nano-hydroxyapatite as a filler for dental composite resins: a systematic review and meta-analysis. J. Mech. Behav. Biomed. Mater. **126**, 105024 (2022)
62. Cellulose Nanocrystals in Dental Composites: A review of properties and applications. Dent. Mater. **37**(12), e45–e56 (2021)

63. Fortunati, E., et al.: Multifunctional nanocellulose-based composites for biomedical applications. Front. Bioeng. Biotechnol. **7**, 303 (2019)
64. Meloa, M.A.S., et al.: Silver nanoparticles in dental adhesive systems: a new strategy against biofilms. ACS Appl. Mater. Interfaces. **11**(27), 24870–24881 (2019)
65. Santiago, S.L., et al.: Chitosan nanoparticles as a promising agent in dental adhesive technology: antibacterial and anti-proteolytic activities. Int. J. Biol. Macromol. **163**, 2215–2226 (2020)
66. Kishen, A.: Advanced therapeutic options for endodontic biofilms. Endod. Top. **22**(1), 99–123 (2010)
67. Jha, P., et al.: Phyto-nanotechnology in endodontics: current findings and future perspectives. J. Funct. Biomater. **13**(3), 112 (2022)
68. Silva, P.V., Guedes, D.F., Pecora, J.D., et al.: Chitosan-based nanoparticles in endodontic therapy: a systematic review. Int. Endod. J. **54**(12), 2141–2155 (2021)
69. Al-Haddad, A.Y., Che Ab Aziz, Z.A.: Bioceramic-based root canal sealers: a review. Int. J. Biomater. 1–13 (2021)
70. Widbiller, M., Eidt, A., Hiller, K.A., et al.: Nanoparticles in endodontics: current status and future prospects. Materials **11**(12), 2453 (2018)
71. Akıncıbay, H., Şenel, S., Yetkin, A.Z.: Application of chitosan nanoparticles in dentistry. Curr. Drug Deliv. **19**(2), 123–35 (2022)
72. Prabhu, S., Poulose, E.K., Kolanthai, E.: Green synthesized chitosan nanoparticles for targeted drug delivery in periodontal therapy. Int. J. Biol. Macromol. **165**, 1740–1748 (2020)
73. Cheng, L., Zhang, K., Melo, M.A., et al.: Antibacterial and osteogenic properties of silver-containing titanium oxide coatings on implants. J. Funct. Biomater. **12**(3), 52 (2021)
74. Mendonça, G., Mendonça, D.B., Aragão, F.J., et al.: The combination of micro- and nano-scale topography and titanium chemistry to enhance osteoblast function and osseointegration. Mater. Sci. Eng., C **118**, 111505 (2021)
75. Sivaswamy, V., Neelakantan, P.: Modern approaches to biofilm management on dental implants. In: Neelakantan, P., Princy Solomon, A. (eds.) Dental implants and oral microbiome dysbiosis: an interdisciplinary perspective, pp. 61–73. Springer International Publishing, Cham (2022)
76. Prakash, S., Swarnakar, R., Singh, S.P.: Green synthesis of gold nanoparticles for colorimetric sensing of oral cancer relevant biomarkers. Nanotechnol. Environ. Eng. **6**(2), 25 (2021)
77. Chandran, D., Soman, A., Ramachandran, S.: Green synthesized gold nanoparticle based biosensor for the electrochemical detection of oral cancer biomarker. Mater. Today: Proc. **51**, 2242–2246 (2022)
78. Devi, L., Gupta, R., Kumar, S.: Phytochemical-assisted synthesis of nanoparticles for targeted drug delivery in oral cancer therapy. Curr. Pharm. Biotechnol. **21**(12), 1180–1190 (2020)
79. Khan, A.U., Khan, M., Khan, M.M.: Biosynthesized metallic nanoparticles as nanocarriers for targeted cancer therapy: a comprehensive review. J. Drug Deliv. Sci. Technol. **79**, 104072 (2023)
80. Mikhailova, E.O.: Silver nanoparticles: Mechanism of action and probable bio-application. J. Funct. Biomater. (2021)
81. Moskalevska, I., Faure, V., Haye, L., Mercey-Ressejac, M., Dey, A.K., Chovelon, B., et al.: Intracellular accumulation and immunological response of NIR-II polymeric nanoparticles. Int. J. Pharm. **630**, 122439 (2023)
82. Mukherjee, S., Dasari, M., Priyamvada, S.: A comparative review on conventional and green synthesis of nanoparticles and their toxicity profile. Curr. Pharm. Biotechnol. **23**(3), 379–394 (2022)
83. Valsami-Jones, E., Lynch, I.: How safe are nanomaterials? Science **350**(6259), 388–389 (2015)
84. Ghidan, A.Y., Al, A., T M.: Green synthesis of metallic nanoparticles and their prospective biotechnological applications: an overview. Biol. Trace Elem.T Res. **199**, 344–70 (2020)
85. Jadoun, S., Arif, R., Jangid, N.K., Meena, R.K.: Green synthesis of nanoparticles using plant extracts: a review. Environ. Chem. Lett. **19**(1), 355–374 (2021)
86. Ouay, B.L., Stellacci, F.: Antibacterial activity of silver nanoparticles: a surface science insight. Nano Today **10**(3), 339–354 (2015)

87. Senthilkumar, N., Nandhakumar, E., Priya, P., Soni, D., Vimalan, M., Potheher, I.V.: Synthesis of ZnO nanoparticles using leaf extract of Tectona grandis (L.) and their anti-bacterial, anti-arthritic, anti-oxidant and in vitro cytotoxicity activities. New J. Chem. **41**(18), 10347–56 (2017)
88. Karthiga, P., Rajeshkumar, S., Annadurai, G.: Mechanism of larvicidal activity of antimicrobial silver nanoparticles synthesized using Garcinia mangostana bark extract. J. Cluster Sci. **32**(2), 515–525 (2021)
89. Jeevanandam, J., Kiew, S.F., Boakye-Ansah, S., Lau, S.Y., Barhoum, A., Danquah, M.K., et al.: Green approaches for the synthesis of metal and metal oxide nanoparticles using microbial and plant extracts. Nanoscale **14**(7), 2534–2571 (2022)
90. Dhand, C., Dwivedi, N., Loh, X.J., et al.: Methods and strategies for the synthesis of diverse nanoparticles and their applications: a comprehensive overview. RSC Adv. **5**(105), 105003–105037 (2015)
91. Halamoda, K.B., Baconnier, S., Bastogne, T., et al.: Bridging communities in the field of nanomedicine. Regul. Toxicol. Pharmacol. **106**, 187–96 (2019)
92. Allaker, R.P., Ren, G.: Potential impact of nanotechnology on the control of infectious diseases. Trans. R. Soc. Trop. Med. Hyg. **102**(1), 1–2 (2008)
93. Venkatesan, J., Kim, S.K., Shim, M.S.: Antimicrobial, antioxidant, and anticancer activities of biosynthesized silver nanoparticles using Marine Algae Ecklonia cava. Nanomaterials **6**(12), 235 (2016)
94. Slavin, Y.N., Asnis, J., Häfeli, U.O., et al.: Metal nanoparticles: understanding the mechanisms behind their antibacterial activity. J. Nanobiotechnology. **15**(1), 65 (2017)
95. Banakar, M., Sijanivandi, S., Mousavi, S.M., Hashemi, S.A., Lai, C.W.: Green materials for 3D printing in dentistry. In: Encyclopedia of Green Materials 2024 Dec 27, pp. 833–838. Singapore: Springer Nature Singapore